Industry 4.0 for SMEs

Industry 4.0 for SMEs - Smart Manufacturing and Logistics for SMEs

Special Issue Editors

Erwin Rauch
Manuel Woschank

MDPI • Basel • Beijing • Wuhan • Barcelona • Belgrade • Manchester • Tokyo • Cluj • Tianjin

Special Issue Editors
Erwin Rauch
Free University of Bolzano-Bozen
Italy

Manuel Woschank
Montanuniversitaet Leoben
Austria

Editorial Office
MDPI
St. Alban-Anlage 66
4052 Basel, Switzerland

This is a reprint of articles from the Special Issue published online in the open access journal *Sustainability* (ISSN 2071-1050) (available at: https://www.mdpi.com/journal/sustainability/special_issues/Smart_Manufacturing_Logistics).

For citation purposes, cite each article independently as indicated on the article page online and as indicated below:

LastName, A.A.; LastName, B.B.; LastName, C.C. Article Title. *Journal Name* **Year**, *Article Number*, Page Range.

ISBN 978-3-03936-567-8 (Hbk)
ISBN 978-3-03936-568-5 (PDF)

Cover image courtesy of Erwin Rauch.

Contents

About the Special Issue Editors

Erwin Rauch holds an M.Sc. in Mechanical Engineering from the Technical University Munich (TUM), an M.Sc. in Business Administration from the TUM Business School and obtained his Ph.D. degree in Mechanical Engineering from the University of Stuttgart with summa cum laude. Currently, he is an Assistant Professor for Manufacturing Technology and Systems at the Free University of Bolzano, where he is the Head of the Smart Mini Factory laboratory for Industry 4.0. His current research is on Industry 4.0, social sustainability in production, smart and sustainable production systems, smart shop floor management and engineering/make-to-order systems. He has 10 years of experience as Consultant and later as an Associate Partner in an industrial consultancy firm operating in production and logistics. He is project manager of the EU-funded H2020 research project "SME 4.0 –Industry 4.0 for SMEs" in an international partner consortium. Furthermore, he is author and co-author of more than 150 scientific and non-scientific books, chapters of books, articles and other contributions and has received several awards for his scientific contributions.

Manuel Woschank is a Postdoc Senior Lecturer and Researcher at the Chair of Industrial Logistics at the Montanuniversitaet Leoben and an Adjunct Associate Professor at the Faculty of Business, Management and Economics at the University of Latvia. Dr. Woschank holds a Diploma degree in Industrial Management and a Master's Degree in International Supply Management from the University of Applied Sciences FH JOANNEUM, Austria, and a Ph.D. in Management Sciences with summa cum laude from the University of Latvia, Latvia. He has published a multitude of international, peer-reviewed papers and has conducted research projects with voestalpine AG, Stahl Judenburg—GMH Gruppe, Knapp AG, SSI Schaefer, MAGNA International Europe GmbH, Hoerbiger Kompressortechnik Holding GmbH, voestalpine Boehler Aerospace GmbH & Co. KG, etc. His research interests include the areas of logistics system engineering, production planning and control systems, logistics 4.0 concepts and technologies, behavioral decision making and industrial logistics engineering education.

Preface to "Industry 4.0 for SMEs - Smart Manufacturing and Logistics for SMEs"

In recent years, the industrial environment has been changing radically due to the introduction of concepts and technologies based on the fourth industrial revolution (also known as Industry 4.0). The fourth industrial revolution should extend to the whole production and supply chain and not only, like in the past revolutions, to the mechanical manufacturing process of products and the associated process organization. Production based on the principles of Industry 4.0 creates the conditions necessary to replace traditional structures, which are based on centralized decision-making mechanisms and rigid organizational forms. These structures are replaced by flexible reconfigurable manufacturing and logistics systems, decentralized and collaborative decision-making mechanisms, as well as digitally supported processes.

Additionally, SMEs have moved into the focus of many economies. Due to their flexibility, entrepreneurial spirit, and innovation capabilities, SMEs have proved to be more robust than large and multi-national enterprises. Small organizations are increasingly proactive in improving their operational processes, which is a good starting point for introducing the new concepts of Industry 4.0. The successful implementation of an industrial revolution like Industry 4.0 has to take place not only in large enterprises, but must also show a high potential, especially in SMEs. The readiness of SME-adapted Industry 4.0 concepts and the organizational capability of SMEs to meet this challenge exist only in some areas. This reveals the need for further research and action plans for preparing SMEs in a technical and organizational direction. Therefore, special research and investigations are needed for the implementation of Industry 4.0 technologies and concepts in SMEs. SMEs will only achieve Industry 4.0 by following SME-customized implementation strategies and approaches and realizing SME-adapted concepts and technological solutions.

Thus, this Special Issue represents a collection of theoretical models as well as practical case studies related to the introduction of Industry 4.0 concepts in small- and medium-sized enterprises.

Erwin Rauch, Manuel Woschank
Special Issue Editors

 *sustainability*

Article

The Role of Smart Technology in Sustainable Agriculture: A Case Study of Wangree Plant Factory

Salinee Santiteerakul [1,2], Apichat Sopadang [1,2,*], Korrakot Yaibuathet Tippayawong [1,2] and Krisana Tamvimol [3]

1   Center of Excellence in Logistics and Supply Chain Management, Chiang Mai University, Chiang Mai 50200, Thailand; salinee@eng.cmu.ac.th (S.S.); korrakot@eng.cmu.ac.th (K.Y.T.)
2   Department of Industrial Engineering, Faculty of Engineering, Chiang Mai University, Chiang Mai 50200, Thailand
3   Wangree Health Factory Co., Ltd., Nayok 26000, Thailand; krisana@oi.co.th
*   Correspondence: apichat.s@cmu.ac.th

Received: 31 March 2020; Accepted: 4 June 2020; Published: 5 June 2020

Abstract: Sustainable development is of growing importance to the agriculture sector because the current lacking utilization of resources and energy usage, together with the pollution generated from toxic chemicals, cannot continue at present rates. Sustainability in agriculture can be achieved through using less (or no) poisonous chemicals, saving natural resources, and reducing greenhouse gas emissions. Technology applications could help farmers to use proper data in decision-making, which leads to low-input agriculture. This work focuses on the role of smart technology implementation in sustainable agriculture. The effects of smart technology implementation are analyzed by using a case study approach. The results show that the plant factory using intelligence technology enhances sustainability performance by increasing production productivity, product quality, crop per year, resource use efficiency, and food safety, as well as improving employees' quality of life.

Keywords: smart technologies; sustainable agriculture; plant factory

---

1. Introduction

In the era of the fourth industrial revolution (Industry 4.0), new technologies, such as artificial intelligence, automation systems, and cloud computing have been developed to combine the digital and physical world together. One of the Industry 4.0 benefits is to provide a crossing workspace among machines, infrastructures, and digital platforms. [1]. The methods of Industry 4.0 have been investigated in the area of agriculture. Various innovations, like sensor technology, machine learning, wireless communication, positioning systems, and data visualization tools, have been adopted to create value and increase productivity in the farming sector [2].

Due to the environmental stress on water scarcities, insufficient land use, soil depletion, and greenhouse gas emissions, the demands on sustainable agriculture are rapidly increasing. The sustainable development (SD) concept was developed regarding a reorganization of a threatened future. Many regions face risks of inevitable damage to the human environment. Environmental stress has been seen as the result of the increasing growth of the population, technology growth and development, and the rising living standards among the affluent. Given the importance of agriculture as the crucial provider of food, the sustainable development of this sector is important. Thus, sustainable agriculture needs an innovative system that protects and enhances the natural resource base while increasing productivity.

There are essential research works that deal with aspects of the technology implementation for sustainable agriculture. The authors of [3] Reviewed relevant research works on intelligent agricultural information handling methods. The technological applications related to agricultural aspects were classified into three categories, namely data sources and collection, machine learning (ML) methodologies for agricultural data, and intelligent knowledge acquisition. These applications lead to the development of innovation for precision and optimal farming. The authors of [4] studied artificial devices of closed plant production systems and found that smart plant production systems produce high-quality plants and transplants with minimum use of resources and carbon dioxide emissions as well as environmental pollutants.

This work aims to study the effects of smart technology implementation on sustainability performance by using a case study. The case study company, namely the Wangree Health Factory Company (located in Nakhon Nayok, Thailand), deploys plant factory with artificial lighting technology to cultivate fresh organic vegetables and fruits. This work develops the analysis framework to analyze how deploying plant factories will impact the sustainability of agriculture. This article is organized as follows: first, the literature review provides a background on smart farming and sustainable agriculture. Then, the material and methods section provides a research framework and a case study analysis, and is followed by the results section. Finally, the roles of smart technology are presented and discussed.

2. Literature Review

2.1. Smart Farming

Smart farming is a management concept using modern technology to increase the quantity and quality of agricultural products. Smart farming involves an integration of information and communication technologies into machines, sensors, actuators, and network equipment for use in agricultural production systems [5]. There are several technologies related to smart farming, including sensors, robotics, the Internet of Things (IoT), mapping, decision-making, and statistical processes [5,6].

Ray [6] proposed the detailed framework of IoT-based agriculture. This framework comprises six layers. First, the physical layer contains various types of devices, such as sensors and microcontrollers, to collect, exchange, and process data to other devices. Second, the network layer comprises the Internet and relevant communication technologies. Third, the IoT-based middleware layer performs various tasks, such as device management, interoperation, context awareness, platform portability, and security-related tasks. Fourth, the service layer provides cloud storage and Software-as-a-Service (SaaS). Fifth, the analytics layer performs big data processing to predictive and multi culture analysis. Sixth, the user experience layer facilitates the farmer's communication using social network activities to share and disseminate agricultural knowledge.

Plant factories are one of the smart farming applications. A plant factory is a closed-growing system that enables a farmer to achieve constant and regular production of vegetables throughout the year. There are three types of plant factories: (1) Plant factory with sunlight; (2) Plant factory with sunlight and supplemental light; and (3) Plant factory with artificial lighting. The plant factory with artificial lighting replaces sunlight with artificial intelligence sources of light. This plant factory creates a more consistent light environment for the plants. Ray [6] explained that there are three units for managing plant factory systems: (1) Farm Gate Way (FGW), (2) A data collection/storage/distribution platform, and (3) an application module. A FGW unit is a data collection and control device. It collects growing conditions data in the plant factory (such as temperature and nutrient content) and the crop production equipment data (such as nutrient solution pumps and heat pumps). A cloud-based data collection, storage, and distribution platform is used to provide communication between the data center and the plant factory for the control of growing conditions and equipment. It also helps in the servicing of command and control information management. The application unit is composed of physical devices to perform three operations: sensory data reception, command delivery to the FGW, and response reception from the FGW.

This work employed the integrated IoT-based agriculture and plant factory unit framework (see Figure 1) to analyze the technology implementation of a case study company.

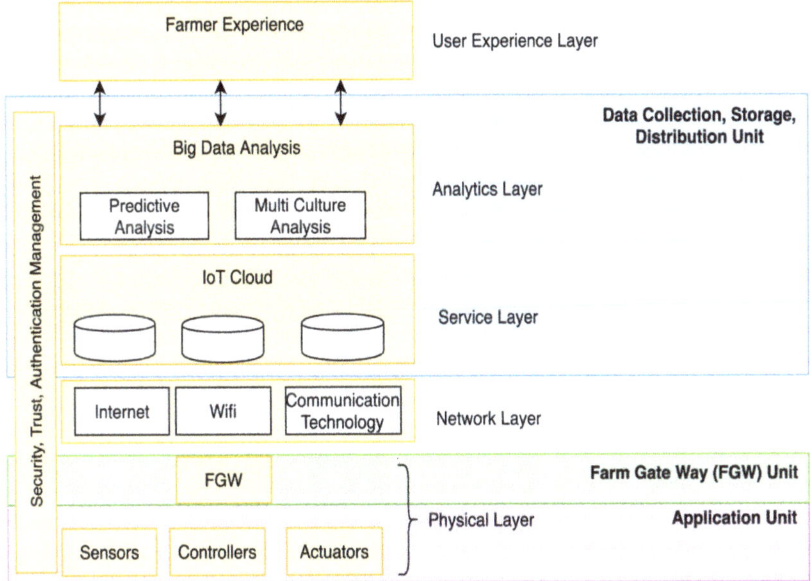

Figure 1. Internet of Things (IoT)-based agriculture and plant factory units framework (modified from [6]).

2.2. Sustainable Agriculture and Sustainability Measurement

Over the past decade, societies have changed rapidly as a result of technology advancement, rapid urbanization, and innovations in production systems. At the same time, enormous changes have occurred, such as climate change, an increasing number of crises and natural disasters, increasing of population dynamics, and economic growth. Crop production is damaged by floods, climate change, drought, and insufficient land resources. The world's population is expected to increase by more than 30% by 2050 (from 7 billion in 2011). Meanwhile the global crop production is expected to grow by more than 90% from higher yields. The world's agricultural production has increased more than three times between 1960 and 2015 [7]. This caused the use of resource conservation technologies and productivity enhancement for agricultural purposes. These trends threaten the sustainability of agricultural systems and undermine the global capacity to meet its needs.

A useful definition of sustainability is that of the World Commission on Environment and Development (1987), which states: "A sustainable economy is one which can meet the needs of the present without compromising the ability of future generations to meet their own needs" [8]. Sustainability as an attribute of agriculture has been gaining recognition globally since the 1970s [9]. The term "sustainable agriculture" refers to an agricultural system that will continue to be productive in the future.

There are many suggested measures of sustainability. Most of the literature uses the triple bottom line (TBL) developed by Elkington [10]. This concept divides sustainability into three categories, including economic, environment and society. The sustainability indicators are developed based on the TBL concept by defining a set of indicators to indicate the performance of each category [11–16].

At the farm level, numerous sustainability assessment methods have been developed to assess the sustainability performance of agricultural systems. Some examples of sustainability assessment methods from the scientific literature are the sustainability assessment of food and agriculture systems (SAFA) [13], sustainability monitoring and assessment routine (SMART) [14], response-inducing sustainability evaluation (RISE) [17], and multi-criteria assessment (MCA) [18].

The indicators of measuring the sustainability of the agriculture system are divided into three main categories. The first category is technical data, which require specific quantitative data, such as energy consumption, water use, soil pH, and farm revenue. The second category is the performance rating indicator, which requires an approach to convert sustainability performance to rating scores. For example, to evaluate working condition performance, the working environment is assessed in terms of a 5-scale rating from worse (score = 1) to excellent (score = 5). The third category is a descriptive indicator, which relates to the driver, pressure, state, impact, or response of sustainability. Quantitative and qualitative descriptive indicators describe the factual situation but do not assess whether this is good or bad—they are, in practical terms, a statement of a fact.

3. Material and Methods

3.1. Research Framework

A framework to analyze the roles of smart technology in sustainable agriculture consists of four steps (see Figure 2).

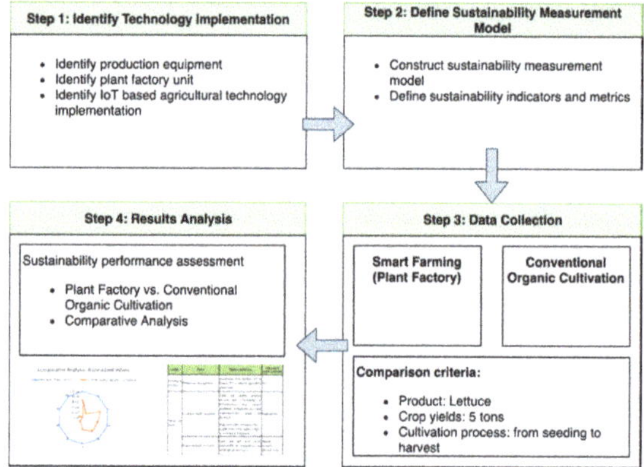

Figure 2. Research Framework.

Step 1: Identify technology implementation of the case study by using the IoT-based agriculture and plant factory unit framework [6].

Step 2: Define a sustainability measurement model by defining sustainability indicators and metrics and constructing the measurement model.

Step 3: Planting data was collected by using the case study. The data on cultivation in a plant factory were collected from Wangree Health Factory Company. The data on conventional organic cultivation were collected from Wangree Organic Farm. The relevant data were collected based on the assumption of equal production outputs, which is one-crop cultivation of 5 tons of product weight. This data collection includes all activities from seeding to harvest process.

Step 4: Assess the sustainability performance of plant factory cultivation compared to conventional organic cultivation.

3.2. Case Study Overview

Wangree Health Factory Company, located in Nakhon Nayok, Thailand, was founded in 2016 with the corporate vision of using modern digital technology to provide fresh organic vegetables and fruits to the Thai market. A plant factory with artificial intelligence light is an indoor farming system connected with a smart control system. The structures of the plant factory separate the plants from the external environment so that the plants are protected from uncertain conditions. These systems permit high-quality and high-yield production year-round under a controlled environment (e.g., light, temperature, humidity, the concentration of carbon dioxide, and culture solution). The IoT-based technologies allow farmers to plan their production by using mobile devices for monitoring and controlling their farming systems.

Wangree Health Factory Company combines a vertical farming system with IoT technologies (see Figures 3 and 4). The growth process is fully automated for watering, lighting, nutrient adding, and temperature controlling. The 173.85 m^2 × 6 m high plant factory produces approximately 50,552 heads of lettuce per month. The main production equipment is composed of electricity supply, an air conditioning system, nutrient solution supply, lighting, CO_2 supply, smart devices, and communication devices.

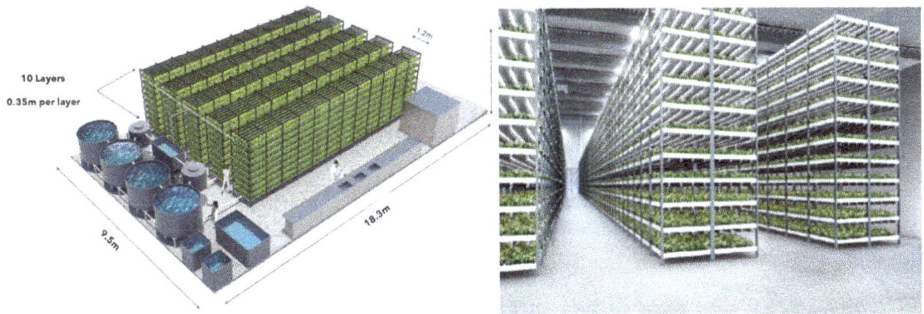

Figure 3. Plant factory, a vertical farming system.

Figure 4. Seedlings grow with light at Wangree Health Factory Company.

Wangree Plant Factory employs IoT-based technology to produce and manage the agriculture production system. The workflow of the system consists of three steps (see Figure 5): (1) Detecting and collecting data from sensor devices; (2) Analysis and control based on a specific AI algorithm; and (3) Data visualization for statistical reporting to help farm owners make decisions.

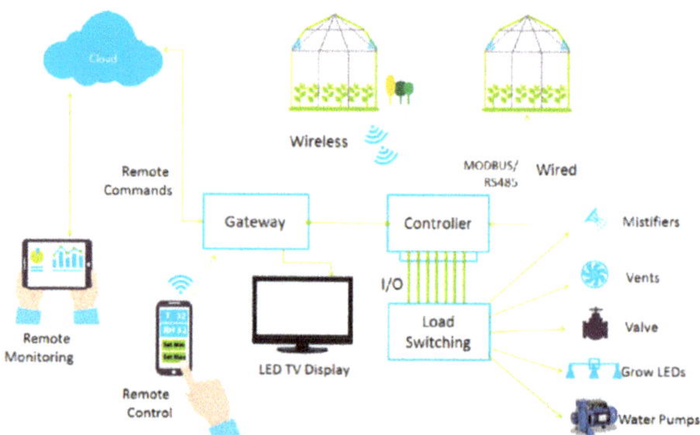

Figure 5. Smart technology implementation at Wangree Health Factory Company.

The lists of equipment for one plant factory are shown in Table 1. The hardware equipment comprises seven categories, i.e., electricity supply, air conditioning, nutrient solution supply, lighting, CO_2 supply, sensors for environmental control, and communication and management equipment.

Table 1. Equipment and environmental sensors typically installed in a plant factory.

Category	Equipment and Environmental Sensors
Electricity supply	Power distribution box and backup power system Power Consumption AC 100 V/240 V 100–150 W (Power Saving Mode), 300–350 W (Full Power Mode)
Air conditioning	Inner units of air conditioners 40,000 BTU * Air circulation fans Air cleaners with filters
Nutrient solution supply	Water and fertilizer system Cultivation controller system Pest protection system Plumbing for clean water supply UV water purifier
Lighting	Light source with reflectors LED lumen 5 cm—11,000 lum and 26 cm—9000 lum
CO_2 supply	Control unit with distribution tubes
Sensors for environmental control	Smart light control Smart air sensors Smart water feeding Temperature and humid sensors CO_2 sensors
Communication and Management	Wireless communication Plant dashboard visualization

* Note: British Thermal Unit (BTU) is a unit of heat. It is defined as the amount of heat required to raise the temperature of one pound of water by one degree Fahrenheit. One BTU is about 1055 Jules.

The software system (see Figure 6) comprises three main functions: (1) a cloud data center system with an intelligence processing system (Big Data and AI) is responsible for storing databases for automatic processing and backup of essential data. (2) A back-end information management system, which is a web-based system, is responsible for managing information in terms of data security, right of data accessing, and data editing. (3) A mobile application system for both iOS and Android platforms is an application module used for managing and controlling operational production. It deploys remote controlling technology and allows users to monitor and manage their operation in real-time from a distance via smartphone and tablet devices.

Figure 6. Software system of Plant Factory at the Wangree Health Factory Company.

Following the framework developed by Ray [6], the IoT-based technology implementation is shown in Figure 7.

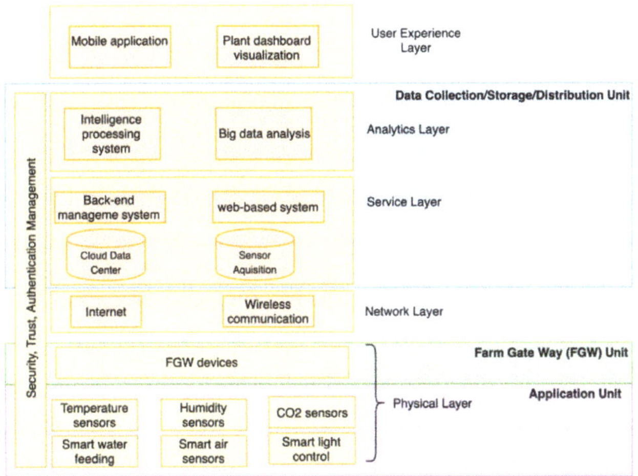

Figure 7. Wangree Health Factory Company technology implementation.

3.3. Sustainability Measurement Model Definition

This work adopts the sustainability performance measurement model proposed by [19] (see Figure 8). The objective of the proposed model is to measure the sustainability performance of sustainable farming. This model is composed of five levels. Level 0 is the top level and identifies the goal, which is sustainable agriculture.

Level 1 is a dimension level dividing sustainability into economic, environmental, and social categories. Level 2 is a sub-dimension level. There are eight sub-dimensions regarding three principal dimensions. For example, raw material, natural resources, and energy are sub-dimensions of the environmental perspective. Level 3 is an indicator level that provides a performance indicator to indicate the performance of each sub-dimension. For example, we could evaluate the impact of natural resources, from resource consumption to pollution and emission. Level 4, the last level, is a metric level that provides a formulation to measure sustainability performance. For example, resource consumption performance is indicated by the quantity of water consumption, land use, and soil quality.

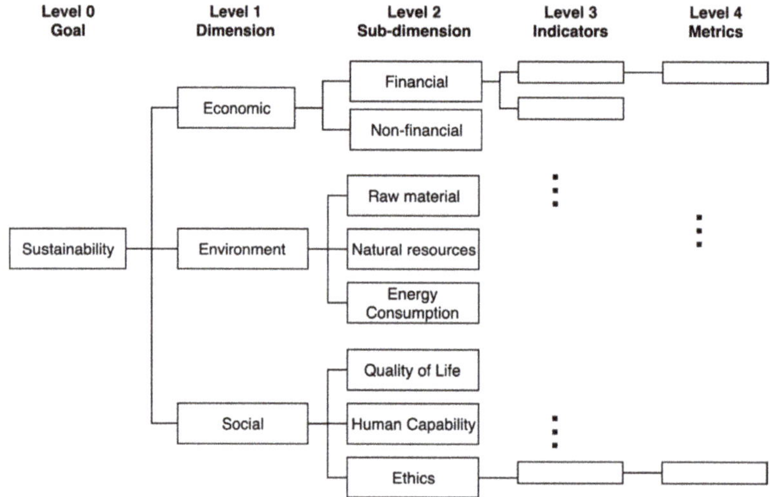

Figure 8. Sustainability performance measurement framework.

To implement the sustainable agriculture performance measurement model, [20] suggested that decision-maker(s) should use level 0 to level 2 as a framework to select specific indicators (level 3) and metrics (level 4) that are appropriate for their application. Therefore, this work will construct the performance indicators and metrics that relate to smart technology implementation in agriculture.

This work selected seven sub-dimension (level 2) and the details are as follows:

- Financial criteria relate to increasing revenue, profit, market share, reducing the cost of operation, and unit cost.
- Non-financial criteria relate to improving product quality, crop per year, and reducing harvest time.
- Raw material criteria relate to conserving and enhancing the raw material by efficient use through the reduce, reuse, and recycle concepts; reducing hazardous material usage; and reducing defects and waste generation.
- Natural resource (water, soils, and land use) criteria relate to conserving and enhancing the natural resource base by efficiently using and reducing environmental emission.
- Quality of life criteria relate to assessing and reducing the potential health impacts of new technologies as well as increasing the well-being of stakeholders.
- Human capability criteria relate to encouraging education to improve human skills and knowledge performance.
- Ethics criteria relate to respecting local and international laws on business and human rights and supporting ethical operating practice issues.

The details of sustainability indicators and metrics are shown in Table 2.

Table 2. Sustainability indicators and metrics.

Dimension	Sub-Dimension	Indicator	Metrics
Economic (EC)	Financial	Cost	Infrastructure cost (EC-1) Fertilizer cost (EC-2) Unit cost (EC-3)
	Non-Financial	Productivity	Harvest time (EC-4) Weight per unit (EC-5) Product quality grade (EC-6) Crop per year (EC-7) Lifetime of cultivation system (EC-8)
Environmental (EN)	Raw material	Defects	Percentage of defects (EN-1)
	Natural resource	Resource consumption	Water use (EN-2) Land use (EN-3)
		Pollution and emission	Wastewater management (EN-4)
Social (SO)	Quality of life	Health and Safety	Consumer health and safety (SO-1) Employee health and safety (SO-2)
	Human capability	Knowledge sharing	Social support by sharing knowledge to the local community (SO-3)
	Ethics	Fair operation practices	Fair salary (SO-4)

4. Results

The results of the sustainability performance of plant factories and conventional organics cultivation are shown in Table 3.

Table 3. Results of comparing sustainability Performance.

Indicator	Metrics	Wangree Plant Factory	Conventional Organics Cultivation
Cost	Infrastructure cost (million baht)	5–6	10–12
	Fertilizer cost (baht)	10,000	50,000
	Unit cost (baht)	4.25	6.00
Productivity	Harvest time (days)	21–30	45–50
	Weight per unit (g)	100–175	75–100
	Product quality grade	Medical grade (post organic)	Good agricultural practice (GAP) or Organic
	Crop per year (crops)	12–15	4–8
	Lifetime of cultivation system (years)	15–30	3–5
	Labors (man)	2	10–20
Defects	Percentage of defects (%)	0.5–1%	30–50%
Resource consumption	Cultivation water use for (liter per month)	30,000	3,000,000
	Land use (m^2)	160	1600
Pollution and emission	Wastewater management	Wastewater from fertilizer will be reused for traditional agriculture production.	N/A
Health and Safety	Consumer health and safety	Clean and healthy products: because of the high controllability of the environmental and sanitary conditions, pesticide-free and other contaminant-free plants are produced. High traceability throughout the supply chain, which enables a high level of risk management.	Pesticide-free
	Employee health and safety	Light and safe work under comfortable air temperature and moderate air movement.	Requires intensive physical work
Knowledge sharing	Social support by sharing knowledge to the local community	Wangree Plant Factory provides knowledge sharing to the local community, such as researchers in academic institutes, governance institutes, and private companies.	N/A
Fair operation practices	Fair salary	Due to the requirements of highly skilled persons and less labor needed (only 2–3 persons compared with 10–15 persons for traditional organic production), the Wangree Plant Factory could pay a higher salary to its employees.	N/A

Wangree Plant Factory has actualized most of the potential benefits in the economic and environmental aspects. Compared with traditional organic cultivation, the Wangree Plant Factory can greatly reduce the consumption of resources than conventional cultivation. The percentage of resources saved and the improvement in produce quality and yield in the plant factory are as follows:

Economic Perspective:

- A 50% reduction of infrastructure cost and a 3–5 times reduction of lifetime cultivation system.
- A 80% reduction of fertilization cost because of lower fertilizer consumption achieved through recycling with little drainage of circulating nutrient solution.
- A 30% reduction of unit cost because of lower resource consumption and higher productivity.
- A 33–75% increase of product weight per unit.
- An increase by 1.8–2 times of amount of a crop per year.
- Better product quality grade.

Environmental Perspective:

- Both the plant factory and organic cultivation are free-pesticide applications. The plant factory keeps the cultivated area clean and free from pest insects.
- A 99% reduction in water consumption.
- A 99% reduction in land use compared to conventional agriculture due to a higher productivity per production area.
- A 30–50% reduction in plant defects.

Social Perspective:

- Increasing demands for fresh food, nutritious food, and functional food for health care and higher quality of life because of high controllability of plant environment. Controlled aerial environmental factors include photosynthetic photon flux density (PPFD), air temperature, CO_2 concentration, light quality, and flow rate of the nutrient solution.
- Light and safe work under comfortable air temperature and moderate air movement.

5. Discussion

The closed environment of a plant factory isolates crop production from an exterior environment. Consequently, the plant growth factors, such as water, light, carbon dioxide concentration, and nutrients are controllable. A plant factory control system can measure the environmental conditions, such as temperature, humidity, and light intensity, and respond by adjusting the plant growth factors via controllers. Measurement and responsiveness capabilities depend on the sensors' data collection and storage and on the processing ability of cloud computing with AI and Big Data analysis. Moreover, the effective management system and data visualization encourage farmers to make good decisions to operate their production.

This work studies how technology affects sustainable agriculture. The sustainability performance is composed of three dimensions: economic, environment, and society. The effects from technology implementation in each dimension are discussed here.

Economic Dimension

- Increasing product quality: precision measurement and a suitable plant growth factor adjustment lead to better product quality. It can be seen that the products from the plant factory are higher in weight per unit, and have a better quality grade and a lower percentage of defects.
- Increasing productivity: the production productivity of the plant factory is higher than the conventional cultivation due to the cost reduction from resource use (water and fertilization), the increase of product weight per unit, and the amount of labor reduction.

- Increasing crop per year: the plant factory has an artificial intelligence system completely closed off the outside environment. It replaces sunlight with controllable lighting sources and controls other plant growth factors, such as humidity, carbon dioxide concentration, temperature, and nutrients by using an artificial intelligence system. Consequently, a plant factory can achieve a year-round production environment. The plant factory produces around twice as much crop per year compared to conventional agriculture and reduces harvest time by around 50 percent compared to traditional cultivation.

Environmental Dimension

- Increasing resource use efficiency: the plant factory enhances crop irrigation water productivity due to a water control system that reduces drained water in the growing area and recycles water vapor into liquid water. The vertical farming of the plant factory increases land use efficiency. It provides a 99% reduction in land use.

Social Dimension

- Increasing food safety: the plant factory gives priority to keeping the growing area free from pests and pesticides. These hygiene conditions create a ready-to-eat product after harvesting. Moreover, the information technology in the plant factory allows customers and stakeholders in the supply chain to trace operational data from producers.
- Increasing employees' quality of life: the controllable working environment in the plant factory is much more desirable than field cultivation, which involves the uncertainty of heat and weather. Further, to work with automatic and high technology systems, the plant factory requires highly skilled workers. It encourages employees to improve their skills and knowledge.

It is clear that the plant factory with artificial system enhances the sustainability performance in all economic, environmental, and social perspectives. The isolated environment of the plant factory provides numerous advantages for improving productivity, improving product quality, increasing resource use efficiency, and increasing food safety. However, plant factories tend to require a greater energy input. Due to the fact that free energy from sunlight has been rejected from this growing system, the plant factory needs new energy to provide a light source to the system. It This leads to much higher costs of lighting in the plant factory. The electricity cost represents nearly one-fourth of the production costs [21] This issue can be solved and further investigated in future research.

6. Conclusions

Technological development and digitalization shape feasible boundaries to increase resource use efficiency. Smart agriculture reduces the negative environmental impacts of farming, increases resilience and soil health, and decreases costs for farmers. The number and types of challenges associated with smart farming expand across various agricultural production systems, and infrastructural limitations apply when it comes to IoT implementation. The plant factory is one of the solutions to solve the problems regarding foods, resources, and the environment. Methodologies have been developed by which the yield and quality of foods are improved, with less consumption of resources and less environmental degradation than the current plant production system. The potential benefits of the plant factory are enhanced economic and environmental sustainability.

Author Contributions: A.S. supervised the work and worked on the selection and application methodology on the real case study. S.S. contributed the whole structure of this paper, worked on the literature review and the development of research framework, and summarized the findings and the conclusion. K.Y.T. contributed on the project coordinating and worked on reviewing and editing manuscript. K.T. contributed the case study data. All authors have read and agreed to the published version of the manuscript.

Funding: This research and the APC were funded by the project "SME 4.0—Industry 4.0 for SMEs" (funded by the European Union's Horizon 2020 R&I program under the Marie Skłodowska-Curie grant agreement No. 734713).

Acknowledgments: This research work was partially supported by Center of Excellence in Logistics and Supply Chain Management, Chiang Mai University and Wangree Health Factory Co., Ltd., Nakhon Nayok, Thailand.

Conflicts of Interest: The authors declare no conflict of interest.

References

1. Saunila, M.; Nasiri, M.; Ukko, J.; Rantala, T. Smart technologies and corporate sustainability: The mediation effect of corporate sustainability strategy. *Comput. Ind.* **2019**, *108*, 178–185. [CrossRef]
2. Braun, A.T.; Colangelo, E.; Steckel, T. Farming in the era of industrie 4.0. *Procedia CIRP* **2018**, *72*, 979–984. [CrossRef]
3. Voutos, Y.; Mylonas, P.; Katheniotis, J.; Sofou, A. A survey on intelligent agricultural information handling methodologies. *Sustainability* **2019**, *11*, 3278. [CrossRef]
4. Kozai, T. Sustainable plant factory: Closed plant production systems with artificial light for high resource use efficiencies and quality produce. *Acta Hortic.* **2013**, 27–40. [CrossRef]
5. Pivoto, D.; Waquil, P.D.; Talamini, E.; Spanhol, C.; Corte, V.F.D.; Mores, G.D.V. Scientific development of smart farming technologies and their application in Brazil. *Inf. Process. Agric.* **2018**, *5*, 21–32. [CrossRef]
6. Ray, P.P. Internet of things for smart agriculture: Technologies, practices and future direction. *J. Ambient Intell. Smart Environ.* **2017**, *9*, 395–420. [CrossRef]
7. FAO. *The Future of Food and Agriculture—Trends and Challenges*; FAO: Rome, Italy, 2017; Available online: http://www.fao.org/3/a-i6583e.pdf (accessed on 15 March 2020).
8. WCED, S.W.S. World commission on environment and development. *Our Common Future* **1987**, *17*, 1–91.
9. Senanayake, R. Sustainable agriculture: Definitions and parameters for measurement. *J. Sustain. Agric.* **1991**, *1*, 7–28. [CrossRef]
10. Elkington, J. *Cannibals with Forks: The Triple Bottom Line of 21st Century Business*; Capstone: Oxford, UK, 1999.
11. Santiteerakul, S.; Sekhari, A.; Bouras, A.; Sopadang, A. Sustainability indicators for evaluating sustainable supply chain. In Proceedings of the International Conference on Green and Sustainable Innovation (ICGSI), Chiang Mai, Thailand, 24–26 May 2012.
12. Veleva, V.; Ellenbecker, M. Indicators of sustainable production: Framework and methodology. *J. Clean. Prod.* **2001**, *9*, 519–549. [CrossRef]
13. FAO. *Sustainability Assessment of Food and Agriculture Systems (SAFA)*; Natural Resources Management and Environment Department Food and Agriculture Organization of the United Nations: New York, NY, USA, 1998.
14. Jawtrusch, J.; Schader, C.; Stolze, M.; Baumgart, L.; Niggli, U. Sustainability monitoring and assessment routine: Results from pilot applications of the FAO SAFA guidelines. In Proceedings of the International Symposium on Mediterranean Organic Agriculture and Quality Signs Related to the Origin, Agadir, Morocco, 2–4 December 2013.
15. Santiteerakul, S.; Sekhari, A.; Bouras, A.; Sopadang, A. Sustainability performance measurement framework for supply chain management. *Int. J. Prod. Dev.* **2015**, *20*, 221. [CrossRef]
16. Sopadang, A.; Wichaisri, S.; Banomyong, R. Sustainable supply chain performance measurement a case study of the sugar industry. In Proceedings of the International Conference on Industrial Engineering and Operations Management, Rabat, Morocco, 11–13 April 2017.
17. Grenz, J.; Thalmann, C.; Stampfli, A.; Studer, C.; Hani, F. RISE—A method for assessing the sustainability of agricultural production at farm level. *Rural Dev. News* **2009**, *1*, 5–9.
18. Kamali, F.P.; Borges, J.A.R.; Meuwissen, M.P.; De Boer, I.J.M.; Lansink, A.G.O. Sustainability assessment of agricultural systems: The validity of expert opinion and robustness of a multi-criteria analysis. *Agric. Syst.* **2017**, *157*, 118–128. [CrossRef]
19. Weerapan, T. Performance Measurement of Organic Vegetables in Sustainable Supply Chain Management. Master's Thesis, Chiang Mai University, Chiang Mai, Thailand, August 2018.

20. Weerapan, T.; Santiteerakul, S. Indicators of performance measurement for agriculture sustainable supply chain. In Proceedings of the International Conference of Simulation and Modeling (SIMMOD), Pattaya, Thailand, 23–25 January 2017; pp. 130–137.

21. Kozai, T.; Niu, G.; Takagaki, M. *Plant Factory: An Indoor Vertical Farming System for Efficient Quality Food Production*; Elsevier: San Diego, CA, USA, 2015.

Article

A Review of Further Directions for Artificial Intelligence, Machine Learning, and Deep Learning in Smart Logistics

Manuel Woschank [1,*], **Erwin Rauch** [2] **and Helmut Zsifkovits** [1]

[1]   Chair of Industrial Logistics, Montanuniversitaet Leoben, 8700 Leoben, Austria;
helmut.zsifkovits@unileoben.ac.at

[2]   Industrial Engineering and Automation (IEA), Faculty of Science and Technology,
Free University of Bozen-Bolzano, 39100 Bolzano, Italy; erwin.rauch@unibz.it

*   Correspondence: manuel.woschank@unileoben.ac.at; Tel.: +43-3842-402-6023

Received: 31 March 2020; Accepted: 29 April 2020; Published: 6 May 2020

Abstract: Industry 4.0 concepts and technologies ensure the ongoing development of micro- and macro-economic entities by focusing on the principles of interconnectivity, digitalization, and automation. In this context, artificial intelligence is seen as one of the major enablers for Smart Logistics and Smart Production initiatives. This paper systematically analyzes the scientific literature on artificial intelligence, machine learning, and deep learning in the context of Smart Logistics management in industrial enterprises. Furthermore, based on the results of the systematic literature review, the authors present a conceptual framework, which provides fruitful implications based on recent research findings and insights to be used for directing and starting future research initiatives in the field of artificial intelligence (AI), machine learning (ML), and deep learning (DL) in Smart Logistics.

Keywords: industry 4.0; artificial intelligence; machine learning; deep learning; smart logistics; logistics 4.0

---

1. Introduction

The fourth industrial revolution (Industry 4.0) comprises a set of concepts and technologies that should be used to strengthen the competitiveness of industrial enterprises by referring to the concepts of interconnectivity, digitalization, and automation [1–4]. In this context, Smart Logistics aims at the successful implementation of intelligent and lean supply chains based on agile and cooperative networks and interlinked organizations. Furthermore, information exchange is established through the usage of modern information and communication technologies (ICT), data networks, actors and sensors, and automatic identification and material tracking technologies. Moreover, automated transport, transition, and storage systems, supported by autonomous transport vehicles, should enable a partial and/or complete self-control of systems [2,4–6].

Furthermore, Smart Logistics can be implemented by using the technological concepts of cyber-physical systems (CPS), the internet of things (IoT), respectively as the industrial internet of things (IIoT), and the physical internet (PI) [2]. Besides the implementation of the technological concepts, the application of artificial intelligence, machine learning, and deep learning concepts can be considered as one of the most important success factors within the process of the digital transformation [7].

In this context, artificial intelligence (AI) can be defined as the science and engineering of intelligent machines with a special focus on intelligent computer programs [8]. Machine learning (ML) is considered as an integral part of AI, which refers to the automated detection of meaningful patterns in datasets. ML tools aim to increase the efficiency of algorithms by ensuring the ability to learn and adapt based on big-data analytics [9]. Moreover, deep learning (DL), is defined as a sub-class of ML within the AI-technologies that explores many layers of non-linear information processing for supervised and/or unsupervised features extraction and transformation, and for pattern analyses and classification [10,11].

In recent years, AI, ML, and DL have gained increasing relevance in a multitude of research fields such as engineering, medicine, economics, and business management as well as in marketing [12–15]. However, to the best of our knowledge, a holistic study on the usage of AI, ML, and DL in the context of Smart Logistics in industrial enterprises is currently missing in the scientific literature. Therefore, the authors conduct a systematic literature review on AI, ML, and DL technologies in the timeframe from 2014 to 2019. The identified studies can be used to provide an overview of research on these emerging topics that can be used as a starting point for further studies in the area of Smart Logistics later on.

This paper is structured as follows. Section 2 describes the selected research methodology and methods of this study as well as the detailed process steps of the systematic literature review. Section 3 presents a descriptive analysis and the content analysis of the identified studies. Section 4 introduces the conceptual framework of AI, ML, and DL approaches in the research area of Smart Logistics in industrial enterprises. Section 5 contains a discussion of the research findings. Section 6 summarizes the implications for both future research initiatives and practical applications, and the limitations of this research study. The final Section 7 briefly concludes by reflecting on the main contributions of this paper.

2. Research Methodology and Methods

In this paper, the authors conduct a systematic literature review (SLR) for the secondary-data-based evaluation of recent research studies regarding the application of AI, ML, and DL approaches in the research area of Smart Logistics in industrial enterprises.

By including the sequential identification, screening, clustering, and evaluation of research-relevant studies in research-related areas, the SLR approach was selected because of its systematic, method driven, and replicable approach [16–18]. Especially for the generation and assessment of new knowledge, the SLR process is considered as a valuable instrument that minimizes various judgment biases by systematically evaluating relevant findings from recent research studies [19–21].

In this context, research offers a multitude of guidelines for SLR, which should be used to ensure high quality in empirical research [16,22–24]. In this study, the authors focus on the SLR guidelines as suggested by Denyer et al. [23] and Hokka et al. [16], which generally can be divided into the following four steps:

- Step 1: Definition of the research objective(s);
- Step 2: Framing of the research subject (conceptual boundaries);
- Step 3: Data collection by using inclusion/exclusion criteria;
- Step 4: Validation of the research results.

2.1. Research Objectives

This paper aims to systematically evaluate possible future directions for AI, ML, and DL in Smart Logistics by analyzing the current knowledge and the state of the research regarding AI, ML, and DL-related topics with a special emphasis on recent developments in industrial research. In particular, the authors want to understand how this research topic evolved during the last years. Moreover, the evaluation results will be used for the identification of key activities for further research as well as practical applications.

2.2. Framing of the Research Subject

This research focuses on the systematic evaluation of artificial intelligence, machine learning, and deep learning approaches for Smart Logistics in industrial enterprises. Therefore, the research subject, i.e., the conceptual boundaries, was defined based on the term "artificial intelligence" and the related terms "machine learning" and "deep learning" in the industrial environment.

2.3. Data Collection by Using a Set of Inclusion and Exclusion Criteria

The SLR further requires a definition of search criteria, databases, search terms, and publication period. In this study, the authors have used Scopus as the main source for the keyword search, because it was identified as the most relevant database for scientific publications in the areas of engineering and management science. An additional review of similar databases (Web of Science, Science Direct, and Emerald) did not lead to significant differences in the resulting research studies. Therefore, the authors have decided to use Scopus as the main database for the evaluation of secondary data in the course of this research study.

The following Figure 1 shows the structured research process [16,23] and the pre-defined inclusion and exclusion criteria, which are based on our overall meta-search query.

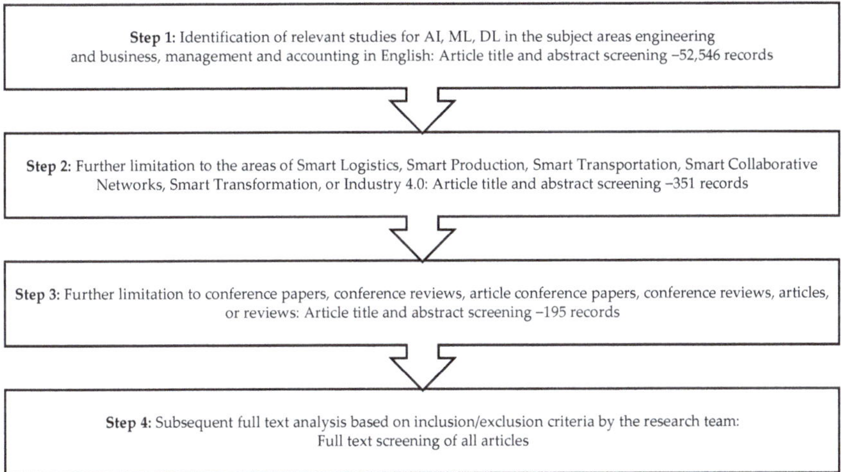

Figure 1. Process steps of the systematic literature review.

In the first step, the authors identified the relevant literature for AI, ML, and DL in the subject areas of engineering and business, management and accounting by screening the article title, abstract, and keywords. In this step, the authors included all kinds of document types and restricted them to the language English. This first approach was primarily used to get a first impression of the current state of research. Therefore, the authors limited the timeframe of the research studies to the period from 1 January 2014 to 1 January 2019. In general, this process resulted in 52,546 identified studies.

In the second step, the authors additionally focused on studies that are related to the areas of Smart Logistics, Smart Production, Smart Transportation, Smart Collaborative Networks, Smart Transformation, or Industry 4.0. Therefore, we limited the previously identified 52,546 papers to the terms Smart Logistics, Smart Production, Smart Transportation, Smart Collaborative Networks, Smart Transformation, or Industry 4.0, which resulted in a total of 351 identified papers. Moreover, the authors have computed a ranking of all identified keywords, which was used to validate the meta-search query for the ongoing database research. No significant keywords were missing in our research strategy.

In the third step, the research was limited to conference papers, conference reviews, articles, or reviews to consider only high-quality studies. In sum, the final meta-search query was formulated as follows: (TITLE-ABS-KEY ("artificial intelligence" OR "machine learning" OR "deep learning") AND TITLE-ABS-KEY ("smart logistics" OR "smart production" OR "smart transportation" OR "smart collaborative networks" OR "smart transformation" OR "industry 4.0")) AND (LIMIT-TO (DOCTYPE, "cp") OR LIMIT-TO (DOCTYPE, "ar") OR LIMIT-TO (DOCTYPE, "cr") OR LIMIT-TO (DOCTYPE, "re")) AND (LIMIT-TO (SUBJAREA, "ENGI") OR LIMIT-TO (SUBJAREA, "BUSI")) AND (LIMIT-TO (PUBYEAR, 2019) OR LIMIT-TO (PUBYEAR, 2018) OR LIMIT-TO (PUBYEAR, 2017) OR LIMIT-TO (PUBYEAR, 2016) OR LIMIT-TO (PUBYEAR, 2015) OR LIMIT-TO (PUBYEAR, 2014)) AND (LIMIT-TO (LANGUAGE, "English")).

The following Table 1 displays the characteristics of the search string.

Table 1. Characteristics of the search string.

Keywords (1)	Keywords (2)	Language	Timeframe	Paper Type [1]
Artificial Intelligence	Smart Logistics	—	—	CP or CR
Machine Learning	Smart Production	English	2014–2019	AR
Deep Learning	Smart Transportation	—	—	RE
—	Smart Collaborative Networks	—	—	—
—	Smart Transformation	—	—	—
—	Industry 4.0	—	—	—

[1] Conference Proceedings (CP), Conference Reviews (CR), Articles (AR), Reviews (RE).

As a result, the authors have identified 195 papers as the basis for the further research process.

2.4. Validation of the Research Results

Based on the quality criteria of empirical research (validity, reliability, objectivity, and generalizability) the authors emphasize that the identification process is one of the most crucial elements within the SLR [17]. Therefore, we ensured the quality of the research results by coding the identified studies with the score "1" (high appropriateness), the score "2" (medium appropriateness), or the score "3" (low appropriateness).

The screening was carried out in two steps by three independent Postdoc-researchers. In the first step, the screening focused only on the title and abstract of the studies. In the second step, the full text of the study was completely evaluated by the research team. Moreover, the reliability was calculated by evaluating significant differences in the scorings. Papers without significant differences were directly included or excluded in/from the research process. Papers with significant differences were reevaluated by the research team to get unambiguous research results.

By reviewing the title and abstract of the 195 papers, we identified 103 relevant papers for the area of Smart Logistics and 148 for the area Smart Production. In the area of Smart Logistics, out of the 103 studies, 33 were assigned as having high appropriateness, whereas in the area of Smart Production, 44 were assigned as having high appropriateness for this research study and the subsequent secondary data analyses. Table 2 displays the total results of the SLR.

Table 2. Total research results of the systematic literature review (SLR).

No./Appropriateness	Smart Logistics	Smart Production
Total studies	103	148
High appropriateness	33	44
Medium appropriateness	52	46
Low appropriateness	18	18

In this paper, we aim to analyze the current state-of-the-art regarding the use of artificial intelligence, machine learning, and deep learning in logistics. Therefore, we consider in the following sections only the papers identified for Smart Logistics.

3. Results

In this section, the authors analyze the descriptive findings of the SLR. Moreover, the full texts of the identified papers will be discussed in the content analysis section.

3.1. Descriptive Analysis

Based on the previously identified total number of 52,546 papers regarding the application of AI, ML, and DL in various scientific disciplines, 103 papers were finally assigned to the area of Smart Logistics in industrial enterprises and therefore rated as relevant for further analysis within this research study. Table 3 shows the distribution of the appropriateness of the identified studies, which was evaluated by screening the title and the abstract of the respective studies.

Table 3. Distribution of appropriateness of studies.

No./Appropriateness	Records	Records (%)
Total studies	103	100.00
High appropriateness	33	32.04
Medium appropriateness	52	50.49
Low appropriateness	18	17.48

Out of the identified 103 full texts in the area of Smart Logistics, 33 papers (32.04%) were classified as papers with high appropriateness, 52 papers (50.49%) were classified as papers with medium appropriateness, and 18 papers (17.48%) were classified as papers with low appropriateness regarding the aim of this research study.

Table 4 shows the distribution of document types based on the identified 103 full papers in the area of Smart Logistics.

Table 4. Distribution of document types.

Type of Document	Records	Records (%)
Conference Proceedings	63	61.17
Conference Reviews	0	0
Articles	38	36.89
Reviews	2	1.94

A total of 63 papers (61.17%) are conference proceedings, 38 (36.89%) are articles, and 2 (1.94%) are reviews. No conference reviews were identified in the final sample. The following Figure 2 shows the development of the relevant research studies from 2014–2019.

Three papers (2.91%) were published in 2014, four papers (3.88%) were published in 2015, six papers (5.83%) were published in 2016, 29 papers (28.16%) were published in 2017, and 61 studies (59.22%) were published in 2018. In general, there is still an increasing trend of relevant studies regarding the application of AI, ML, and DL in the area of Smart Logistics in industrial enterprises.

The following Table 5 displays the distribution of sources of the identified studies. Most of the studies were published in Lecture Notes in Computer Science, Advances in Intelligent Systems and Computing, IFAC-Papers Online, and IFIP Advances in Information and Communication Technology.

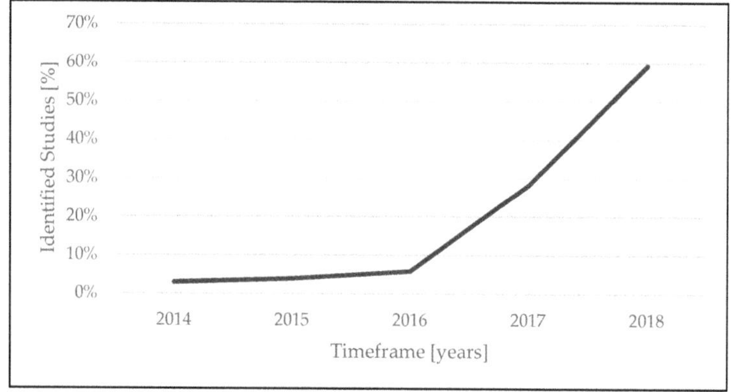

Figure 2. Development of the relevant research studies 2014–2018.

Table 5. Distribution of the identified studies.

Source	Records	Records (%)
Lecture Notes in Computer Science	8	7.77
Advances in Intelligent Systems and Computing	5	4.85
IFAC-Papers Online	5	4.85
IFIP Advances in Information and Communication Technology	4	3.88
Journal of Manufacturing Systems	3	2.91
Procedia CIRP	3	2.91
Procedia Manufacturing	3	2.91
CEUR Workshop Proceedings	2	1.94
Computers in Industry	2	1.94
IEEE Intelligent Systems	2	1.94
Journal of Big Data	2	1.94
Manufacturing Letters	2	1.94
IEEE Transactions on Industrial Informatics	2	1.94
2018 IEEE Global Communications Conference	2	1.94
Others	58	56.31

The following Figure 3 shows an overview of the identified research collaborations based on the systematic literature review.

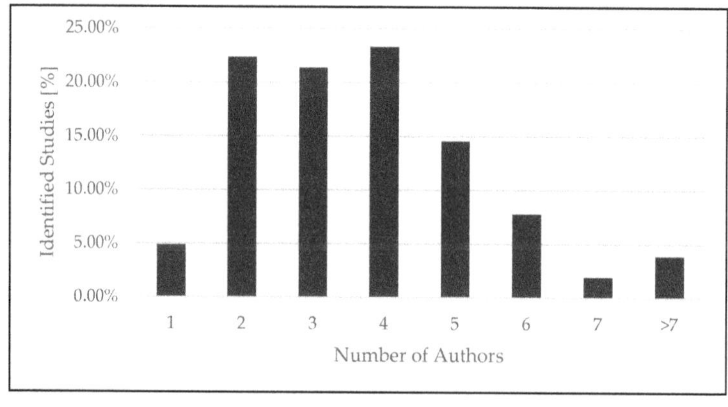

Figure 3. Overview of the identified research collaborations.

Five papers (4.85%) were published by one author, 23 papers (22.33%) were published by two authors, 22 papers (21.36%) were published by three authors, 24 papers (23.30%) were published by four authors, 15 papers (14.56%) were published by five authors, eight papers (7.77%) were published by six authors, two papers (1.94%) were published by seven authors, and four papers (3.88%) were published by more than seven authors.

In the next step, the authors performed a keyword analysis by extracting the author keywords and index keywords of the relevant papers from the Scopus database. The results are displayed in the following Figure 4.

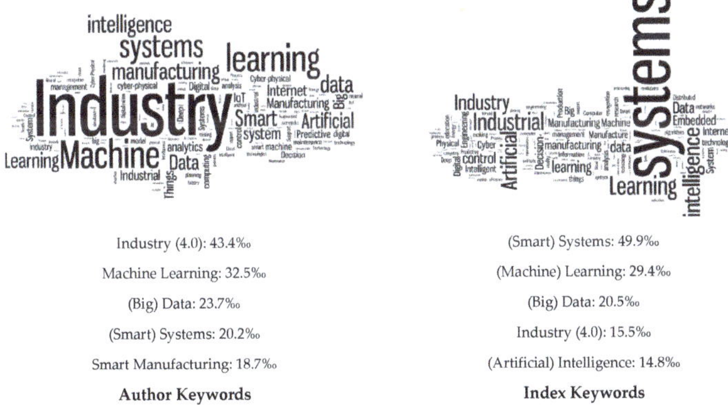

Industry (4.0): 43.4‰

Machine Learning: 32.5‰

(Big) Data: 23.7‰

(Smart) Systems: 20.2‰

Smart Manufacturing: 18.7‰

Author Keywords

(Smart) Systems: 49.9‰

(Machine) Learning: 29.4‰

(Big) Data: 20.5‰

Industry (4.0): 15.5‰

(Artificial) Intelligence: 14.8‰

Index Keywords

Figure 4. Analysis of keywords.

The keywords were analyzed by using the software packages IBM SPSS Statistics 24 (IBM, New York, NY, USA) and Wordle (www.wordle.net, California, CA, USA). Thereby, the size of the words and letters reflects the frequency of the keywords from the final sample of 103 full papers. As expected, the most important author keywords are "Industry (4.0)", "Machine Learning", "(Big) Data", "(Smart) Systems", and "Smart Manufacturing". Moreover, the top-ranked index keywords are "(Smart) Systems", "(Machine) Learning", "(Big) Data", "Industry (4.0)", and "(Artificial) Intelligence".

The following Figure 5 displays the classification of the relevant research study by the type of study, divided into Case Study (CS), Conceptual Approach (CA), Experiment (EX), and Literature Review (LR). The classification is based on the content analysis and subsequent assignment of the works to one main type of study by the research team.

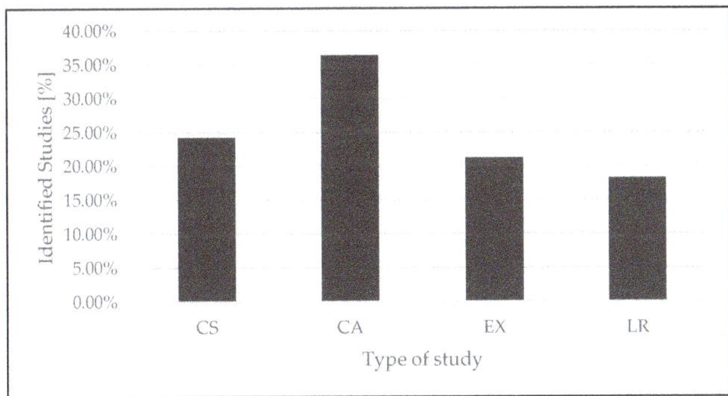

Figure 5. Classification of the relevant research studies per type of study—Case Study (CS), Conceptual Approach (CA), Experiment (EX), and Literature Review (LR).

From the identified 33 articles with high appropriateness, eight (24.24%) articles can be classified as case articles (CS), 12 (36.36%) articles can be classified as conceptual approaches (CA), seven (21.21%) articles can be classified as an experiment (EX), and six (18.18%) articles can be classified as literature review (LR).

3.2. Content Analysis

In this section, the authors analyze the full texts of the identified papers. Therefore, Tables 6 and 7 display the comprehensive results from the systematic literature review by briefly summarizing the clusters, the main references, the number of records, and the percentage of records in Table 6, as well as the authors, the type of the study, the clusters, and the main content of each identified work in Table 7. The research team conducted a profound content analysis to categorize the selected literature into similar clusters.

Table 6. Classification of the identified studies.

No.	Cluster	Main References	Records	Records (%)
1	Strategic and Tactical Process Optimization	[25–28]	4	12.12%
2	Cyber-Physical Systems in Logistics	[29–35]	7	21.21%
3	Predictive Maintenance	[36–41]	7	21.21%
4	Hybrid Decision Support Systems	[42–44]	3	9.09%
5	Production Planning and Control Systems	[45–49]	4	12.12%
6	Improvement of Operational Processes in Logistics	[50–53]	4	12.12%
7	Intelligent Transport Logistics	[54–57]	4	12.12%

The results were further aggregated to the following five clusters:

(1)   Strategic and tactical process optimization;
(2)   Cyber-physical systems in logistics;
(3)   Predictive maintenance;
(4)   Hybrid decision support systems;
(5)   Production planning and control systems;
(6)   Improvement of operational processes in logistics;
(7)   Intelligent transport logistics.

The following Table 6 displays the results of the classification of five clusters regarding the application of AI, ML, and DL in the area of Smart Logistics in industrial enterprises.

In the following paragraphs, we will describe and summarize the content of the identified studies for each of the defined clusters.

3.2.1. Strategic and Tactical Process Optimization

A total of 12.12% of the identified articles were assigned to the cluster "strategic and tactical process optimization". Gursch et al. review learning systems for management support in the area of condition monitoring and process control of enterprises. Furthermore, the authors state that ML can provide solutions in the fields of optimization, automation, and human support by handling complex problems that cannot be solved via static and non-adaptive computer programs [25]. Brodsky et al. focus on the application of decision analytics, for example, for the development of decision guidelines in strategic and tactical reasoning. Furthermore, the guidelines can be used to perform monitoring, analysis, planning, and optimization tasks within manufacturing and logistics processes [26]. Qu et al. propose a general computational reasoning and learning framework under realistic conditions. The approach explains how to realize smart processes from a formal and data-orientated methodological point of view [27]. Bonino and Vergori introduce a concept and architecture for connecting inner and outer value chains in modern factories by using ML, AI, and IoT technologies. Thereby, the authors highlight how modern IT-systems will innovate and optimize daily processes in production and logistics by implementing multi-agent systems and ML technologies [28].

3.2.2. Cyber-Physical Systems in Logistics

A total of 21.21% of the identified articles were assigned to the cluster "frameworks for CPS in logistics". In this context, Peres et al. introduce an IDARTS framework for intelligent data analysis and real-time supervision as a guideline for the implementation of scalable, flexible, and pluggable analyses within production environments. The framework can be used to translate data into business advantages in fields such as predictive maintenance, quality control, production, and logistics through the implementation of CPS based on cloud computing. The framework further includes technologies of data acquisition, ML, and run-time reasoning for production and logistics processes [29]. Guo et al. describe a method for learning from imbalanced machinery data by transferring visual elements detectors. The authors propose a methodology to organize collected sensor data into image form and utilize visual element detectors based on convolutional neuronal networks (CNN) [30]. Ferrer et al. present a concept for the implementation the adaptation of CPS by focusing on the production line, logistics, and facilities based on advanced data analytics, machine-to-machine IoT protocols, machine learning, knowledge representation algorithms, and cloud-based platforms in (smart) manufacturing environments. Moreover, they present a scenario for the implementation of CPS in the course of a pilot study [31].

Thalmann et al. describe new analytic methods and tools for the optimization of industrial processes. The projects aim to develop a set of tools, e.g., algorithms, analytic machinery, planning approaches, and visualization for industrial process improvement based on data analytics [32]. Morozov et al. present an adaptation approach for CPS, namely the Formal Concept Analysis (FCA), which includes AI and ML for the structuring of knowledge and the optimization of the CPS interoperability [33]. Marella and Mecalla and Marella et al. introduce a process management system (PMS) for cyber-physical processes (CPPs) and smart process management (SmartPM). The system can be used to automatically adapt processes by using AI and cognitive computing based on a large amount of data [34,35].

3.2.3. Predictive Maintenance

A total of 21.21% of the identified articles were assigned to the cluster "predictive maintenance". Thereby, Subakti and Jiang propose the design, development, and implementation of an augmented reality system for machines in smart factories. A deep learning image detection module identifies different machines and IoT allows the machines to report machine settings and machine states to a cloud-based server [36]. Susto et al. describe a methodology to derive the health factor of machines by applying a Monte Carlo approach based on particle filtering techniques to a real industrial predictive maintenance problem in the semiconductor industry [37]. Wang and Wang discuss the impact of AI on the future of predictive maintenance by focusing on DL technology. Thereby, they state a list of state-of-the-art algorithms in the field of predictive maintenance, e.g., deep feedforward networks, long short-term memory, convolutional networks, deep belief networks, etc., which can be considered as key success factors for future industries [38].

Klein and Bergman propose an approach for the generation of predictive maintenance data for ML investigations by using a Fischertechnik model factory equipped with several sensors. The dataset will be published and can be used in future investigations and for the further development of ML and DL approaches as well [39]. Cho et al. introduce a hybrid ML approach for predictive maintenance in smart factories. The approach combines unsupervised learning with semi-supervised learning to provide data-driven decision support [40]. Wuest et al. review the application of ML techniques in manufacturing in the areas of monitoring and image recognition. Thereby, they distinguish between supervised learning, unsupervised learning, and reinforcement learning, which can be used as powerful toolsets within manufacturing environments [41].

3.2.4. Hybrid Decision Support Systems

A total of 9.09% of the identified articles were assigned to the cluster "decision support systems and man-machine interaction". In this context, Terziyan et al. introduce Pi-Mind technology as a mixture of human-expert-driven decision-making and AI-driven decision-making approaches for smart manufacturing processes based on AI and ML technologies. In many cases, this hybrid approach outperforms rational decision-making processes by human operators [42]. Venkatapathy et al. and Zeidler et al. describe a basic concept of a hybrid network for the usage of human and machine synergies in the intra-logistics. Thereby, they focus on optical reference systems, radio reference systems, laser project systems, virtual reality systems, robot systems, LR-WPAN, and other wireless network systems, and networked computational systems [43,44].

3.2.5. Production Planning and Control Systems

A total of 12.12% of the identified articles were assigned to the cluster "production planning and control systems". Lolli et al. present an approach for a multi-criteria-based classification of inventory based on machine learning techniques. Thereby, the authors achieve a reduction of simulation effort by using supervised classifiers from the field of ML [45]. He et al. review a set of multi-objective swarm intelligence algorithms for flow shop scheduling problems. Besides conventional algorithms, new ML-based approaches are also presented in this paper. The performance of the various algorithms can be evaluated by the minimum distance to the Pareto-optimal front, the goodness of the distribution, and the maximum spread [46]. Zhang et al. systematically review the literature on job shop scheduling research in the context of Industry 4.0, based on ML, AI, and DL technologies. The algorithms were further classified in exact optimization methods, e.g., efficient rule approach, mathematical programming approach, branch and bound methods, and approximate methods (e.g., constructive methods, artificial intelligence methods, local search methods, and meta-heuristic methods) [47].

Gomes et al. develop an ambient intelligent-based decision support system for production planning and control based on ML that assists in the creation of standard work procedures that assure production quantity and efficiency by using ambient intelligence, optimization heuristics, and ML [48]. Luetkehoff et al. develop a self-learning production control system by using algorithms of AI. The developed approach presents a new platform-based concept to collect and analyze data by using self-learning algorithms. The patterns can be used for a prediction of future system behavior [49].

3.2.6. Improvement of Operational Processes in Logistics

A total of 12.12% of the identified articles were assigned to the cluster "improvement of operational processes in logistics". Wen et al. discuss the application of swarm robotics in the area of Smart Logistics by outlining some possibilities and fields of application, e.g., smart warehousing, smart delivery, route tracking, precise supply chains, green logistics, and smart ICTs, in the next area of modern logistics [50]. Laux et al. describe a reflection-based sound localization system that is based on ML approaches and can be used for indoor localization and object tracking [51].

Ademujimi et al. review the current literature on ML techniques in manufacturing by focusing on techniques, e.g., Bayesian networks, the artificial neural network, the support vector machine, and the hidden Markov model for the optimization of manufacturing fault diagnosis [52]. Teschemacher and Reinhart develop an ant colony optimization algorithm to enable dynamic milk runs in logistics to reduce the number of necessary vehicles in long-term optimization approaches [53].

Table 7. Systematic literature review (SLR) on artificial intelligence (AI), machine learning (ML), and deep learning (DL) in Smart Logistics.

No.	Ref. No.	Author(s) and Year	Type of Study	Cluster	Content
1	[29]	Peres et al., 2018	Case Study	Cyber-Physical Systems in Logistics	IDARTS framework for intelligent data analysis and real-time supervision as a guideline for the implementation of scalable, flexible, and pluggable analyses within production environments based on cyber-physical systems (CPS) and cloud computing.
2	[30]	Guo et al., 2018	Experiment	Cyber-Physical Systems in Logistics	Methodology to organize collected sensor data into image form and utilize visual element detectors based on convolutional neuronal networks (CNN).
3	[31]	Ferrer et al., 2018	Case Study	Cyber-Physical Systems in Logistics	Concept and exemplary case study for the implementation of CPS by focusing on the production line, logistics, and facilities in (smart) manufacturing companies.
4	[32]	Thalmann et al., 2018	Conceptual Approach	Cyber-Physical Systems in Logistics	Concept and resp. discussion of a set of tools, e.g., algorithms, analytic machinery, planning approaches, and visualization for industrial process improvement based on data analytics.
5	[50]	Wen et al., 2018	Conceptual Approach	Improvement of Operational Processes	Application of swarm robotics in the area of Smart Logistics (smart warehousing, smart delivery, route tracking, precise supply chains, green logistics, and smart information and communication technologies (ICTs), etc.).
6	[42]	Terziyan et al., 2018	Conceptual Approach	Hybrid Decision Support Systems	Pi-Mind as a mixture of human-expert-driven and AI-driven decision-making approaches based on AI and ML.
7	[36]	Subakti & Jiang, 2018	Case Study	Predictive Maintenance	Usage of AR and ML for machine detection and the provision of machine data in smart factories.
8	[51]	Laux et al., 2018	Experiment	Improvement of Operational Processes	Reflection-based sound localization system based on ML approaches.
9	[37]	Susto et al., 2018	Experiment	Predictive Maintenance	Predictive maintenance model based on a numerical Monte Carlo approach with particle filtering techniques.
10	[38]	Wang & Wang, 2018	Literature Review	Predictive Maintenance	Conceptual discussion of AI and DL in predictive maintenance applications.
11	[39]	Klein & Bergmann, 2018	Case Study	Predictive Maintenance	Approach for the generation of predictive maintenance data for ML investigations by using a Fischertechnik model factory.
12	[45]	Lolli et al., 2019	Experiment	Production Planning and Control Systems	Approach for a multi-criteria-based classification of inventory based on machine learning techniques.
13	[33]	Morozov et al., 2018	Conceptual Approach	Cyber-Physical Systems in Logistics	Adaptation approach for CPS, namely the formal concept analysis (FCA), which includes AI and ML for the structuring of knowledge and the optimization of the CPS interoperability.
14	[46]	He et al., 2018	Literature Review	Production Planning and Control Systems	Review swarm intelligence algorithms for multi-objective flow shop scheduling problems.
15	[40]	Cho et al., 2018	Case Study	Predictive Maintenance	Hybrid ML approach for predictive maintenance, which combines unsupervised learning with supervised learning in smart factories.

Table 7. *Cont.*

No.	Ref. No.	Author(s) and Year	Type of Study	Cluster	Content
16	[34]	Marrella & Mecella, 2018	Conceptual Approach	Cyber-Physical Systems in Logistics	Process management system (PMS) for cyber-physical processes (CPPs) and smart process management (SmartPM) based on AI and cognitive computing.
17	[35]	Marrella et al., 2018	Case Study	Cyber-Physical Systems in Logistics	Process management system (PMS) for cyber-physical processes (CPPs) and smart process management (SmartPM) based on AI and cognitive computing.
18	[43]	Venkatapathy et al., 2017	Conceptual Approach	Hybrid Decision Support Systems	Concept of a hybrid network for the usage of human and machine synergies in the intra-logistics.
19	[44]	Zeidler et al., 2017	Conceptual Approach	Hybrid Decision Support Systems	Concept of a hybrid network for the usage of human and machine synergies in the intra-logistics.
20	[28]	Bonino & Vergori, 2017	Case Study	Strategic and Tactical Process Optimization	Architecture for connecting inner and outer value chains in modern factories by using ML, AI, and IoT technologies.
21	[47]	Zhang et al., 2019	Literature Review	Production Planning and Control Systems	Review of literature on job shop scheduling research in the context of Industry 4.0.
22	[52]	Ademujimi et al., 2017	Literature Review	Improvement of Operational Processes	Review of literature on ML techniques in manufacturing.
23	[25]	Gursch et al., 2016	Literature Review	Strategic and Tactical Process Optimization	Review of learning systems for management support in the areas of condition monitoring and process control, scheduling, and predictive maintenance.
24	[48]	Gomes et al., 2017	Conceptual Approach	Production Planning and Control Systems	Ambient intelligent-based decision support system for production planning and control based on ML.
25	[49]	Luetkehoff et al., 2017	Conceptual Approach	Production Planning and Control Systems	Development of a self-learning production control system by using algorithms of AI.
26	[53]	Teschemacher & Reinhart, 2017	Experiment	Improvement of Operational Processes	Development of an ant colony optimization algorithm to enable dynamic milk runs in logistics.
27	[41]	Wuest et al., 2016	Literature Review	Predictive Maintenance	ML techniques in manufacturing in the areas of monitoring, and image recognition.
28	[26]	Brodsky et al., 2014	Conceptual Approach	Strategic and Tactical Process Optimization	Application of decision analytics in smart manufacturing.
29	[27]	Qu et al., 2014	Conceptual Approach	Strategic and Tactical Process Optimization	General computational reasoning and learning framework for smart manufacturing.
30	[54]	Li et al., 2018	Experiment	Intelligent Transport Logistics	Intelligent transport system based on AI, ML, and DL technologies.
31	[55]	Cheng et al., 2017	Case Study	Intelligent Transport Logistics	Fuzzy-group-based control for smart transportation processes.
32	[56]	Li et al., 2015	Experiment	Intelligent Transport Logistics	Analyses and optimization of traffic conditions based on internet of things (IoT) techniques.
33	[57]	Edelstein, 2014	Conceptual Approach	Intelligent Transport Logistics	Smart transport management systems through ICT.

3.2.7. Intelligent Transport Logistics

A total of 12.12% of the identified articles were assigned to the cluster "intelligent transport logistics". Li et al. (2018) introduce an intelligent transport system that combines information technology, data communication technology, electronic sensing technology, global positioning technology, geographical information system technology, computer processing technology, and system engineering technology to a real-time, accurate, efficient and intelligent transportation management system based on deep belief network models (DBN) and support vector regression classifier (SVR) [54]. Cheng et al. (2017) discuss a fuzzy-group-based control for smart transportations, which reduces waiting time and improves the performance by up to 40% [55]. Li et al. (2015) analyze traffic conditions based on IoT techniques by applying numerical techniques from data mining and machine learning approaches [56]. Edelstein (2014) outlines the potentials of smarter transport management through ICT in terms of operational efficiency, reduced costs, and an enhanced entrepreneurial environment for the staff to develop and implement process innovations by using integrated multimodal transport systems [57].

4. Conceptual Framework for the Application of AI, ML, and DL in Smart Logistics

4.1. Framework for the Application AI, ML, and DL in Smart Logistics

In this paper, the authors have identified five clusters that can be amalgamated to a conceptual framework for the application of AI, ML, and DL in Smart Logistics in industrial enterprises. Figure 6 illustrates the developed clusters, namely, "frameworks for CPS in logistics", "predictive maintenance", "decision support systems and man-machine interaction", "production planning and control systems", and "improvement of operational processes in logistics". According to the results of the content analysis, the research team mapped to each of the identified clusters of relevant methods and instruments indicated in the literature.

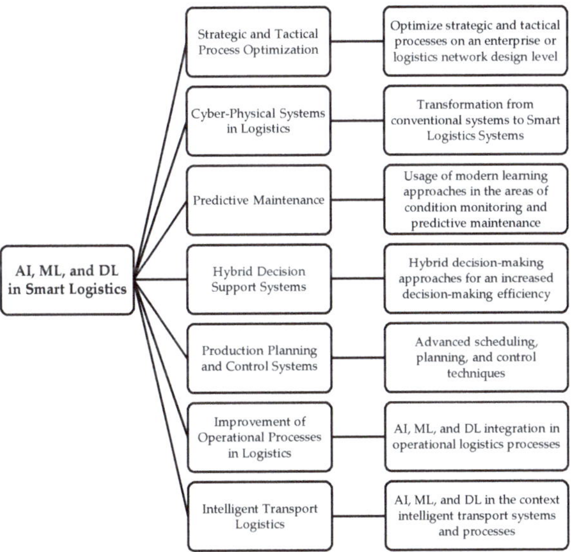

Figure 6. Conceptual framework for AI, ML, and DL in Smart Logistics.

Cluster 1, "strategic and tactical process optimization", concentrates on the application of AI, ML, and DL methods to optimize strategic and tactical processes on an enterprise or logistics network design level. Based on learning systems for management support, so-called management information systems can be enriched with intelligence, therefore providing not only data, information, and key performance indicators but also preparing and make decisions on a strategic and tactical level [25–28]. Further progress in such strategic and tactical process optimization will be of significant importance for large companies operating in different countries.

Cluster 2, "cyber-physical systems in logistics", describes opportunities for the development from a conventional logistics system to a Smart Logistics system based on actors and sensors for real-time data analysis and enhanced knowledge management through state-of-the-art learning approaches. CPS can be used to improve the quality of production and logistics processes, which further affects the overall efficiency of the industrial enterprise. Moreover, CPS will lead to increased connectivity, consistent digitalization, better modeling techniques, more flexibility, and higher versatility and reusability of systems and systems' components [29–35].

Cluster 3, "predictive maintenance", focuses on the usage of learning approaches in the areas of condition monitoring and predictive maintenance. Thereby, a multitude of studies suggests the application of recent AL, ML, and DL approaches for the continuous reporting of machine settings, machine states, and quality parameter settings. Based on real-time data, enhanced knowledge can be used for further predictive analysis regarding a strategic and pro-active plant maintenance strategy for production and logistics processes [36–41].

Cluster 4, "hybrid decision support systems", describes the improvement of non-automated and therefore human-centered decision-making processes by using learnable support systems. In this context, decision-relevant information will be automatically collected, aggregated, and pre-analyzed by AI, ML, and DL technologies. In many cases, these hybrid decision-making processes outperform purely rational decision-making processes [42–44,58].

Cluster 5, "production planning and control systems", introduces new opportunities for advanced planning and control approaches in the research fields of inventory management, flow shop problems, traditional job shop scheduling problems, production process optimization, and the self-learning abilities of modern production planning and control systems based on the application of AI, ML, and DL technologies [45–49].

Cluster 6, "improvement of operational processes in logistics", outlines various possibilities for the enhancement of operational processes in logistics by applying AI, ML, and DL technologies. Thereby, swarm robotics can be used to optimize smart warehouses, sound location systems can increase the efficiency of identification and tracking approaches, AI-based algorithms ensure holistic manufacturing fault diagnosis, and ant colony optimization approaches enable the optimization of milk runs in logistics problems [50–53].

Cluster 7, "intelligent transport logistics", analyzes the application of AI, ML, and DL technologies in intelligent transport systems and intelligent transport processes. In this context, the performance of transport logistics can be increased by applying AI-based methods in combination with state-of-the-art approaches based on the usage of information technology, data communication technology, electronic sensing technology, global positioning technology, geographical information system technology, computer processing technology, and system engineering technology [54–57].

4.2. Practical Examples

In this section, we looked for practical examples for each of the identified clusters or applications of AI, ML, and DL in logistics. This analysis should provide an overview of how the proposed framework can be transferred into practice and what are the practical implications for managers of companies.

Aitheon presents on his website one of the first AI-driven project management tools. The project management AI generates a straightforward list for those who want to be in the general work process without the need to understand or learn the depths of project management tools. The tool generates prioritized and grouped tasks by artificial intelligence. An AI-based sketchboard visually organizes information to show the relationships amongst the pieces of the whole project. This helps to create a vision of complex ideas, concepts, or projects, which is generated into tasks viewable in Gantt or Kanban views [59].

DHL research describe their vision of digital twins in logistics. Artificial intelligence has given digital twins and cyber-physical systems a big push in creating new value. Today, all the data DHL has from sensors, historical performance, and inputs about behavior lends itself to being linked to spatial models and to predict future behavior by changing different inputs. The data and prediction capabilities of AI make the spatial model come alive [60].

Presenso is a predictive maintenance software using machine learning and deep learning algorithms to drive precise and continuous failure prediction. With the use of software, logistics companies can achieve operational savings through full predictive maintenance aimed at yield optimization. The system collects large amounts of data at high speed and streams the data to the Cloud in real-time. Using unique, proprietary deep neural-network architectures, Presenso's analytic engine autonomously interlinks events with components within the machines and ultimately predicts evolving failures. In addition, it provides valuable information about the remaining time to failure and its origin within the machine [61].

According to a study of PWC, 67% of business leaders believe that AI and automation will impact negatively on stakeholder trust levels in their industry in the next five years. The central challenge is that many of the AI applications operate within black boxes, offering little if any discernible insight into how they reach their outcomes. The greater the confidence in AI, the faster and more widely can it be deployed [62].

SkyPlanner APS is an advanced planning and scheduling system using AI. The software includes AI that instantly optimizes work order, in which production is most efficiently scheduled. The AI algorithm also takes into account details that can further enhance production efficiency. For example, in many productions, it is advisable to schedule jobs that use the same materials or tools in succession [63]. AI is also able to more accurately estimate the time it takes to complete the different jobs.

Swarm Logistics is a deep-tech software technology company specializing in the development of intelligent, autonomous transportation systems. The Auto-Dispatcher of Swarm Logistics is based on a complex algorithm that is constantly improving itself with the use of artificial intelligence. This software was tested in a case study and compared to previous planning. The results of the comparison showed cost savings of 25% and 35% faster delivery for the transportation company [64].

Siemens Mobility is testing its ITS Digital Lab applications and services for smarter management of road traffic, fleets such as e-bikes, and intermodal mobility. Connected vehicles sending data in real-time, infrastructure systems transmitting their status to Siemens' internet-of-things platform MindSphere and road users who are connected with their smartphones all produce an immense amount of data. This rich and growing source of data is changing the types of services of mobility that are feasible. Siemens Mobility is working on solutions for a Balanced Intermodal Mobility Ecosystem, which manages not only the road network but also specific fleets within the network and ultimately the traveler across different modes of transportation [65].

5. Discussion

In this paper, the authors identified the following five clusters for the application of AI, ML, and DL in Smart Logistics: (1) frameworks for cyber-physical systems (CPS) in logistics; (2) AI, ML, and DL in the context of (predictive) maintenance; (3) decision support systems and man-machine interaction; (4) production planning and control systems, and (5) improvement of operational processes in logistics. In the above-described conceptual framework, we analyzed the main clusters where AI, ML, or DL are applied today or where the potential for application is already studied.

Looking at the first cluster in the proposed framework, AI and related technologies and methods will be used in the near future to do better forecasting and to better understand users and customers. This will have a significant impact on the optimization of the forecast on a strategic/tactical level and may be also used to determine new customer demand. In the future, AI may be used also in higher-level processes to detect fraud, prevent cybersecurity threats, and generally optimize higher-level processes. Possibilities of the future use of AI can also be in project management tasks, reducing the failure rate of projects thanks to predictive analysis and to more accurate project management. In addition, if AI algorithms can win chess games, they will also be able to be used in the generation of corporate strategies.

Taking a more detailed look at the second identified cluster dealing with the application of AI methods to create CPS in logistics, we can observe that already there are many research activities for realizing CPS and mono-directional information flow from the physical object to the digital model (digital shadow), and in a further step a bidirectional information flow between both models to control the physical system through a so-called digital twin [66]. CPS are receiving large amounts of different datasets and data from the logistics systems. Today, most of the practitioners face the challenge that they do not have the right tools to exploit the amount of data in the right manner. Here AI methods can complete the vision of a digital-twin-based control of logistics systems by analyzing big data from CPS, determining data patterns and therefore automating workflows and processes for more immediate and more qualitative control of logistics systems. Although some works deal with this topic, we find that, especially for logistics systems, this field is still an emerging research field with much need for further investigation and case study applications.

The third cluster deals with AI methods for realizing what we call predictive maintenance as an enhancement of preventive maintenance, which is well-known from lean management. Predictive maintenance as a concept has been well-discussed since the beginning of the new century, but looking at the Scopus database using this keyword, research on this concept has "exploded" in the years after 2017 when progress was made in the application of AI-based methods for realizing predictive maintenance. As AI methods depend a lot on the availability of a large amount of data, the field of predictive maintenance is a highly important research direction due to new generations of logistics systems equipped with sensors and providing large amounts of data that can be used for data analytics. Seeing this development, we consider predictive maintenance as one of the more consolidated fields of research for AI in logistics, although we are conscious of the fact that there is still a huge need for investigation especially for applied research and case study research.

The fourth cluster of "hybrid decision support systems" follows the current trend of human-centered engineering. Despite all of the potential for automation and digitalization in production and logistics, we have understood that many processes still depend from skilled workers and that the human will also in the future play a big and important role in the industry. What we can do is to support workers with cognitive assistance systems to enhance their capabilities in decision-making as well as taking advantage of a large amount of data. AI-based methods play a big role in this ongoing development of such digital aid systems for our workers. As indicated also in the analyzed literature, it will be important for the future not to only offer pure rational decision-making assistance tools, as this would decrease the role of the human from an intelligent individuum to a "stupid" executor of commands and decisions taken from a "super-intelligent system". It will become important to develop solutions using an approach based on so-called Explainable AI [67] and thus a hybrid way where the human is still able to understand the result through explainable interfaces. We state the hypothesis that only such hybrid systems will succeed in being applied also in practice, as the human being is still a skeptical individuum that would not accept assistance systems based on pure rational decision-making. While the use of AI in man–machine interactions has already been studied for many years, we consider therefore the development of decision-making systems based on explainable AI as one of the main future challenges in research.

The fifth cluster of "production planning and control systems" can be considered again as one of the already more developed and perhaps more consolidated fields compared to the other clusters identified in the proposed conceptual framework. For many years now production planning and control deal with the optimization of the planning when it comes to conduct advanced planning and scheduling tasks and to transfer this results in meaningful control mechanisms for production control.

We can already find several works dealing with the use of simulation-based methods for scheduling since the early 1990s [68]. On the other hand, we postulate that there will be a "renaissance" in this field due to the ongoing trend of AI, opening an emerging field towards real-time planning. A human planner would never be able to process information as fast as AI does based on sensors and seamless vertical data integration. Through the combination of a steady increase of computational power and storage capacity for a large amount of data with sensor devices for collecting real-time data, real-time planning comes within reach in the near future.

The sixth cluster deals with improvements of operational processes in logistics that can often be deduced from nature, as the examples in the framework description have shown (see also [50–53]). The topic of biological transformation is an already emerging theme promoted, e.g., by the Fraunhofer research institutes [69] and other researchers [70], and describes the change from solely bioinspired manufacturing or logistics systems towards those that are bio-integrated and bio-intelligent by combining what we can deduce and learn from the observation of processes in nature and leaping computational power as well as progress in AI, ML, and DL.

The seventh cluster of "intelligent transport logistics" has been making enormous progress in data-based traffic optimization and the development of autonomous vehicles for years. With the latest technologies in object recognition, roads and important traffic junctions can be monitored even better in real-time, reducing possible delays in processing time. A great challenge for the future will be the investigation and finding of appropriate solutions for reducing or eliminating the risk of cyber-attacks and biased decisions about transport as well as discussing ethical questions regarding liability for the decisions taken by artificial intelligence in the place of humans. In this regard, for example, the EU is taking important steps to adapt its regulatory framework to these developments so that it supports innovation while at the same time ensuring respect for fundamental values and rights [71].

The following Table 8 summarizes the main research actions that can be deduced from the discussion of the identified clusters for applying AI, ML, and DL in Smart Logistics. In addition, it addresses also the experts/stakeholders that should be involved in research activities, as logistics is a broad field in which several different subjects, perspectives, and types of expertise converge. This should guarantee that further research is target-oriented and follows the needs of practitioners in the field of logistics. The listed research actions should motivate researchers of basic and applied research to conduct further research to exploit the full potential of AI, ML, and DL in the field of logistics.

Table 8. Research actions and experts/stakeholders to be involved.

No.	Research Action	Cluster	Experts/Stakeholder
1	Further research on the automated generation of corporate strategies	C1—Strategic and Tactical Process Optimization	Top management
2	Further research for AI support in project management to reduce failures and increase the success of projects	C1—Strategic and Tactical Process Optimization	Project managers
3	Case study research on applications for AI-based CPS in logistics	C2—Cyber-Physical Systems in Logistics	Users and experts from all fields of logistics
4	Case study research on applications of predictive maintenance	C3—Predictive Maintenance	Maintenance managers of logistics systems and vehicle fleets
5	Further research on explainable AI for AI-based hybrid decision support systems	C4—Hybrid Decision Support Systems	Users and experts from all fields of logistics
6	Real-time planning and (re-)scheduling	C5—Production Planning and Control Systems	Production planners and intralogistics experts
7	Learning from nature through bio-intelligence in logistics	C6—Improvement of Operational Processes in Logistic	Operations managers
8	Development of low-cost real-time monitoring devices/solutions	C7—Intelligent Transport Logistics	Traffic Planners, hardware providers
9	Clarify questions and provide solutions for reducing the risk of cyber-attacks and regarding ethical/liability questions	C7—Intelligent Transport Logistics	Experts in ethics and cybersecurity

6. Limitations and Implications

6.1. Limitations of This Study

The study conducted a review of the scientific literature from 2014 to January 2019, the date of data extraction from the Scopus scientific database. A limit of this research is that during the period of the screening, selection and analysis and development of the framework, further research on AI in logistics may have been performed and published. We are aware that the current rapid development of research in this area is a limit for the results of this review. For this reason, in the following we have summarized the most important work that has been done on this topic in the period from the beginning of 2019 to the beginning of 2020, to analyze whether further research directions have emerged. Firstly, this analysis confirmed the proposed framework, as also after the period investigated by the authors, many works deal with the clusters identified. In addition, a scientific discussion on some new topics has been started. The following short overview does not present an exhaustive list of new research topics, but some examples of the most interesting ones.

Giusti et al. address synchromodal logistics as an emergent topic in logistics and transportation where AI-based methods may open new possibilities. Synchromodality is an emerging and attractive concept in logistics, developed and established in the Benelux region during the last decade. The main purpose of synchromodality is reducing costs, emissions, and delivery times while maintaining the quality of supply chain service through the smart utilization of available resources and synchronization of transport flows [72].

In Le et al. neural network-based approaches are used for example to estimate, predict, and optimize fuel consumption for container ships in Korea. The developed model has thus been applied to confirm the effectiveness of the slow-steaming method for achieving energy efficiency [73].

Liu et al. examine the effects of context-awareness on the ambient intelligence of logistics service quality, as context-awareness makes it possible to provide personalized service for each client. Context-awareness-based ambient intelligence predicts users' intention-to-use depending on the contexts they provide. By applying the prediction to logistics services, it can provide customized service to keep clients satisfied. A key issue in user-centered services is how to detect each user-specific situation and choose a certain service that meets users' requirements the best, and then to provide support for real-time decision-making [74].

According to Ceyhun, using AI will contribute to the prevention of ship-related accidents by anticipating future cases by using pinpoint calculations [75].

Another use of AI-based methods will be to develop intelligent road inspection systems for smart mobility and transportation maintenance systems [76].

6.2. Implications for Academia and Practitioners

From our point of view, this research is interesting for both academics and practitioners. While there are now some literature review articles on artificial intelligence, machine learning, or deep learning in manufacturing or business processes, there is a great lack of such review articles in the field of logistics. With this paper, researchers can find a comprehensive review of the current state of research. In addition, the paper presents a framework for the use of AI, ML, or DL in logistics. This framework, as well as the research actions and fields collected in the discussion Section, should contribute to and encourage researchers to carry out further research in these areas and thus take an important step towards consolidating these areas.

More and more managers and practitioners from logistics companies or logistics sectors are interested in the potential of AI-based methods in logistics. However, the complexity of these topics makes it very difficult for practitioners to estimate their use or even to get an overview of the current status. Therefore, it was our concern after the creation of the framework, which is informative for academics as well as for practitioners, to show practical examples from management literature and case studies. This will provide practitioners with an overview of the current methods as well as new input to plan and start initiatives for the use of AI in their logistics processes.

7. Conclusions

In the context of Smart Logistics, the application of AI, ML, and DL technologies is still in an early stage of development. Most of the identified studies are concepts, laboratory experiments, or in a very early testing phase. Mature industrial applications are still missing. However, the continuous reporting of machine settings, machine states, quality parameter settings, predictive maintenance, decision-making support systems, advanced scheduling, planning, and control approaches in the research fields of inventory management, flow shop problems, traditional job shop scheduling problems, production process optimization, and the improvement of operational logistics processes, e.g., identification and tracking approaches, can be seen as promising areas within the Smart Logistics framework.

The findings of this research study should be used as a starting point for future investigations regarding the application of AI, ML, and DL technologies in the area of Smart Logistics in industrial enterprises, and provide a framework for practitioners in industrial companies for the successful implementation of state-of-the-art technologies as well. Therefore, it is important to integrate different research areas, e.g., information technology, logistics, mechanical engineering, industrial engineering, mathematics, and statistics, into future research projects.

Author Contributions: H.Z. led and supervised the research project. M.W. and E.R. organized, planned, and conducted the SLR. All authors have read and agreed to the published version of the manuscript.

Funding: This research was funded by the project "SME 4.0—Industry 4.0 for SMEs" (funded by the European Union's Horizon 2020 R&I program under the Marie Skłodowska-Curie grant agreement No. 734713).

Conflicts of Interest: The authors declare no conflict of interest.

References

1. Matt, D.T.; Modrák, V.; Zsifkovits, H. *Industry 4.0 for SMEs*; Springer: Berlin/Heidelberg, Germany, 2020.
2. Zsifkovits, H.; Woschank, M. Smart Logistics–Technologiekonzepte und Potentiale. *BHM Berg Hüttenmännische Monatshefte* **2019**, *164*, 42–45. [CrossRef]
3. Bosch, G.; Bromberg, T.; Haipeter, T.; Schmitz, J. Industrie und Arbeit 4.0: Befunde zu Digitalisierung und Mitbestimmung im Industriesektor auf Grundlage des Projekts "Arbeit 2020". In *IAQ-Report: Akutelle Forschungsergebnisse aus dem Institut Arbeit und Qualifikation*; IAQ: Duisburg Essen, Germany, 2017. Available online: http://www.iaq.uni-due.de/iaq-report/2017/report2017-04.pdf (accessed on 1 March 2020).
4. Kagermann, H.; Anderl, R.; Gausemeier, J.; Schuh, G.; Wahlster, W. *Industrie 4.0 in a Global Context: Strategies for Cooperating with International Partners*; Herbert Utz Verlag: Munich, Germany, 2016.
5. Dallasega, P.; Woschank, M.; Ramingwong, S.; Tippayawong, K.; Chonsawat, N. Field study to identify requirements for smart logistics of European, US and Asian SMEs. In Proceedings of the International Conference on Industrial Engineering and Operations Management, Bangkok, Thailand, 5–7 March 2019; pp. 844–855.
6. Dallasega, P.; Woschank, M.; Zsifkovits, H.; Tippayawong, K.; Brown, C.A. Requirement Analysis for the Design of Smart Logistics in SMEs. In *Industry 4.0 for SMEs*; Matt, D.T., Modrák, V., Zsifkovits, H., Eds.; Springer: Berlin/Heidelberg, Germany, 2020; pp. 147–162.
7. Cioffi, R.; Travaglioni, M.; Piscitelli, G.; Petrillo, A.; de Felice, F. Artificial Intelligence and Machine Learning Applications in Smart Production: Progress, Trends, and Directions. *Sustainability* **2020**, *12*, 1–24. [CrossRef]
8. Mc Carthy, J.; Minsky, M.L.; Rochester, N.; Shannon, C.E. A Proposal for the Dartmouth Summer Research Project on Artificial Intelligence. *AI Mag.* **2006**, *27*, 12–14.
9. Shalev-Shwartz, S.; Ben-David, S. *Understanding Machine Learning: From Theory to Algorithms*; Cambridge University Press: Cambridge, UK, 2014.
10. Diez-Olivan, A.; Del Ser, J.; Galar, D.; Sierra, B. Data Fusion and Machine Learning for Industrial Prognosis: Trends and Perspectives towards Industry 4.0. *Inf. Fusion* **2019**, *50*, 92–111. [CrossRef]

11. Dey, N.; Ashour, A.S.; Nguyen, G.N. Deep Learning for Multimedia Content Analysis. In *Mining Multimedia Documents*; Karaa, W.B.A., Dey, N., Eds.; CRC Press: Boca Raton, FL, USA, 2017; pp. 193–203.

12. Salehi, H.; Burgueño, R. Emerging artificial intelligence methods in structural engineering. *Eng. Struct.* **2018**, *171*, 170–189. [CrossRef]

13. Hamet, P.; Tremblay, J. Artificial intelligence in medicine. *Metab. Clin. Exp.* **2017**, *69S*, 36–40. [CrossRef]

14. Baryannis, G.; Validi, S.; Dani, S.; Antoniou, G. Supply chain risk management and artificial intelligence: State of the art and future research directions. *Int. J. Prod. Res.* **2018**, *57*, 2179–2202. [CrossRef]

15. Wirth, N. Hello marketing, what can artificial intelligence help you with? *Int. J. Mark. Res.* **2018**, *60*, 435–438. [CrossRef]

16. Hokka, M.; Kaakinen, P.; Polkki, T. A systematic review: Non-pharmacological interventions in treating pain in patients with advanced cancer. *J. Adv. Nurs.* **2014**, *70*, 1954–1969. [CrossRef]

17. Töpfer, A. *Erfolgreich Forschen: Ein Leitfaden für Bachelor-, Master-Studierende und Doktoranden*, 3rd ed.; Springer: Berlin/Heidelberg, Germany, 2012.

18. Booth, A.; Sutton, A.; Papaioannou, D. *Systematic Approaches to a Successful Literature Review*, 2nd ed.; SAGE: London, UK, 2016.

19. Ali, N.B.; Usman, M. Reliability of search in systematic reviews: Towards a quality assessment framework for the automated-search strategy. *Inf. Softw. Technol.* **2018**, *99*, 133–147. [CrossRef]

20. Palmarini, R.; Erkoyuncu, J.A.; Roy, R.; Torabmostaedi, H. A systematic review of augmented reality applications in maintenance. *Robot. Comput. Manuf.* **2018**, *49*, 215–228. [CrossRef]

21. Tranfield, D.; Denyer, D.; Smart, P. Towards a Methodology for Developing Evidence-Informed Management Knowledge by Means of Systematic Review. *Br. J. Manag.* **2003**, *14*, 207–222. [CrossRef]

22. Petticrew, M.; Roberts, H. *Systematic Reviews in the Social Sciences: A Practical Guide*; Blackwell Publisher: Malden, MA, USA, 2012.

23. Denyer, D.; Tranfield, D. Producing a systematic review. In *The Sage Handbook of Organizational Research Methods*; Buchanan, D.A., Bryman, A., Eds.; Sage Publications Inc.: Thousand Oaks, CA, USA, 2011; pp. 671–689.

24. Durach, C.F.; Kembro, J.; Wieland, A. A New Paradigm for Systematic Literature Reviews in Supply Chain Management. *J. Supply Chain Manag.* **2017**, *53*, 67–85. [CrossRef]

25. Gursch, H.; Wuttei, A.; Gangloff, T. Learning Systems for Manufacturing Management Support. 2016. Available online: http://ceur-ws.org/Vol-1793/paper5.pdf (accessed on 1 March 2020).

26. Brodsky, A.; Krishnamoorthy, M.; Menasce, D.A.; Shao, G.; Rachuri, S. Toward smart manufacturing using decision analytics. In Proceedings of the 2014 IEEE International Conference on Big Data (Big Data), Washington, DC, USA, 27–30 October 2014; pp. 967–977.

27. Qu, S.; Jian, R.; Chu, T.; Wang, J.; Tan, T. Computational reasoning and learning for smart manufacturing under realistic conditions. In Proceedings of the 2014 International Conference on Behavior, Economic and Social Computing (BESC), Shanghai, China, 30 October–2 November 2014; pp. 1–8.

28. Bonino, D.; Vergori, P. Agent Marketplaces and Deep Learning in Enterprises: The COMPOSITION Project. In Proceedings of the 2017 IEEE 41st Annual Computer Software and Applications Conference (COMPSAC), Turin, Italy, 4–8 July 2017; pp. 749–754.

29. Peres, R.S.; Dionisio Rocha, A.; Leitao, P.; Barata, J. IDARTS–Towards intelligent data analysis and real-time supervision for industry 4.0. *Comput. Ind.* **2018**, *101*, 138–146. [CrossRef]

30. Guo, Q.; Miyamae, Y.; Wang, Z.; Taniuchi, K.; Yang, H.; Liu, Y. Senvis-Net: Learning from Imbalanced Machinery Data by Transferring Visual Element Detectors. *Int. J. Mach. Learn. Comput.* **2018**, *8*, 416–422.

31. Ferrer, B.R.; Mohammed, W.M.; Martinez Lastra, J.L.; Villalonga, A.; Beruvides, G.; Castano, F.; Haber, R.E. Towards the Adoption of Cyber-Physical Systems of Systems Paradigm in Smart Manufacturing Environments. In Proceedings of the 2018 IEEE 16th International Conference on Industrial Informatics (INDIN), Porto, Portugal, 18–20 July 2018; pp. 792–799.

32. Thalmann, S.; Mangler, J.; Schreck, T.; Huemer, C.; Streit, M.; Pauker, F.; Weichhart, G.; Schulte, S.; Kittl, C.; Pollak, C.; et al. Data Analytics for Industrial Process Improvement A Vision Paper. In Proceedings of the 2018 IEEE 20th Conference on Business Informatics (CBI), Vienna, Austria, 11–13 July 2018; pp. 92–96.

33. Morozov, D.; Lezoche, M.; Panetto, H. Multi-paradigm modelling of Cyber-Physical Systems. *IFAC-PapersOnLine* **2018**, *51*, 1385–1390. [CrossRef]

34. Marrella, A.; Mecella, M. Cognitive Business Process Management for Adaptive Cyber-Physical Processes. In *Business Process Management Workshops*; Teniente, E., Weidlich, M., Eds.; Springer: Berlin/Heidelberg, Germany, 2018; pp. 429–439.

35. Marrella, A.; Mecella, M.; Sardiña, S. Supporting adaptiveness of cyber-physical processes through action-based formalisms. *AI Commun.* **2018**, *31*, 47–74. [CrossRef]

36. Subakti, H.; Jiang, J.-R. Indoor Augmented Reality Using Deep Learning for Industry 4.0 Smart Factories. In Proceedings of the 2018 IEEE 42nd Annual Computer Software and Applications Conference (COMPSAC), Tokyo, Japan, 23–27 July 2018; pp. 63–68.

37. Susto, G.A.; Schirru, A.; Pampuri, S.; Beghi, A.; De Nicolao, G. A hidden-Gamma model-based filtering and prediction approach for monotonic health factors in manufacturing. *Control. Eng. Pract.* **2018**, *74*, 84–94. [CrossRef]

38. Wang, K.; Wang, Y. How AI Affects the Future Predictive Maintenance: A Primer of Deep Learning. In *Advanced Manufacturing and Automation VII*; Wang, K., Wang, Y., Strandhagen, J.O., Yu, T., Eds.; Springer: Berlin/Heidelberg, Germany, 2018; pp. 1–9.

39. Klein, P.; Bergmann, R. Data Generation with a Physical Model to Support Machine Learning Research for Predictive Maintenance. In Proceedings of the LWDA 2018, Mannheim, Germany, 22–24 August 2018; pp. 179–190.

40. Cho, S.; May, G.; Tourkogiorgis, I.; Perez, R.; Lazaro, O.; de La Maza, B.; Kiritsis, D. A Hybrid Machine Learning Approach for Predictive Maintenance in Smart Factories of the Future. In *Advances in Production Management Systems. Smart Manufacturing for Industry 4.0*; Moon, I., Lee, G.M., Park, J., Kiritsis, D., von Cieminski, G., Eds.; Springer: Berlin/Heidelberg, Germany, 2018; pp. 311–317.

41. Wuest, T.; Weimer, D.; Irgens, C.; Thoben, K.-D. Machine learning in manufacturing: Advantages, challenges, and applications. *Prod. Manuf. Res.* **2016**, *4*, 23–45. [CrossRef]

42. Terziyan, V.; Gryshko, S.; Golovianko, M. Patented intelligence: Cloning human decision models for Industry 4.0. *J. Manuf. Syst.* **2018**, *48*, 204–217. [CrossRef]

43. Venkatapathy, A.K.R.; Bayhan, H.; Zeidler, F.; Hompel, M. Human Machine Synergies in Intra-Logistics: Creating a Hybrid Network for Research and Technologies. In Proceedings of the 2017 Federated Conference on Computer Science and Information Systems, Prague, Czech Republic, 3–6 September 2017; pp. 1065–1068.

44. Zeidler, F.; Bayhan, H.; Ramachandran Venkatapathy, A.K.; ten Hompel, M. Referenzfeld zur Erforschung und Entwicklung neuartiger hybrider Formen der Zusammenarbeit von Menschen und Maschinen in der Logistik. *Logist. J. Proc.* **2017**, *1*, 1–8. Available online: https://www.logistics-journal.de/proceedings/2017/4593/zeidler_2017.pdf (accessed on 1 March 2020). [CrossRef]

45. Lolli, F.; Balugani, E.; Ishizaka, A.; Gamberini, R.; Rimini, B.; Regattieri, A. Machine learning for multi-criteria inventory classification applied to intermittent demand. *Prod. Plan. Control.* **2018**, *30*, 76–89. [CrossRef]

46. He, L.; Li, W.; Zhang, Y.; Cao, J. Review of Swarm Intelligence Algorithms for Multi-objective Flowshop Scheduling. In *Internet and Distributed Computing Systems*; Xiang, Y., Sun, J., Fortino, G., Guerrieri, A., Jung, J.J., Eds.; Springer: Berlin/Heidelberg, Germany, 2018; pp. 258–269.

47. Zhang, J.; Ding, G.; Zou, Y.; Qin, S.; Fu, J. Review of job shop scheduling research and its new perspectives under Industry 4.0. *J. Intell. Manuf.* **2017**, *30*, 1809–1830. [CrossRef]

48. Gomes, M.; Silva, F.; Ferraz, F.; Silva, A.; Analide, C.; Novais, P. Developing an Ambient Intelligent-Based Decision Support System for Production and Control Planning. In *Intelligent Systems Design and Applications*; Madureira, A.M., Abraham, A., Gamboa, D., Novais, P., Eds.; Springer: Berlin/Heidelberg, Germany, 2017; pp. 984–994.

49. Luetkehoff, B.; Blum, M.; Schroeter, M. Self-learning Production Control Using Algorithms of Artificial Intelligence. In *Collaboration in a Data-Rich World*; Camarinha-Matos, L.M., Afsarmanesh, H., Fornasiero, R., Eds.; Springer: Berlin/Heidelberg, Germany, 2017; pp. 299–306.

50. Wen, J.; He, L.; Zhu, F. Swarm Robotics Control and Communications: Imminent Challenges for Next Generation Smart Logistics. *IEEE Commun. Mag.* **2018**, *56*, 102–107. [CrossRef]

51. Laux, H.; Bytyn, A.; Ascheid, G.; Schmeink, A.; Kurt, G.K.; Dartmann, G. Learning-based indoor localization for industrial applications. In Proceedings of the 15th ACM International Conference, New York, NY, USA, 24–29 September 2018; pp. 355–362.

52. Ademujimi, T.T.; Brundage, M.P.; Prabhu, V.V. A Review of Current Machine Learning Techniques Used in Manufacturing Diagnosis. In *Advances in Production Management Systems. The Path to Intelligent, Collaborative and Sustainable Manufacturing*; Lödding, H., Riedel, R., Thoben, K.-D., von Cieminski, G., Kiritsis, D., Eds.; Springer: Berlin/Heidelberg, Germany, 2017; pp. 407–415.

53. Teschemacher, U.; Reinhart, G. Ant Colony Optimization Algorithms to Enable Dynamic Milkrun Logistics. *Procedia CIRP* **2017**, *63*, 762–767. [CrossRef]

54. Li, D.; Deng, L.; Cai, Z.; Franks, B.; Yao, X. Intelligent Transportation System in Macao Based on Deep Self-Coding Learning. *IEEE Trans. Ind. Inform.* **2018**, *14*, 3253–3260. [CrossRef]

55. Cheng, J.; Wu, W.; Cao, J.; Li, K. Fuzzy group-based intersection control via vehicular networks for smart transportations. *IEEE Trans. Ind. Inform.* **2017**, *13*, 751–758. [CrossRef]

56. Li, B.Y.S.; Yeung, L.F.; Tsang, K.F. Analysing traffic condition based on IoT technique. In Proceedings of the 2014 IEEE International Conference on Consumer Electronics (ICCE-C 2015), Shenzhen, China, 10–13 January 2015; pp. 1–4.

57. Edelstein, R. Smarter transportation management through its. In Proceedings of the 21st World Congress on Intelligent Transport Systems, ITSWC 2014: Reinventing Transportation in Our Connected World, Detroit, MI, USA, 7–11 September 2014; pp. 1–4.

58. Woschank, M. The Impact of Decision Making Maturity on Decision Making Efficiency. Ph.D. Thesis, University of Latvia, Riga, Latvia, 2018.

59. Aitheon. Project Management. Available online: https://aitheon.com/project-management (accessed on 19 April 2020).

60. DHL Research. Digital Twins in Logistics. A DHL Perspective on the Impact of Digital Twins on the Logistics Industry. Available online: https://www.dhl.com/content/dam/dhl/global/core/documents/pdf/glo-core-digital-twins-in-logistics.pdf (accessed on 19 April 2020).

61. Presenso. Presenso Announces the Production Release of Its Predictive Maintenance Solution. Available online: https://www.prnewswire.com/news-releases/presenso-announces-the-production-release-of-its-predictive-maintenance-solution-839580485.html (accessed on 19 April 2020).

62. PWC. Explainable AI Driving Business Value through Greater Understanding. Electronically. Available online: https://www.pwc.co.uk/audit-assurance/assets/explainable-ai.pdf (accessed on 19 April 2020).

63. Skycode. SkyPlanner APS-AI Production Planning and Scheduling. Available online: https://skycode.fi/en/skyplanner-aps-ai-production-planning-and-scheduling/ (accessed on 19 April 2020).

64. Swarm Logistics. Available online: https://www.swarmlogistics.net/proof-of-concept-validates-swarm-logistics-system.html (accessed on 19 April 2020).

65. Siemens Mobility. AI-Based Traffic and City Mobility Solutions. Available online: https://www.mobility.siemens.com/global/en/portfolio/road/digital-lab.html (accessed on 29 April 2020).

66. Kritzinger, W.; Karner, M.; Traar, G.; Henjes, J.; Sihn, W. Digital Twin in manufacturing: A categorical literature review and classification. *IFAC-PapersOnLine* **2018**, *51*, 1016–1022. [CrossRef]

67. Rai, A. Explainable AI: From black box to glass box. *J. Acad. Mark. Sci.* **2019**, *48*, 137–141. [CrossRef]

68. Aytug, H.; Bhattacharyya, S.; Koehler, G.; Snowdon, J. A review of machine learning in scheduling. *IEEE Trans. Eng. Manag.* **1994**, *41*, 165–171. [CrossRef]

69. Miehe, R.; Bauernhansl, T.; Schwarz, O.; Traube, A.; Lorenzoni, A.; Waltersmann, L.; Full, J.; Horbelt, J.; Sauer, A. The biological transformation of the manufacturing industry–envisioning biointelligent value adding. *Procedia CIRP* **2018**, *72*, 739–743. [CrossRef]

70. Matt, D.T.; Riedl, M.; Rauch, E. Die Natur als Inspiration. *ZWF Zeitschrift Wirtschaftlichen Fabrikbetrieb* **2020**, *115*, 158–161. [CrossRef]

71. European Parliament. Artificial Intelligence in Transport. Available online: https://www.europarl.europa.eu/RegData/etudes/BRIE/2019/635609/EPRS_BRI(2019)635609_EN.pdf (accessed on 29 April 2020).

72. Giusti, R.; Manerba, D.; Bruno, G.; Tadei, R. Synchromodal logistics: An overview of critical success factors, enabling technologies, and open research issues. *Transp. Res. Part. E Logist. Transp. Rev.* **2019**, *129*, 92–110. [CrossRef]

73. Le, L.T.; Lee, G.; Park, K.S.; Kim, H. Neural network-based fuel consumption estimation for container ships in Korea. *Marit. Policy Manag.* **2020**, *1*, 1–18. [CrossRef]

74. Liu, C.; Park, E.; Jiang, F. Examining effects of context-awareness on ambient intelligence of logistics service quality: User awareness compatibility as a moderator. *J. Ambient. Intell. Humaniz. Comput.* **2018**, *11*, 1413–1420. [CrossRef]
75. Ceyhun, G.Ç. Recent Developments of Artificial Intelligence in Business Logistics: A Maritime Industry Case. In *Digital Business Strategies in Blockchain Ecosystems*; Springer: Berlin/Heidelberg, Germany, 2020; pp. 343–353.
76. Karballaeezadeh, N.; Zaremotekhases, F.; Shamshirband, S.; Mosavi, A.; Nabipour, N.; Csiba, P.; Várkonyi-Kóczy, A.R. Intelligent Road Inspection with Advanced Machine Learning; Hybrid Prediction Models for Smart Mobility and Transportation Maintenance Systems. *Energies* **2020**, *13*, 1718–1741. [CrossRef]

Article

Digital Twin of Experimental Smart Manufacturing Assembly System for Industry 4.0 Concept

Kamil Židek [1,*], Ján Piteľ [1], Milan Adámek [2], Peter Lazorík [1] and Alexander Hošovský [1]

[1] Department of Industrial Engineering and Informatics, Faculty of Manufacturing Technologies with a Seat in Presov, Technical University of Kosice, Bayerova 1, 08001 Prešov, Slovakia; jan.pitel@tuke.sk (J.P.); peter.lazorik@tuke.sk (P.L.); alexander.hosovsky@tuke.sk (A.H.)

[2] Department of Security Engineering, Faculty of Applied Informatics, Tomas Bata University in Zlín, Nad Stráněmi 4511, 76005 Zlín, Czech Republic; adamek@utb.cz

* Correspondence: kamil.zidek@tuke.sk; Tel.: +421-903-137961

Received: 31 March 2020; Accepted: 26 April 2020; Published: 1 May 2020

Abstract: This article deals with the creation of a digital twin for an experimental assembly system based on a belt conveyor system and an automatized line for quality production check. The point of interest is a Bowden holder assembly from a 3D printer, which consists of a stepper motor, plastic components, and some fastener parts. The assembly was positioned in a fixture with ultra high frequency (UHF) tags and internet of things (IoT) devices for identification of status and position. The main task was parts identification and inspection, with the synchronization of all data to a digital twin model. The inspection system consisted of an industrial vision system for dimension, part presence, and errors check before and after assembly operation. A digital twin is realized as a 3D model, created in CAD design software (CDS) and imported to a Tecnomatix platform to simulate all processes. Data from the assembly system were collected by a programmable logic controller (PLC) system and were synchronized by an open platform communications (OPC) server to a digital twin model and a cloud platform (CP). Digital twins can visualize the real status of a manufacturing system as 3D simulation with real time actualization. Cloud platforms are used for data mining and knowledge representation in timeline graphs, with some alarms and automatized protocol generation. Virtual digital twins can be used for online optimization of an assembly process without the necessity to stop that is involved in a production line.

Keywords: Industry 4.0; smart manufacturing; digital twin; cloud platform

---

1. Introduction and Relevant Previous Works

The Industry 4.0 concept is referred to as the fourth industrial revolution and is the current trend in automation, monitoring, and data mining from manufacturing processes. The main technologies to realize these tasks are [1]:

- IoT (internet of things);
- CPS (cyber-physical systems);
- CP (cloud platforms);
- CC (cognitive computing);
- AR/VR (augmented reality/virtual reality) devices; and
- other related disciplines.

The main tasks are digitization of data, analysis, and knowledge extraction. In these areas many papers have been published with research results aimed towards the use of cyber-physical systems within the Industry 4.0 concept, big data processing [2], and the combination of CPS with

IoT systems [3]. The next task of the Industry 4.0 concept is product customization, because current customers want unique products. This requirement is a big challenge for the Industry 4.0 concept to ensure production with minimal cost and high customization.

Generally, manufacturing systems, production processes, and products are nowadays suitable for full digitalization. The first task in process digitalization is the specification of suitable technologies. There can be defined two groups of devices, the first for permanent installation in product, which must be low cost and the second group for machine and production process monitoring.

Perspective technologies for data digitization and collection from machines, processes, and products are:

- RFID (radio-frequency identification) labels for wireless part identification, and RFID transceivers for fixture monitoring in the production process;
- MEMS sensors integrated into the product for contactless data measuring and also integrated in the production process;
- IoT devices with independent wireless communication for data upload to cloud platforms for next processing; and
- Cloud platforms with data mining for knowledge extraction and data representation in timelines and alarms.

A modern approach to product identification is using RFID UHF (ultra high frequency) technology, because it reaches higher distances and reliability like LF and HF (low and high frequency) RFID technology. Some research results concerning RFID systems were published in articles about security of tags [4], detection of missing tags [5], and new searching protocols [6].

MEMS sensors have minimal power consumption and can be powered from battery during a product's lifetime. They can be used, for example, for monitoring of product overheating and vibration during operation by the customers, and are generally used for continuous monitoring of the product environment. Some research on the usability of MEMS sensor devices was published in the articles about the monitoring of mechatronic systems by MEMS [7], vibration analyses by MEMS [8], and structural monitoring by MEMS [9].

Modern IoT systems are based on specialized communication technologies for data isolation from standard Wi-Fi (Wireless-Fidelity) or Bluetooth networks. The main IoT communication standard is LPWAN (low-power wide-area network) and includes solutions such as LoRa/LoRaWAN and Sigfox. Other technologies are using modifications of GSM (Global System for Mobile Communication) networks for low data transfer. Some research results in this area of interest are described in articles about energy-efficient communications in the internet of things [10], novel chaining encryption algorithm for LPWAN IoT networks [11], evaluation of next-generation low-power communication technology to replace GSM in IoT applications [12], performance assessment of long-range and Sigfox protocols with mobility support [13], and a new effective way to boost LoRaWAN network capacity [14].

Knowledge extraction (data mining) and data analysis are the main tasks of a cloud system. Cloud platforms can provide data in user-friendly form by timelines, day/weeks, or monthly automated reports. There can be also integrated an alarm system for critical production status, usually as a message by email or SMS (short message service). Some new research trends in cloud platforms and data mining are described in the papers concerning clouds and big data connection [15], cloud robotics data [16], clouds in industrial automation [17], and chaos theory combined with cloud systems [18]. The practical aspects are described in the papers about industrial process simulation and data digitization in Tecnomatix software [19,20].

Applying a virtual model for remote monitoring is a new trend of the Industry 4.0 concept and can represent the real manufacturing system, production process, or product. Such virtual models digitally replicate all aspects of real devices and they are called digital twins, which continuously learn and update themselves from multiple data sources. The articles and case studies aimed at digital twins are, for example, about digital twins with ergonomic optimization [21], digital twin methodology to commentary [22], the learning of new experiences by digital twins [23], automatic generation of simulation for digital twins [24], digital twins' implementation for legacy systems [25], new possibilities of digital twin technology [26], and its application for Smart Industry [27].

Our research idea was to create a unified methodology for full data transfer from an experimental manufacturing assembly system to its digital twin, with real-time simulation and long-time data storage in a cloud platform with adequate synchronization.

The main purpose of this article is to present an architectural concept for a digital twin design with cloud platform and methodology of full data digitalization for smart manufacturing, which was implemented and tested on the experimental assembly system based on a belt conveyor system and an automatized line for quality production check.

2. Problem Specification and Proposed Work

The basic problem in implementation of an Industry 4.0 concept is data acquisition from the manufacturing environment. So, the first task is to select suitable technologies for data collection from production machines, production processes, and products.

Data from production processes can be provided by integrated position check sensors, for example: inductive proximity sensors, optical sensors, and laser sensors. These sensors can provide the actual status of the production, but no customized data about products; for example, they cannot provide information about where an exact order is currently positioned within the production process.

The next task is implementation of technology for exact part localization in production. The useful technology for exact part localization in production is RFID technology. An RFID reader, with an RFID transceiver on fixture and RFID labels on final assembly or parts, can provide exact information about the product status and its history.

The next important task in production is quality control. It can be divide into these levels:

- standard part identification;
- part presence check-in;
- dimension check in defined tolerance; and
- assembly check for product completeness.

If there is need for automatic feedback from a realized product, some autonomous monitoring equipment has to be implemented. The solution is implementation of IoT devices combined with MEMS sensors. This provides a reliable and low-cost solution with long-term operation. The data collection must be separated from standard communication networks like Wi-Fi or Bluetooth. The new IoT communication technologies provide this feature also with roaming possibilities.

The last task involes data storage and knowledge representation in some readable form. The standard databases cannot be used, because industrial systems now usually produce big data. The solution is to use industrial cloud platforms for data storage and visualization from all production processes. Open source timeline databases with graphical visualization are also suitable for data collection from IoT devices.

The proposed work is divided into four phases:

1. Creation of minimalized physical experimental assembly system with technologies for data collection;
2. Design of data digitization from all integrated technologies;
3. Synchronization of the actual status of physical devices with a digital twin; and
4. Data accumulation and representation in a cloud platform.

The main novelty of the research is its methodology for full digitalization of different inspection and identification technologies used in the experimental smart manufacturing assembly system. The basis is an architectural concept design of the digital twin of the experimental system and its synchronization with the cloud platform.

2.1. The Concept of Experimental Smart Manufacturing Assembly System

The concept of the experimental smart manufacturing assembly system is shown in Figure 1 and partial physical realization for research and educational purposes is shown in Figure 2. This concept includes all necessary technologies for digital data acquisition: RFID technology, vision systems, and IoT devices. All data from these technologies must be transformed to standardized industrial format. The PLC (programmable logic controller) is the main data collector for the next data processing and transfer to cloud system. Standardization of data is reached by the open platform communications (OPC) server, which also provides data distribution to the digital twin based on Siemens Tecnomatix plant simulator and the cloud platform MindSphere.

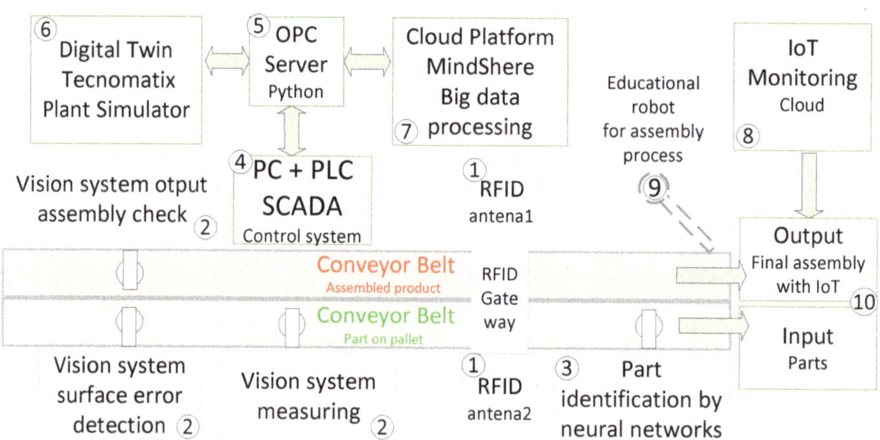

Figure 1. The concept of experimental smart manufacturing assembly system with data transfer to digital twin and cloud.

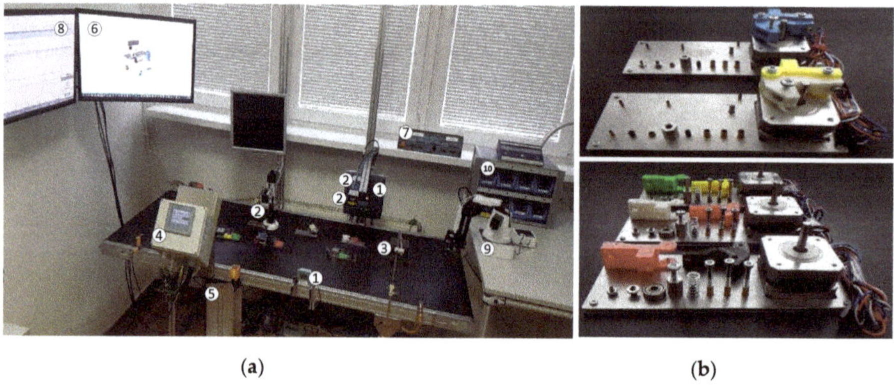

(a)   (b)

Figure 2. (**a**) The experimental assembly system; (**b**) the fixture with parts and finalized assembly.

2.2. RFID Part Identification System

The RFID reader (Siemens SIMATIC R685R with one integrated and second external antenna SIMATIC RF615A) was used for product identification. The first antenna checked the part presence by RFID labels tags. The second antenna checked the assembly completeness and wrote data for the fixture's RFID tag transponder. Implementation of RFID technology created an identification gate for input parts and output for finalized assembly. Any data acquired from vision system can be written to a part or assembly RFID tag as unique data. The used RFID system with two antennas, tested tags, and RFID tag implementation for some assembling parts and the fixture is shown in Figure 3.

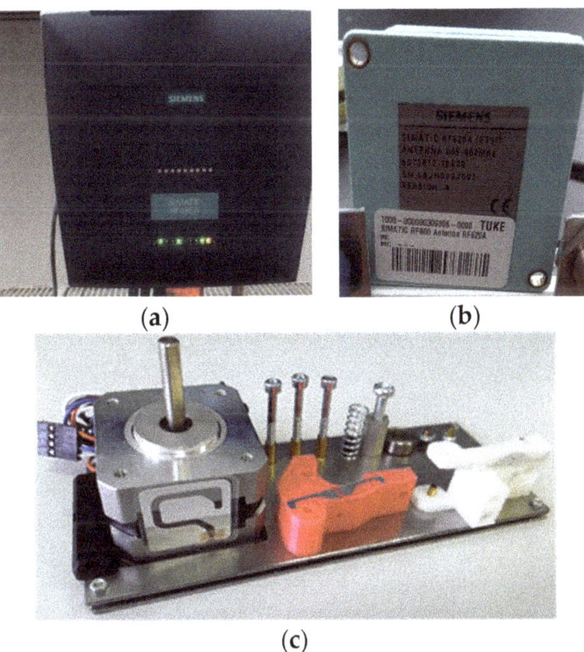

Figure 3. (a) Radio-frequency identification (RFID) gateway for assembly identification; (b) RFID gateway for parts localization; (c) parts with RFID tags.

Data from the RFID system were acquired as EPCID (Electronic Product Code ID) numbers and collected in PLC system, as shown in Figure 4.

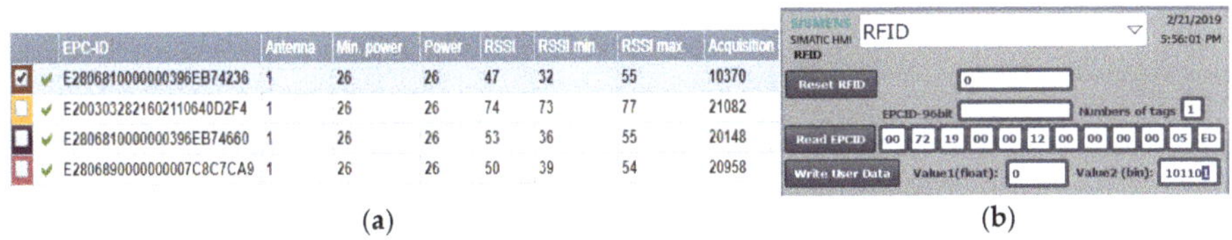

Figure 4. (a) RFID data from integrated tags on fixture and parts; (b) RFID data in the programmable logic controller (PLC).

2.3. Vision Systems for Part and Assembly Inspection

The used industrial vision systems, and specialized vision system with integrated trained convolutional neural network for standard parts (nuts, screws, washers) recognition [28], are shown in Figure 5. Examples of using vision systems for part checks, dimension measuring, assembly completeness, and standard part (nut) identification are shown in Figure 6.

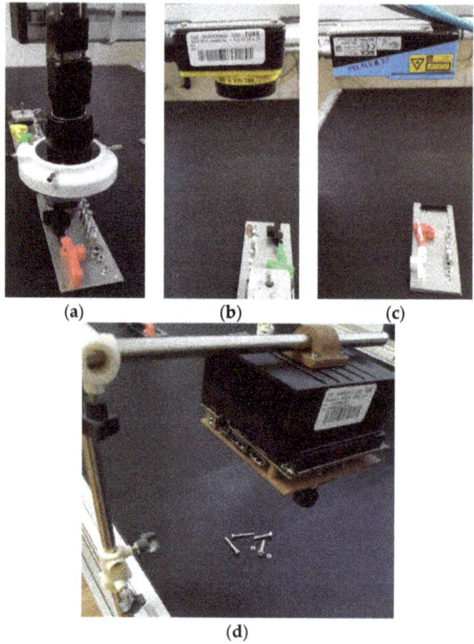

Figure 5. Industrial vision systems: (**a**) Omron FZ3-L355 with 5Mpix camera sensor; (**b**) Cognex with Insight-Explorer software; (**c**) Sick Inspector I10 with SOPAS Engineering tools; and (**d**) the specialized vision system Nvidia Jetson AGX Xavier Developer Kit with e-con Systems e-CAM130 CUXVR dual 13Mpix camera sensors using neural networks for parts identification.

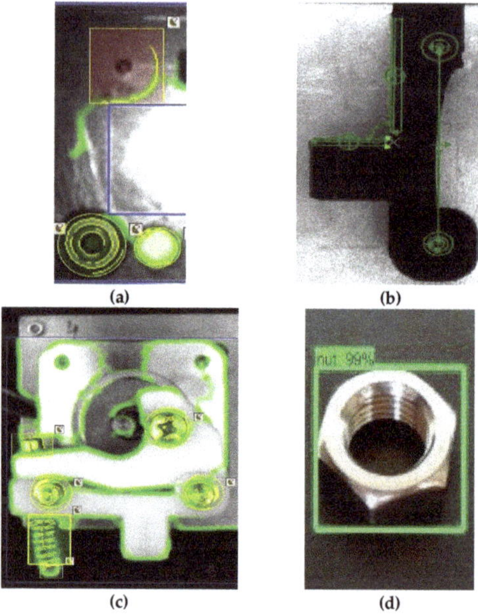

Figure 6. Examples of using vision systems for: (**a**) part checks; (**b**) dimension measuring; (**c**) assembly completeness; and (**d**) standard part (nut) identification.

An example of digital data output from a vision system for measured dimension and its transfer to the OPC server is shown in Figure 7.

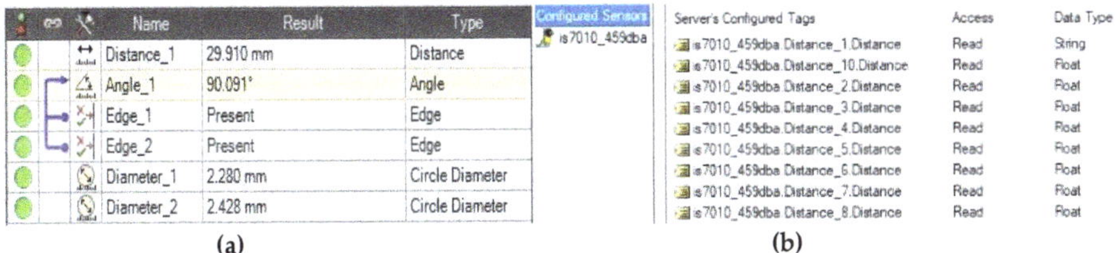

(a)   (b)

Figure 7. Digital data from the vision system: (**a**) in the graphical table; (**b**) in the open platform communications (OPC) server.

2.4. IoT Devices for Long-term Product Monitoring by MEMS Sensors

Three IoT communication technologies (GSM technology, Sigfox, and LoRaWAN) which can provide isolated data transfer were selected for long-term product monitoring. The experimental system combines these IoT technologies to one module with data collection to an open source system, as shown in Figure 8.

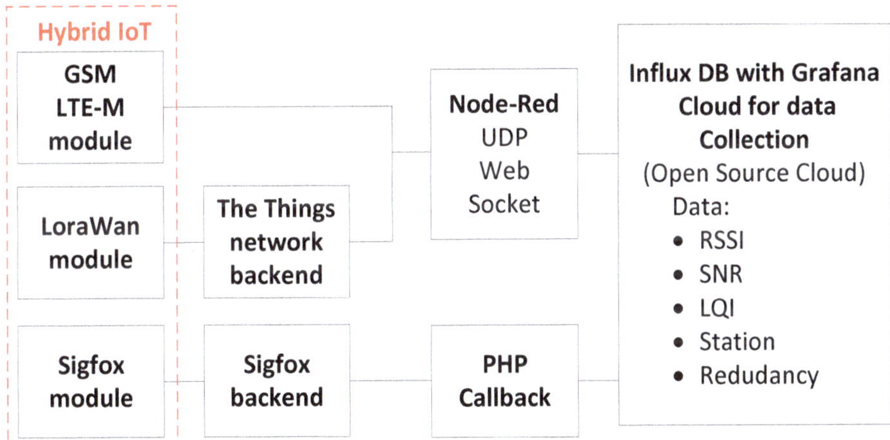

Figure 8. The principle scheme of the experimental combined IoT device.

Examples of the tested components for such combined system are shown in Figure 9. The software was programmed by Pymakr IDE inside Visual Studio and transferred to an FiPy module. The multiple IoT communication was operated by the FiPy module, which was connected to a MEMS sensor to collect vibration information.

These devices were used for experimental signal quality measurements, and for real implementation they must be reduced in dimension for integration to final assembly product.

An example of digital data collected from the hybrid IoT device by LoRaWAN network user interface "The things network" (TTN) is shown in Figure 10.

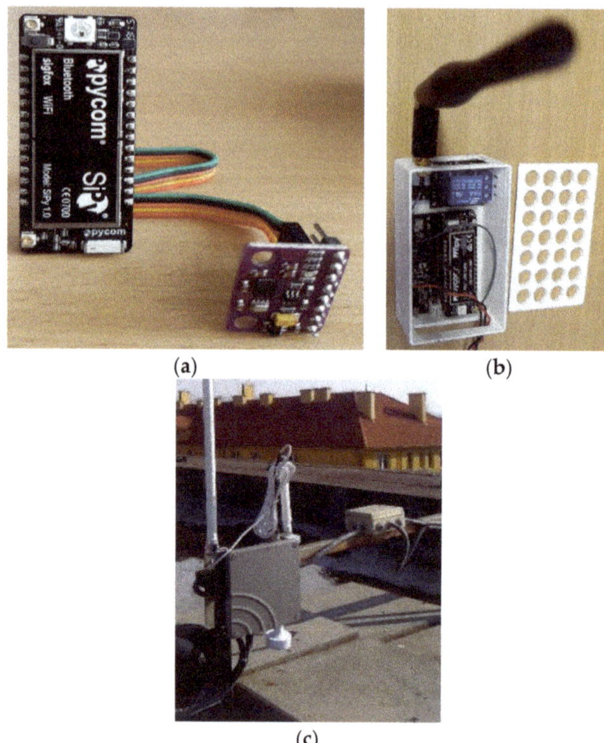

Figure 9. (**a**) IoT device FiPy and 9-axis MEMS sensor MPU-6050—accelerometer, gyroscope, magnetometer; (**b**) integration of the FiPy module from PyCom to development board; (**c**) LoRaWAN Gateway "The Things Outdoor Gateway—868Mhz version" with Omni antenna SIRIO GP 868 C.

(**a**)   (**b**)

Figure 10. Digital data from: (**a**) LoRaWAN TTN network; (**b**) Sigfox network.

The limitations of the tested IoT communication technologies for industrial use are summarized in Table 1.

Table 1. Comparison of IoT technologies for industrial tasks.

Technology	Nr. of Uplink MSG/Day	Nr. of Downlink MSG/Day	Bytes	Delay
LoRaWAN (TTN)	1440	max 10	10	30s
Sigfox	140	max 4	12	30s
GSM	not defined	not defined	UDP/TCP	-

The simple principle of MEMS system calibration in production is to acquire the reference vibration of product during normal working condition. Then it is possible to identify increased vibration or critical states during operating status. An example of the calibration and the critical status identification is shown in Figure 11. The reference status in this case is vibration amplitude in a stepper motor running at nominal speed. So, there is need to acquire this reference vibration status as the value range and write the offset value to the MCU (MicroController Unit) as a critical level. Data acquisition from MEMS sensors has been tested in Arduino IDE (Integrated Development Environment). An example of the vibration with filtration based on Kalman filters in x-axis and cumulated data from MEMS sensor decoded by Node Red interface is shown in Figure 12.

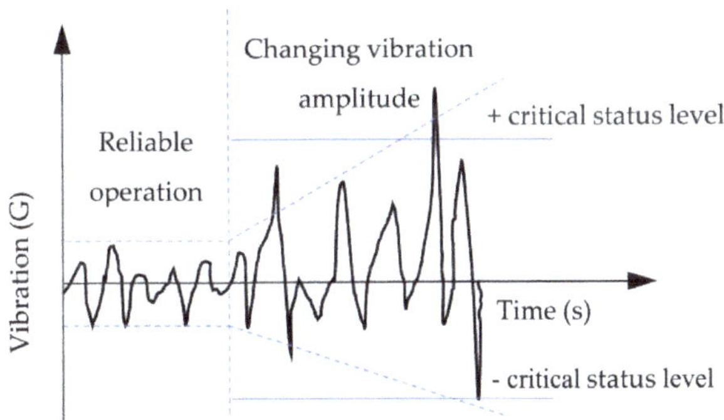

Figure 11. The principle of MEMS sensor calibration.

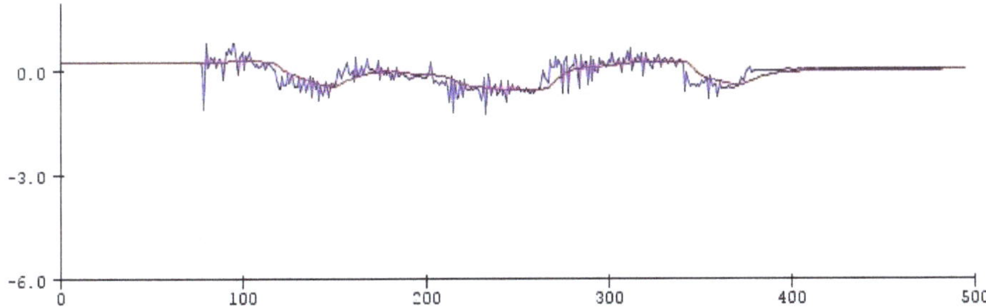

Figure 12. An example of data acquired from an accelerometer in x-axis with filter in Arduino IDE.

3. Digital Twin

A digital twin is a digital replica (a virtual model) for several hierarchy levels, like a sensor, an actuator, a production unit, a plant. It consists of 3D models grouped to assemblies with the possibility of remote monitoring and data synchronization with a real system and offline simulation.

The basis for digital twin creation is 3D virtual models of empty fixture, a fixture with parts, a fixture with assembly (Figure 13), and a 3D virtual model of a belt conveyor system with equipment for data acquisition (Figure 14).

Optimization of the production systems and their processes have so far usually been executed offline, mainly before the production starts or before it had to be stopped for optimization. The main improvement in the suggested approach is online optimization within the digital twin and synchronization with the real system. In the presented setup, a Python OPC UA (Unified Architecture) server (FreeOpcUa framework) is responsible for data transfer from the digital twin to the PLC system (Siemens PLC S7-1200 with HMI 4,3″ TFT KTP 400).

So, it is possible to synchronize simulation (digital twin of the system) with control system of the real manufacturing system and update process variables in a very short time by user interface. The designed 2D and 3D digital twins of the experimental assembly system are shown in Figure 15.

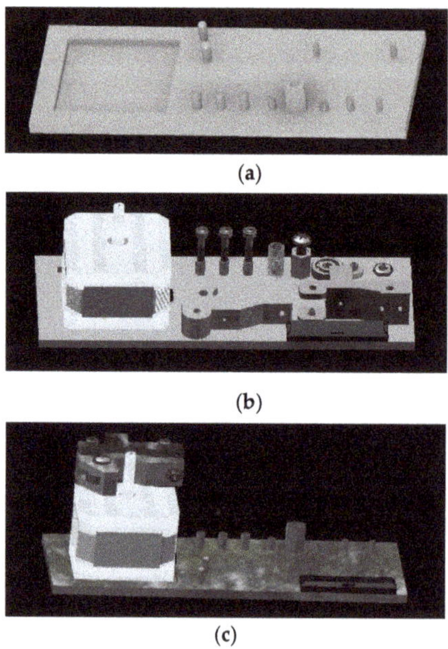

Figure 13. (**a**) 3D model of the fixture; (**b**) 3D model of the fixture with parts; (**c**) 3D model of the assembly.

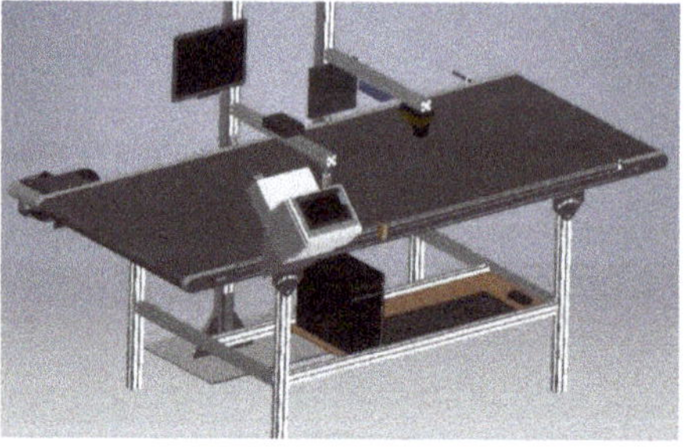

Figure 14. 3D model of the belt conveyor system with equipment for data acquisition.

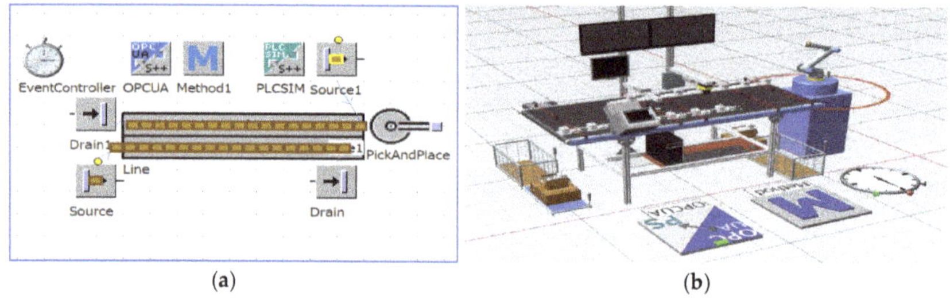

(**a**)   (**b**)

Figure 15. Simulation in Tecnomatix plant simulator with OPC communication: (**a**) 2D model; (**b**) 3D model.

The digital twin was transferred to a virtual reality device—HTC Vive Pro virtual reality glasses. This equipment can be used in quality control to remotely monitor the actual status of production. The main task is the synchronization of data from real production to the 3D virtual digital twin with simple visualization. An example of a digital twin of product identification and customization (a) and transfer (b) of Technomatix simulation to virtual reality by HTC Vive Pro is shown in Figure 16.

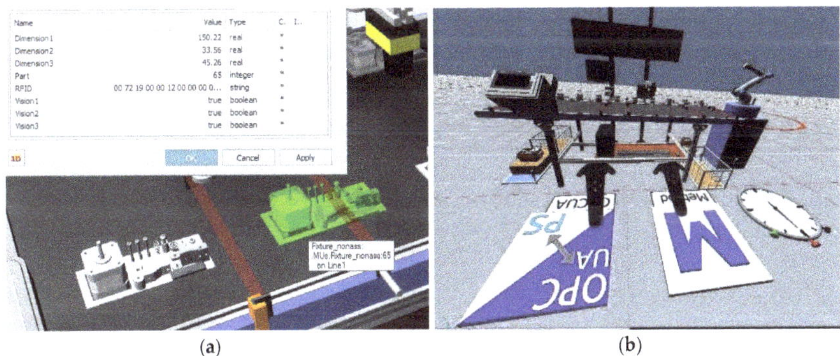

Figure 16. (**a**) 3D model in plant simulator; (**b**) digital twin inspection in virtual reality (VR).

VR devices (HTC Vive Pro, Oculus Rift, Valve Index VR) for communication with virtual scenes use hand controllers, but in assisted assembly there is the need to have both hands free. This problem was solved by using a Leap motion sensor to detect hand and finger position. Another problem in assisted assembly is that recognized parts must be detected in any position, rotation, and scale, but this is unsolvable by standard industrial vision systems. This task we solved by using deep learning techniques; all assembly parts were trained in deep neural networks by Google TensorFlow frameworks [29]. The trained network was transformed to OpenCV library DNN (Deep Neural Network) for Android and executed in an AR device inside a Unity 3D engine. An example of hand detection by Leap motion installed on a VR device (HTC Vive Pro with SteamVR software) is shown in Figure 17a, and standardized assembly parts recognized by the AR device (Epson Moverio BT350 with the development software Unity engine) in Figure 17b [28].

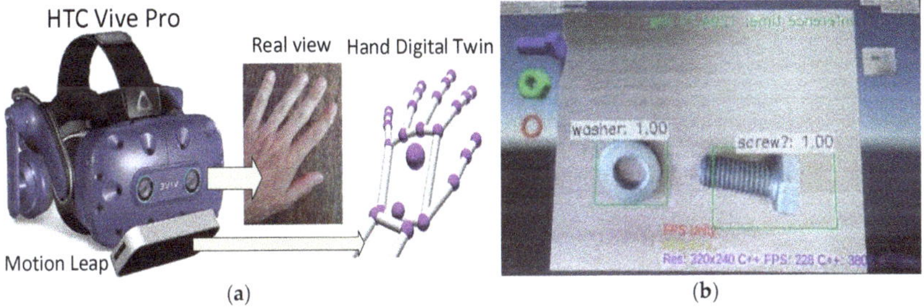

Figure 17. (**a**) Hand detection; (**b**) implementation of standardized part recognition for assisted assembly.

4. Data Transfer to Cloud System

To isolate the plant from the internet, we used IoT gateway MindConnect (connected to Amazon Web Services Cloud access) with separated network interfaces: the first one for internet connection, the second one for data transfer from the isolated manufacturing process by a Python OPC Server. So, security of all critical digital data transfer from the experimental manufacturing system to the cloud platform was solved by Mindsphere network unit MindConnect, which works like a hardware firewall and isolates the production process (plant) from the internet. The extracted data sent to the cloud platform Mindsphere were encrypted. The encryption for OPC server data is possible, but all OPC data were transferred in local networks, so there was no problem with security.

The used communication devices for data transfer to the MindSphere cloud platform are shown in Figure 18.

Figure 18. MindConnect devices as gateway to MindSphere cloud.

An example of data timelines' visualization from the vision system for all measured dimensions in the MindSphere cloud platform is shown in Figure 19.

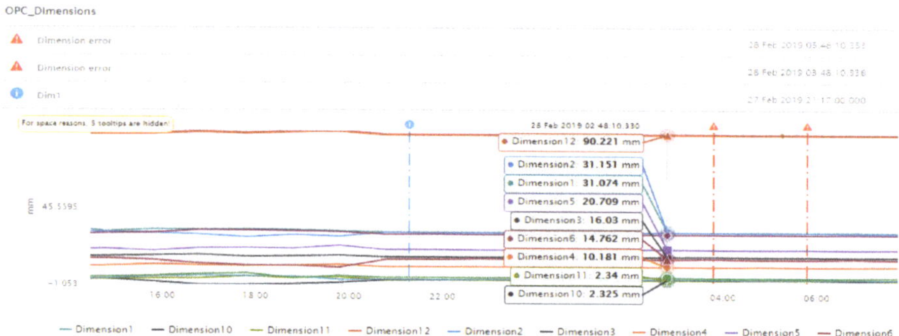

Figure 19. Cloud digital data from vision system.

Some typical IoT data from MEMS sensors as acceleration in two axis X and Y in the cloud platform Thinger.io, and signal quality monitoring by Grafana visualization, are shown in Figure 20.

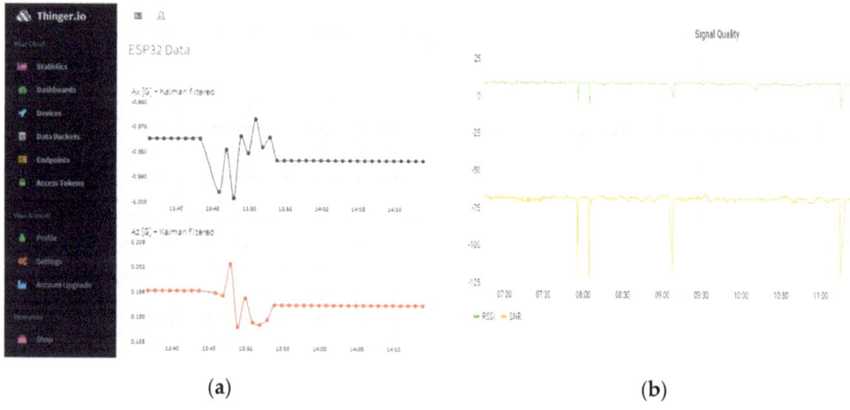

(a)   (b)

Figure 20. (**a**) IoT data from MEMS sensor (acceleration) in Thinger IO platform; (**b**) quality signal monitoring by Grafana.

5. Discussion

All data acquisition technologies, which were integrated into the experimental assembly system for identification (RFID tags) and inspection (vision systems, MEMS sensors, IoT devices) have their own limitations, which must be solved before data is stored into a cloud platform.

Identification systems used in manufacturing based on UHF RFID devices cannot be used for very small parts, because the tag size is limited by antenna length. RFID readers working with low frequency (LF) or high frequency (HF) tags are the solutions for tagging smaller parts, but they are not primarily developed for industrial parts (mainly for the food industry) and they reach lower distances for stable communication. The main limitation of UHF RFID technology in experimental assembly systems is UHF RFID tag size. The small parts—for example, screws, washers, and nuts—cannot be tagged. One of the possible solutions could be the combination of UHF with LF or HF RFID technologies, but this would have the disadvantage of shorter distance detection of assembly parts.

Smart industrial vision systems are usually closed source systems without the capability to modify an algorithm. This limitation arises especially in surface errors in recognizing very complicated parts. The next limitation of industrial vision systems for assembly operation is that detected parts must be in the fixed distance and position from camera, with only limited rotation. A new approach and solution could be the use of convolutional neural networks (CNN) with deep learning training techniques described in details in research articles about automated training of deep learning networks by 3D virtual models [28] and recognition of assembly parts by convolutional neural networks [29]. This new approach was also tested in the experimental smart manufacturing assembly system using the Tensor Flow framework, with good results for its application in real production. The main advantage of these approaches is a relatively simple integration into assisted assembly operation with AR/VR devices.

The data acquisition from MEMS sensors is based on some physical theorems which provide data with limited long-term precision (LTP), because they are physically isolated from measured objects and therefore must acquire data contactless. This is main reason why the MEMS sensors' data from accelerometers or gyroscopes have to be filtered by some filters, for example a Kalman filter, because their output signal is very noisy. The position of MEMS sensors in the manufacturing system must be with minimum interferences from RFID system. MEMS sensors integrated into the product are not influenced by RFID signal because they start to operate after the production process, during the lifetime of products. The next problem is high frequency output bandwidth generated by MEMS sensors, which is not compatible with currently used cloud platforms. A standard cloud platform minimal data acquisition time is one second. So, there is a need to use some buffers for data accumulation. The problem of how to accumulate more data in the one second interval is described in the research article about data optimization for communication between wireless IoT devices and Cloud platforms in production process [30].

The significant problem in the usability of clouds platforms is the lack of support to store a digital twin and its simulation and visualization. Industry 4.0 uses both technologies: cloud platforms for simple data storage, and digital twins for actual status visualization and simulation. So, there is a need to have some connection between these technologies.

Cloud platforms are primarily focused on data collection and basic graphical representation via timelines and data knowledge extraction, with critical status alarm systems or user-defined periodical reports: daily, weekly, monthly, and yearly. Available cloud platforms currently offer for storage simple variables like integer, float, bool, data time, and string. All digital data are usually stored in predefined frames and defined only by strings or numbers. It is not possible to include advanced structures with combined blob (Binary Large OBject) data, for example images, or binary data combined with numerical/string variables.

In the currently available cloud platforms, it is not possible to store customized 3D virtual digital twin models of products with adequate digital variable data.

The first solution could be to create a separate database for the customized digital twin 3D model, with a hyperlink to standard digital data stored in a cloud platform. This solution could link data from a digital twin to a cloud platform in one direction of synchronization. The main advantage of this approach would be increased security for critical data of products, because customized digital twin models of products would be stored in local databases.

The second solution could be the modification of an open source cloud platform for visualization and storage of customized digital twin models and image data, with simple variables, in one place. The main advantage of this approach would be very low operating costs, because whole solutions could be integrated into factory server infrastructure.

The main disadvantage of the designed and tested architecture, with the digital twin and cloud platform after implementation to real device, is the necessity of trained staff to modify, optimize, and extract knowledge, which is specific to every production process.

6. Conclusions

Full digitalization of production processes is a very important part of the Industry 4.0 concept to increase the effectivity of SME (Small and Medium-sized Enterprise) manufacturing. Suitable technologies for fast digitalization, data transfer, data storage, and finally for data mining have to be used. In this paper, we described an approach for how to capture data with contactless technologies in a smart manufacturing assembly process, and to transfer this data from the production process to the extended digital twin 3D model.

To build a bridge between the real production system and its digital twin, an OPC technology for data synchronization in both directions was implemented using OPC UA Server (OPC DCOM –Distributed Component Object Mode). The OPC server must ensure three communications: the first to the digital twin model, the second to the cloud platforms, and the third to the PLC system. For that reason, the customized OPC server written in Python Programming Language was designed and implemented. The IoT gateway MindConnect was used for data transfer to the MindSphere cloud platform.

The vision system, consisting of three camera modules, was implemented into the experimental manufacturing assembly process to check the shapes and surfaces of assembly parts and to measure their dimensions. An RFID system was used to localize parts on the conveyor line by RSSI (radio signal strength indication) signal from tags. The RFID gate wrote unique information to the main assembly RFID tag label of every product, as for example, acquired dimension data and quality data. So, in this way a smart identification system was applied to the experimental manufacturing assembly system.

Furthermore, the MEMS sensor data acquisition in combination with IoT communication technologies was tested. The product vibrations had been measured by an integrated accelerometer. The IoT data were processed by Node Red data conversion technology to the specialized NoSQL (Non-Structured Query Language) database with Grafana visual interface. Independent IoT communication technologies Sigfox and LoRaWAN were used for data transfer to a cloud platform.

After the implementation of the architectural concept of the digital twin with the cloud platform to the real manufacturing system with full digitalization, the finding is that there are some limitations which arise from currently available technologies. The main limitation is the cloud platform because it does not support storage of customized digital twins and it provides a minimum delay of about one second in data transfer.

The realized experimental smart manufacturing assembly system will serve for further research (for example, to extend the digital twin via data from articulated assembly robot) and also for educational purposes. Using advanced technologies based on the Industry 4.0 concept by educated students and workers can help to develop sustainable production of SMEs. The designed and tested digital twin architecture on the experimental manufacturing system with full data digitalization provides a universal digital model which can be used as a template for real manufacturing of SMEs. Implementation of digital twins into manufacturing is a necessary condition for product lifecycle management (PLM) to ensure sustainable production.

Author Contributions: Methodology, K.Ž.; software, K.Ž.; validation, J.P. and M.A.; formal analysis, A.H.; investigation, M.A. and A.H.; resources, K.Ž. and P.L.; data curation, P.L.; writing—original draft preparation, K.Ž.; writing—review and editing, J.P.; visualization, P.L.; supervision, M.A. and A.H.; project administration, K.Ž and J.P.; funding acquisition, J.P. All authors have read and agreed to the published version of the manuscript.

Funding: This research was funded by the European Union's Horizon 2020 research and innovation program under the Marie Skłodowska-Curie, grant number 734713, by the Ministry of Industry and Trade of the Czech Republic project No. FV20419 and also by the projects VEGA 1/0700/20, 055TUKE-4/2020 granted by the Ministry of Education of the Slovak Republic.

Conflicts of Interest: The authors declare no conflict of interest.

References

1. Xu, L.D.; Duan, L. Big data for cyber physical systems in industry 4.0: A survey. *Enterp. Inf. Syst.* **2019**, *13*, 148–169. [CrossRef]
2. Li, G.; Tan, J.; Chaudhry, S.S. Industry 4.0 and big data innovations. *Enterp. Inf. Syst.* **2019**, *13*, 145–147. [CrossRef]
3. Sanin, C.; Haoxi, Z.; Shafiq, I.; Waris, M.M.; Silva de Oliveira, C.; Szczerbicki, E. Experience based knowledge representation for Internet of Things and Cyber Physical Systems with case studies. *Future Gener. Comput. Syst.* **2019**, *92*, 604–616. [CrossRef]
4. Han, W.; Liu, W.; Zhang, K.; Li, Z.; Liu, Z. A protocol for detecting missing target tags in RFID systems. *J. Netw. Comput. Appl.* **2019**, *132*, 40–48. [CrossRef]
5. Lee, C.-C.; Chen, S.-D.; Li, C.-T.; Cheng, C.-L.; Lai, Y.-M. Security enhancement on an RFID ownership transfer protocol based on cloud. *Future Gener. Comput. Syst.* **2019**, *93*, 266–277. [CrossRef]
6. Liu, C.-G.; Liu, I.-H.; Lin, C.-D.; Li, J.-S. A novel tag searching protocol with time efficiency and searching accuracy in RFID systems. *Comput. Netw.* **2019**, *150*, 201–216. [CrossRef]
7. Židek, K.; Hošovský, A. Wireless Device Based on MEMS Sensors and Bluetooth Low Energy (LE/Smart) Technology for Diagnostics of Mechatronic Systems. *Appl. Mech. Mater.* **2013**, *460*, 13–21. [CrossRef]
8. Varanis, M.; Silva, A.; Mereles, A.; Pederiva, R. MEMS accelerometers for mechanical vibrations analysis: A comprehensive review with applications. *J. Braz. Soc. Mech. Sci. Eng.* **2018**, *40*, 527. [CrossRef]
9. Dumont, M.; Wolf, D. Usage of MEMS capacitive acceleration sensors for structural monitoring. In *Dynamics of Civil Structures, Volume 2*; Springer: Cham, Switzerland, 2019; pp. 77–89.
10. Deepak, G.C.; Bouhafs, F.; Raschellà, A.; Mackay, M.; Shi, Q. Radio resource management framework for energy-efficient communications in the Internet of Things. *Trans. Emerg. Telecommun. Technol.* **2019**, *30*. [CrossRef]
11. Bidgoly, A.J.; Bidgoly, H.J. A Novel Chaining Encryption Algorithm for LPWAN IoT Network. *IEEE Sens. J.* **2019**, *19*, 7027–7034. [CrossRef]
12. Durand, T.G.; Visagie, L.; Booysen, M.J. Evaluation of next-generation low-power communication technology to replace GSM in IoT-applications. *IET Commun.* **2019**, *13*, 2533–2540. [CrossRef]
13. Oliveira, L.; Rodrigues, J.J.P.C.; Kozlov, S.A.; Rabêlo, R.A.L.; Furtado, V. Performance assessment of long-range and Sigfox protocols with mobility support. *Int. J. Commun. Syst.* **2019**, *32*, e3956. [CrossRef]
14. Centenaro, M.; Vangelista, L. Time-Power Multiplexing for LoRa-Based IoT Networks: An Effective Way to Boost LoRaWAN Network Capacity. *Int. J. Wirel. Inf. Netw.* **2019**, *26*, 308–318. [CrossRef]
15. Lu, Y.; Xu, X. Cloud-based manufacturing equipment and big data analytics to enable on-demand manufacturing services. *Robot. Comput. Integr. Manuf.* **2019**, *57*, 92–102. [CrossRef]
16. Aissam, M.; Benbrahim, M.; Kabbaj, M.N. Cloud Robotic: Opening a New Road to the Industry 4.0. In *Development and Manufacturing of Recombinant Antibodies and Proteins*; Springer Nature Singapore Pte Ltd.: Singapore, 2019; pp. 1–20.
17. Mahmoud, M.S. Architecture for Cloud-Based Industrial Automation. In *Third International Congress on Information and CommunicationTechnology*; Springer Nature Singapore Pte Ltd.: Singapore, 2019; pp. 51–62.
18. Hu, Y.; Zhu, F.; Zhang, L.; Lui, Y.; Wang, Z. Scheduling of manufacturers based on chaos optimization algorithm in cloud manufacturing. *Robot. Comput. Integr. Manuf.* **2019**, *58*, 13–20. [CrossRef]
19. Hrehova, S. Predictive model to evaluation quality of the manufacturing process using Matlab tools. *Procedia Eng.* **2016**, *149*, 149–154. [CrossRef]

20. Lazár, I.; Husár, J. Validation of the serviceability of the manufacturing system using simulation. *J. Effic. Responsib. Educ. Sci.* **2012**, *5*, 252–261. [CrossRef]

21. Caputo, F.; Greco, A.; Fera, M.; Macchiaroli, R. Digital twins to enhance the integration of ergonomics in the workplace design. *Int. J. Ind. Ergon.* **2019**, *71*, 20–31. [CrossRef]

22. Tomko, M.; Winter, S. Beyond digital twins – A commentary. *Environ. Plan. B Urban Anal. City Sci.* **2019**, *46*, 395–399. [CrossRef]

23. David, J.; Lobov, A.; Lanz, M. Learning experiences involving digital twins. In Proceedings of the IECON 2018—44th Annual Conference of the IEEE Industrial Electronics Society, Washington, DC, USA, 21–23 October 2018; IEEE: New York, NY, USA, 2018; pp. 3681–3686.

24. Martinez, G.S.; Sierla, S.; Karhela, T.; Vyatkin, V. Automatic generation of a simulation-based digital twin of an industrial process plant. In Proceedings of the IECON 2018—44th Annual Conference of the IEEE Industrial Electronics Society, Washington, DC, USA, 21–23 October 2018; IEEE: New York, NY, USA, 2018; pp. 3084–3089.

25. Khan, A.; Dahl, M.; Falkman, P.; Fabian, M. Digital Twin for legacy systems: Simulation model testing and validation. In Proceedings of the 2018 IEEE 14th International Conference on Automation Science and Engineering (CASE), Munich, Germany, 20–24 August 2018; IEEE: New York, NY, USA, 2018; pp. 421–426.

26. Shubenkova, K.; Valiev, A.; Shepelev, V.; Tsiulin, S.; Reinau, K.H. Possibility of digital twins technology for improving efficiency of the branded service system. In Proceedings of the 2018 Global Smart Industry Conference (GloSIC), Chelyabinsk, Russia, 13–15 November 2018; IEEE: New York, NY, USA, 2018; pp. 1–7.

27. Luściński, S. Digital Twinning for Smart Industry. In Proceedings of the 3rd EAI International Conference on Management of Manufacturing Systems, Dubrovnik, Croatia, 6–8 November 2018; pp. 1–9.

28. Židek, K.; Lazorík, P.; Piteľ, J.; Hošovský, A. An automated training of deep learning networks by 3D virtual models for object recognition. *Symmetry* **2019**, *11*, 496. [CrossRef]

29. Židek, K.; Hosovsky, A.; Piteľ, J.; Bednár, S. Recognition of assembly parts by convolutional neural networks. In *Advances in Manufacturing Engineering and Materials*; Lecture Notes in Mechanical Engineering; Springer: Cham, Switzerland, 2019; pp. 281–289.

30. Zidek, K.; Janáčová, D.; Hošovský, A.; Piteľ, J.; Lazorik, P. Data optimization for communication between wireless IoT devices and Cloud platforms in production process. In Proceedings of the 3rd EAI International Conference on Management of Manufacturing Systems, Dubrovnik, Croatia, 6–8 November 2018; pp. 1–8.

Article

The Advantages of Industry 4.0 Applications for Sustainability: Results from a Sample of Manufacturing Companies

Riccardo Brozzi [1], David Forti [1,2], Erwin Rauch [2,* and Dominik T. Matt [1,2]

[1] Fraunhofer Italia Research s.c.a.r.l., A.-Volta-Str. 13/A, 39100 Bolzano, Italy; riccardo.brozzi@fraunhofer.it (R.B.); david.forti@fraunhofer.it or david.forti@natec.unibz.it (D.F.); dominik.matt@fraunhofer.it or dominik.matt@unibz.it (D.T.M.)

[2] Industrial Engineering and Automation (IEA), Faculty of Science and Technology, Free University of Bozen-Bolzano, Universitätsplatz 5, 39100 Bolzano, Italy

* Correspondence: erwin.rauch@unibz.it; Tel.: +39-0471-017111

Received: 30 March 2020; Accepted: 25 April 2020; Published: 1 May 2020

Abstract: Far from being exclusively related to economic considerations, the advantages of Industry 4.0 applications also include environmental and social concerns. An increasing amount of scientific publications relate the implementation of the fourth industrial revolution paradigm to sustainability. Several studies reported opportunities of Industry 4.0 implementation particularly to the environmental dimension of sustainability, e.g., through improved logistics streams and lowered waste from production. The present research aims at providing evidence on whether manufacturing companies consider Industry 4.0 implementation as an advantage contributing to environmental and social sustainability in terms of lower environmental impact of production, as well as higher physical relief for workers and flexibility of work organisation. The results were an attempt to study such relations with company sizes, industry sectors, turnover and self-assessed levels of digitalization varying. The sample encompasses 65 companies located in the Marche region (Italy). The results show that overall the perception of economic opportunities prevail, while the association of a beneficial impact of Industry 4.0 on environmental sustainability is rather low across companies, regardless of their size, turnover and digital level. As for the statistically significant variables, the results suggest a strong association of the size and the digital level to specific Industry 4.0 related advantages, referring to the social and economic dimension of sustainability, respectively.

Keywords: Industry 4.0; SME; sustainability; assessment; field study; small and medium sized enterprises

---

1. Introduction

The core of Industry 4.0 lies in the real-time and smart connection of people, machines and objects for the management of production systems [1]. The introduction of Cyber-Physical Systems (CPS) will enable the digitalisation of the entire value chains, reshaping relations between producers, consumers and suppliers [2]. Consequently, a large scale and heterogeneity of components is emerging also driven by the Key Enabling Technologies (KET) of Industry 4.0, encompassing, among other things, advance manufacturing solutions, additive manufacturing, augmented and virtual reality, simulation, Internet of Things (IoT), cloud, big data and cyber security. Industry 4.0 solutions exhibit an increasing application potential across several businesses and industries. However, the success of Industry 4.0 and the approaches under it will ultimately depend on whether small and medium sized enterprises (SMEs) can adopt and implement these technologies, as they represent the backbone of several productive systems worldwide [3]. The process leading SMEs shifting their production and organisational mindsets towards Industry 4.0, as well as the implementation of digital projects, introduces several opportunities and challenges [4–6]. To support digital transformation in SMEs in the last years, several EU countries launched innovation policies. Most of them foresee measures encouraging research and development actions which have been considered to play an important role in the decision to launch investments in digital technologies [7]. The introduction of technologies and concepts inherited from the Industry 4.0 paradigm is considered strategic also for small and medium sized enterprises (SME) to ensure competitiveness, increasing productivity and flexibility of production systems [8,9], while matching the growing market demand for customised products and services [10]. In recent times, business, research and institutions have been increasingly focusing their attention on the sustainability implications of Industry 4.0, in terms of economic, environmental and social impacts of manufacturing and digitalisation, requiring the scientific community to further explore such dynamics [11].

Several scholars agree that Industry 4.0 can provide, beyond economic advantages, numerous opportunities for environmental and social sustainability [12–14]. The expected advantages of Industry 4.0 for sustainability target the overall promotion of sustainable development [15] through the achievement of higher resource efficiency, waste reduction [16] and the realisation of a more favourable production environment for workers, e.g., in terms of higher physical relief and safety [17,18]. The consideration of these growing series of aspects calls for shifting the focus of analysis, adopting the Triple Bottom Line (TBL) approach, that is, extending the assessment of relevant factors among companies, for the implementation of Industry 4.0 to the economic, environmental and social dimensions of sustainability [19]. Within the ongoing scientific debate on this topic, notably Müller et al. have acknowledged the importance of understanding the underlying dynamics of the implementation of Industry 4.0, the predominant focus of existing literature on the analysis of technical motivations and the need to adopt differentiated approaches considering, e.g., company size as an explanatory variable to adequately investigate the emerging relations [20]. Beier et al. affirm the lack in research investigating the impact of digitalized industry on the relevance of sustainability aspects [21].

In this work we conduct a study to investigate the benefits of Industry 4.0 applications for sustainability. We use the results from a sample of manufacturing companies, which were asked by means of a questionnaire. In detail, we investigate to what extent manufacturing companies see Industry 4.0 as an advantage for economic, environmental and social sustainability. Furthermore, the study gives an overview of whether there are different opinions depending on turnover or company size. The main findings of our study indicate that the consideration of economic opportunities prevails over environmental and social ones. Environmental opportunities are not considered relevant by companies. The overall perception of the beneficial impact of Industry 4.0 on environmental sustainability is that it is low, and this trend is consistent among companies of different sizes, yearly turnover and digital level. The social opportunity relating to flexible organisation of work emerges as a driver in large companies, more so than SMEs. The economic opportunity, relating to improved logistics, results were strongly implicated for companies with a higher digital level.

The paper is structured as follows. After a short introduction the state of the art regarding the main relations between Industry 4.0 and sustainability will be outlined. The methodology employed to collect data and typologies of the analysis conducted are presented. To test the hypothesis emerging from the literature review indicating Industry 4.0 holding opportunity for realizing sustainable industrial value creation [12,20,22], the present paper derived evidence from the results of a digital self-assessment tool, distributed among a sample of manufacturing companies, encompassing both SMEs and large companies. For this reason, the presentation of results will provide general information on the characteristics of the sample and specific analysis of the subject of this research, namely the existing relations regarding sustainability and Industry 4.0 among the considered groups of responding companies. The results will be discussed considering the scientific publications about the topic included and concrete implications for practitioners and academia. Finally, conclusions will be drawn paving the way for future studies based on the emerging evidence and limits of this research.

2. Theoretical Background

2.1. Literature Review

The "Brundtland" report published in 1987 by the World Commission coined the term "sustainable development", giving rise to an increasing societal awareness towards the environmental impact of industrial manufacturing [23]. The emerging environmental concerns such as availability of non-renewable resources and stricter legislation, as well as consumer preference for environmentally friendly products, require manufacturing companies to consider sustainability in their supply chain [24]. From a sustainability perspective, focusing exclusively on the maximisation of profits without considering additional stakeholder concerns is becoming less tolerable [25]. Nevertheless, it is logical to think that the economic profitability of the company is an indispensable element for its existence [26]. The consideration of aspects other than the economic implications of sustainability has the potential to boost innovation and enhance the competitiveness of companies [27]. Against this background, an increasing number of scholars argue that Industry 4.0 holds great opportunity for realizing sustainable industrial value creation on all three dimensions of sustainability: Economic, social and environmental [12].

The terms Industry 4.0 and fourth industrial revolution were used for the first time at the Hannover Fair in 2011, referring to a smart real time connection between people, machines and objects with the aim of managing a production system [1]. Industry 4.0 has been defined as the convergence of Internet technologies in the manufacturing environment [28]. The concept of Industry 4.0 includes several technologies, which are defined as the key enabling technologies of this strategic initiative; these are: Advanced manufacturing solutions, additive manufacturing, augmented and virtual reality, simulation, horizontal and vertical data integration, industrial Internet, cloud computing, cyber security, big data and analytics. Internet of things (IoT) and Cyber Physical Systems (CPS) make it possible to create networks incorporating the entire manufacturing processes converting factories into smart environments in which the physical and the virtual world cooperate [29]. Over the last few years the attention that has been given and the number of conferences and studies dedicated to the topic of Industry 4.0 have grown exponentially [30], not only in the academic environment but also from managers who recognize the potential of this paradigm-shift impacting the entire concept of production, the processes involved as well as the dynamics among suppliers and customers. The fourth industrial revolution has created several opportunities and challenges for SMEs [31,32], which can take advantage of these technologies to increase their flexibility, productivity and competitiveness [9,28].

From a sustainability point of view, Stock and Selinger [12] identified Industry 4.0 as a step towards a more sustainable industrial value creation, arguing that industrial value creation must be geared towards sustainability and Industry 4.0, providing immense opportunities for the realization of sustainable manufacturing. By providing detailed information on each point of the production process, resource and energy use can be optimized over the entire value network [22]. The scientific debate considers the Triple Bottom Line construct [23], consisting of the analysis of environmental, economic and social aspects of sustainability. Authors mentioning such an approach included in their analysis combinations of these dimensions such as environmental [33], economic and social [24] as well as the entire TBL construct [14]. A comprehensive review of the state of the art in terms of environmental and economic aspects reported that research activities are clearly unbalanced and focus mainly on energy as specific input mean [24]. In fact, Morelli [34], relates the use of environmentally responsible and sustainable energy sources and the effort of improving energy efficiency to the concept of sustainability. Vance et al. [35] show that a reduction of energy consumption can be considered as a step towards sustainability. Bonilla et al. [36] provided a systematic overview of scientific publications relating Industry 4.0 to environmental sustainability, drawing conclusions about the uncertainty of the long-run impact of digital transformation on environmental sustainability. The impacts on sustainability have been assessed in relation to specific key enabling technologies on the sustainable value creation of a manufacturing company such as IoT on reshoring. In logistics [12], a reduction in the number of transport processes as well as unnecessary material flows can be achieved [22]. Decentralised production, which can be translated into reduced transport distances based on relocating production closer to where products and production equipment are purchased, leads to a decrease in both logistics costs and environmental impact [5,37]. This can be achieved with the use of new production technologies such as additive manufacturing that can help to shorten and reduce the size of the value chain [38]. Furthermore, the number of wrong deliveries, unnecessary waiting time and damaged products can be reduced by data transparency throughout the entire supply chain [39,40]. In other words technologies associated with the concept of Industry 4.0 are considered to have the unique potential to unlock environmentally sustainable manufacturing [33], and positive relations to environmental sustainability are detected in terms of load balancing optimisation, leading to a reduction in energy consumption [41,42]. Further aspects which are related to environmental sustainability are product usage, waste, end-life products and information [36].

Additional sources relate Industry 4.0 to the concept of sustainability from an ecological point of view and at the same time link it to the social perspective. Since social sustainability is necessary to the fulfilment of the three dimensions of the TBL, it deserves a specific and detailed analysis [43]. The increasing level of automation [44] and collaboration between the workers and the machines [45] will improve people's working conditions. This is particularly advantageous, since the proportion of older workers will increase due to the age shift [22]. It implements collaborative robots into the workstations, which are non-ergonomically designed and physically demanding and preserves workers' health and consequently the productivity of the production system in the long term [18,46]. Smart worker assistance systems will support operators in performing tasks that are monotonous and repetitive, resulting in higher employee satisfaction and motivation [47]. These approaches meet the current demographic challenges since the working environment needs to be designed age-appropriately [28]. These are just examples of the impact Industry 4.0 has on sustainability. From the literature review, it has been shown that even sustainability has a positive impact on the implementation of Industry 4.0, by working as an incentive for the implementation of the related technologies. As Kiel et al. [14] state, the environmental and social opportunities which Industry 4.0 has to offer have a positive impact on the companies' tendency to start the implementation process of Industry 4.0. In this regard, literature identifies sustainability also as the implementation and usage of tools, methods, processes, approaches and practices that aim at improving the company's contribution to sustainable development [48], which is the main focus of the present study.

2.2. Research Question

The literature analysis shows numerous advantages of the introduction of technologies and concepts of Industry 4.0 for sustainability. In the literature, however, there is currently a limited number of studies analysing and reporting on the differences between the perceived opportunities of Industry 4.0 implementation on sustainability in larger companies and SMEs. We see here a large gap in scientific research, which should be tackled, increasing targeted research in this field. Therefore, in this paper we focus on the influence of Industry 4.0 on sustainability, adopting a TBL approach, among manufacturing SMEs and large companies. In detail, the present work will answer the research question of whether manufacturing companies perceive the economic, environmental and social sustainability opportunities driven from the implementation of Industry 4.0. More specifically the following research questions (RQ) have been formulated:

RQ1: To what extent do manufacturing companies consider Industry 4.0 as an advantage related to economic, social and environmental sustainability?

RQ2: Which variables might explain differences in the perception of economic, social and environmental sustainability among companies?

To answer these questions, the advantages and challenges related to the implementation of Industry 4.0 technologies and concepts perceived by companies were analysed, considering the

TBL of sustainability, which covers different aspects such as the economic, environmental and social dimensions [23,49,50]. Furthermore, the paper investigated whether specific characteristics of companies, such as size and digital level of the company, are associated with differences in the perception of specific dimensions of sustainability (economic, social or environmental) as an advantage driven by Industry 4.0 implementation. This approach is aligned to the study of Müller et al., which effectively examined differences in size, industry sector and the company's role as an Industry 4.0 provider or user in determining the Industry 4.0-related opportunities and challenges [20]. Following this comprehensive approach investigating the three dimensions of sustainability, we formulated specific hypotheses related to RQ1 and RQ2, aiming at deepening the focus on the perception of environmental sustainability. In this regard, we expect a relatively low perception of Industry 4.0 implementation to be an advantage for the sustainability performance of the companies, regardless of their size and digital level:

Hypothesis 1 (H1). *Industry 4.0 is not considered a main advantage related to environmental sustainability.*

Hypothesis 2 (H2). *Companies do not perceive environmental sustainability as an advantage, regardless of their size in terms of both number of employees and turnover.*

Hypothesis 3 (H3). *Companies do not perceive environmental sustainability as an advantage, regardless of their digital level.*

RQ1 is linked to H1, while RQ2 entails an in-depth discussion following H1 and H2. The next section will provide more details on the methods employed to test the hypotheses, the approach used to collect data and the metrics used to derive the digital level of companies in the considered sample.

3. Research Methodology

3.1. The Online Self-Assessment Tool "Digital Check"

The assessment of the digital level of companies and the perceived advantages relating to the implementation of Industry 4.0 was part of the survey, distributed among a sample of manufacturing companies associated to Confindustria Marche Nord, the industrial representative association located in Central Italy Marche Region. An online tool named "Digital Check" allowed local companies to self-assess their digital level and readiness for the Industry 4.0 paradigm. The survey was defined considering the general structure and objectives of existing self-assessment tools [51,52] and adapted according to the challenges SMEs face in initiating digital transformation, considering recent scientific works on this topic carried out, among others, by the authors of the present work [31]. The "Digital Check" comprises 26 questions across the following dimensions: (1) Strategy, (2) processes, (3) Industry 4.0, (4) employees, (5) Information Technology (IT) and data security. The current digital level perceived by the individual companies responding to the survey was calculated considering the average of selected questions (see those marked with •) within each dimension as indicated in Table 1.

Table 1. Structure of the Digital Check.

Dimension	Question	Type
D1 *Strategy*	Q-1 Degree of definition of a strategy towards Industry 4.0?	Likert •
	Q-2 Have you planned intervention related to Industry 4.0 over the next two years?	Likert
	Q-3 Degree of collaboration with external experts (e.g., universities, research centres and consulting agencies) on Industry 4.0 topics?	Likert •
	Q-4 Expected collaborations with external experts over the next two years?	Likert
D2 *Processes*	Q-5 Use of key enabling technologies.	Likert
	Q-6 Integration of IT in production systems.	Likert •
	Q-7 Flexibility of production systems.	Likert •
	Q-8 Digitalisation of organisation and processes.	Likert •
	Q-9 Impact of Industry 4.0 on the organisation of the company.	Likert
D3 *Industry 4.0*	Q-10 Knowledge about Industry 4.0.	Likert •
	Q-11 Perceived need to increase knowledge about Industry 4.0 over the next two years.	Likert
	Q-12 Relevance of Industry 4.0 for the company.	Likert
	Q-13 Contribution of Industry 4.0 to the development of new products/services and business models.	Likert •
	Q-14 Expected contribution of Industry 4.0 to the development of new products/services and business models over the next two years.	Likert
	Q-15 Challenges related to Industry 4.0.	Multiple response
	Q-16 Advantage related to Industry 4.0.	Multiple response
	Q-17 Areas requiring support to local companies.	Multiple response
D4 *Collaborators*	Q-18 Adequacy of digital skills of collaborators.	Likert •
	Q-19 Perceived need to increase digital competences of collaborators over the next two years.	Likert
D5 *Information* *Technology*	Q-20 Degree of use of data from production	Likert
	Q-21 Use of cyber-security systems	Likert
	Q-22 Degree of data-security	Likert •
D6 *Company profile*	Q-23 Sector.	Multiple choice
	Q-24 Number of employees.	Multiple choice
	Q-25 Yearly turnover (optional question).	Multiple choice
	Q-26 Role of the respondent in the company.	Multiple choice

• Rating considered to calculate average values determining the digital level of companies.

Respondents could rate the questions within each dimension of the survey along a five levels Likert Scale ranging from 1 (low implementation level) to 5 (high implementation level), according to the specific topic of analysis, as shown in the example in Table 2.

Table 2. Exemplary scale according to topic and dimension in the survey.

Dimension	D1 Strategy	
Question topic	To what extent has your company defined a strategy toward Industry 4.0	
Scale	1	No strategy has been formulated or planned
	2	Planning to develop a strategy in the short term
	3	Development of the strategy currently ongoing
	4	The strategy is formulated but not yet implemented
	5	The strategy is fully implemented

The overall digital level resulted from the average values assigned by companies to each item object of evaluation across the different dimensions considered. The results of the survey enabled to assign one category portraying the self-assessed digital level of each company taking part to the survey, namely into so called (i) "Digital Newcomers", (ii) "Companies in Transition" and (iii) "Top Performers" (Table 3).

Table 3. Description of digital levels.

Category	Description
Digital Newcomers (*Low digital level*)	Companies belonging to this level do not consider the transition to Industry 4.0 relevant to their business or are testing the first pilot projects in isolated areas to test possible effects. A concrete strategy for digitization is still under development or not yet at an advanced stage. Few or none of the production processes support IT integration. The sharing of corporate data and information is still limited. The skills of employees are still inadequate in the Industry 4.0 perspective.
Companies in Transition (*Medium digital level*)	Companies belonging to this level demonstrate that they have already incorporated the concept of Industry 4.0 into their strategic orientation. They are taking the first steps towards sharing information both at company level and with business partners. IT services are supported and sometimes incorporated into production processes. In certain areas the skills of employees are proving to be enough to apply/test Industry 4.0 concepts. This type of company is not new to collaborations with research institutes and/or strategic consulting firms.
Top-Performer (*High digital level*)	Companies of this level are in the advanced stage of implementation of Industry 4.0. A strategy has already been implemented. Investments in multiple areas are either in the planning stage or already underway. The sharing of information at company level between departments and with suppliers is fully or largely integrated. Production processes are integrated with IT systems thanks to which production data are collected and used for monitoring and optimisation of the processes themselves. This type of company has concluded several collaboration experiences with different partners (e.g., applied research institutes, consulting companies, system integrators).

The variable representing the digital level of companies will be employed to test the RQ2 of the present study, namely which characteristics of companies may be associated with different perceptions of sustainability, together with other attributes that were subjects of the survey such as size (number of employees) and turnover (EUR) as summarised in Table 4.

Table 4. Variables.

Category	Description	Metrics	Source
Size	1 Small	10–49 employees	[53]
	2 Medium	50–249 employees	
	3 Large	Over 250 employees	
Turnover	1 Small	Up to 10 million EUR	[53]
	2 Medium	11–49 million EUR	
	3 Large	Over 50 million EUR	
Digital level	1 Digital Newcomers (low level)	Own elaboration	Own elaboration
	2 Companies in Transition (medium level)		
	3 Top Performer (high level)		

A specific question in the survey (Q-16) assesses the advantages of Industry 4.0 perceived by companies, considering the TBL as identified in prior studies and systematically outlined in Table 5.

Table 5. Structure of Q-16: Advantages of Industry 4.0 perceived by companies.

ID	Advantage	Triple Bottom Line	Reference
Log	*Optimization in logistics and warehousing*	Economic	[54–57]
TS	*Time savings*	Economic	[58,59]
Flex	*Flexible organisation of work*	Social	[60,61]
Qual	*Quality (reduction of errors)*	Economic	[62–64]
PS	*Lower physical stress*	Social	[65–68]
Work	*Reduction of workforce*	Economic	[69,70]
Cost	*Reduction of costs*	Economic	[71,72]
Sust	*Sustainability (lower environmental impact)*	Environment	[20,73]

Most advantages included in the survey relate to the economic line that turned out to be relatively more covered by previous related studies in the recent scientific literature. Being a multiple-response question, respondents could express more preferences (in the specific case up to three questions): The typology of the question was considered for the choice of the proper statistical test searching for an association between attributes of the companies such as size, turnover and digital level with the perception of the different advantages related to Industry 4.0 (Section 3.3).

3.2. Data Collection and Sample

The data derived with Digital Check were collected via an online survey available to be filled out from 15 October 2018 to 18 November 2018. The survey was managed by Confindustria Marche Nord (regional unit of the General Confederation of Italian Industry) and distributed among a sample of companies operating in the Marche Region (Italy). The sample consists of 65 companies, mainly operating in the machine construction (40%), chemical (13%) and footwear (7%) sectors. Approximately 70% of the respondent companies could be classified as SMEs employing less than 250 collaborators and having less than 50 million Euro of yearly turnover (see Table 6).

Table 6. Sample of participating companies according to size.

Size	Number	Percent
Small	23	35.4%
Medium	23	35.4%
Large	19	29.2%
Total	**65**	**100.0%**

Figure 1 shows the results of the survey and the categorization of the sample into the three digital levels as explained before. A positive result of the survey is that few companies ended up as digital newcomers (4.6% of total) and therefore at the outset of introducing Industry 4.0 and digitalization into their company. While 20.0% of the respondents belong to the digital top performers, most of the participating companies are in transition and therefore on an average digital level (75.4%). Taking a further look only to SMEs we can see that especially those that are newcomers are almost all SMEs, while top performers seem to be almost all medium to large companies. It is interesting to note that the companies in the transitional phase consist nearly equally of small, medium-sized and large companies.

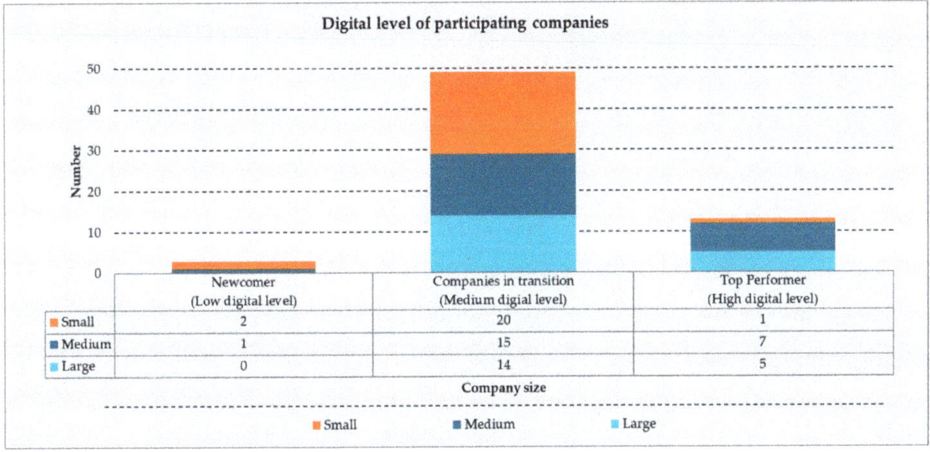

	Newcomer (Low digital level)	Companies in transition (Medium digial level)	Top Performer (High digital level)
Small	2	20	1
Medium	1	15	7
Large	0	14	5

Figure 1. Digital level of participating companies.

3.3. Data Analysis

The structure and the results of the Digital Check provide the opportunity to answer the research questions of the present study, and particularly to test whether an association between variables exists. In the analysis, we employed the multiple response to question 16 (Q-16) assigned by companies ordered in different variables, namely size, turnover and digital level, to study whether there are differences in the perceptions of sustainability among a group of companies. The analysis and modelling of categorical variables have been the focus of significant research activities such as [74]. Considering the multiple-response nature of the data, the work of Agresti and Liu and Thomas and Decady were considered [75,76]. Pearson Chi-squared test for nominal (categorical) data was performed. For the data analysis, the statistical software SPSS v.19 was employed.

4. Results

This section presents the main results of the analysis. Section 4.1 answers RQ1 and the related H1. Sections 4.2–4.4 deal with RQ2. In this regard, while the Sections 4.2 and 4.3 provide evidence to assess H2, Section 4.4 exclusively addresses H3.

4.1. Advantages Related to Industry 4.0 Implementation

As shown in Figure 2 the main advantages linked to the adoption of Industry 4.0 concern the possibility of increasing the quality of products through the reduction of errors (63.1%). Advantages in the management of logistics (53.8%) and time savings (50.8%) are also considered major benefit following the implementation of Industry 4.0 concepts. Within the social dimension of sustainability, less than half of the respondents consider Industry 4.0 beneficial to lower physical stress of collaborators (40.0%). The percentage decreases with consideration to the advantage related to higher flexibility of work (24.6%). A rather low number of respondents consider Industry 4.0 applications as an opportunity to improve sustainability, e.g., reducing negative environmental impact of manufacturing processes (6.2%). Regarding RQ1, we conclude overall that companies consider economic sustainability at a higher extent compared to the social and environmental dimension. The consideration of environmental sustainability as an advantage among companies is very low, hence H1 is supported.

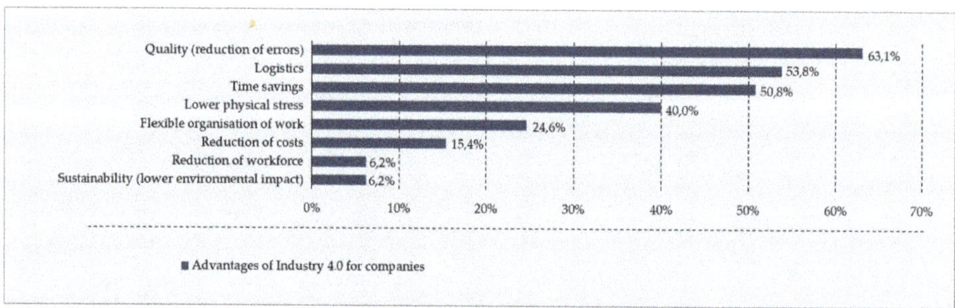

Figure 2. Advantages of Industry 4.0 for companies.

4.2. Company Size as Variable for Different Associations in the Perception of Sustainability

The first categorical variable tested for a potential association with the perception of Industry 4.0 as beneficial for sustainability in manufacturing is the size of the company, which considers the number of employees. There is a strong association ($\chi^2 = 7.093$, $p < 0.019$) between the size of the company and the beneficial perception of Industry 4.0 in achieving more flexible organisation of work (Table 7). Indeed, only a very marginal percentage of small companies (4.3%) perceive this aspect of social sustainability as beneficial, against a substantial 34.8% and 36.8% of medium and large companies, respectively.

Table 7. Summary table of the joint distributions of size of the companies (rows) and each investigated advantage of Industry 4.0 (columns).

Size	Advantages							
N = 65	*Qual*	*Log*	*TS*	*LPS*	*Flex*	*Cost*	*Work*	*Sust*
Small	56.5% 13	60.9% 14	60.9% 14	43.5% 10	4.3% 1	13% 3	4.3% 1	8.7% 2
Medium	69.6% 16	56.5% 13	39.1% 9	43.5% 10	34.8% 8	21.7% 5	4.3% 1	8.7% 2
Large	63.2% 12	42.1% 8	52.6% 10	31.6% 6	36.8% 7	10.5% 2	10.5% 2	0% 0
χ^2	0.840	1.577	2.212	0.793	7.903	1.155	0.889	1.761
Sig.	0.657	0.455	0.331	0.673	0.019	0.561	0.641	0.415

Considering the variable environmental sustainability as advantage of Industry 4.0, the available data show no significant association ($\chi^2 = 1.761$, $p < 0.415$). As for the statistically significant variables referring to RQ2, results suggest a strong association between the size of the company and the advantage of higher flexibility of work organization within the social dimension of sustainability. None of the other advantages related to other dimensions of sustainability (economic and environmental) show a significant association with size. Consequently, H2 is supported, since environmental sustainability is not perceived as an advantage related to Industry 4.0 regardless of the size of the company.

4.3. Turnover of the Company as Variable for Different Association in the Perception of Sustainability

Following the same approach, the categorical variable referring to the yearly turnover of the company (EUR) was tested against the hypothesis of independence with the perception of Industry 4.0 as beneficial for specific economic, social and environmental dimensions concerning sustainability in manufacturing (see Table 8). In this case the number of respondents covers 92.3% of the entire sample (N = 60) since five companies opted for not disclosing their category of yearly turnover. The Chi-Squared statistics indicate the existence of no significant association across the economic, social and environmental advantages proposed, such as cost savings ($\chi^2 = 0.051$, $p < 0.0975$), lower physical stress ($\chi^2 = 0.008$, $p < 0.996$) and environmental impact ($\chi^2 = 4.058$, $p < 0.131$), respectively. The answer to RQ2 is that the variable turnover does not show any significant association to specific advantages related to the economic, social and environmental dimensions of sustainability. Therefore, H2 is supported.

Table 8. Summary table of the joint distributions of turnover of the companies (rows) and each investigated advantage of Industry 4.0 (columns).

Turnover	Advantages							
N = 60	*Qual*	*Log*	*TS*	*LPS*	*Flex*	*Cost*	*Work*	*Sust*
Small	63.6% 14	54.5% 12	59.1% 13	40.9% 9	9.1% 2	13.6% 3	9.1% 2	4.5% 1
Medium	68.4% 13	63.2% 12	42.1% 8	42.1% 8	31.6% 6	15.8% 3	31.6% 6	15.8% 3
Large	63.2% 12	47.4% 9	42.1% 8	42.1% 8	36.8% 7	15.8% 3	36.8% 7	0% 0
χ^2	0.144	0.960	1.610	0.008	4.829	0.051	0.328	4.058
Sig	0.930	0.619	0.447	0.996	0.089	0.975	0.849	0.131

4.4. Digital Level as Variable for Different Association in the Perception of Sustainability

Finally, a test was performed to investigate whether the digital level of the company is associated with advantages related to Industry 4.0 applications (see Table 9). The Chi-Squared statistics indicate the existence of a significant association between the advantage of Industry 4.0 for logistics and the digital level of the company ($\chi^2 = 8.035$, $p < 0.018$). This provides an additional element to answer RQ2 since the variable digital level results associated to a specific advantage of Industry 4.0 related to the economic dimension of sustainability and related to the optimisation of logistics and warehousing. The association of Industry 4.0 as beneficial for environmental sustainability remains low ($\chi^2 = 0.250$), regardless of the digital level of companies. Consequently, H2 is supported.

Table 9. Summary table of the joint distributions of the digital level of the companies (rows) and each investigated advantage of Industry 4.0 (columns).

Digital Level	Advantages							
N = 65	*Qual*	*Log*	*TS*	*LPS*	*Flex*	*Cost*	*Work*	*Sust*
Low	66.7% 2	0% 0	33.3% 1	66.7% 2	0% 0	0% 0	0% 0	0% 0
Medium	61.2% 30	63.3% 31	53.1% 26	40.8% 20	22.4% 11	14.3% 7	22.4% 11	6.1% 3
High	69.2% 9	30.8% 35	46.2% 6	30.8% 4	38.5% 5	23.1% 3	38.5% 5	7.7% 1
χ^2	0.300	8.035	0.579	1.364	2.447	1.182	2.447	0.250
Sig	0.861	0.018	0.749	0.506	0.249	0.554	0.294	0.882

5. Discussion

5.1. General Discussion of Results

The present research studied whether differences exist in the perception of the advantages related to the implementation of Industry 4.0 within a sample of manufacturing companies located in the Marche Region (Italy). In particular, the existence of differences in the association of Industry 4.0 to specific economic, social and environmental sustainability concerns was investigated considering as the explanatory variable "company specific characteristics" such as number of employees, yearly turnover and digital level. With respect to the RQ1 formulated in the present paper, the overall analysis of the sample indicates that companies in general rate economic advantages among the most relevant factors, compared to other social and environmental benefits, resulting from Industry 4.0. This evidence is aligned to the existing literature, indicating a consistent number of economic opportunities at the base of the implementation of Industry 4.0 technologies and concepts, which can also be gained in the context of SMEs [77,78]. Similar conclusions were drawn from other studies in relation to the most recurrent advantages rated in the sample object of the present research, such as quality (reduction of errors) and logistics [79,80]. Furthermore, the relatively low perception in terms of reduction of workforce, following Industry 4.0, complies with the current literature on the topic, portraying unified perspectives on potential gain or loss in the number of employees in industry [81]. Social aspects of sustainability, such as more flexible organisation of work and lower physical stress, contributing to improved working environment for employees, have been rated overall of secondary importance, although previous studies acknowledged such aspects as drivers for Industry 4.0 implementation [20]. Environmental opportunities emerge as irrelevant drivers relating to the implementation of Industry 4.0, within the considered sample. A very limited number of respondents declared considering Industry 4.0 as an opportunity to improve their companies. This evidence differs from the findings of previous studies [82], indicating environmental opportunities as relevant drivers for a tendency towards Industry 4.0 [20,28]. Another relevant study on this field showed with a higher degree of details that green production in general, as well as energy and resource management, resulted as high priority within the surveyed sample of companies [22]. The results of this analysis offered enough elements to confirm the proposed H1, acknowledging the absence of sustainable opportunities influencing the tendency for Industry 4.0 implementation. Consequently, at the aggregate level of the sample, economic opportunities are considered prominent.

Answering RQ2 implied studying the opportunities associated to Industry 4.0 according to different characteristics of the sample such as size, turnover and digital level. With respect to company size in terms of employees, the findings for RQ2 revealed the existence of a stronger association for large companies regarding the perceived social opportunity, related to the higher flexibility in the organisation of work. The latter is acknowledged as being enabled by Industry 4.0 [58] and in terms of reshaping organisations is considered a driver for sustainability [83]. In the literature, SMEs are recognised for their higher degree of flexibility compared to large companies [84], due to specific characteristics already enabling them to react to increasingly volatile market conditions [85]. This aspect may explain the higher relevance attributed by larger companies to Industry 4.0 implementation as potentially nurturing their demand for higher flexibility. Considering the advantage of cost, our research did not find any association with size. In this respect a recent study identified a greater relevance of this advantage for larger companies, such as multinational enterprises, compared to SMEs due to their higher cost reduction and profitability expectations [86]. There is no strong association of Industry 4.0 to environmental sustainability, regardless of the size of the companies. For this reason, the proposed H2 is confirmed.

The analysis considering yearly level of turnover did not show significant associations to specific Industry 4.0 related advantages. This confirmed our initial hypothesis (H2), related to the turnover as a potential variable resulting in different associations to environmental sustainability among companies. Several scholars argue that the lack of financial resources in general may represent a main constraint to the implementation of Industry 4.0 concepts, particularly across SMEs [87–89]. In this respect, the results of our study could not indicate clear directions on whether large companies, due to their higher investment capabilities, rate specific advantages differently compared to SMEs.

The results obtained analysing the digital level against the different advantages relating to Industry 4.0 do not show any significance association for environmental and social sustainability. Therefore, companies of the sample, regardless of their digital level, as proposed in H3, do not associate significantly environmental sustainability as a driving advantage for application of Industry 4.0. In this regard, the results indicate for RQ2 a rather stronger association for economic sustainability, such as the opportunity for optimization of logistics and warehousing, among companies rated with a higher digital level, namely companies in transition and top-performer. Reviewing the scientific literature on the topic, we noted that the association between digital level and the perceived advantages relating to Industry 4.0 implementation appears have limited assessment. While the existence of assessment of the adoption of specific typologies of technologies according to the maturity level of implementation of Industry 4.0 is acknowledged [90], the review of existing literature did not yield to the identification of studies analysing the aforementioned association between digital level and advantages of Industry 4.0. While not relating to social or environmental sustainability, we observe that companies with a higher digital level consider the advantages of Industry 4.0 for efficient logistics and warehousing more relevant, compared to less digitalised companies. Considering the limited number of similar studies on this specific matter, it is challenging to draw further conclusions in comparison with existing literature. A partial explanation for this association can be suggested considering the general large size of companies which achieved the highest digital level (top performer) within the considered sample. In this respect, large companies may consider Industry 4.0 applications more relevant for logistics, being more advanced in integrating information technology systems into their production facilities compared to SMEs facing more obstacles to their integration in digital supply chain [91,92].

5.2. Implications for Practitioners and Academia

The results of this study are relevant for researchers as well as for practitioners from industry and especially managers of SMEs. As already seen in the literature review there is a lack of investigation concerning the current status as well as opportunities for the introduction of Industry 4.0 and digitalization in companies. To be precise, the literature review showed that there are lots of contributions to presenting different concepts and technologies for Industry 4.0, but not that many works that give an insight into already ongoing projects to introduce Industry 4.0 or its results. With this work and survey, we provide information from the perspective of companies themselves and how they see the impact and advantages of Industry 4.0 regarding economic, social and environmental aspects.

The study provides valuable information for practitioners and mangers of SMEs. First of all, the study used a self-assessment tool adapted to the needs of SMEs to assess the current status of companies in introducing Industry 4.0. The results show that almost all companies started to deal with the introduction of Industry 4.0. This is an important result for all those companies that did not yet start to implement any kind of measure, as it would mean that they are already far behind most companies and thus that intervention measures are compelling. In addition, this kind of benchmarking data shows also that for most of the companies there is still a long way to go to be fully digitalized and to become a top-performer.

An unfortunately sobering result of this study is that most of the companies (including SMEs as well as large enterprises) do not perceive Industry 4.0 as a chance to enhance sustainability (especially environmental and social sustainability). This should be an important input for researchers working in the area of Industry 4.0 and sustainability, as it shows that there is still a need to raise awareness of the opportunities arising from the introduction of Industry 4.0 to increase sustainability and performance aspects as already shown by recent studies [73,93,94]. In this respect, these results may suggest that, e.g., regional business and industrial associations update accordingly the programs offered on digital transformation and Industry 4.0 toward the topics increasingly relevant but currently less mastered by companies, such as the social and environmental opportunities of Industry 4.0. In addition, there seems to be a general lack of enterprises in perceiving sustainability as an important aim for the future, combining long-term economic aspects with social and environmental aspects. Therefore, it will be important for the future to conduct, e.g., case study research in order to provide information of real case studies where Industry 4.0 contributes to increasing sustainability and at the same time to improving the overall performance of the company. Targeted actions could also consist in knowledge transfer, providing more information about the relationship between Industry 4.0 and sustainability. As the lack of perception of advantages of Industry 4.0 concerning sustainability seems to be evident not only for SMEs, but also for large companies, further studies should also motivate large companies with higher financial and human resources to take a first step in implementing measures. Table 10 summarizes all identified implications in a sort of action plan.

Table 10. Sample of participating companies according to size.

N°	Result of this study	Derived Actions	Interested Parties
1	SMEs participating in this study learned a lot from conducting the self-assessment.	SMEs should take advantage of offered self-assessment tools to identify the current status in introducing Industry 4.0.	Practitioners in SMEs
2	Almost all companies started to deal with the introduction of Industry 4.0.	Intervention measures are compelling for all those SMEs that have not yet started to introduce Industry 4.0.	Practitioners in SMEs
3	Most of the companies are not yet top performer.	SMEs need to define a long-term oriented strategy with the aim to become top performer.	Practitioners in SMEs
4	Most of the companies do not perceive Industry 4.0 as a chance to enhance sustainability.	Revise programs and offer of initiatives to increase the awareness of social and environmental opportunities of Industry 4.0.	Regional and Industrial Associations
5	Most of the companies do not perceive Industry 4.0 as a chance to enhance sustainability.	Conduct research to raise awareness of the opportunities of Industry 4.0 to increase performance as well as sustainability aspects.	Researchers in Academia
6	Lack of information from real case studies where Industry 4.0 contributed to increase sustainability.	Conduct case study research to provide best practice examples	Researchers in Academia together with practitioners from industry
7	Missing knowledge transfer regarding the relationship between Industry 4.0 and sustainability	Conduct research on the relationship of Industry 4.0 and sustainability	Researchers in Academia

5.3. Limitations

We want to note that our study might have certain limitations that need to be considered. While we consider the size of the sample with 65 companies as sufficient to derive generalized results and hypotheses, we are conscious of the fact that all companies are located in one specific geographical region in Italy. The study is therefore representative for all rural regions in Italy or Europe with the similar characteristics of industry sectors and company size, while there might be differences in the perception between the participating companies in this study and, e.g., companies in urban regions and heavily industrialized areas or large cities.

Further, the study was mainly targeted to assess the current state in the introduction of Industry 4.0 in the region of Marche in Italy with only a partial focus on sustainability aspects. Therefore, future research should be conducted to focus and further investigate the relationship between Industry 4.0 and sustainability. This will enable defining more specific relations between sustainability and Industry 4.0 in specific domains and to explore innovative ways to enhance the improvement of either economic or social and environmental performance and contribution of companies. Future research directions, deepening the discourse about sustainability in SMEs approached in the present study, should extend the focus of the analysis to the realm of circular economy and Industry 4.0. This foresees the development of targeted tools assessing the degree of circular economy, adapted to the requirements of SMEs, similar to the methodological work and considerations undertaken to define the structure and metrics of the self-assessment tool for digital readiness employed in the present study.

We exclude any subjective influence on the part of the research team, as data has been taken directly and without any subjective assessment or modification and evaluated using statistical methods and software.

6. Conclusions and Outlook

This work deals with the investigation of the advantages in the application of Industry 4.0 for sustainability. It presents the partial results of a larger survey-based study with 65 companies from the Marche region in Italy. Two research questions were set up for the study and answered based on the results. The first research question deals with whether Industry 4.0 is seen as beneficial for sustainability, considering the economic, environmental and social dimension. The results confirm the initial hypothesis (H1) that currently Industry 4.0 is not seen as beneficial for achieving environmental sustainability by most of the companies. The advantages relating to the economic dimensions of sustainability prevail also over social considerations of sustainability. The second research question deals with whether there are different views on this aspect depending on factors such as size, turnover or the digital level of the companies. Two hypotheses were also put forward that were also confirmed by the data. The first of these two hypothesis (H2) is that companies, regardless of their size (small, medium, large) and level of turnover (small, medium, large), do not consider Industry 4.0 to be advantageous for achieving environmental sustainability. The other hypothesis (H3) is that Industry 4.0 is not considered beneficial for environmental sustainability, regardless of the current digital level of companies.

While it has been shown that some recent research indicates positive links between Industry 4.0 and environmental sustainability, the results of this study show that companies are not fully aware of them. Overall, it is evident that issues such as social and environmental sustainability are often neglected by companies to the benefit of economic sustainability. We therefore see a great need for actions for knowledge transfer in the future, but also to prove a positive correlation between Industry 4.0, sustainability and the performance of companies through applied research and real industrial case studies.

Author Contributions: R.B. developed the overall research design; D.F. worked on the systematic literature review and data analysis; E.R. contributed validating the findings considering the ongoing scientific debate and related implications for practitioners and academia; D.T.M. supervised the correct framing of the proposed concept within the scientific context of the topics discussed. All authors have read and agreed to the published version of the manuscript.

Funding: The APC was funded by the project "SME 4.0—Industry 4.0 for SMEs" (European Union's Horizon 2020 R&I programme under the Marie Skłodowska-Curie grant agreement No 734713).

Acknowledgments: This work belongs to the project "SME 4.0—Industry 4.0 for SMEs" (funded in the European Union's Horizon 2020 R&I programme under the Marie Skłodowska-Curie grant agreement No 734713) and the project "SCENA" (CIG: 7417905858D9 funded by Fondirigenti). The authors would like to thank Confindustria Marche, Federmanager and all the participating local institutions for the fruitful collaboration.

Conflicts of Interest: The authors declare no conflict of interest.

References

1. Platform Industry 4.0. Available online: https://www.plattform-i40.de/PI40/Navigation/EN/Home/home.html (accessed on 7 December 2019).
2. Pfohl, H.-C.; Yahsi, B.; Kurnaz, T. The Impact of Industry 4.0 on the Supply Chain. Innovations and Strategies for Logistics and Supply Chains. Proocedings of the International Conference of Logistics (HICL), Hamburg, Germany, 24–25 September 2015.
3. Andulkar, M.; Le, D.T.; Berger, U. A multi-case study on Industry 4.0 for SME's in Brandeburg, Germany. In Proceedings of the 51st International Conference on System Sciences, Honolulu, Hawaii, 3–6 January 2018; pp. 4544–4553.
4. Schröder, C. *The Challenges of INDUSTRY 4.0 for Small and Medium-sized Enterprises*; Friedrich-Ebert-Stiftung: Bonn, Germany, 2017.
5. Rauch, E.; Dallasega, P.; Matt, D.T. Critical Factors for Introducing Lean Product Development to Small and Medium Sized Enterprises in Italy. *Procedia CIRP* **2017**, *60*, 362–367. [CrossRef]
6. Bauer, W.; Schulund, W.; Marrenbach, D.; Ganshar, O. *Industrie 4.0-Volkswirttschafliches Potenzial fur Deutschland*; BITKOM: Berlin, Germany, 2014.

7. Lasinio, C.J.; Lucchetti, F.; Perani, G.; Rinaldi, M.; Zeli, A.; Zurlo, D. *Il Piano Nazionale Impresa 4.0: Prime Valutazioni*; ISTAT: Rom, Italy, 2018.

8. Spath, D.; Ganschar, O.; Gerlach, S.; Hämmerle, M.; Krause, T.; Schlund, S. *Manufacturing Work of the Future–Industry 4.0*; Fraunhofer Verlag: Stuttgart, Germany, 2013.

9. Wenking, M.; Benninghaus, C.; Friedli, T. Umsetzungsbarrieren und -lösungen von Industrie 4.0: Welche Faktoren limitieren die Produktion der Zukunft? *ZWF* **2016**, *111*, 847–850. [CrossRef]

10. Baum, G. Innovation as the basis for the next industrial revolution. In *Industry 4.0-Control of Industrial Complexity with SysLM (in German)*; Springer: Berlin/Heidelberg, Germany, 2013.

11. Ghobakhloo, M. Industry 4.0, Digitization, and Opportunities for Sustainability. *J. Clean. Prod.* **2020**, *252*, 119869. [CrossRef]

12. Stock, T.; Seliger, G. Opportunities of Sustainable Manufacturing in Industry 4.0. *Procedia CIRP* **2016**, *40*, 536–541. [CrossRef]

13. Prause, G. Sustainable business models and structures for Industry 4.0. *J. Secur. Sustain. Issues* **2015**, *5*, 159–169. [CrossRef]

14. Kiel, D.; Müller, J.M.; Arnold, C.; Voigt, K.-I. Sustainable industrial value creation: Benefits and challenges of industry 4.0. *Int. J. Innov. Mgt.* **2017**, *21*, 1740015. [CrossRef]

15. Jensen, J.P.; Remmen, A. Enabling Circular Economy through Product Stewardship. *Procedia Manuf.* **2017**, *8*, 377–384. [CrossRef]

16. Tortorella, L.G.; Fettermann, D. Implementation of Industry 4.0 and lean production in Brazilian manufacturing companies. *Int. J. Prod. Res.* **2018**, *56*, 2975–2987. [CrossRef]

17. Ender, J.; Wagner, J.C.; Kunert, G.; Guo, F.B.; Larek, R.; Pawletta, T. Concept of a self-learning workplace cell for worker assistance while collaboration with a robot within the self-adapting-production-planning-system. *IAPGOS* **2019**, *9*, 4–9. [CrossRef]

18. Hirsch-Kreinsen, H. Smart production systems: A new type of industrial process innovation. In Proceedings of the 2014 DRUID Society Conference, Copenhagen, Denmark, 16–18 June 2014.

19. Littig, B.; Griessler, E. Social Sustainability: A Catchword between Political Pragmatism and Social Theory. *Int. J. Sust. Dev.* **2005**, *8*, 65–78. [CrossRef]

20. Müller, J.M.; Kiel, D.; Voigt, K.-I. What Drives the Implementation of Industry 4.0? The Role of Opportunities and Challenges in the Context of Sustainability. *Sustainability* **2018**, *10*, 247. [CrossRef]

21. Beier, G.; Niehoff, S.; Ziems, T.; Xue, B. Sustainability Aspects of a Digitalized Industry–A Comparative Study from China and Germany. *Int. J. Pr. Eng. Man.-GT.* **2017**, *4*, 227–234. [CrossRef]

22. Gabriel, M.; Pessel, E. Industry 4.0 and sustainability impacts: Critical discussion of sustainability aspects with a special focus on future of work and ecological consequences. *Int. J. Eng.* **2016**, *1*, 131–136.

23. Elkington, J. Towards the Sustainable Corporation: Win-Win-Win Business Strategies for Sustainable Development. *Calif. Manage. Rev.* **1994**, *36*, 90–100. [CrossRef]

24. Giret, A.; Trentesaux, D.; Prabhu, V. Sustainability in manufacturing operations scheduling: A state of the art review. *J. Manuf. Syst.* **2015**, *37*, 126–140. [CrossRef]

25. Mcwilliams, A.; Parhankangas, A.; Coupet, J.; Welch, E.; Barnum, D. Strategic Decision Making for the Triple Bottom Line. *Bus. Strateg. Environ.* **2014**. [CrossRef]

26. Markley, M.; Davis, L. Exploring Future Competitive Advantage through Sustainable Supply Chains. *Int. J. Phys. Distr. Log.* **2007**, *37*, 763–774. [CrossRef]

27. Zampou, E.; Plitsos, S.; Karagiannaki, A.; Mourtos, I. Towards a framework for energy-aware information systems in manufacturing. *Comput. Ind.* **2014**, *65*, 419–433. [CrossRef]

28. Kagermann, H.; Wahlster, W.; Helbig, J. *Reccomendations for Implementing the Strategic Initiative Industrie 4.0*; acatec -National Academy of Science and Engineering: München, Deutschland, 2013.

29. Schwab, K. *The Fourth Industrial Revolution*; Crown Business: New York, NY, USA, 2016.

30. Liao, Y.; Deschamps, F.; Rocha Loures, E.; Ramos, L. Past, present and future of Industry 4.0-a systematic literature review and research agenda proposal. *Int. J. Prod. Res.* **2017**, *55*. [CrossRef]

31. Brozzi, R.; D'Amico, R.D.; Pasetti Monizza, G.; Marcher, C.; Riedl, M.; Matt, D. Design of Self-Assessment Tools to Measure Industry 4.0 Readiness. A Methodological Approach for Craftsmanship SMEs. In *Product Lifecycle Management to Support Industry 4.0*; Chiabert, P., Bouras, A., Noël, F., Ríos, J., Eds.; Springer: Berlin/Heidelberg, Germany, 2018; pp. 566–578.

32. Matt, D.; Rauch, E. Chancen Zur Bewältigung Des Fachkräftemangels in KMU Durch Die Urbane Produktion von Morgen (Opportunities to address shortages in qualified staff in SMEs through the Urban Production of tomorrow). In *Industrie 4.0 Wie Intelligente Vernetzung und Kognitive Systeme Unsere Arbeit Verändern*; Lödding, H., Kersten, W., Koller, H., Eds.; Gito Verlag: Berlin, Germany, 2014; pp. 155–176.

33. de Sousa Jabbour, A.B.L.; Jabbour, C.J.C.; Foropon, C.; Godinho Filho, M. When titans meet–Can industry 4.0 revolutionise the environmentally-sustainable manufacturing wave? The role of critical success factors. *Technol. Forecast. Soc.* **2018**, *132*, 18–25. [CrossRef]

34. Morelli, J. Environmental Sustainability: A Definition for Environmental Professionals. *J. Environ. Sustain.* **2011**, *1*, 1–10. [CrossRef]

35. Vance, L.; Eason, T.; Cabezas, H. Energy sustainability: Consumption, efficiency, and environmental impact. *Cleam. Techn. Environ. Policy* **2015**, *17*, 1781–1792. [CrossRef]

36. Bonilla, S.; Silva, H.; Terra da Silva, M.; Franco Gonçalves, R.; Sacomano, J. Industry 4.0 and Sustainability Implications: A Scenario-Based Analysis of the Impacts and Challenges. *Sustainability* **2018**, *10*, 3740. [CrossRef]

37. Müller, J.; Dotzauer, V.; Voigt, K. Industry 4.0 and its Impact on Reshoring Decisions of German Manufacturing Enterprises. In *Supply Management Research. Advanced Studies in Supply Management*; Bode, C., Bogaschewsky, R., Eßig, M., Lasch, R., Stölzle, W., Eds.; Springer: Berlin/Heidelberg, Germany, 2017; pp. 165–179.

38. Gebler, M.; Schoot Uiterkamp, A.J.M.; Visser, C. A global sustainability perspective on 3D printing technologies. *Energy Policy* **2014**, *74*, 158–167. [CrossRef]

39. Qiu, X.; Luo, H.; Xu, G.; Zhong, R.; Huang, G.Q. Physical assets and service sharing for IoT-enabled Supply Hub in Industrial Park (SHIP). *Int. J. Prod. Econ.* **2015**, *159*, 4–15. [CrossRef]

40. Hermann, M.; Pentek, T.; Otto, B. Design Principles for Industrie 4.0 Scenarios. In Proceedings of the 49th Hawaii International Conference on Systems Sciences (HICSS), Honolulu, HI, USA, 5–8 January 2016.

41. Ding, K.; Jiang, P.; Zheng, M. Environmental and economic sustainability-aware resource service scheduling for industrial product service systems. *J. Intell. Manuf.* **2017**, *28*, 1303–1316. [CrossRef]

42. Fysikopoulos, A.; Pastras, G.; Alexopoulos, T.; Chryssolouris, G. On a generalized approach to manufacturing energy efficiency. *Int. J. Adv. Manuf. Technol.* **2014**, *73*, 1437–1452. [CrossRef]

43. Pfeffer, J. Building Sustainable Organizations: The Human Factor. *Acad. Manag. Perspect.* **2010**, *24*. [CrossRef]

44. Gualtieri, L.; Rojas, R.; Carabin, G.; Palomba, I.; Rauch, E.; Vidoni, R.; Matt, D.T. Advanced Automation for SMEs in the I4.0 Revolution: Engineering Education and Employees Training in the Smart Mini Factory Laboratory. In Proceedings of the 2018 IEEE International Conference on Industrial Engineering and Engineering Management (IEEM), Bangkok, Thailand, 16–19 December 2018; pp. 1111–1115.

45. Rojas, R.; Rauch, E.; Dallasega, P.; Matt, D.T. Safe Human-Machine Centered Design of an Assembly Station in a Learning Factory Environment. In Proceedings of the International Conference on Industrial Engineering and Operations Management, Bandung, Indonesia, 6–8 March 2018; pp. 403–411.

46. Gualtieri, L.; Rauch, E.; Rojas, R.; Vidoni, R.; Matt, D. Application of Axiomatic Design for the Design of a Safe Collaborative Human-Robot Assembly Workplace. *MATEC Web Conf.* **2018**, *223*, 01003. [CrossRef]

47. Mark, B.G.; Hofmayer, S.; Rauch, E.; Matt, D.T. Inclusion of Workers with Disabilities in Production 4.0: Legal Foundations in Europe and Potentials Through Worker Assistance Systems. *Sustainability* **2019**, *11*, 5978. [CrossRef]

48. Schulte, J.; Hallstedt, S.I. Self-Assessment Method for Sustainability Implementation in Product Innovation. *Sustainability* **2018**, *10*, 4336. [CrossRef]

49. Elkington, J. Partnerships from Cannibals with Forks: The Triple Bottom Line of 21st-Century Business. *Environ. Qual. Manag.* **1998**, *8*, 37–51. [CrossRef]

50. Norman, W.; MacDonald, C. Getting to the Bottom of "Triple Bottom Line". *Bus. Ethics Q.* **2004**, *14*, 243–262. [CrossRef]

51. Schumacher, A.; Erol, S.; Sihn, W. A Maturity Model for Assessing Industry 4.0 Readiness and Maturity of Manufacturing Enterprises. *Procedia CIRP* **2016**, *52*, 161–166. [CrossRef]

52. Matt, D.T.; Rauch, E.; Riedl, M. Knowledge Transfer and Introduction of Industry 4.0 in SMEs: A Five-Step Methodology to Introduce Industry 4.0. In *Analyzing the Impacts of Industry 4.0 in Modern Business Environments*; Jirsák, P., Bureš, V., Eds.; IGI Global: Pennsylvania, PA, USA, 2018; pp. 256–282.

53. European Commission. Available online: https://ec.europa.eu/growth/smes/business-friendly-environment/sme-definition_en (accessed on 12 March 2020).

54. Barreto, L.; Amaral, A.; Pereira, T. Industry 4.0 Implications in Logistics: An Overview. *Procedia Manuf.* **2017**, *13*, 1245–1252. [CrossRef]

55. Schmidt, R.; Möhring, M.; Härting, R.-C.; Reichstein, C.; Neumaier, P.; Jozinović, P. Industry 4.0-Potentials for Creating Smart Products: Empirical Research Results. In *Business Information Systems*; Abramowicz, W., Ed.; Springer: Berlin/Heidelberg, Germany, 2015; pp. 16–27.

56. Schuhmacher, J.; Hummel, V. Decentralized Control of Logistic Processes in Cyber-Physical Production Systems at the Example of ESB Logistics Learning Factory. *Procedia CIRP* **2016**, *54*, 19–24. [CrossRef]

57. Zhong, R.Y.; Huang, G.Q.; Lan, S.; Dai, Q.Y.; Chen, X.; Zhang, T. A Big Data Approach for Logistics Trajectory Discovery from RFID-Enabled Production Data. *Int. J. Prod. Econ.* **2015**, *165*, 260–272. [CrossRef]

58. Tirabeni, L.; De Bernardi, P.; Forliano, C.; Franco, M. How Can Organisations and Business Models Lead to a More Sustainable Society? A Framework from a Systematic Review of the Industry 4.0. *Sustainability* **2019**, *11*, 6363. [CrossRef]

59. Davies, R.; Coole, T.; Smith, A. Review of Socio-Technical Considerations to Ensure Successful Implementation of Industry 4.0. *Procedia Manuf.* **2017**, *11*, 1288–1295. [CrossRef]

60. Peukert, B.; Benecke, S.; Clavell, J.; Neugebauer, S.; Nissen, N.F.; Uhlmann, E.; Lang, K.-D.; Finkbeiner, M. Addressing Sustainability and Flexibility in Manufacturing Via Smart Modular Machine Tool Frames to Support Sustainable Value Creation. *Procedia CIRP* **2015**, *29*, 514–519. [CrossRef]

61. Brettel, M.; Klein, M.; Friederichsen, N. The Relevance of Manufacturing Flexibility in the Context of Industrie 4.0. *Procedia CIRP* **2016**, *41*, 105–110. [CrossRef]

62. Rüßmann, M.; Lorenz, M.; Gerbert, P.; Waldner, M.; Justus, J.; Engel, P.; Harnisch, M. *Industry 4.0: The Future of Productivity and Growth in Manufacturing*; BCG: Boston, MA, USA, 2015.

63. Oliff, H.; Liu, Y. Towards Industry 4.0 Utilizing Data-Mining Techniques: A Case Study on Quality Improvement. *Procedia CIRP* **2017**, *63*, 167–172. [CrossRef]

64. Shibin, K.T.; Gunasekaran, A.; Papadopoulos, T.; Childe, S.J.; Dubey, R.; Singh, T. Energy Sustainability in Operations: An Optimization Study. *Int. J. Adv. Manuf. Technol.* **2016**, *86*, 2873–2884. [CrossRef]

65. Rönick, K.; Kremer, T.; Wakula, J. Evaluation of an Adaptive Assistance System to Optimize Physical Stress in the Assembly. In Proceedings of the 20th Congress of the International Ergonomics Association (IEA 2018), Florence, Italy, 26–30 June 2018; Bagnara, S., Tartaglia, R., Albolino, S., Alexander, T., Fujita, Y., Eds.; Springer: Berlin/Heidelberg, Germany, 2019; pp. 576–584.

66. Romero, D.; Bernus, P.; Noran, O.; Stahre, J.; Fast-Berglund, Å. The Operator 4.0: Human Cyber-Physical Systems & Adaptive Automation towards Human-Automation Symbiosis Work Systems. In *Advances in Production Management Systems, Initiatives for a Sustainable World*; Nääs, I., Vendrametto, O., Mendes Reis, J., Gonçalves, R.F., Silva, M.T., von Cieminski, G., Kiritsis, D., Eds.; Springer: Berlin/Heidelberg, Germany, 2016; Volume 488, pp. 677–686.

67. Rauch, E.; Linder, C.; Dallasega, P. Anthropocentric Perspective of Production before and within Industry 4.0. *Comput. Ind. Eng.* **2020**, *139*, 105644. [CrossRef]

68. Rodrigues, V.P.; Pigosso, D.C.A.; McAloone, T.C. Process-Related Key Performance Indicators for Measuring Sustainability Performance of Ecodesign Implementation into Product Development. *J. Clean. Prod.* **2016**, *139*. [CrossRef]

69. Frey, C.B.; Osborne, M.A. The Future of Employment: How Susceptible Are Jobs to Computerisation? *Technol. Forecast. Soc.* **2017**, *114*, 254–280. [CrossRef]

70. Zhang, J.; Ding, G.; Zou, Y.; Qin, S.; Fu, J. Review of Job Shop Scheduling Research and Its New Perspectives under Industry 4.0. *J. Intell. Manuf.* **2019**, *30*, 1809–1830. [CrossRef]

71. Dalenogare, L.S.; Benitez, G.B.; Ayala, N.F.; Frank, A.G. The Expected Contribution of Industry 4.0 Technologies for Industrial Performance. *Int. J. Prod. Econ.* **2018**, *204*, 383–394. [CrossRef]

72. Fettermann, D.C.; Cavalcante, C.G.S.; de Almeida, T.D.; Tortorella, G.L. How Does Industry 4.0 Contribute to Operations Management? *J. Ind. Prod. Eng.* **2018**, *35*, 255–268. [CrossRef]

73. Kamble, S.S.; Gunasekaran, A.; Gawankar, S.A. Sustainable Industry 4.0 Framework: A Systematic Literature Review Identifying the Current Trends and Future Perspectives. *Process Saf. Environ.* **2018**, *117*, 408–425. [CrossRef]

74. Agresti, A. Logistic regrission models using cumulative logits. In *Analysis of Categorical Data*, 2nd ed.; John Wiley & Sons, Inc.: Hoboken, NJ, USA, 2010; pp. 9–43.

75. Agresti, A.; Liu, L.M. Modeling a categorical variable allowing arbitrarily many category choices. *Biometrics* **1999**, *55*, 936–943. [CrossRef] [PubMed]

76. Thomas, R.D.; Decady, Y.J. Testing for Association Using Multiple Response Survey Data: Approximate Procedures Based on the Rao-Scott Approach. *Int. J. Test.* **2004**, *4*, 43–59. [CrossRef]

77. Rauch, E.; Vickery, A.R.; Brown, C.A.; Matt, D.T. SME Requirements and Guidelines for the Design of Smart and Highly Adaptable Manufacturing Systems. In *Industry 4.0 for SMEs: Challenges, Opportunities and Requirements*; Matt, D.T., Modrák, V., Zsifkovits, H., Eds.; Springer: Berlin/Heidelberg, Germany, 2020; pp. 39–72.

78. Moeuf, A.; Pellerin, R.; Lamouri, S.; Tamayo-Giraldo, S.; Barbaray, R. The Industrial Management of SMEs in the Era of Industry 4.0. *Int. J. Prod. Res.* **2018**, *56*, 1118–1136. [CrossRef]

79. Eleftheriadis, R.J.; Myklebust, O. A guideline of quality steps towards zero defect manufacturing in industry. In Proceedings of the 2016 International Conference on Industrial Engineering and Operations Management, Detroit, MI, USA, 23–25 September 2016.

80. Büchi, G.; Cugno, M.; Castagnoli, R. Smart factory performance and Industry 4.0. *Technol. Forecast. Soc.* **2020**, *150*, 119790.

81. Bonekamp, L.; Sure, M. Consequences of Industry 4.0 on Human Labour and Work Organisation. *J. Bus. Media Psychol.* **2015**, *6*, 33–40.

82. Lins, T.; Rabelo Oliveira, R.A. Energy efficiency in industry 4.0 using SDN. In Proceedings of the 2017 IEEE 15th International Conference on Industrial Informatics (INDIN), Emden, Germany, 24–26 July 2017.

83. Peruzzini, M.; Gregori, F.; Luzi, A.; Mengarelli, M.; Germani, M. A social life cycle assessment methodology for smart manufacturing: The case of study of a kitchen sink. *J. Ind. Inf. Integr.* **2017**, *7*, 24–32. [CrossRef]

84. Matt, D.T.; Rauch, E.; Fraccaroli, D. Smart Factory for SMEs. *ZWF* **2016**, *111*, 52–55. [CrossRef]

85. Matt, D.T.; Rauch, E. SME 4.0: The Role of Small- and Medium-Sized Enterprises in the Digital Transformation. In *Industry 4.0 for SMEs: Challenges, Opportunities and Requirements*; Matt, D.T., Modrák, V., Zsifkovits, H., Eds.; Springer: Berlin/Heidelberg, Germany, 2020; pp. 3–36.

86. Horváth, D.; Szabó, R.Z. Driving Forces and Barriers of Industry 4.0: Do Multinational and Small and Medium-Sized Companies Have Equal Opportunities? *Technol. Forecast. Soc.* **2019**, *146*, 119–132.

87. Müller, J.; Voigt, K.-I. Industrie 4.0 für kleine und mittlere Unternehmen. Welche spezifischen Probleme werden bei der Einführung von Industrie 4.0 von kleinen und mittleren Unternehmen gesehen? *Product. Manag.* **2016**, *3*, 28–30.

88. Erol, S.; Jäger, A.; Hold, P.; Ott, K.; Sihn, W. Tangible Industry 4.0: A Scenario-Based Approach to Learning for the Future of Production. *Procedia CIRP* **2016**, *54*, 13–18. [CrossRef]

89. Dassisti, M.; Panetto, H.; Lezoche, M.; Merla, P.; Semeraro, C.; Giovannini, A.; Chimienti, M. Industry 4.0 paradigm: The viewpoint of the small and medium enterprises. In Proceedings of the 7th International Conference on Information Society and Technology, Kopaonik, Serbia, 12–15 March 2017.

90. Frank, A.G.; Dalenogare, L.S.; Ayala, N.F. Industry 4.0 Technologies: Implementation Patterns in Manufacturing Companies. *Int. J. Prod. Econ.* **2019**, *210*, 15–26. [CrossRef]

91. Luco, J.; Mestre, S.; Henry, L.; Tamayo, S.; Fontane, F. Industry 4.0 in SMEs: A Sectorial Analysis. In *Advances in Production Management Systems. Production Management for the Factory of the Future, IFIP Advances in Information and Communication Technology*; Ameri, F., Stecke, K.E., von Cieminski, G., Kiritsis, D., Eds.; Springer: Berlin/Heidelberg, Germany, 2019; pp. 357–365.

92. Bär, K.; Herbert-Hansen, Z.N.L.; Khalid, W. Considering Industry 4.0 Aspects in the Supply Chain for an SME. *Prod. Eng.* **2018**, *12*, 747–758. [CrossRef]

93. Dev, N.K.; Shankar, R.; Qaiser, F.H. Industry 4.0 and Circular Economy: Operational Excellence for Sustainable Reverse Supply Chain Performance. *Resour. Conserv. Recy.* **2020**, *153*, 104583. [CrossRef]

94. Jena, M.C.; Mishra, S.K.; Moharana, H.S. Application of Industry 4.0 to Enhance Sustainable Manufacturing. *Environ. Prog. Sustain. Energy* **2020**, *39*, 13360. [CrossRef]

Article

Management of Product Configuration Conflicts to Increase the Sustainability of Mass Customization

Vladimir Modrak * and Zuzana Soltysova

Faculty of Manufacturing Technologies, Technical University of Kosice, Bayerova 1, 080 01 Presov, Slovakia; zuzana.soltysova@tuke.sk
* Correspondence: vladimir.modrak@tuke.sk; Tel.: +421-55-602-6449

Received: 30 March 2020; Accepted: 28 April 2020; Published: 29 April 2020

Abstract: An important role in product variety management is finding an accurate variety extent to which the product matches the consumer's expectations. In principle, customers prefer to have more rather than less versions of a product from which to choose. This motivates producers to offer a richer variety of goods. As a consequence, it brings a large amount of manufacturing complexity, and configuration conflicts may frequently occur. In order to avoid a situation in which a customer will select mutually incompatible components, product configurators usually recommend corrective actions for generating valid configurations. Nevertheless, the presence of infeasible configurations in customer options are negatively perceived by customers, and therefore it has an unfavorable impact on the sustainability of mass customization. One way to solve this problem is to eliminate, or at least reduce, mutually incompatible components. When considering the fact that eliminating all incompatible components may cause a rapid decrease in product variety, then the reduction of incompatible components can help to solve the product configuration problem. The proposed method aims to find a trade-off solution between minimizing configuration conflicts and maintaining a sufficient level of mass customization. Moreover, two supplementary methods for the determination of infeasible product configurations are proposed in this paper. The applicability and effectiveness of the proposed methods are demonstrated by two practical examples.

Keywords: positive complexity; negative complexity; infeasible configurations; product platform; customer's perception

---

1. Introduction

A great challenge facing original equipment manufacturers (OEMs) today is learning about which mass customization strategy is appropriate for specific product categories and markets, which large product variety should be offered, and how to manage product variety and its resulting complexity. Commonly, regarding the extent of customizable products, it is perceived by customers that the larger the product variety, the better [1]. This fact stimulates manufacturers to offer more and more individualized products in order to meet customer requirements [2]. It is obvious that such an approach may result in unexpected turbulence in their manufacturing systems, leading to, for example, higher direct production costs [3]. However, the higher costs themselves are not an issue if a company offers customization at acceptable prices. On the contrary, product configuration conflicts are becoming a serious problem that appear mostly when a high variety of products is offered. The presence of product configuration conflicts results in infeasible configurations, which are negatively perceived by customers due to uncertainty in customer choice. In order to solve this problem, different approaches are offered, recommending that designers take corrective actions such as removing selected components or adding new components [4,5]. In spite of this, so called ex-post interventions are limited. Moreover, Fraizer and Wells [6] showed that eliminating infeasible options through product configurators is not always viable. Therefore, product designers might reconsider such constraints because they trigger adverse reactions and/or customer disappointment. A promising solution to solve this problem is to eliminate, or at least reduce, mutually incompatible components in the early phase of the design process.

The main scope of this paper is to outline the necessary steps that are needed to solve this problem by changing the balance between infeasible product configurations and feasible product configurations. For this purpose, two methods for the determination of viable product configurations are proposed in order to find trade-off solution between minimizing configuration conflicts and maintaining a sufficient level of mass customization. These methods will be described and their feasibility will be demonstrated by two realistic examples. Finally, a discussion and conclusion with a summary drawn from current and future research relating to product configuration conflicts and sustainability of mass customization is presented.

2. Related Work

Mass customization as a business strategy can be very successful when sufficient varieties of products are offered at competitive prices with reasonable delivery times [7–10]. However, mass customization is an emerging paradigm, and may not be the panacea for all organizations [11]. According to Hadzistevic and Moraca [12], typical high product variety environments tend to come with a higher number of differentiated products that have a higher number of design changes, but Whitley et al. [13] argue that "consumers' perceptions of how many choices they prefer change depending on whether they intend to use an item for pleasure or to meet a functional need". Therefore, consumers motivated by pleasure prefer a large assortment, and customers looking only for functional products will be satisfied by a smaller assortment from which to choose. In this context, our research follows the behavior of the first group of customers. Aydogdu et al. [14] emphasized that product configuration conflicts cause enormous problems in the product or service design. Krus [15] proposed expressing the quality of a modular design through the rest of the design space that is outside the constrained design space by the term 'waste information entropy'. His approach inspired us to think in this way and adopt his idea with respect to the propriety of this complexity attribute.

Several studies, those by [16–19] for example, have underlined the fact that with the increasing demand for individual products and variants the transition from mass customization (MC) towards a personalized customization becomes more and more realistic. Some studies have considered advanced methods and techniques that lead to formal approaches for the design of entire product families in terms of MC. For example, a formal computer-assisted approach that addresses the requirements for the design of product family architecture products has been proposed by Bonev et al. [20]. A dynamic constraint satisfaction approach for configuring structural products under MC was presented in the studies of Huang et al. [21] and Yang et al. [22]. Matt and Rauch [23] focused their research on designs for the modular manufacturing of mass customized goods. High product variety in MC is often associated with complexity problems [24,25]. The positive impact of so called variety-induced complexity on sales performance has been presented, e.g., in works by Zhou et al. [26] and ElMaraghy et al. [27]. Tseng and Piller [28] developed a generic product variety structure consisting of the common product structure, variety parameters, and configuration constraints. A CAD-based approach for the automatic variation of three-dimensional product structure by changing the combination of parts, selecting the assembly method, and rearranging the assembly sequence was proposed by Chu et al. [29]. Such techniques allow product designers to determine the optimum level of the feasible rate of product customization.

Moreover, our research has been motivated by previous findings that restricted options are not perceived positively by individual users. In this context, authors, e.g., Mailharro [30], Antonelli et al. [31], Aldanondo et al. [32], Helo et al. [33], and Paul et al. [34] argue that infeasible configurations might be eliminated or minimized to avoid customer disappointment.

3. Methodological Framework

In general, it is assumed that a large scale of product variety and its resulting complexity are inherently tied together. In order to establish a working definition for product variety-induced complexity, it is first supposed that complexity can be separated into structural and dynamic complexity concepts. Product variety-induced complexity belongs to the category of structural properties that describes only the static structure of a system by the number of system elements and their interrelationships [35]. Abdelkafi [36] developed a comprehensive framework explaining the term 'variety–induced complexity', assuming that this aspect of complexity is directly associated with the extent of product variety and not with the sates of the system. He emphasized that such a view on complexity differs a great deal from its meaning in cybernetics. Hu et al. [37] proposed a unified measure of complexity by combining both product variety and assembly process information, and developed models for evaluating complexity in assembly systems. Trattner et al. [38] argued that there is no strong relationship between the complexity of the products produced and any measure of manufacturing operational performance. Therefore, taking into account that every effort to find an adequate definition for product variety-induced complexity is not trivial, we further assume that variety-induced complexity relates to the amount of product variants.

Second, it is important to distinguish between an external customers' views on variety-induced complexity problems and the internal viewpoints of manufacturers. In this context, internal complexity can be defined as the amount of input, information, or energy that the manufacturing system receives from its environment [39]. External complexity relates to customers experience during the product selection process in extensive variety environments [40]. Our focus, in this research work, is only on the external complexity that reflects the number of realistic product configurations, N_{real}, offered to customers as well as the number of infeasible product variants, N_{inf}, that can occur during product configuration by the customers.

The first question when dealing with infeasible product configurations due to incompatible components concerns the distinction between customer perceptions of available product configurations and infeasible product configurations. The number of available product configurations are perceived by customers positively, i.e., the larger the better [20]. Therefore, uncertainty related to product variety reflecting only the quantity of realistic product configurations can be viewed as a positive source of variety-induced complexity. By contrast, infeasible product configurations are undesirable for customers [41]. Thus, the smaller the better criterion can be taken as a preliminary assessment of the presence of infeasible product configurations. Accordingly, the related variety-based complexity reflecting relative rate of N_{inf} in the total amount of product configurations , $N_{total,}$ is attributed here as a negative source of variety-induced complexity.

Based on the above, product configuration conflicts negatively affect the sustainability of mass customization. When following the words of Lord Kelvin, "If you cannot measure it, you cannot improve it" the subsequent challenge is to measure this impact in a way that is quantifiable as well as qualitative. For this purpose, it is reasonable to employ Axiomatic Design Theory [42] for selecting the best design parameters (DPs) to satisfy the functional requirements (FRs). The second axiom states that when several alternative designs satisfy the first axiom, then the best design is the one with minimal additional information. In other words, design is good if its additional information content equals zero. Quantity of information content can be constructed by using the probability density function, as shown in Figure 1, where the design range represents the extent of the FRs, the system range corresponds with the DPs, and the common range is the mutual area between the design range and the system range.

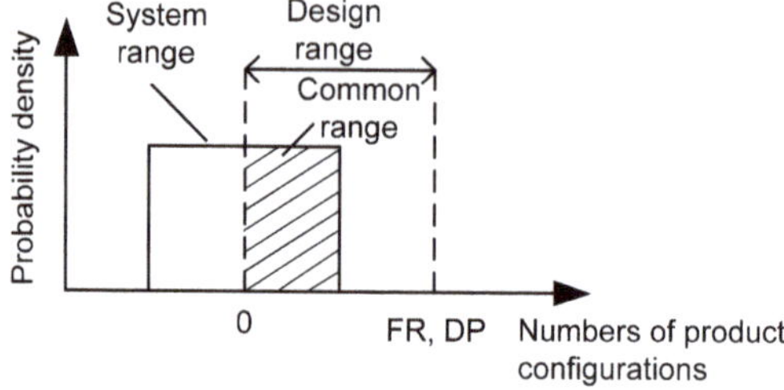

Figure 1. Probability density function of uniform distribution. Note: functional requirements (FR), design parameters (DP).

According to information theory, the information content of an event, *E*, is expressed as follows:

$$I(E) = -log_2(p(E)) = log_2\left(\frac{1}{p(E)}\right) \tag{1}$$

Then, additional information content of the design space can be calculated by formula [43]:

$$I = log_2\left(\frac{System\ range}{Common\ range}\right) \tag{2}$$

In our specific case we consider that:

Design range is represented by the N_{total};

Common range is represented by the N_{real};

System range is identical to the design range because, if $N_{real} = N_{total}$, then $N_{inf} = 0$, i.e., information content is optimal because I = 0, and if $N_{real} \neq N_{total}$, then $N_{inf} \neq 0$; i.e., information content is not optimal because $I \neq 0$.

Based on that amount of the additional information content, i.e., negative complexity of the product design variety, (C_N) can be expressed as:

$$C_N = log_2 N_{total} - log_2 N_{real} \tag{3}$$

Moreover, gradual categorization of the notion of negative complexity is shown in Figure 2.

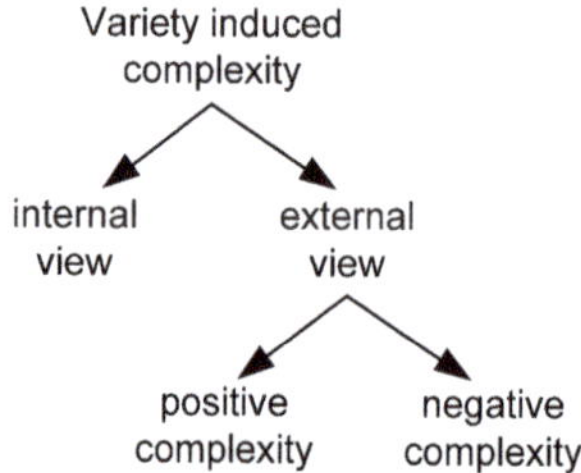

Figure 2. Classification of variety-induced complexity in terms of mass customization.

In order to apply this methodology to product design projects, figures for N_{total} and N_{real} are first determined and, subsequently, figures for N_{inf} are enumerated. N_{total} can be calculated using common combinatorial rules, while standard procedures for the enumeration of N_{real} are not available. Therefore, the two different methods for determining N_{real} are presented in the next section by using the realistic Case 1. Consecutively, the realistic Case 2 demonstrates the applicability of the methodological framework to measure the negative complexity of the product design variety.

4. Application of Methodology on Real Case Studies

As mentioned above, one of the aims of this section is to show how to calculate the number of realistic product configurations when restrictions between product components occur. Prior to this, it is beneficial to classify the main types of product components (PCs), which are as follows: Stable components (S), which are comprised in all possible configuration; Optional components (O), which are chosen according to the customer's requirements and are alterable in any combination, including the case where customers do not choose any of the offered options; Restricted optional components (RO) that can be selected from a set of different components according to exactly specified restriction conditions. Restrictions can be determined using at least three types of the volition rules: minimum, maximum, and particular requirements on selection of optional components.

The particular volition rule can be specified in a simple way by combinatorial number $\binom{k}{l}$, where l defines the required number of selected RO from the available number, k, of RO, while $1 \leq l < k$.

Selection by minimum volition rule $\binom{k}{lmin}$ should be limited to a certain number or more.

Selection by maximum volition rule $\binom{k}{lmax}$ should be limited to a certain number or less.

It should be noted that the number of possible product configurations depends practically on the number of optional components on the number of restricted optional components described by volitions rules for their selection.

As mentioned earlier, two methods concerning how to enumerate N_{real} will be presented. The first is based on graphical presentation using multidimensional matrices, whereas the second one uses proposed calculator software, which enumerates N_{real} using two-dimensional matrices between two complementary types of optional components and their variants. Both of the mentioned methods will be applied to Case 1, and the software based method will be applied to Case 2.

4.1. Case 1

The ways to enumerate N_{real} using the graphical method and the calculator-based method with component restrictions are described through the example of a customizable mobile phone containing one stable component, namely Calls; two optional components, specifically GPS and Media; and one restricted optional component, i.e., Screen with specific volition rule $\binom{3}{1}$ and defined restrictions, as shown in Figure 3.

4.1.1. Enumeration of Possible Product Configurations Using the Graphical Method

To enumerate the number of N_{inf} using the graphical method, the following procedure is proposed. Let us start, for example, with the restricted optional component, i.e., Screen, for which we select, e.g., Color screen (CSc.). Other options are Basic screen (BSc.) and High resolution screen (HRSc.). It is then possible to construct an incidence sub-matrix for the Color screen option and a stable component—Calls. Because there is no restriction, Color screen as an option can be combined with the Calls component (see Figure 4a). Then, a three-dimensional matrix is created with relations between the Color screen component, the Calls component, and the optional GPS component. Note that a mobile phone configuration without GPS can also be designed (marked in the matrix as without GPS). There are no restrictions in this case (see Figure 4b), therefore, a four-dimensional matrix of relations is subsequently constructed to determine the number of restricted product configurations with Color screen by adding a fourth dimension—Media, with two optional components. It is possible to create four choices here, such as only the MP3 component, only the Camera component, the MP3 and Camera components together, or none of these components. Only one restricted option occurs in this four-dimensional matrix and, accordingly, Camera must be combined with a compatible type of screen (Camera requires High resolution screen in its product configuration). Subsequently, it is possible to identify two options for Media—the presence of MP3 in the final product configuration or no Media components (see Figure 4c).

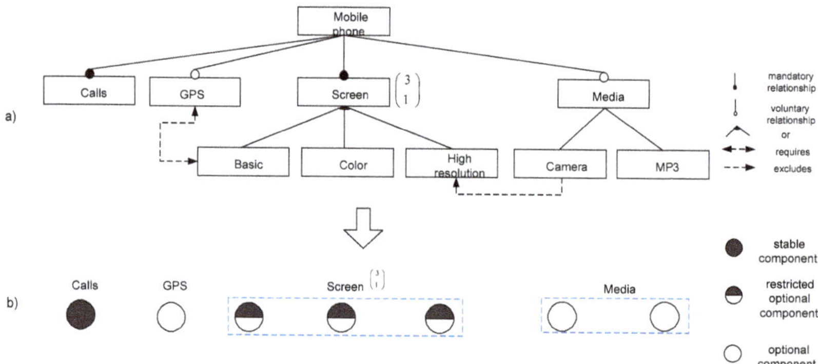

Figure 3. (**a**) Feature diagram of mobile phone; (**b**) Product component structure.

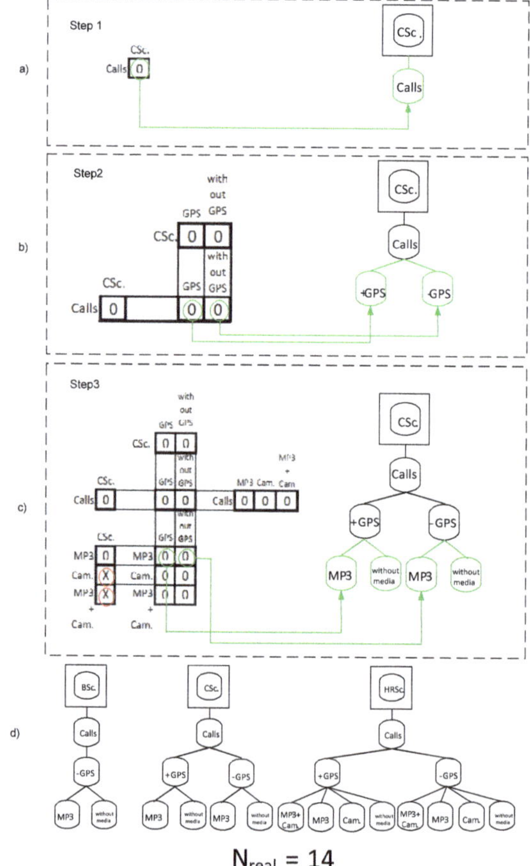

Figure 4. Procedure for transforming incidence matrices into product component structure with the initial component Color screen: (**a**) Step 1; (**b**) Step 2; (**c**) Step 3; (**d**) Generated N_{real} respecting defined restrictions.

Consecutively, the sub-procedure depicted in Figure 4a–c has to be repeated for the rest of the components, namely, Basic screen and High resolution screen. Then one can obtain 14 realistic product configurations of the mobile phone in total (see Figure 4d).

4.1.2. Enumeration of Possible Product Configurations Using Software

The second method generates N_{real} using the proposed online calculator [44]. As this method is based on two-dimensional matrices, it allows the calculation of N_{real} between two complementary types of optional components, while restrictions can occur. Let us use the previous example—a mobile phone—to calculate N_{real}. Firstly, we start with two components; Screen and Media. To obtain the correct number of N_{real}, one more variant of Media has to be considered, i.e., the option *without media*. Media variations are as follows: MP3 (V22), Camera (V21), MP3 + Camera (V23), or configuration without media (V24). The initial step of how to use the online calculator for N_{real} is to define "Value Problems", known as an $m \times n$ matrix, which has m vertical columns and n horizontal rows. In our case, m presents Screen options (BSc. (V11), CSc. (V12), and HRSc. (V13)) and n presents Media options, then $m \times n = 3 \times 4$. Subsequently, it is possible to assign defined restrictions (BSc. and CSc) in columns V21 and V23 of this matrix by checking the white boxes, which turn red, as can be seen in Figure 5a. Finally, the online calculator generates N_{real}, respecting the restrictions. In the case of the generation of N_{total} without restrictions, we leave the white boxes unchecked.

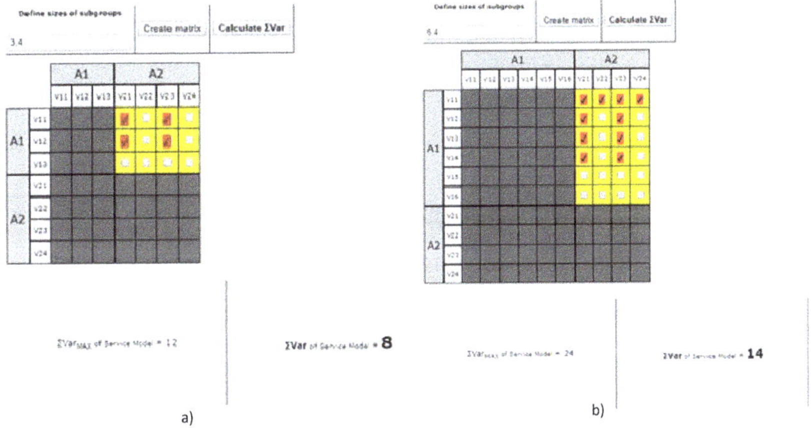

Figure 5. (**a**) Calculator software window with calculation of preliminary N_{total} and N_{real}. (**b**) Calculator software window with calculation of final N_{total} and N_{real}.

To calculate the final N_{real}, when restrictions are included for all the components of the mobile phone, it is necessary to join together Screen and Media into one component type with its variations (BSc. with GPS (V21); BSc. without GPS (V22); CSc. with GPS (V23); CSc. without GPS (V24); HRSc. with GPS (V25), and HRSc. without GPS (V26)), which creates the rows of the new two-dimensional matrix. The columns of this matrix will form the Media component type with its variants. The component *Calls*, being the stable one, is included in each configuration, and it does not affect the number of N_{real}. For this reason, only the mentioned components and their variants are included in this procedure. Then, it is necessary to define in the software window "Define sizes of subgroups" $m \times n = 6 \times 4$, and to assign restrictions in the new matrix, which are as follows: GPS cannot be in configuration with a BSc., Camera and MP3 + Camera cannot be in configuration with a BSc. or a CSc. Finally, we obtain the N_{real} and the N_{total} (see Figure 5b).

According to the obtained results by the proposed calculator, the number of N_{real} is 14, which is the same as that gained by using the graphically based method. Finally, N_{inf} is enumerated as $N_{total} - N_{real} = 10$.

4.2. Case #2

The purpose of this section is twofold: (1) To describe the calculator-based method of enumerating N_{real} on a realistic case, and (2) to demonstrate the applicability of the methodology for identification and reduction of negative complexity in designing product platforms and sub-platforms.

4.2.1. Description of the Calculator-Based Method

The calculator-based procedure will be described using the two inter-operating parts of a front drivetrain bicycle module, which can be found on every model of bicycle. This product sub-platform consists of nine types of gear component and two types of front drivetrain parts each with a different chain stay angle (CSA). Each of the nine groups differs from the others in the specific number of related components to be combined with the selected alternative of the front drivetrain (FD). For example, Gear type 1 can be combined with six Front Crank sets (FC11–FC16). In Table 1, elements of the matrix noted with "X" stand for a restriction/incompatibility.

With this range of components, we can construct an incidence matrix with sizes of subgroups $m \times n = 38 \times 19$ using the same principle as described in Section 4.1.2. First, we add into the software window the names of the columns marked as V21 to V219 and the names of the rows marked as V11 to V138, representing the bicycle components. Next we assign restrictions between the components by checking the white boxes. Subsequently, by clicking on "Calculate $\sum$Var" we obtain $N_{real} = 239$ product configurations, as depicted in Figure 6.

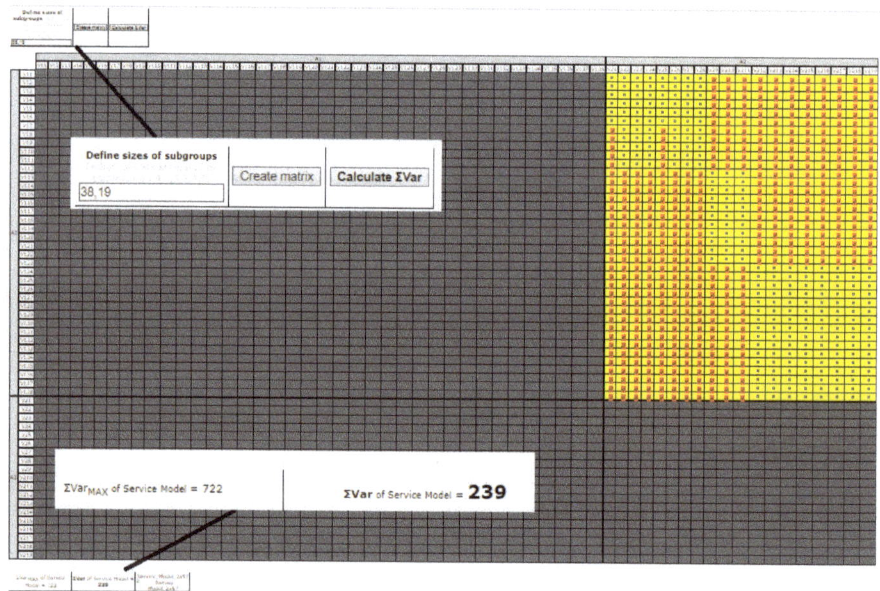

Figure 6. Generated number of N_{total} and N_{real} using software.

Finally, the number of infeasible product configurations is enumerated as $N_{total} - N_{real} = 483$.

Table 1. Bicycle components presented by compatibility platform D_0 for gears and front drive train.

Sub-Platform D_0		Front Drivetrain (FD)																		
		FD Type 1														FD Type 2				
Chain Stay Angle (CSA)		CSA 1								CSA 2						CSA 3				
Gears	Front Crankset (FC)	FD11	FD12	FD13	FD14	FD15	FD16	FD17	FD18	FD21	FD22	FD23	FD31	FD32	FD33	FD34	FD35	FD36	FD37	FD38
Gear type 1	FC11									X	X	X	X	X	X	X	X	X	X	X
	FC12									X	X	X	X	X	X	X	X	X	X	X
	FC13									X	X	X	X	X	X	X	X	X	X	X
	FC14									X	X	X	X	X	X	X	X	X	X	X
	FC15									X	X	X	X	X	X	X	X	X	X	X
	FC16									X	X	X	X	X	X	X	X	X	X	X
Gear type 2	FC21	X				X				X	X	X	X	X	X	X	X	X	X	X
	FC22	X				X				X	X	X	X	X	X	X	X	X	X	X
	FC23	X				X				X	X	X	X	X	X	X	X	X	X	X
	FC24	X				X				X	X	X	X	X	X	X	X	X	X	X
	FC25	X				X				X	X	X	X	X	X	X	X	X	X	X
Gear type 3	FC31	X	X	X	X	X	X	X	X				X	X	X	X	X	X	X	X
	FC32	X	X	X	X	X	X	X	X				X	X	X	X	X	X	X	X
	FC33	X	X	X	X	X	X	X	X				X	X	X	X	X	X	X	X
	FC34	X	X	X	X	X	X	X	X				X	X	X	X	X	X	X	X
	FC35	X	X	X	X	X	X	X	X				X	X	X	X	X	X	X	X
	FC36	X	X	X	X	X	X	X	X				X	X	X	X	X	X	X	X
	FC37	X	X	X	X	X	X	X	X				X	X	X	X	X	X	X	X
	FC38	X	X	X	X	X	X	X	X				X	X	X	X	X	X	X	X
Gear type 4	FC41	X	X	X	X	X	X	X	X				X	X	X	X	X	X	X	X
	FC42	X	X	X	X	X	X	X	X				X	X	X	X	X	X	X	X
	FC43	X	X	X	X	X	X	X	X				X	X	X	X	X	X	X	X
Gear type 5	FC5	X	X	X	X	X	X	X	X	X	X	X								
Gear type 6	FC6	X	X	X	X	X	X	X	X	X	X	X								

Table 1. *Cont.*

Sub-Platform D_0		Front Drivetrain (FD)																		
		FD Type 1													FD Type 2					
Chain Stay Angle (CSA)		CSA 1								CSA 2						CSA 3				
Gears	Front Crankset (FC)	FD11	FD12	FD13	FD14	FD15	FD16	FD17	FD18	FD21	FD22	FD23	FD31	FD32	FD33	FD34	FD35	FD36	FD37	FD38
Gear type 7	FC71	X	X	X	X	X	X	X	X	X	X	X								
	FC72	X	X	X	X	X	X	X	X	X	X	X								
	FC73	X	X	X	X	X	X	X	X	X	X	X								
	FC74	X	X	X	X	X	X	X	X	X	X	X								
	FC75	X	X	X	X	X	X	X	X	X	X	X								
Gear type 8	FC81	X	X	X	X	X	X	X	X	X	X	X								
	FC82	X	X	X	X	X	X	X	X	X	X	X								
	FC83	X	X	X	X	X	X	X	X	X	X	X								
	FC84	X	X	X	X	X	X	X	X	X	X	X								
	FC85	X	X	X	X	X	X	X	X	X	X	X								
Gear type 9	FC91	X	X	X	X	X	X	X	X	X	X	X								
	FC92	X	X	X	X	X	X	X	X	X	X	X								
	FC93	X	X	X	X	X	X	X	X	X	X	X								
	FC94	X	X	X	X	X	X	X	X	X	X	X								

4.2.2. Resolving Product Configuration Conflicts

In order to resolve the product configuration conflicts, one can see the two possible ways that can be used to eliminate or at least reduce mutually incompatible components. When considering the fact that eliminating all incompatible components may cause a rapid decrease in product variety, then reduction of incompatible components seems to be a more useful approach. Therefore, it appears to be reasonable to generate derived alternatives of the original product sub-platforms, in which selected incompatible components will be omitted. For this purpose, this method can be used to generate three other alternative product platforms, D_1–D_3, in which selected incompatible components will be gradually removed from the initial design platform, D_0, where the complete design space (N_{total}) is defined by 722 configurations, and the restricted design space (N_{real}) is represented by 239 configurations.

In this task, in order to propose a possible concurrent product platforms at once, let us remove Gear type 3, which includes 8 crank sets (FC31–FC38), from D_0. This group of components was selected to be excluded from D_0, using the criterion of the highest density of restrictions. In this way we obtained platform D_1, which is defined by N_{total1} = 570 drivetrain configurations, from which 215 are viable.

Afterwards, for determination of the platform, D_2, we proceed towards the reduction of Gear type 4, which includes three crank sets (FC41, FC42, and FC43). We obtained platform D_2 with N_{total2} = 513 drivetrain configurations and N_{real} = 206 restricted (viable) product configurations.

In order to provide the next alternative product platform, D_3, two FDs, namely FD11 and FD15, can be removed from D_2 due to the high number of restrictions related to these two components. Then, D_3 is represented by N_{total} = 459 drivetrain configurations and N_{real} = 194 feasible product configurations.

In the next step, the obtained number of configurations are used to quantify and compare the values of negative complexity of the concurrent product sub-platforms, as depicted in Table 2.

Table 2. Comparison of the product sub-platforms.

	D_0	D_1	D_2	D_3
N_{total}	722 conf.	570 conf.	513 conf.	459 conf.
N_{real}	239 conf.	215 conf.	206 conf.	194 conf.
N_{inf}	483 conf.	355 conf.	307 conf.	265 conf.
Entropy of design space (without constraints)	9.50 bits	9.16 bits	9.00 bits	8.84 bits
Entropy of constrained design space	7.90 bits	7.75 bits	7.69 bits	7.60 bits
C_N	1.60 bits	1.41 bits	1.31 bits	1.24 bits

5. Discussion

Firstly, as regards the two proposed computational methods to calculate N_{real}, it is important to note that the graph-based method approach using more dimensional matrices is not suitable for large numbers of components, while the software-based method is intended to enumerate numbers of module variations and/or product configurations with large numbers of optional product components. In addition, the graph based method can be used as the theoretical basis for developing an effective concurrent software programming technique that could be used for the same purpose.

Coming back to the results obtained by generating the alternative product platforms D_1–D_3, it is evident that D_3 is has the smallest value of negative complexity. In other words, this product platform has a minimal negative effect on the customers' perception of the company product offer.

6. Conclusions

Summarily, it is worth noting that sustainable mass customization projects are oriented on the building of trust with customers. This also means that each potential negative perception of risk must be eliminated or reduced because, otherwise, it could cause customers to feel reluctant to invest their time and effort in mass customization and may prefer a standard product instead. It is quite obvious that a universal solution to this problem is rather counterintuitive. Theoretically, the best way is to eliminate all configuration constraints by product design changes, if possible. Alternatively, the next effort could be focused on eliminating infeasible options through product configurators. Computational experiments have clearly shown that designers can generate alternative product platforms and identify their negative complexity through the proposed method. However, only designers of specific products can decide which product platform is the best for them, according to the given criterion.

According to related experts, each different MC sector requires a specific approach to solve configuration conflicts, and therefore elimination of infeasible options through product configurators is significantly difficult. In this context, experts recommend that this type of configurators should be capable of analyzing inconsistencies between components during the selection process and propose to the customers the minimum changes required to achieve a resolution. A complementary solution to the previous one is offered in the present study.

The proposed method has been theoretically verified through the investigation of several realistic problems, two of which have been demonstrated in this study. The next stage of research will be focused on the willingness of manufacturers to reduce the number of incompatible components in their product platforms, and the exploration of the potential benefits of such decisions.

Finally, it can be stated that the proposed method combines known elements of knowledge in a novel way.

Author Contributions: V.M. proposed the whole structure of this paper, provided the introduction and the related work methodology framework, and summarized the findings in the conclusion section. Z.S. worked on the selection and application of methodology on real case studies. All authors have read and agreed to the published version of the manuscript.

Funding: This research was funded by the European Union's Horizon 2020 research and innovation program under the Marie Skłodowska-Curie, grant number 734713 and by the KEGA project Nr. 025TUKE-4/2020 granted by the Ministry of Education of the Slovak Republic.

Conflicts of Interest: The authors declare no conflict of interest.

References

1. Da Silveira, G.; Borenstein, D.; Fogliatto, F.S. Mass customization: Literature review and research directions. *Int. J. Prod. Econ.* **2001**, *20*, 1–13. [CrossRef]
2. Schuh, G.; Reuter, C.; Hauptvogel, A. Increasing collaboration productivity for sustainable production systems. *Procedia CIRP* **2015**, *29*, 191–196. [CrossRef]
3. Gilmore, J.H.; Pine, B.J. The four faces of mass customization. *Harv. Bus. Rev.* **1997**, *75*, 91–101.
4. Yang, D.; Dong, M. A constraint satisfaction approach to resolving product configuration conflicts. *Adv. Eng. Inform.* **2012**, *26*, 592–602. [CrossRef]
5. Liu, Y.; Liu, Z. Multi-objective product configuration involving new components under uncertainty. *J. Eng. Des.* **2010**, *21*, 473–494. [CrossRef]
6. Fraizer, S.; Wells, R. System for Presenting Customer Constrained Purchase Choices in an On-Line Store. WO Patent 2001091019 A1 filed, 29 November 2001.
7. Cox, W.M.; Alm, R. *The Right Stuff—America's Move to Mass Customization*; Federal Reserve Bank of Dallas Annual Report; Federal Reserve Bank of Dallas: Dallas, TX, USA, 1998; pp. 3–26.
8. Funke, M.; Ruhwedel, R. Product variety and economic growth: Empirical evidence from the OECD countries. *IMF Staff. Pap.* **2001**, *48*, 225–242. [CrossRef]
9. Modrak, V.; Bednar, S.; Soltysova, Z. Application of axiomatic design-based complexity measure in mass customization. *Procedia CIRP* **2016**, *50*, 607–612. [CrossRef]

10. Forza, C.; Salvador, F. Application support to product variety management. *Int. J. Prod. Res.* **2008**, *46*, 817–836. [CrossRef]
11. Pollard, D.; Chuo, S.; Lee, B. Strategies for mass customization. *J. Bus. Econ. Res. (JBER)* **2008**, *6*, 77–86.
12. Hadzistevic, M.; Moraca, S. Networks and Quality Improvement. *Int. J. Qual. Res.* **2009**, *3*, 353–361.
13. Whitley, S.C.; Trudel, R.; Kurt, D. How many versions of a product do consumers really want? *Harv. Bus. Rev.* **2018**. Available online: https://hbr.org/2018/06/how-many-versions-of-a-product-do-consumers-really-want, (accessed on 27 June 2018).
14. Aydogdu, A.A.; Saka, M.P. Design optimization of real-world steel space frames using artificial bee colony algorithm with Levy flight distribution. *Adv. Eng. Softw.* **2016**, *92*, 1–14. [CrossRef]
15. Krus, P. Design space configuration for minimizing designinformation entropy. In *ICoRD'15—Research into Design AcrossBoundaries Volume 1: Theory, Research Methodology, Aesthetics, Human Factors and Education, vol. 34 of Smart Innovation, Systems and Technologies*; Springer: New Delhi, India, 2015; pp. 51–60.
16. Matt, D.; Rauch, E.; Dallasega, P. Trends towards distributed manufacturing systems and modern forms for their design. *Procedia CIRP* **2015**, *33*, 185–190. [CrossRef]
17. Modrak, V.; Marton, D.; Bednar, S. The influence of mass customization strategy on configuration complexity of assembly systems. *Procedia CIRP* **2015**, *33*, 538–543. [CrossRef]
18. Wang, Y.; Ma, H.S.; Yang, J.H.; Wang, K.S. Industry 4.0: A way from mass customization to mass personalization production. *Adv. Manuf.* **2017**, *5*, 311–320. [CrossRef]
19. Zhang, X.; Ming, X.; Liu, Z.; Qu, Y.; Yin, D. State-of-the-art review of customer to business (C2B) model. *Comput. Ind. Eng.* **2019**, *132*, 207–222. [CrossRef]
20. Bonev, M.; Hvam, L.; Clarkson, J.; Maier, A. Formal computer-aided product family architecture design for mass customization. *Comput. Ind.* **2015**, *74*, 58–70. [CrossRef]
21. Huang, Y.; Liu, H.; Ng, W.K.; Song, B.; Li, X. Automating knowledge acquisition for constraint-based product configuration. *J. Manuf. Technol. Manag.* **2008**, *19*, 744–754. [CrossRef]
22. Yang, D.; Dong, M.; Chang, X.K. A dynamic constraint satisfaction approach for configuring structural products under mass customization. *Eng. Appl. Artif. Intell.* **2014**, *25*, 1723–1737. [CrossRef]
23. Matt, D.T.; Rauch, E. Design of a Network of Scalable Modular Manufacturing Systems to Support Geographically Distributed Production of Mass Customized Goods. *Procedia CIRP* **2013**, *12*, 438–443. [CrossRef]
24. Behunova, A.; Soltysova, Z.; Behun, M. Complexity Management and Its Impact on Economy. *TEM J.* **2018**, *7*, 324–329.
25. Dima, I.C.; Gabrara, J.; Modrak, V.; Piotr, P.; Popescu, C. Using the expert systems in the operational management of production. In *11th WSEAS International Conference on Mathematics and Computers in Business and Economics (MCBE'10)*; WSEAS Press: Iasi, Romania, 2010; pp. 307–312.
26. Zhou, Y.M.; Wan, X. Product variety, sourcing complexity, and the bottleneck of coordination. *Strateg. Manag. J.* **2017**, *38*, 1569–1587. [CrossRef]
27. ElMaraghy, H.; Schuh, G.; ElMaraghy, W.; Piller, F.; Schönsleben, P.; Tseng, M.; Bernard, A. Product variety management. *CIRP Ann. Manuf. Technol.* **2013**, *62*, 629–652. [CrossRef]
28. Tseng, M.M.; Piller, F. The Customer Centric Enterprise. In *The Customer Centric Enterprise: Advances in Mass Customization and Personalization*; Springer Science & Business Media: Berlin, Germany, 2011; pp. 3–16.
29. Chu, C.H.; Luh, Y.P.; Li, T.C.; Chen, H. Economical green product design based on simplified computer-aided product structure variation. *Comput. Ind.* **2009**, *60*, 485–500. [CrossRef]
30. Mailharro, D. A classification and constraint-based framework for configuration. *Artif. Intel. Eng. Des. Anal. Manuf.* **1998**, *12*, 383–397. [CrossRef]
31. Antonelli, D.; Pasquino, N.; Villa, A. Mass-customized production in a SME network. *IFIP Int. Fed. Inf. Process.* **2007**, *246*, 79–86.
32. Aldanondo, M.; Vareilles, E.; Djefel, M. Towards an association of product configuration with production planning. *Int. J. Mass Cust.* **2010**, *3*, 316–332. [CrossRef]
33. Helo, P.T.; Xu, Q.L.; Kyllonen, S.J.; Jiao, R.J. Integrated vehicle configuration system connecting the domains of mass customization. *Comput. Ind.* **2010**, *61*, 44–52. [CrossRef]
34. Paul, P.; Aldanondo, M.; Vareilles, E. Concurrent product configuration and process planning: Some optimization experimental results. *Comput. Ind.* **2014**, *65*, 610–621.

35. Mueller, D.; Graefenstein, J.; Scholz, D.; Henke, M. Complexity-Oriented Evaluation of Production Systems for Online-Switching of Autonomous Control Methods. In *Interdisciplinary Conference on Production, Logistics and Traffic 2019*; Springer: Cham, Switzerland, 2019; pp. 246–264.

36. Hu, S.J.; Zhu, X.; Wang, H.; Koren, Y. Product variety and manufacturing complexity in assembly systems and supply chains. *CIRP Ann.* **2008**, *57*, 45–48. [CrossRef]

37. Trattner, A.L.; Hvam, L.; Herbert-Hansen, Z.N.L.; Raben, C. Product variety, product complexity and manufacturing operational performance: A systematic literature review. In Proceedings of the 24th International Annual EurOMA Conference, Edinburgh, UK, 1–5 July 2017; pp. 1–10.

38. Abdelkafi, N. *Variety Induced Complexity in Mass Customization: Concepts and Management*; Erich Schmidt Verlag GmbH & Co KG: Berlin, Germany, 2008; Volume 7, pp. 1–314.

39. He, J. Complexity in Adaptive Systems. In *Encyclopedia of Machine Learning*; Sammut, C., Webb, G.I., Eds.; Springer: Boston, MA, USA, 2011; pp. 194–198.

40. Blecker, T.; Friedrich, G.; Kaluza, B.; Abdelkafi, N.; Kreutler, G. *Information and Management Systems for Product Customization*; Springer Science & Business Media Inc.: New York, NY, USA, 2004; Volume 7.

41. Felfernig, A.; Friedrich, G.E.; Jannach, D. UML as domain specific language for the construction of knowledge-based configuration systems. *Int. J. Softw. Eng. Know. Eng.* **2000**, *10*, 449–469. [CrossRef]

42. Suh, N.P. Axiomatic design theory for systems. *Res. Eng. Des.* **1998**, *10*, 189–209. [CrossRef]

43. Suh, N.P. Complexity in engineering. *CIRP Ann. Manuf. Technol.* **2005**, *54*, 46–63. [CrossRef]

44. Online Calculator. Available online: http://web.tuke.sk/fvt-mms/complexity/calc.php (accessed on 1 September 2019).

Article

Design of Human-Centered Collaborative Assembly Workstations for the Improvement of Operators' Physical Ergonomics and Production Efficiency: A Case Study

Luca Gualtieri, Ilaria Palomba, Fabio Antonio Merati, Erwin Rauch and Renato Vidoni *

Industrial Engineering and Automation (IEA), Faculty of Science and Technology, Free University of Bozen-Bolzano, Piazza Università 5, 39100 Bolzano, Italy; luca.gualtieri@unibz.it (L.G.); ilaria.palomba@unibz.it (I.P.); fabioantonio.merati@natec.unibz.it (F.A.M.); erwin.rauch@unibz.it (E.R.)
* Correspondence: renato.vidoni@unibz.it; Tel.: +39-0471-017111

Received: 11 March 2020; Accepted: 27 April 2020; Published: 29 April 2020

Abstract: Industrial collaborative robotics is one of the main enabling technologies of Industry 4.0. Collaborative robots are innovative cyber-physical systems, which allow safe and efficient physical interactions with operators by combining typical machine strengths with inimitable human skills. One of the main uses of collaborative robots will be the support of humans in the most physically stressful activities through a reduction of work-related biomechanical overload, especially in manual assembly activities. The improvement of operators' occupational work conditions and the development of human-centered and ergonomic production systems is one of the key points of the ongoing fourth industrial revolution. The factory of the future should focus on the implementation of adaptable, reconfigurable, and sustainable production systems, which consider the human as their core and valuable part. Strengthening actual assembly workstations by integrating smart automation solutions for the enhancement of operators' occupational health and safety will be one of the main goals of the near future. In this paper, the transformation of a manual workstation for wire harness assembly into a collaborative and human-centered one is presented. The purpose of the work is to present a case study research for the design of a collaborative workstation to improve the operators' physical ergonomics while keeping or increasing the level of productivity. Results demonstrate that the achieved solution provides valuable benefits for the operators' working conditions as well as for the production performance of the companies. In particular, the biomechanical overload of the worker has been reduced by 12.0% for the right part and by 28% for the left part in terms of manual handling, and by 50% for the left part and by 57% for the right part in terms of working postures. In addition, a reduction of the cycle time of 12.3% has been achieved.

Keywords: industry 4.0; collaborative robotics; physical ergonomics; human-robot collaboration; human-centered design; assembly; SME; small and medium sized enterprise

---

1. Introduction

Nowadays, production systems are shifting from mass production to mass customization [1]. As a consequence, in order to remain competitive and profitable, modern companies need further production flexibility, efficiency, and sustainability in terms of lot sizes, variants, and time-to-market. These conditions require the integration of adaptable, reconfigurable, and agile manufacturing systems and technologies characterized by a scalable degree of automation. The term "Industry 4.0", also known as the fourth industrial revolution, was initially introduced by a German government strategic initiative in 2011 [2] and represents the current transformation, which is affecting a large part of worldwide industrial companies.

Process sustainability is one of the drivers of the fourth industrial revolution, and it is defined as a multi-dimensional concept encompassing environmental, social, and economic aspects [3]. Operators' occupational health and safety (OHS) as well as wellbeing and satisfaction are crucial parts of social sustainability in industry. Manufacturing companies should consider the human element as their core and a valuable part by improving work conditions and developing human-centered production systems. In this context, a major role is played by work-related ergonomics. Even if ergonomics includes three different categories dealing with the physical, cognitive, and organizational aspects of the interaction between humans and systems, this paper mainly focuses on the aspects related to physical one. The International Ergonomics Association (IEA) defines physical ergonomics as the scientific discipline concerned with human anatomical, anthropometric, physiological, and biomechanical characteristics as they relate to physical activity [4].

A possible and concrete solution to improve the social sustainability in terms of physical ergonomics without neglecting the production efficiency is represented by the collaborative robotics [5,6], as demonstrated by the increasing scientific literature on such a topic. Castro et al. [7] developed an integrated robot simulation and digital human modeling aimed to be a tool to create and confirm successful ergonomic collaborative workstations. Lietaert et al. [8] proposed a model-based methodology to aid the layout design of a collaborative workcell by considering feasibility, reachability, safety, and ergonomics as constraints. Mateus et al. [9] provided an information-based generic methodology for collaborative assembly solution development, starting from CAD models. Petruck et al. [10] designed an ergonomic human-robot collaboration (HRC) workstation where intelligent sensors are used to adapt the system behavior to the way the working person works. Dombrowski et al. [11] presented an interactive and real-time physics simulation as a tool for workcell validation and optimization. Makrini et al. [12] developed a modular architecture used to enhance human-robot interaction (HRI) in assembly tasks. Schmidtler et al. [13] presented a study about the optimization of working conditions by adapting collaborative assistance systems in terms of human acceptance and well-being. Basically, all these works aimed to finalize design tools and methodologies for the realization of a new collaborative workstation. Conversely, just few works have addressed the problem of the conversion of an existing manual workstation into a collaborative one. In particular, Heydaryan et al. [14] developed an HRC workstation for the assembly of an automotive brake disc by implementing a hierarchical task analysis to allocate operational tasks to humans and robots based on productivity, quality, human fatigue, and safety considerations.

This paper presents the conversion of an industrial manual workstation for wire harness assembly into a collaborative one and its realization and validation. The conversion aims at improving the operators' physical ergonomics, in terms of reduction of biomechanical overload, and the company's productivity, in terms of cycle time. Before starting with the workstation conversion, the preliminary steps are introduced by proposing two algorithms. One is addressed to the evaluation of the physical ergonomics conditions, while the other is aimed at assessing the feasibility of using collaborative robotics. The first one is a framework based on three key questions that guide the evaluator towards the most suitable ergonomics assessment methods to be used. The second is a simplified version of the methodology proposed by Gualtieri et al. [15]. It allows for an easy and quick identification of the potentials for the adoption of collaborative solutions in terms of ergonomics and safety, product and process quality, and economics.

The paper is organized as follows. Section 2 describes the methodology adopted for transforming a manual assembly workstation into a collaborative one. Section 3 describes the industrial case study employed. Section 4 collects all the results. Finally, Section 5 discusses the technical measures designed for the workstation re-design according to ergonomics and production efficiency requirements and draws the possible future improvements.

2. Materials and Methods

In order to design a new collaborative assembly workstation starting from an existing one by considering the physical ergonomics and production efficiency as main conversion goals, the following sequential steps, described in detail in the next subsections, were applied:

a)   Analysis of the current situation in terms of physical ergonomics;
b)   Evaluation of the potentials for collaborative robotics;
c)   Workstation re-design for physical ergonomics and production efficiency enhancement.

2.1. Analysis of the Current Situation in Terms of Physical Ergonomics

A crucial part of the ergonomic re-design of an assembly workstation is the physical ergonomics assessment. A detailed analysis is needed to identify the tasks that present unsuitable conditions from the biomechanical point of view. To this end, a framework to assess the manual assembly activities that are typically carried out in a workstation is presented.

The idea was to evaluate the current conditions related to the assembly by considering the potentials of collaborative robots in terms of physical assistance. The proposed framework was inspired by the approach presented in the technical report ISO TR 12295 [16]. It was formulated for the evaluation of typical manual assembly tasks, which could be possibly supported by a fixed-position anthropomorphic collaborative robot. For this reason, other ergonomic evaluations related to two-handed whole-body pushing and pulling of loads were ignored [17]. The suggested framework, shown in Figure 1, mainly focuses on the evaluation of the biomechanical overload of neck, trunk, and/or upper limbs related to manual assembly tasks and is based on a key question analysis [16]. In particular, three key questions were adopted, focusing on: the manual handling of heavy objects, the repetitive tasks of upper limbs, and the awkward working postures. According to the proposed framework, for each positive key question a further physical ergonomics analysis must be performed to understand whether or not there are the needs for an ergonomics re-design by eventually integrating a collaborative robot. On the other hand, if all the key questions are negative, no problems are highlighted, and changes will be not required. The presented key questions were based on the following methodologies, respectively:

—the National Institute for Occupational Safety and Health (NIOSH) lifting equation for manual lifting of objects [18–20];

—the Occupational Repetitive Action (OCRA) for repetitive tasks (in check-list form) [21,22];

—the Rapid Upper Limb Assessment (RULA) for working postures [23].

The first two methods are discussed and partially developed into the technical report ISO TR 12295 and related ISO standards; the third method is an evaluation procedure mainly focused on working postures which differs from the one proposed into the technical report. It is worth noticing that the suggested framework does not aim to calculate occupational risks, but to evaluate which tasks of the manual assembly cycle are not suitable from the point of view of the biomechanical overload. This will be useful to identify which tasks could potentially be improved by integrating a collaborative robot to be used as a physical assistance system for the enhancement of operators' work conditions.

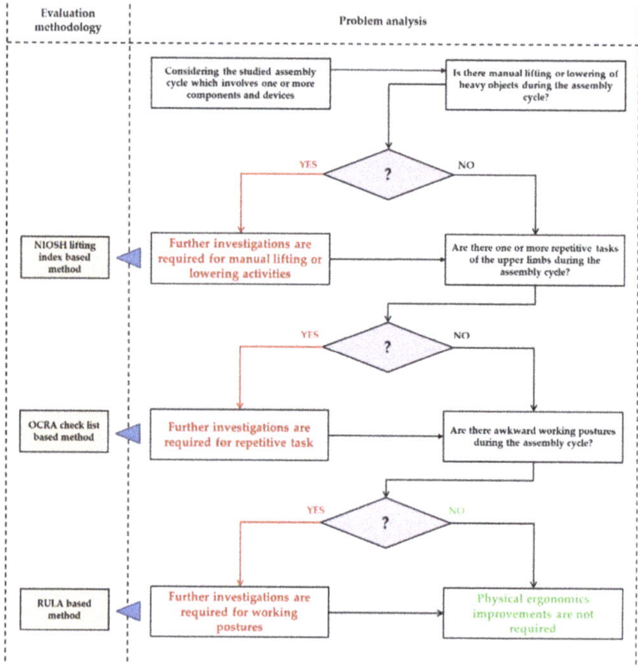

Figure 1. Framework for the identification of not-ergonomic tasks which could be improved by integrating a collaborative robot to be used as a physical assistance system.

2.2. Evaluation of the Potential for Collaborative Robotics

A so-called Quick Assessment algorithm is proposed to identify if the current assembly cycle has the potential to be performed (or not) in collaboration with a robot. The sense of such a Quick Assessment algorithm is to propose a simple and easy to use assessment methodology that can also be applied by small and medium-sized enterprises (SMEs). This approach is a simplified version of the study presented in [15]. Basically, the simplification is addressed to better support the technicians without specific skills and experiences by providing a preliminary tool for a quick and simple evaluation. Firstly, it helps SMEs in understanding their weaknesses with respect to a certain assembly process. This will be useful to define priorities and understand where to focus improvement interventions. Secondly, this will support them in quantifying the potential benefits of using collaborative robots to fill these gaps. This should overcome the technological barriers related to the lack of technical and organizational knowledge about a proper implementation of collaborative robots. The evaluation is based on the analysis of five process critical issues (PCIs) ranging from safety and ergonomics, product and/or process quality to economics (see the first and second columns of Table 1). In accordance with the Industry 4.0 principles, which aim to promote sustainable production systems, the algorithm is organized to provide more relevance to the operators' occupational safety and wellbeing. After that, the procedure considers the relevance of process and product quality, which is a crucial matter for SMEs. Finally, the last consideration refers to economic factors. The algorithm requires the scoring of the five PCIs for each task of the assembly cycle. The score is an integer from 0 to 3, where 0 means that there are no possible improvements for the analyzed task and 3 that they are extremely recommendable.

Table 1. Quick Assessment of the actual manual assembly of wire harness.

Category	Process Critical Issue (PCI)	Weight (W_i)	Score (S_{ij})
Safety and ergonomics	PCI 1: Are there physical ergonomic problems related to: —lifting/lowering or carrying of objects OR —repetitive tasks of the upper limbs characterized by repeated work cycles OR —static or awkward working postures?	3	0–3
	PCI 2: Are there occupational risks for the operator's safety which are not properly managed OR occupational risk for operator's health which are not properly managed? (not yet considered in the previous point)	3	0–3
	PCI 3: Is there high work monotonousness OR very low requirements in terms of task qualification of the manual work?	3	0–3
Product and Process Quality	PCI 4: Are there not-constant/satisfactory product quality OR unsuitable process quality levels according to the nominal values (i.e., characterized by high variability/low standardization)?	2	0–3
Economics	PCI 5: Is there an inefficient use of time and/or resources without a real advancement of production, which means tasks which cannot generate value to the final costumers (not value-added activities) OR Is there low/not satisfactory process productivity and/or production efficiency?	1	0–3
	(*) Score meaning: 0—Improvements are not required for that activity / the problem does not exists 1—The achievable improvements could be moderate for that activity 2—The achievable improvements could be good for that activity 3—Improvements could be very significant for that activity		

The potential value (*Pval$_j$*) in terms of HRI for the *j*-th task is calculated as a weighted sum of the scores (S_{ij}) given to the five PCIs:

$$Pval_j = \sum_{i=1}^{5} S_{ij} * W_i; \qquad j = 1, \ldots, n_t.$$ (1)

In Equation (1), n_t is the number of tasks of the assembly process. The PCI weights *Wi* (stated in the third column of Table 1) are introduced in order to follow the abovementioned considerations. The index *Pval$_j$*, computed according to Equation (1) ranges from 0 to 36. The resulting value describes the potential for a successful adoption of collaborative solutions in the analyzed manual tasks by using the following five classes:

—No potential: Pvalj = 0;

—Low potential: 1 ≥ Pvalj > 9;

—Modest potential: 9 ≥ Pvalj > 18;

—Good potential: 18 ≥ Pvalj > 27;

—High potential: 27 ≥ Pvalj ≥ 36.

2.3. Workstation Re-Design for Physical Ergonomics and Production Efficiency Enhancement

If the physical ergonomics assessment highlights unsuitable work conditions, the process and the workstation shall be improved. According to the (positive) Quick Assessment results, this can be done by exploiting the benefits of the integration of a collaborative robot in lowering the biomechanical overload for the operator. In addition to the introduction of the collaborative robot, another important aspect concerns the development of an ergonomically optimized layout of the workspaces. The proposed re-design follows the guidelines presented in the EN ISO 14738 [24]. Basically, this standard establishes the principles for dimension derivation from anthropometric measurements and their application to the design of workstations.

On the other hand, improvement of the collaborative workstation in terms of productivity (and hence of cycle time) is made by properly setting the layout of the new workstation. In particular, the new layout is chosen on the basis of the following production variables:

—Number of robots (R);

—Number of kinds of workpieces (P);

—Number of workers (W);

—Number of working areas (A).

By combining these four production variables, it is possible to identify 16 different layouts for the collaborative workstation, as shown in Table 2. The more convenient layout will depend on the specific values assumed by such variables for the analyzed assembly process.

Table 2. Classification of the possible layouts for collaborative workstations according to the following variables: Number (Nr) of robots (R), Nr of different kind of workpieces (P) and related assembly tasks which are intended to be allocated to the robot, Nr of workers (W), and Nr of working areas (A).

Possible HRI Workstation Layouts		Nr. of robots R = 1		Nr. of robots R = n > 1	
		Nr. of kind of workpieces P = 1	Nr. of kind of workpieces P = m > 1	Nr. of kind of workpieces P = 1	Nr. of kind of workpieces P = m > 1
Nr. of workers W = 1	Nr of working area A = 1	RPWA-1111 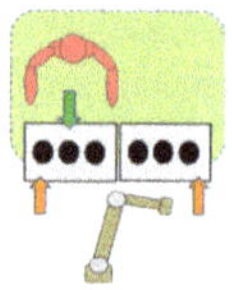	RPWA-1m11 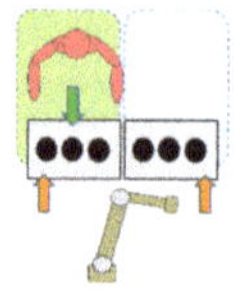	RPWA-n111 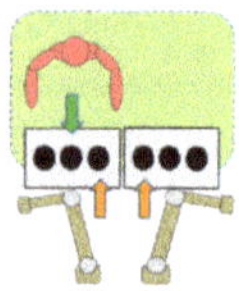	RPWA-nm11 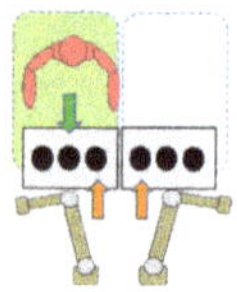
	Nr of working areas A = s > 1	RPWA-111s 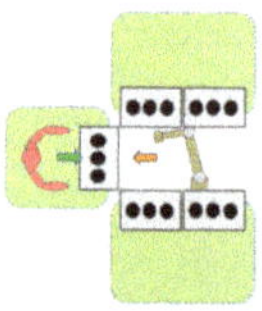	RPWA-1m1s 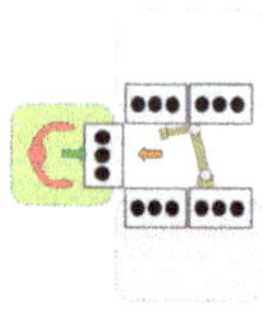	RPWA-n11s 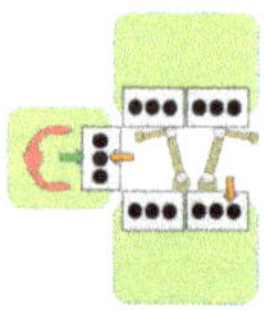	RPWA-nm1s 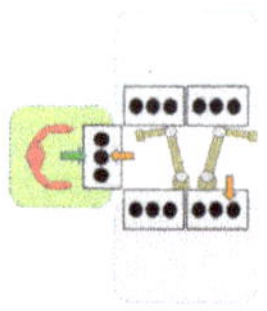
Nr. of workers W = q > 1	Nr of working areas A = 1	RPWA-11p1 	RPWA-1mq1 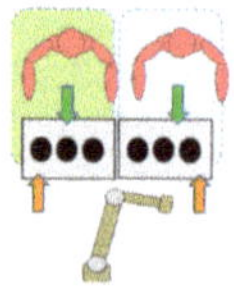	RPWA-n1q1 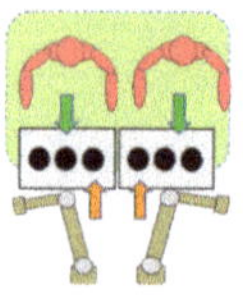	RPWA-nmq1 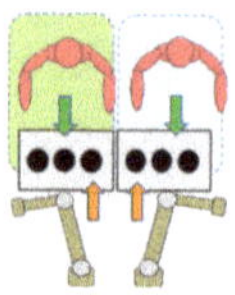
	Nr of working areas A = s > 1	RPWA-11qs 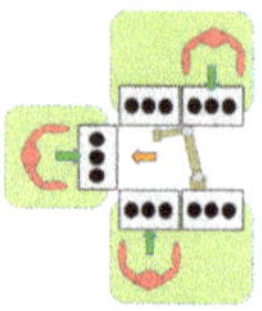	RPWA-1mqs 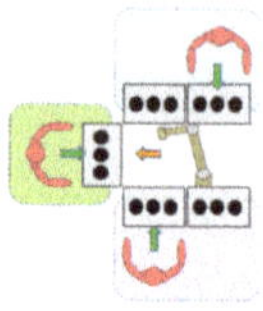	RPWA-n1qs 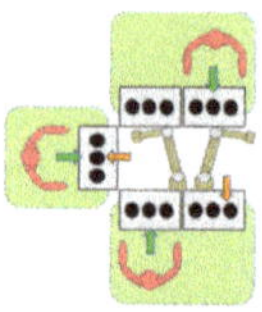	RPWA-nmqs 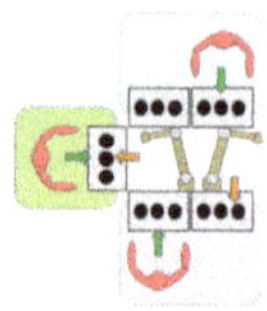

3. Test Case Study Description

The investigated case study was proposed by the company ELVEZ d.o.o. [25], a manufacturer of cable harnesses and processer of plastic components located in Višnja Gora, Slovenia. The company asked for a robotic collaborative system able to fulfill the assembly of wire harnesses for the automotive sector while interacting with an operator in a safe, ergonomic, and efficient way. Such a case study is one of the manufacturing challenges proposed in the first call of ESMERA (European SMEs Robotics Applications), the European Union's "Horizon 2020" project for research and innovation [26]. The solution to the challenge outlined in this paper is part of the project Wire Cobots, which is funded by ESMERA and developed by the research group of the Smart Mini Factory Lab (SMF) of the Free University of Bolzano-Bozen [27] and Carretta s.r.l. [28], a company specializing in the design and installation of industrial automation solutions, located in Quinto di Treviso, Italy.

The wire harnesses are currently assembled on a manual assembly line made by two stations working on average six days per week with three shifts (8h/shift) per day. The average total year production of the line is 900,000 pieces, computed as follows:

500 pcs/shift × 3 shifts × 6 days/week × 50 weeks × 2 lines = 900,000 pcs/year.

It can vary between 750,000 and 1,050,000 pcs/year depending on the number of workdays per week as a result of market interest. An efficiency of 40 s/unit is considered for each station.

The assembly process consists of taping together three groups of wires with connectors by means of a taping pistol. The wires are sequentially inserted by the worker into dedicated assembly jigs (see Figure 2 [29] and Figure 3) and then fastened together by means of the isolating tape in seven different spots (Figure 2b). The assembly process consists of 19 elementary tasks, listed in Table 3 together with the average time. Since the case study refers to an industrial challenge provided by a company, the publication of some sensitive data (e.g., the standard deviation of the task time) is not possible. The total assembly cycle lasts on average 40 s. All the tasks can be conceptually grouped into two groups: the group of operations related to wire insertion (tasks 1–5, 9,10,19 in Table 3—see Figure 2a) and the one related to the tapings (tasks 6–8, 11–18 in Table 3—see Figure 2b). The total cycle time is roughly divided in half between these two groups of tasks.

The company Elvez d.o.o. wants to make this manual assembly workstation collaborative in order to improve the following main metrics:

—Physical ergonomics: the new process should provide better physical work conditions to the operators;

—Productivity: the new cycle time should be the same as—or shorter than—the current manual process.

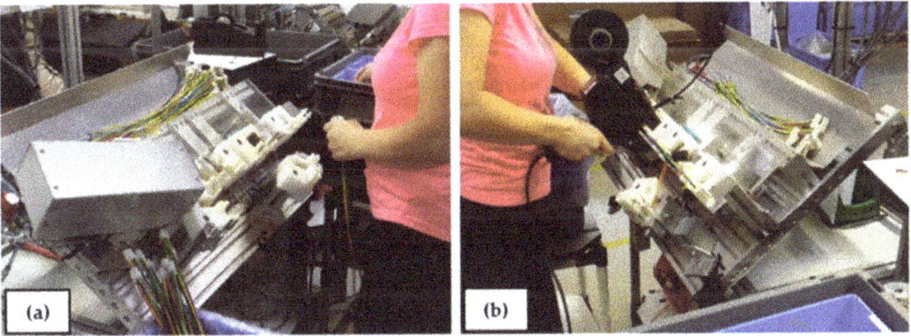

Figure 2. Wire harness manual assembly workstation at the ELVEZ d.o.o. company [29]: tasks related to wire insertion (**a**) and tasks related to the tapings (**b**).

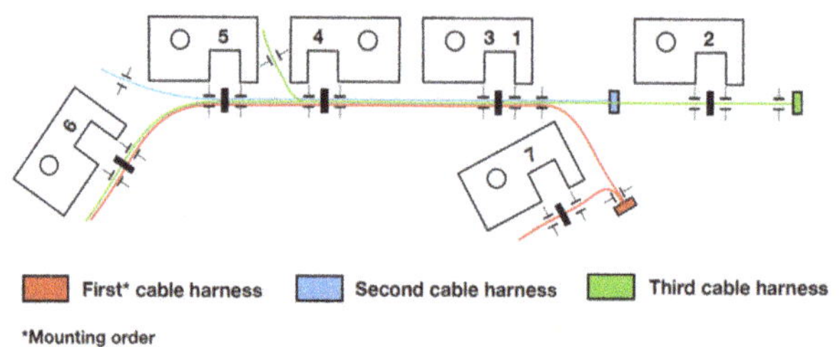

Figure 3. Layout of the assembly jigs and related position of wires and tape spots.

Table 3. Cable harness manual assembly sequence and related average task time.

Nr.	Task	Average Task Time (s)
1	Taking 1st wire harness	2
2	Positioning 1st wire harness	1
3	Taking 2nd wire harness	3
4	Positioning 2nd wire harness	3
5	Adjusting 1st and 2nd wire harness	2
6	Taking taping pistol	2
7	Performing 1st taping	1
8	Depositing taping pistol	1
9	Taking 3rd wire harness	3
10	Positioning 3rd wire harness	5
11	Taking taping pistol	2
12	Performing 2nd taping	
13	Performing 3rd taping	
14	Performing 4th taping	10
15	Performing 5th taping	
16	Performing 6th taping	
17	Performing 7th taping	
18	Depositing taping pistol	1
19	Storing the wire harnesses	4
	Total assembly time (s)	40

4. Results

4.1. Analysis of the Current Situation in Terms of Physical Ergonomics

A physical ergonomics assessment of the manual assembly workstation was carried out by means of the framework proposed and discussed in Section 2.1. The application of the proposed approach to our case study is shown in Table 4. The results show that the key questions related to repetitive upper limbs tasks and awkward working postures were positive. Therefore, further ergonomics evaluations on these aspects are needed. An OCRA check list and a RULA analysis were performed and discussed in Section 4.1.1 and Section 4.1.2., respectively. Conversely, since the assembly tasks did not present manual lifting or lowering of objects of 3 kg or more, the NIOSH-based evaluation was not necessary.

Table 4. Preliminary identification of unsuitable work conditions related to physical ergonomics for the case study of manual assembly cycle (inspired by ISO TR 12295 [16]).

HRI Physical Assistance	Key Question	Answer	Motivation
Manual handling of heavy objects	Is there manual lifting or lowering of heavy objects during the assembly cycle?	NO	The cables are very light (weight $<<$ 3kg). The taping pistol weight is less than 3 kg and is supported.
	(Note: it is possible to consider a "heavy object" a weight $\geq$ 3 kg [18].)		
Repetitive upper limb task	Are there one or more repetitive tasks of the upper limbs during the assembly cycle?	YES	The operator performs the same assembly all the day. The average assembly cycle is 40s. In that time, the operator performs 19 tasks. This means that the total amount of daily manual tasks performed by an operator is about 10,945 per shift (it is supposed a shift duration of 8 hours and an effective working time of 80%).
	(Note: it is possible to consider a "repetitive task" an activity with a total duration of one hour or more per shift and characterized by repeated work cycles or tasks during which the same working actions are repeated for more than 50% of the cycle time [19].)		
Awkward working postures	Are there awkward working postures during the assembly cycle?	YES	According to a first visual ispection, there are different human body districts which do not maintain a suitable posture during the work. The operator performs the same assembly all day.
	(Note: the awkward working postures are related to neck, trunk, and/or upper limbs and repeated for a significant part of the working time [23].)		

4.1.1. The OCRA Analysis Results

The analysis of the effects provided by manual handling of low loads at high frequency was based on the OCRA check list [21,22]. For the reader's convenience the OCRA method is recalled in "Annex A". Table 5 summarizes all the variables requested by the OCRA check list for the evaluation of the biomechanical overload and the related motivations for those choices. The table also shows the final estimation of the workstation risk value: "medium" for both the right and the left part. Major technical problems come from the static frequency of actions (which causes a multiplier equal to 4.5 for both the sides) and from awkward postures, especially for the hands (multiplier equal to 8 for both the sides). Since it was behind the aim of this work to provide organizational improvements, the ergonomics re-design of the workstation was focused on the technical aspects of the abovementioned multipliers. Therefore, only the frequency of technical actions and the awkward postures were considered in the new workstation proposal (force, stereotype, and additional factors are not a problem in the current assembly process).

4.1.2. The RULA Analysis Results

The preliminary key questions underline the need for further investigations about the working postures. In order to identify which tasks of the entire assembly cycle are less suitable from the postural point of view, the RULA analysis was performed for each task summarized in Table 3. This subdivision is useful to identify the least suitable manual tasks that are the best candidates to be potentially performed by the collaborative robot. Further theoretical information about the methodology and the evaluation are provided in "Annex A". Muscle use and force/load scores were ignored in this case, since the analysis was related to postures. Nevertheless, the related values were calculated for all the tasks and they were all equal to zero. The procedure was repeated in detail for each working task. The final RULA values related to the assembly cycle (the so-called overall section "C") are summarized in Table 6. A detailed analysis of each body district for all the assembly tasks is provided in Tables A1 and A2 of "Appendix A".

Table 5. Summary of the multipliers and final estimation of the workstation risk (manual workstation).

Multiplier		Right Part Value	Left Part Value	Motivation
Recovery multiplier		1.33		The shift (8h) is interspersed with a break (1h). As a result, the operator works for 4 hours without a recovery period.
Constant of frequency (dynamic actions)		4	4	The movements of the arms are rapid (~40 action/min), interruptions are infrequent and uneven.
Constant of frequency (static actions)		4.5	4.5	The operator is handling the taping pistol and the cables for all the assembly cycle (there is a pinch for more than the 80% of the time for both the right and the left hands).
Force multiplier		0	0	According to the operator interviews, the assembly tasks do not require the use of force (Borg scale values lower than 3).
Posture and movements multiplier	Shoulder	6	1	The movements and related postures were studied according to the guidelines provided in the check list.
	Elbow	2	6	
	Wrist	0	0	
	Hand	8	8	
Stereotype		0	0	No stereotype.
Additional factors score		0	0	No additional factors.
Multiplier for net duration		0.95		The total net duration of the repetitive tasks was estimated to 361–420 min/shift, which takes into account the unplanned interruptions that may occur.
Final Check-list values		15.8	15.8	Medium red
Final risk estimation		**Medium**	**Medium**	

Table 6. Final Rapid Upper Limb Assessment (RULA) values (overall section "C" table) for the analyzed assembly cycle.

Nr.	Task	Left Value	Right value
1	Taking 1st wire harness	5	4
2	Positioning 1st wire harness	3	3
3	Taking 2nd wire harness	5	6
4	Positioning 2nd wire harness	6	7
5	Adjusting 1st and 2nd wire harness	6	7
6	Taking taping pistol	5	5
7	Performing 1st taping	6	7
8	Depositing taping pistol	6	5
9	Taking 3rd wire harness	4	4
10	Positioning 3rd wire harness	6	7
11	Taking taping pistol	5	5
12	Performing 2nd taping	6	7
13	Performing 3rd taping	6	7
14	Performing 4th taping	6	7
15	Performing 5th taping	6	7
16	Performing 6th taping	6	7
17	Performing 7th taping	5	7
18	Depositing taping pistol	3	3
19	Storing the wire harnesses	6	7
Max values (Overall values)		**6**	**7**

As for the OCRA check list approach, this representation of the results makes it possible to quickly and efficiently understand the needs for an ergonomic re-design of the layout and/or process. The workstation final values (left and right side) were equal to the maximum values calculated for all the analyzed tasks. In general, with the exception of tasks 2, 9, and 18, all the others required important improvements in terms of working postures. From the technical point of view, this could be achieved by re-designing the workspaces in a more ergonomic way and by adding a collaborative robot for performing the most stressful tasks.

4.2. Evaluation of the Potential for Collaborative Robotics

With reference to the studied assembly process, the proposed algorithm was not applied to each of the 19 identified tasks (Table 3), since many of them are the same operation repeated in different moments. In particular, this is the case of the taking and the positioning of cable harness (tasks 1–4, 9–10), and the picking and the dropping off the taping pistol, as well as the related taping tasks (6–8, 11–18). It is worth noticing that although the grouped tasks have a different duration and involve different components, for the purpose of assessing their potentials in HRI they behave like they are the same task. Additionally, the grouping of the harness picking/placing as well as of picking/placing the taping pistol and taping avoids introducing additional tasks (e.g., the transfer of the harness/pistol from human to robot or vice versa), since each task group is performed by the same human/robot resource.

The scoring of the five PCIs for each of the four groups of the studied assembly cycle is shown in Table 7. Considering the results of the physical ergonomics analysis presented in Section 4.1, the first and third groups received the highest possible value for the first process critical issue (PCI 1). In addition, according to a preliminary risk analysis, none of the groups received a value for the aspects related to the occupational risks (PCI 2). On the other hand, due to the repetitiveness of the assembly cycle, all the groups received a value equal to two for the process critical issue related to work monotonousness (PCI 3). Since the taping tasks often require to be re-performed, the third group (only) received a value equal to two for the process critical issue related to the product and process quality (PCI 4). Finally, the first and the last groups received a value equal to three for the last process critical issue since they are classified as not value-added (PCI 5). The third group received a value equal to one due to the fact that only the picking and dropping off of the taping pistol are tasks of the group classified as not value-added.

Table 7. Quick Assessment of the actual manual assembly of wire harness.

Process Critical Issue	Score (S_{ij})			
	Task 1–4, 9–10	Task 5	Task 6–8, 11–18	Task 19
PCI 1	3	2	3	2
PCI 2	0	0	0	0
PCI 3	2	2	2	2
PCI 4	0	0	2	0
PCI 5	3	0	1	3
Potential value (Pval$_j$)	18	12	20	15
Potential class	good	good	good	good

According to the resulting *Pval* indexes (see Table 7), there are good potentials for the implementation of HRI solutions for all the identified groups of tasks. In particular, the highest values were related to the first group (*Pval*$_{G1}$ equal to 18) and to the third group (*Pval*$_{G3}$ equal to 20). If in addition to the evaluated PCIs the technical feasibility for the robot execution was considered, the tasks of the third group were by far the ones with more potential, since the picking and handling (group 1) of flexible and tangle components, such as wires, are very complex and expensive tasks (both in terms of time as well as investments) for a robotic system.

4.3. Workstation Re-Design for Physical Ergonomics and Production Efficiency Enhancement

The results in Section 4.1 show that the physical ergonomics of the manual workstation need to be improved, while the ones in Section 4.2 demonstrate that there is a good potential for collaborative robotics. The highest score in terms of potential for collaborative robots was the one of the taping tasks, which are the ones that mostly contribute to the biomechanical overload for the operator. In addition, in order to provide flexible and cost-effective solutions, there is the necessity to improve the productivity of the current assembly cycle without using fully automated solutions. This means that a sharing of tasks and workspaces between human and automated solutions could be a suitable solution. These conditions make the use of the collaborative robot as a physical assistance system very attractive, especially if properly combined with an ergonomic re-design of the workspace. Based on such results, the manual assembly workstation was re-designed by introducing a collaborative robot supporting the worker in the tapings. In order to further improve the physical ergonomics of the new workstation and its productivity, both the workstation and the assembly cycle were re-organized.

A layout RPWA-1112 (one robot, one type of workpiece, one operator, two working areas), among the ones suggested in Section 2.3, was chosen. The main considerations that led to the adoption of this layout with two working areas were:

—the need to assemble only one kind of product;

—the need to improve productivity without increasing company costs;

—the small dimension of the assembly panel (about 500 × 300 mm), that might lead the operator and the robot to hinder each other preventing a safe sharing of the working area;

—the sequentiality of most of the tasks to be performed, that does not make it possible to parallelize operations between human and robot.

The adoption of two working areas, i.e., a double-panel workstation, overcomes the issues stressed above. Indeed, while the operator is working on one panel, the robot works on the other, and vice-versa (basically they work in parallel).

The results of the physical ergonomics assessment (see Section 4.1.) highlight that the current assembly process leads the worker to assume awkward postures as well as to perform unsuitable repetitive movements. A summary of the highlighted problems as well as the related solutions to improve the manual handling and the postural conditions is given in Table 8.

Table 8. Potential ergonomic improvements according to the physical ergonomics assessment.

Involved Body Part, Posture, and Movements	Involved Task	Possible Solutions
upper arm position and movements	tapings	Use of a collaborative robot as support for tapings
wrist position and twist and movements	all tasks	Re-design of the worktable areas according to main anthropometric requirements [24]
neck position	all tasks	Re-design of the worktable areas according to main anthropometric requirements [24]
trunk position	all tasks	
Involved work-related feature	**Involved task**	**Possible solutions**
activity rhythm	all tasks	Use of a collaborative robot as support for the reduction of manual frequency
overload balance	all tasks	Use of a collaborative robot as support for the balance of the overload between left and right body sides

With reference to the working area, it was designed according to the guidelines defined in the EN ISO 14738 [24]. Since for productivity purposes the new workstation will have two working areas, i.e., two side by side assembly panes, the optimal working area was designed by properly matching two single optimal zones. The resulting working area is shown in Figure 4. The same figure also shows the new layout of the working area. According to such a layout, the assembly panel is completely included in the optimal area as well as the picking zone of the boxes for cables supply. Therefore, during the assembly process, the operator moves in a zone that optimizes the working postures.

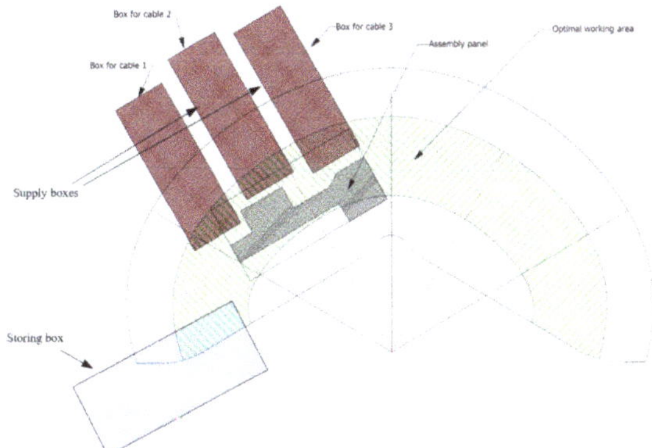

Figure 4. New assembly workstation working area and related layout.

As far as the workstation is concerned, its re-design was based on the EN ISO 14738 guidelines too. In particular, the following features have been provided:

—the working area is inclined 30° with respect to the horizontal, that is, the orientation that minimize the operator's wrist twist;

—the angle with respect to the vertical axis between the two assembling panels is 120°. Such an arrangement of the assembling panels allows the operator to work on both the panels with a minimal trunk twist and without colliding with the adjacent panel;

—the workstation height as well as the panel positions are adjustable, so that they can fit the anthropometric measure of the operator working on it.

A preliminary numerical validation of the physical ergonomics of the re-designed working area and workstation was carried out in "Siemens Tecnomatix Process Simulate" [30]. Such a simulation software makes it possible to perform postural analysis (RULA index) for manufacturing tasks. This preliminary evaluation showed that a great improvement in terms of operators' working postures can be obtained with the proposed workstation. Indeed, the maximum values of the RULA scores were decreased from 6 for the left arm and 7 for the right arm of the old workstation to 3 for both the arms.

The collaborative robot has been placed at the back side of the workstation, while the worker at the front one, i.e., they operate one in front of the other, with the assembling panels interposed. The robot is located on the floor level and moves towards the panel from the underside. In order to equally and easily reach the taping zones of both the panels, the robot has been placed in the middle of the two panels. Such a placement of the robot drastically reduces the possibility of collision with the operator's head. However, since the robot moves under the workstation, if the operator is in a seated position, collisions between the robot and the operator's legs could happen. In order to avoid such a risk, the operator will use a high stool that permits an upright position or will directly take a standing position. The suggested design for the conversion of the manual assembly workstation for the wire harness assembly has been implemented in the laboratory prototype shown in Figure 5.

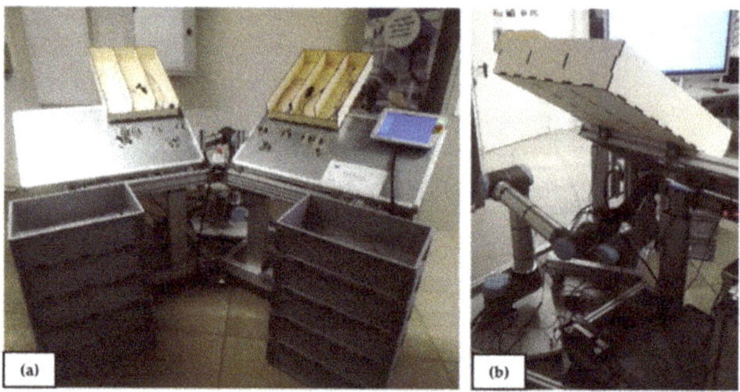

Figure 5. Laboratory prototype of the new collaborative workstation: front (**a**), back (**b**) view.

4.4. Collaborative Workstation Assessment

In order to experimentally assess the physical ergonomics and production efficiency improvements of the proposed solution, several wire-assembly tests were carried out on the developed prototype (see Figure 6). The laboratory tests involved two persons with no knowledge of robotics. The age, gender, anthropometric features, and work skills of the two candidates were very similar and reasonably represent the "real" workers of the company. The ergonomic assessments (OCRA index and RULA analysis) of the new workstation were carried out by the members of the SMF lab with expertise in biomechanical overload analysis and collaborative robotics.

Figure 6. Pictures of experiments with the new assembly cycle.

4.4.1. Manual Handling Improvements

Table 9 summarizes the results of the OCRA check-list carried out on the new workstation. By comparing the final check list values in Table 9 with the ones of Table 5 (related to the manual workstation), it is evident that the operator's work conditions in terms of manual handlings are considerably improved in the new workstation. The overall values were reduced by 12.0% for the right part and by 28% for the left part. In particular, the final indexes changed from 15.8 to 13.9 for the right side and from 15.8 to 11.4 for the left side. This reduction will provide important benefits in terms of biomechanical overload related to manual handlings. In particular, the use of the collaborative robot considerably reduced the multipliers related to the dynamic and static constant of frequency, the posture, and the movements for different body regions. This is possible since the parallel work allows a reduction of the work rhythm. In addition, the use of the robot for the taping tasks reduces the operator's average awkward postures and movements.

Table 9. Summary of the other multipliers and final estimation of the workstation risk (collaborative workstation). The percentage of indexes reduction between manual and collaborative workstation is highlighted in the brackets.

Multiplier		Right Part Value	Left Part Value	Motivation
Recovery multiplier		1.33		It was supposed that the shift (8h) is interspersed with a break (1h). As a result, the number of hours without recovery period is 4, which means a multiplier equal to 1.33
Constant of frequency (dynamic actions)		3 (−25%)	3 (−25%)	The movements of the arms are rapid (~40 action/min). There is the possibility of short interruptions. This means a multiplier equal to 3.
Constant of frequency (static actions)		2.5 (−44%)	2.5 (−44%)	The operator handles the taping pistol and the cables for more than half of the assembly cycle (there is a pinch for about the 70% of the time for both the right and the left hands). This means a multiplier equal to 2.5.
Force multiplier		0	0	According to the operator interviews, the assembly tasks do not require the use of force (Borg scale values lower than 3).
Posture and movements multiplier	Shoulder	1 (−83%)	0 (−100%)	The movements and related postures were studied according to guidelines provided in the check list.
	Elbow	8 (+300%)	6	
	Wrist	0	0	
	Hand	0 (−25%)	0 (−25%)	
Stereotype		0	0	According to the definition, there is no stereotype
Additional factors score		0	0	According to the definition, there are no additional factors
Multiplier for net duration		0.95		It was supposed that some working interruptions sometimes occur. As a result, the total net duration of repetitive tasks is 361–420 min/shift, which means a multiplier equal to 0.95
Final Check-list values		13.9	11.4	Yellow
Final risk estimation		light	light	
Reduction percentage		−12.0%	−28%	

4.4.2. Postural Improvements

The postural assessment results are summarized in Table 10. The large part of the RULA values for different body regions were lower than the ones of the manual workstation. The operator's postures were considerably improved in the new workstation. The ergonomic benefits provided by changing the workstation from the manual to the collaborative one in terms of RULA values are shown in Table 11. The overall workstation values were reduced by 50% for the left part and by 57% for the right part (which are the combination of the left or right arm and related wrist analysis with the neck, trunk, and leg analysis). In particular, they changed from 6 to 3 for the left side and from 7 to 3 for the right side. This reduction will provide important benefits in terms of working postures. In particular, the use of the collaborative robot and the new layout reduced considerably the impact of the postures for different body regions.

4.4.3. Production Efficiency Improvements

The cycle time of the proposed solution is about 35 s/part, it was computed as an average value of the cycle times of the different experiments conducted with the prototype assembly station. In particular, such a value was obtained by dividing by two the overall assembly time which is needed for the parallel fabrication of two harnesses (one for each panel). The original cycle time was around 40 s/part for each workstation. Therefore, the new assembly cycle is more efficient than the original one: the cycle time is reduced by 12.3%, which means about 1460 h/year in terms of "saved manual labor".

Table 10. RULA results analysis of the new workstation.

Nr	Task	Final Value Left Side	Final Value Right Side
1	Picking the 1st wire harness the box	2	2
2	Positioning the 1st wire harness into the frame	3	3
3	Picking the 2nd wire harness from the box	2	2
4	Positioning the 2nd wire harness into the frame	2	2
5	Adjusting the first cable group on the second cable group	2	2
6	Pushing the button (robot call)	3	3
7	Moving from Panel 1 to Panel 2	2	2
8	Taking the previous final assembly from the frame	3	3
9	Storing the previous final assembly into the box	3	3
10	Moving from Panel 2 to Panel 1	2	2
11	Picking the 3rd wire harness from the box	2	3
12	Positioning the 3rd wire harness into the frame	3	3
13	Pushing the button (robot call)	3	3
14	Moving from Panel 1 to Panel 2	2	2
	Perform the same tasks (from 1 to 6) on Panel 2 and then come back to Panel 1 ***		
15	Taking final assembly from the frame	3	3
16	Storing the final assembly into the box	3	3
	Max values (Overall values)	**3**	**3**

*** Since the layout of Panel 2 is specular to Panel 1 and since the assembly sequence is the same, no other or different tasks are required for harness assembly on Panel 2. For this reason, from a postural point of view, an additional RULA analysis of the assembly tasks related to Panel 2 was not required (because the results will be comparable to the one calculated for Panel 1 tasks).

Table 11. RULA values for each analyzed body region.

		RULA Analysis Results (Max Values for Each Posture)					
		Actual Workstation		Collaborative Workstation		Index Variation	
		Left Side	Right Side	Left Side	Right Side	Left Side	Right Side
Arm and Wrist Analysis	Upper Arm Posture Scores	4	6	2	2	−2	−4
	Lower Arm Posture Scores	3	3	2	2	−1	−1
	Wrist Posture Scores	3	4	2	2	−1	−2
	Wrist Twist Posture Scores	2	2	2	2	0	0
	Muscle Use Scores	0	0	0	0	0	0
	Force/Load Scores	0	0	0	0	0	0
Neck, Trunk, and Leg Analysis	Neck Posture Scores	4		3		−1	
	Trunk Posture Scores	4		2		−2	
	Leg Posture Scores	1		1		0	
	Muscle Use Scores	0		0		0	
	Force/Load Scores	0		0		0	
	Max values (overall values)	**6**	**7**	**3**	**3**	**−3**	**−4**
	Improvements (reduction % of overall workstation values)					**50%**	**57%**

5. Discussion and Conclusions

In this work, a manual workstation for wire harness assembly was transformed into a collaborative one. According to the formal definition, the proposed result is a co-existence solution, since the robot is supporting the human without providing a real hand-by-hand physical interaction. This was possible by properly introducing a collaborative robot and by changing the working layout and the related assembly cycle. The main driver for the conversion was the improvement of operators' physical ergonomics. To this end, a framework was proposed to perform an accurate ergonomics assessment of the starting situation. Such a framework is based on ergonomics standards and state-of-the-art-methods for the ergonomic analyses. It is suitable for analyzing the manual assembly activities that are typically carried out in a workstation.

In order to preliminarily evaluate if the introduction in the manual assembly of a collaborative robot is beneficial, a methodology called Quick Assessment algorithm was exploited. It considered the benefits that the integration of a collaborative robot could provide to the manual process in terms of ergonomics and safety, product and process quality, and economics.

After that, the guidelines followed to design the new collaborative assembly workstation, starting from the existing one by particularly focusing on physical ergonomics and production efficiency, were outlined and applied.

Finally, the developed solution was implemented in the SMF laboratory through a TRL 4 prototype and its performance assessed. The achieved results show that the ergonomics indexes related to manual handlings (OCRA check-list) and postures (RULA values) were considerably reduced in the new workstation. In particular, the biomechanical overload of the worker was reduced by 12.0% for the right part and by 28% for the left part in terms of manual handling and by 50% for the left part and by 57% for the right part in terms of working postures. In addition, the estimated cycle time was reduced by 12.3%. Considering the annual productivity, this improvement means a potential annual reduction of 1460 working hours.

The re-design approach introduced in this work is based on solutions that are effectively and simply utilizable by SMEs. The use of the framework for the identification of not-ergonomic tasks which could be improved by integrating a collaborative robot, the Quick Assessment algorithm as well as the virtual simulations are smart and easy-to-use tools for a proper design of a collaborative workstation starting from a manual one. The implementation of a prototype in a learning factory laboratory based on a real industrial case study provided by a manufacturing SME confirmed the effectiveness of the proposed solutions.

Although the results were satisfactory, both the employed methodologies and the developed collaborative workstation could be further improved, as discussed in the below.

1.   Physical ergonomic assessment:

The framework presented for the physical ergonomics could be improved and generalized by using a different methodology for the assessment of the working postures, even if the RULA method is quick and easy to use; indeed, it is suggested to use such an approach only for preliminary postural assessment. A possible improvement relies on replacing the RULA method with the standardized guidelines included in ISO 11226 [31], as for the other methodologies mentioned in the framework.

2.   Ergonomic improvements:

The achieved results are more than satisfactory in terms of reduction of work-related biomechanical overload. Even if the project goal was successfully achieved, the final OCRA check-list and RULA values were slightly over the optimal zone, i.e., the green one. Better results could be achieved by integrating the proposed technical solutions (which usually have the major impact) with organizational solutions. This should provide further benefits and a real possibility to achieve the green zone.

3.   Cycle time:

Even though the achieved results were satisfactory, the cycle time could be shortened by optimizing the robot trajectories and/or developing a new taping pistol (specially designed for robotic applications).

In both the cases, the time taken by the robot for performing the assembly cycle would be reduced and the cycle time improved. Indeed, the bottleneck (in terms of cycle time) of the new collaborative workstation is the second set of tapings (from the second to the seventh). A minimum speeding up of such tasks can bring enormous benefits to the annual productivity: a reduction of about the 5% of the robot cycle time ($\approx$1 s less than the current overall cycle time) results in a potential reduction of 292 annual working hours.

In addition to the economic considerations, a taping pistol which is specially designed for robotic applications could provide several benefits also in terms of product quality.

4. Future research directions:

Future research should focus on the possibility to autonomously and dynamically adapt the workstation elements according to operators' psychophysical work conditions, needs, and wants [32]. This will surely improve both physical and cognitive ergonomics and therefore have a good impact on productivity. These improvements would be possible by properly integrating the robot systems with specific monitoring sensors and wearable devices. In addition, the possibility to increase the system flexibility by rapidly adapting the assembly to different product variants should be investigated.

Author Contributions: Conceptualization, all the authors.; Methodology, L.G.; Software, F.A.M.; Validation, L.G., I.P.; Formal Analysis, L.G.; Writing—Original Draft Preparation, L.G.; Writing—Review and Editing, I.P., E.R. and R.V.; Supervision, E.R. and R.V.; Project Administration, I.P. All authors have read and agreed to the published version of the manuscript.

Acknowledgments: The research leading to these results can be framed within the Wire Cobots project, a cascade funding project which receives funding from the European Commission for organizing the Robotics Application Oriented Research Experiment, under the ESMERA Funding Agreement (agreement No 780265). Further this research and the development of the Quick Assessment methodology belongs to the project "SME 4.0—Industry 4.0 for SMEs" and has received funding from the European Union's Horizon 2020 R&I programme under the Marie Skłodowska-Curie grant agreement No 734713.

Conflicts of Interest: The authors declare no conflict of interest.

Appendix A

Appendix A.1. NIOSH Lifting Equation

A large proportion of occupational back disorders originate from improper manual lifting activities that provide operators' biomechanical and psychological overload as well as psychosocial factors. For these reasons, NIOSH proposed a practical evaluation procedure for the analysis of the physical demands of two-handed manual lifting [18]. The methodology is composed of the assessment of two main factors: the recommended weight limit and the lifting index. The former is calculated through the analysis of main task features, such as the geometry of the hand location, frequency of lifting, work duration, and type of hand coupling required for the task. The latter estimates the physical demand related to the task and is defined as the ratio between the actual weight of the load lifted and the recommended weight limit for the same activity [20].

Appendix A.2. OCRA Check List

Manual handling activities could be potentially harmful for operators if proper physical ergonomics expedients are not correctly implemented. In general, it is advisable to avoid or reduce as much as possible these operations through work enlargements, job rotation, and/or automation. Especially in the case of repetitive tasks characterized by manual handling of low loads at high frequency, industrial robotics could be an efficient solution [22]. The OCRA check list is a quantitative methodology for the evaluation of the biomechanical overload related to upper limbs in the case of manual handling of low loads at high frequency. This method is a simplified form of the so called "OCRA index" and is part of the ISO standard 12228-3 [22] about manual handling ergonomics and considers the following risk factors (for the right and left parts): frequency of technical actions, repetitiveness, awkward postures (for shoulders, elbows, wrists, and hands), force, additional factors, lack of recovery periods, duration of repetitive task. The method is based on the combination of the following multipliers [22]:

—Recovery multiplier: it evaluates the recovery time, which is defined as any time in which the upper limb is primarily physically inactive;

—Constant of frequency: it evaluates the contribution of the movement frequency to the development of a muscle-tendon disorder. The technical action may be dynamic (characterized by movement) or static (characterized by holding a single posture, such as when a worker must hold an object in their hand);

—Force multiplier: it evaluates the operator's use of force during the manual tasks by means of interviewing them and asking to describe the subjectively perceived muscular effort made when they are carrying out a repetitive task;

—Posture and movements multiplier: it evaluates the contribution of awkward postures and movements of shoulder, elbow, wrist, and hand (type of grip and finger movements) required during the activity to the work-related muscle-tendon disorders;

—Additional factors score: it evaluates the further contribution of additional physico-mechanical factors and/or additional socio-organizational factors. These additional factors may increase the risk if they are present and, as a consequence, should be carefully considered;

—Multiplier for net duration: it aims to adapt the risk value by weighing the final index for the real duration of the repetitive work.

As output, it provides the intrinsic risk of a certain work situation or workplace by providing an exposure level classified through a three-zone system. These zones are represented by different colors, as following:

—Green zone (check-list value ≤ 7.5): if the results of the analysis of a certain activity are within this zone, the overall risk is considered to be acceptable;

—Yellow zone (7.5 < check-list value ≤ 11.0): if the results of the analysis of a certain activity are within this zone, a more detailed risk assessment is needed or else remedial action should be taken to reduce the risk to the green zone;

—Red zone (11 < check-list value ≤ 14.0 for light red, 14.0 < check-list value ≤ 22.5 for medium red, and check-list value > 22.5 for high red): if the results of the analysis of a certain activity are within this zone, then the work is judged to be harmful for the operators.

Appendix A.3. RULA Method

RULA is a guided survey method for simple and fast evaluation of the musculoskeletal system through the analysis of postures. It involves the analysis of the neck, trunk, and upper limbs, along with muscle function and the external loads experienced by the body. Basically, the method allows the study of the upper body activities by using body part diagrams integrated with code for joint angles, body postures, load/force, coupling, and muscle activity [23]. The evaluation is done on a single worksheet page composed by different sequential tables to fulfill. Scores are then entered based on both the two central regions: section "A" for wrist and arm and section "B" for trunk and neck (and eventually legs). By combining the results of these two first sections, it is possible to calculate a final score (section "C") which is related to a proposed a scale of action levels:

—Green zone (RULA value between 1 and 2): if the results of the analysis of a certain activity are within this zone, then the work is acceptable and no improvement action are required;

—Pale yellow zone (RULA value between 3 and 4): if the results of the analysis of a certain activity are within this zone, further investigations are necessary and changes may be required;

—Dark yellow zone (RULA value between 5 and 7): if the results of the analysis of a certain activity are within this zone, further investigations are necessary and changes are required soon;

—Red zone (RULA value equal or higher than 7): if the results of the analysis of a certain activity are within this zone, further investigations are necessary and changes are required immediately.

These action levels should provide a guide to the level of risk and need for action to conduct more detailed assessments. Basically, they evaluate whether the posture is acceptable or requires further investigation and change, which means the possibility to introduce the collaborative robot for physical ergonomics improvement.

Even if the method was originally proposed for the investigations of work-related upper limb disorders, the results are usually not sufficient for a comprehensive evaluation of the biomechanical overload related to the assembly activities. Due to the fact that RULA is easy to use by enabling a quick study without requiring a higher degree in ergonomics or even the need for expensive equipment, this method is often chosen for the postural analysis.

Appendix A.4. RULA Evaluation of the Current Assembly Process

Tables A1 and A2 explain in details the RULA evaluation for each body district involved in the manual assembly (Table 3). The summary of the final results is provided in Table 6.

Table A1. Summary of the RULA analysis related to the arm and wrist values.

| | A—Arm and Wrist Analysis | | | | | | | | | | | |
| | Upper Arm Posture Scores | | Lower Arm Posture Scores | | Wrist Posture Scores | | Wrist Twist Posture Scores | | Look-Up Posture Index in Table A | | Find row in Table C | |
Task nr.	Left	Right	Left	Right	Left	Right	Left	Right	Left	Right	Left	Right
1	4	3	3	1	2	3	2	2	5	4	5	4
2	1	1	2	2	3	2	2	2	3	2	3	2
3	1	2	2	3	1	3	2	2	3	4	3	4
4	3	3	2	2	3	4	2	2	4	5	4	5
5	3	4	1	2	3	3	2	2	4	5	4	5
6	2	2	2	1	2	2	2	1	3	3	3	3
7	3	6	1	1	3	4	2	1	4	8	4	8
8	2	2	2	1	3	2	2	1	4	3	4	3
9	4	4	1	2	3	3	2	2	5	5	5	5
10	3	3	2	2	3	4	2	2	4	5	4	5
11	2	2	2	1	2	2	2	1	3	3	3	3
12	3	5	2	2	3	3	2	1	4	6	4	6
13	3	6	1	1	3	3	2	1	4	7	4	7
14	2	6	1	2	3	4	2	1	4	9	4	9
15	2	6	1	2	3	4	2	1	4	9	4	9
16	2	6	1	2	3	3	2	2	4	9	4	9
17	1	6	2	1	3	4	2	2	3	9	3	9
18	2	2	1	1	2	2	2	1	3	3	3	3
19	3	3	2	3	3	4	2	2	4	5	4	5

Table A2. Summary of the RULA analysis related to the neck, trunk, and leg values.

| | B—Neck, Trunk, and Leg Analysis | | | |
Task nr.	Neck Posture Scores	Trunk Posture Scores	Leg Posture Scores	Lookup Posture Index in Table B	Find Column in Table C
1	3	3	1	4	4
2	3	1	1	3	3
3	4	3	1	6	6
4	4	3	1	6	6
5	4	3	1	6	6
6	4	3	1	6	6
7	4	4	1	7	7
8	4	3	1	6	6
9	3	2	1	3	3
10	4	3	1	6	6
11	4	3	1	6	6
12	4	3	1	6	6
13	4	4	1	7	7
14	4	4	1	7	7
15	4	4	1	7	7
16	4	4	1	7	7
17	4	3	1	6	6
18	3	2	1	3	3
19	4	3	1	6	6

References

1. Pedersen, M.R.; Nalpantidis, L.; Andersen, R.S.; Schou, C.; Bøgh, S.; Krüger, V.; Madsen, O. Robot skills for manufacturing: From concept to industrial deployment. *Robot. Comput. Integr. Manuf.* **2016**, *37*, 282–291. [CrossRef]
2. Kagermann, H.; Helbig, J.; Hellinger, A.; Wahlster, W. *Recommendations for Implementing the Strategic Initiative INDUSTRIE 4.0: Securing the Future of German Manufacturing Industry*; Final Report of the Industrie 4.0 Working Group; Forschungsunion: Acatech, Munich, 2013.
3. Braccini, A.M.; Margherita, E.G. Exploring organizational sustainability of industry 4.0 under the triple bottom line: The case of a manufacturing company. *Sustainability* **2018**, *11*, 36. [CrossRef]
4. International Ergonomics Association. 2019. Definition and Domains of Ergonomics. Available online: https://iea.cc/what-is-ergonomics/ (accessed on 12 December 2019).
5. Zhong, R.Y.; Xu, X.; Klotz, E.; Newman, S.T. Intelligent manufacturing in the context of industry 4.0: A review. *Engineering* **2017**, *3*, 616–630. [CrossRef]
6. Gualtieri, L.; Palomba, I.; Wehrle, E.; Vidoni, R. The Opportunities and Challenges of SME Manufacturing Automation: Safety and Ergonomics in Human–Robot Collaboration. In *InDustry 4.0 for SMEs Challenges, Opportunities and Requirements*; Matt, D.T., Modrak, V., Zsifkovits, H., Eds.; Palgrave Macmillan: Basingstoke, UK, 2019.
7. Castro, P.R.; Högberg, D.; Ramsen, H.; Bjursten, J.; Hanson, L. Virtual simulation of human-robot collaboration workstations. In *Congress of the International Ergonomics Association*; Springer: Cham, Switzerland, 2018; pp. 250–261.

8. Lietaert, P.; Billen, N.; Burggraeve, S. Model-based Multi-Attribute Collaborative Production Cell Layout Optimization. In Proceedings of the 2019 20th International Conference on Research and Education in Mechatronics (REM), Wels, Austria, 23–24 May 2019; pp. 1–7.

9. Mateus, J.C.; Claeys, D.; Limère, V.; Cottyn, J.; Aghezzaf, E.H. A structured methodology for the design of a human-robot collaborative assembly workplace. *Int. J. Adv. Manuf. Technol.* **2019**, *102*, 2663–2681. [CrossRef]

10. Petruck, H.; Faber, M.; Giese, H.; Geibel, M.; Mostert, S.; Usai, M.; Mertens, A.; Brandl, C. Human-robot collaboration in manual assembly–A collaborative workplace. In *Congress of the International Ergonomics Association*; Springer: Cham, Switzerland, 2018; pp. 21–28.

11. Dombrowski, U.; Stefanak, T.; Perret, J. Interactive simulation of human-robot collaboration using a force feedback device. *Procedia Manuf.* **2017**, *11*, 124–131. [CrossRef]

12. El Makrini, I.; Merckaert, K.; Lefeber, D.; Vanderborght, B. Design of a collaborative architecture for human-robot assembly tasks. In Proceedings of the 2017 IEEE/RSJ International Conference on Intelligent Robots and Systems (IROS), Vancouver, BC, Canada, 24–28 September 2017; pp. 1624–1629.

13. Schmidtler, J.; Knott, V.; Hölzel, C.; Bengler, K. Human Centered Assistance Applications for the working environment of the future. *Occup. Ergon.* **2015**, *12*, 83–95. [CrossRef]

14. Heydaryan, S.; Suaza Bedolla, J.; Belingardi, G. Safety design and development of a human-robot collaboration assembly process in the automotive industry. *Appl. Sci.* **2018**, *8*, 344. [CrossRef]

15. Gualtieri, L.; Rauch, E.; Vidoni, R.; Matt, D.T. An Evaluation Methodology for the Conversion of Manual Assembly Systems into Human-Robot Collaborative Workcells. In Proceedings of the 2019 International Conference in Flexible Automation and Intelligent Manufacturing (FAIM 2019), Limerick, Ireland, 24–28 June 2019.

16. International Organization for Standardization. ISO-TR 12295—Ergonomics—Application Document for ISO Standards on Manual Handling (ISO 11228-1, ISO 11228-2 and ISO 11228-3) and Evaluation of Static Working Postures (ISO 11226) (ISO-TR12295: 2014). 2014. Available online: https://www.iso.org/standard/51309.html (accessed on 12 December 2019).

17. International Organization for Standardization. ISO 11228-2—Ergonomics—Manual Handling—Part 2: Pushing and Pulling (ISO 11228-2:2007). 2007. Available online: https://www.iso.org/standard/26521.html (accessed on December 2019).

18. Waters, T.R.; Putz-Anderson, V.; Garg, A.; Fine, L.J. Revised NIOSH equation for the design and evaluation of manual lifting tasks. *Ergonomics* **1993**, *36*, 749–776. [CrossRef] [PubMed]

19. International Organization for Standardization. ISO 11228-1—Ergonomics—Manual handling—Part 1: Lifting and Carrying (ISO 11228-1:2003). 2003. Available online: https://www.iso.org/standard/26520.html (accessed on December 2019).

20. Waters, T.R.; Lu, M.L.; Piacitelli, L.A.; Werren, D.; Deddens, J.A. Efficacy of the revised NIOSH lifting equation to predict risk of low back pain due to manual lifting: Expanded cross-sectional analysis. *J. Occup. Environ. Med.* **2011**, *53*, 1061–1067. [CrossRef] [PubMed]

21. Colombini, D. *Risk Assessment and Management of Repetitive Movements and Exertions of Upper Limbs: Job Analysis, Ocra Risk Indicies, Prevention Strategies and Design Principles (Vol. 2)*; Elsevier: Amsterdam, The Netherlands, 2002.

22. International Organization for Standardization. (2007). ISO 11228-3—Ergonomics—Manual Handling—Part 3: Handling of Low Loads at High Frequency. International Organization for Standardization. (ISO 11228-3:2007). 2007. Available online: https://www.iso.org/standard/26522.html (accessed on 13 December 2019).

23. McAtamney, L.; Corlett, E.N. RULA: A survey method for the investigation of work-related upper limb disorders. *Appl. Ergon.* **1993**, *24*, 91–99. [CrossRef]

24. International Organization for Standardization. (2002). ISO 14738:2002—Safety of Machinery—Anthropometric Requirements for the Design of Workstations at Machinery. International Organization for Standardizatio. (ISO 14738:2002). 2002. Available online: https://www.iso.org/standard/27556.html (accessed on 8 December 2019).

25. The Company. 2019. Available online: https://elvez.si/en/ (accessed on 12 December 2019).

26. ESMERA Open Calls. 2020. Available online: http://www.esmera-project.eu/open-calls/ (accessed on 3 April 2020).

27. Gualtieri, L.; Rojas, R.; Carabin, G.; Palomba, I.; Rauch, E.; Vidoni, R.; Matt, D.T. Advanced automation for SMEs in the I4. 0 revolution: Engineering education and employees training in the smart mini factory laboratory. In Proceedings of the 2018 IEEE International Conference on Industrial Engineering and Engineering Management (IEEM), Bangkok, Thailand, 16–19 December 2018; pp. 1111–1115.

28. Carretta Automations. 2019. Available online: http://www.carrettaautomazioni.it/en/ (accessed on 20 December 2019).
29. ESMERA Elvez Challenge. 2020. Available online: https://www.youtube.com/watch?v=DA1nueKqQt8 (accessed on 8 January 2020).
30. Tecnomatix. 2020. Available online: https://www.plm.automation.siemens.com/global/en/products/tecnomatix/ (accessed on 8 January 2020).
31. International Organization for Standardization. ISO 11226: 2000/COR 1:2006. Ergonomics—Evaluation of Static Working Postures—Technical Corrigendum 1. International Organization for Standardizatio. (ISO 11226: 2000/COR 1:2006). 2006. Available online: https://www.iso.org/standard/44143.html (accessed on 12 December 2019).
32. Gualtieri, L.; Rojas, R.A.; Ruiz Garcia, M.A.; Rauch, E.; Vidoni, R. Implementation of a Laboratory Case Study for Intuitive Collaboration between Man and Machine in SME Assembly. In *Industry 4.0 for SMEs Challenges, Opportunities and Requirements*; Matt, D.T., Modrak, V., Zsifkovits, H., Eds.; Palgrave Macmillan: Basingstoke, UK, 2019.

Article

A Maturity Level-Based Assessment Tool to Enhance the Implementation of Industry 4.0 in Small and Medium-Sized Enterprises

Erwin Rauch [1], Marco Unterhofer [1], Rafael A. Rojas [1,*], Luca Gualtieri [1], Manuel Woschank [2] and Dominik T. Matt [1,3]

[1]   Industrial Engineering and Automation (IEA), Faculty of Science and Technology, Free University of Bozen-Bolzano, 39100 Bolzano, Italy; erwin.rauch@unibz.it (E.R.); marco.unterhofer@unibz.it (M.U.); luca.gualtieri@unibz.it (L.G.); dominik.matt@unibz.it (D.T.M.)

[2]   Chair of Industrial Logistics, Monanuniversitaet Leoben, 8700 Leoben, Austria; manuel.woschank@unileoben.ac.at

[3]   Innovation Engineering Center (IEC), Fraunhofer Italia Research s.c.a.r.l., 39100 Bolzano, Italy

*   Correspondence: rafael.rojas@unibz.it

Received: 30 March 2020; Accepted: 22 April 2020; Published: 27 April 2020

Abstract: Industry 4.0 has attracted the attention of manufacturing companies over the past ten years. Despite efforts in research and knowledge transfer from research to practice, the introduction of Industry 4.0 concepts and technologies is still a major challenge for many companies, especially small and medium-sized enterprises (SMEs). Many of these SMEs have no overview of existing Industry 4.0 concepts and technologies, how they are implemented in their own companies, and which concepts and technologies should primarily be focused on future Industry 4.0 implementation measures. The aim of this research was to develop an assessment model for SMEs that is easy to apply, provides a clear overview of existing Industry 4.0 concepts, and supports SMEs in defining their individual strategy to introduce Industry 4.0 in their firm. The maturity level-based assessment tool presented in this work includes a catalog of 42 Industry 4.0 concepts and a norm strategy based on the results of the assessment to support SMEs in introducing the most promising concepts. For testing and validation purposes, the assessment model has been applied in a field study with 17 industrial companies.

Keywords: industry 4.0; small and medium-sized enterprises; SME; assessment model; sustainability

1. Introduction

The proclamation of the fourth industrial revolution with the aim of a digital transformation of companies changed the industrial world. While in the past a focus was given to the introduction of lean production [1,2], almost every modern enterprise aspired in recent years to become a proper smart factory according to the 'Industry 4.0' principles. In the industrial environment, a curious feeling grew. At first glance, Industry 4.0 was not really perceived as an opportunity, but more as a challenge. Since the first use of the term Industry 4.0 in 2011 [3], many studies have been conducted to investigate the impact of Industry 4.0, its significance for the sustainable competitiveness of companies, the related technologies, as well as methods and strategies for the introduction and implementation of Industry 4.0 in industrial enterprises.

Industry 4.0 is considered to be another word for the fourth industrial revolution [3]. After mechanization, electrification, and computerization, the fourth stage of industrialization aims to introduce concepts like cyber-physical systems (CPS), internet of things (IoT), automation, human–machine interaction (HMI), as well as advanced manufacturing technologies in an intelligent and digitalized factory environment [4]. A distinctive feature of the fourth industrial revolution is the ability to combine the digital and the physical world, affecting a broad spectrum of industrial disciplines [5]. The term was introduced in 2011 by a German group of scientists during the Hannover Fair event, which symbolized the beginning of this fourth industrial revolution [6]. Since then, the term Industry 4.0, or also its synonym smart manufacturing [7], has been one of the most popular manufacturing topics among industry and academia in the world [3,8,9].

Considering recent developments in terms of the industrial progression to smart factories and the fourth industrial revolution, companies are overwhelmed and seem to be incapable of developing appropriate implementation strategies [10]. Each company has to analyze its situation and its individual needs and then choose those Industry 4.0 concepts that promise the best prospects for achieving the goals set [11]. There is a need to develop suitable models and instruments to assess the current status or maturity of applied technologies in industrial companies and to implement Industry 4.0 concepts based on the appropriateness for the individual company.

One of the remarkable works, titled 'Guideline Industry 4.0—Guiding principles for the implementation of Industry 4.0 in small and medium-sized businesses', presented by the VDMA (German Engineering Federation), proposes an 'Industry 4.0-toolbox' that should provide first guidance in the challenging change process [12]. The supplied toolbox was perceived as an acceptable starting base to assess one's own company regarding the implementation of Industry 4.0 technologies and concepts, but is limited to the evaluation of a few concepts for product design and manufacturing.

The need for instruments for the assessment of one's own company regarding the level of Industry 4.0 implementation motivated researchers to develop further models. Schumacher et al. [13] developed an assessment model for the determination of the readiness for Industry 4.0-related measures. Compared with former works [14], the degree of completeness of the model is enhanced through the inclusion and combination of technological and organizational aspects, which are categorized into nine dimensions and single specific assessment items.

However, there is still room for further improvements in the development of models for the evaluation of companies introducing Industry 4.0. Therefore this work aimed to enrich the evolution of Industry 4.0 assessment models giving a specific focus to small and medium-sized enterprises. Further, it opens doors to new and enriched evaluation possibilities that are no longer addressing a problem of readiness of enterprises, but one affiliated to the maturity level associated to the Industry 4.0 implementation. Small and medium-sized enterprises (SMEs) are usually less informed about Industry 4.0 concepts. Therefore, an assessment model must clearly describe the concepts and also show the companies which different possibilities of concepts and technologies Industry 4.0 has to offer. Subsequently, such a model should support SMEs simply and pragmatically to show their status of implementation and the potential for future Industry 4.0 projects. Thus, the model's objective was (i) to inform SMEs about the existing Industry 4.0 concepts, (ii) to assess the current progress in the implementation and application of these concepts, and (iii) to signalize to SMEs which of the Industry 4.0 concepts are the most important ones for the individual company.

The paper is structured as follows: after a first introduction to the topic in Section 1, the authors provide in Section 2 the research methodology for this work. Section 3 follows a literature review of existing assessment and maturity models, as well as for the identification of Industry 4.0 concepts and technologies. Section 4 provides a detailed insight into the structure and development of the proposed maturity level-based assessment tool. Section 5 describes the evaluation and testing of the developed model in several industrial companies. Section 6 provides a discussion about the proposed model, the implications for practitioners, as well as its limitations. Finally, the paper ends with a summary and a look at further research needed in the future.

2. Research Methodology

In this work, we applied a research approach in three phases (see also Figure 1): (i) in the first phase a literature review was conducted in 2018 and Industry 4.0 concepts and technologies were identified; (ii) in the second phase the assessment tool for SMEs was developed in 2018–2019; (iii) in the third phase the developed assessment model was tested and validated by using a field study with industrial companies in 2019.

In the first phase (see also Section 3), we conducted a literature review to analyze already existing assessment models and their suitability for SMEs. In addition, we looked into the scientific literature for Industry 4.0 concepts and technologies to assess their implementation in SMEs in the assessment model.

In the second phase (see also Section 4), we developed the maturity level-based assessment tool for Industry 4.0 to be used in SMEs. In the first step, we defined the maturity levels for each of the Industry 4.0 concepts and technologies. The assessment model foresees the evaluation of the current implementation of Industry 4.0 in SMEs as well as the future target implementation. The difference between current and target state defines the gap that needs to be overcome by SMEs. Further, the assessment model should include also the assessment of the potential of each concept/technology for the single SME. Based on the gap and the potential of Industry 4.0 concepts/technologies, a norm strategy matrix indicates to SMEs a meaningful prioritization of final implementation measures.

In the third phase (see also Section 5), we conducted a field study with 17 industrial companies from Italy, Austria, Slovakia, and the United States to test and validate the developed assessment model. First, we will describe the structure of the field study describing the dimensions of the analyzed companies, their origin, as well as the approach of the field study. Then we present the results of the field study for the application of the assessment model and discuss the lessons learned to focus on feedback from SMEs.

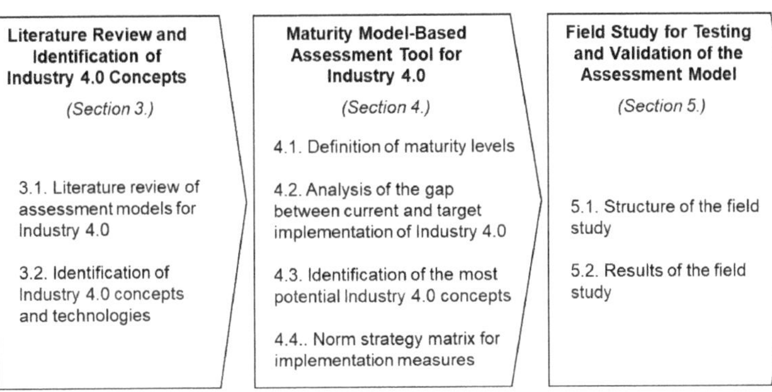

Figure 1. Research methodology used in this work.

3. Literature Review and Identification of Industry 4.0 Concepts and Technologies

3.1. Overview of Existing Assessment and Maturity Models for SMEs

The analysis of existing assessment and/or maturity models for Industry 4.0 was based on a literature search in SCOPUS database, as this is one of the leading and most complete scientific databases for industrial and manufacturing engineering. The research team used for the literature analysis the following research query: (TITLE-ABS-KEY ("Industry 4.0") AND TITLE-ABS-KEY (assessment OR maturity) AND TITLE-ABS-KEY (sme OR "small and medium sized")) AND (LIMIT TO (LANGUAGE,"English")), with an output of 32 search hits. In the search, only works in the English language were considered, excluding 1 work in German. The remaining 31 works were published mainly as conference papers (22 works) and conference reviews (2 works), while 7 works were articles/reviews in journals and 1 had been published as a book chapter. This shows that the development of assessment models for Industry 4.0 in SMEs is a quite new topic and is already mainly discussed in scientific conferences and not yet so much in peer-reviewed journals. Also, the year of publication confirmed this as 2 works were published in 2016, 14 works in 2018 and 16 works in 2019. In the first screening of title and abstract, a total of 19 works were encoded as firmly pertinent. In a second screening reading all remaining works, a total of 13 research papers were considered for a further content analysis.

In the following, a summary of the content analysis of the remaining works (see Table 1) will be given to explain the current status of assessment and maturity models with a special focus on SMEs.

The first group of works focused on maturity-based assessment models determining the maturity/implementation of concepts and technologies from Industry 4.0. The authors in [15] were focused more on the maturity levels of the general implementation of an Industry 4.0 strategy but did not go into detail regarding single Industry 4.0 concepts. The model in [16] was based on five maturity levels (novice, beginner, learner, intermediate, expert) as well as five dimensions (finance, people, strategy, process, product). Also in this model, the assessment provided only a rough overview of seven categories of Industry 4.0 so-called toolboxes (manufacturing/fabrication, design and simulation, robotics and automation, sensors and connectivity, cloud/storage, data analytics, and business management), but not of the single Industry 4.0 concepts and technologies. The authors in [17] presented a study of the maturity of Industry 4.0 in German companies, focusing on six dimensions (product development process, steering, and control, manufacturing and operation, smart services, process organization, big data). The authors in [18] presented a maturity level assessment with five fields of action and 29 subordinated action elements. The action elements showed only areas of implementation (e.g., production logistics or communication), thus the results gave only an overview of a general assumption of the implementation of Industry 4.0 in each action element, but not based on single Industry 4.0 concepts and technologies. In [19] the authors proposed a maturity model based on five dimensions (strategy, technology, production, products, people) and a Likert scale from 1 to 5. The study used a Likert scale from 0 to 6. In [20] the authors used five dimensions (strategy, people, processes, technology integration) with a Likert scale from 1 to 5. Also in the latter cases, the maturity assessment did not go in deep on the concept and technology level, but instead remained on a very superficial level of exploration.

The second group of works concentrated on the assessment model for evaluating the level of readiness of Industry 4.0. [21] determined the readiness for Industry 4.0 based on six dimensions (1—strategy and organization; 2—smart factory; 3—smart products; 4—data-driven services; 5—smart operations; 6—employees) based on a Likert scale from 0 to 5. The works did not provide any further insight into the readiness regarding detailed Industry 4.0 concepts. The work described in [22] presented a readiness self-assessment for craftsmanship companies with the three dimensions production/operations, digitalization, and ecosystem. In these three dimensions, a total of 23 items (e.g., data security, perception of digitalization, quality of internet connection) were assessed with a Likert scale from 1 to 5. Also, in [23] the model provided only a rough overview of two dimensions (smart factory, strategy, and culture). The authors in [24] provided a readiness assessment based on five dimensions (manufacturing and operations, people capability, technology-driven process, digital support, and business and organization strategy) and 43 subdimensions with a Likert scale from 0 to 4. Summarizing, most of the models discussed did not go very in depth and followed a different scope to the one in this work as they assessed the readiness for Industry 4.0 and not the maturity of applied Industry 4.0 concepts and technologies.

The third group of works dealt with a specific assessment model for Industry 4.0 in SMEs. In [25] the following process modules of intralogistics were assessed: incoming goods, internal transport, storage, order picking, packaging, and outgoing goods. In [26], four dimensions (data, communication, processes, and intellectual capital) were assessed based on a Likert scale from 1 to 5. A characteristic of this work was that each maturity level was described in detail with examples, which makes it easier for SMEs to validate the right maturity level. The work in [27] referred to nine characteristics of SMEs (e.g., ability to produce customized products) and assessed the maturity of lean and Industry 4.0 components). In summary, this group of assessment models dealt only with specific areas of an SME, like logistics, or with additional concepts from lean manufacturing and is therefore not suitable for the scope of the work proposed in this article.

Table 1. Overview of existing assessment and maturity models for small and medium-sized enterprises (SMEs).

No	Ref.	Authors	Year	Title	Assessment of
1	[15]	Ganzarain J., Errasti N.	2016	Three-stage maturity model in SMEs towards industry 4.0	maturity
2	[16]	Mittal S., Romero D., Wuest T.	2018	Towards a smart manufacturing maturity model for SMEs (SM3E)	maturity
3	[17]	Bittighofer D., Dust M., Irslinger A., Liebich M., Martin L.	2018	State of Industry 4.0 Across German Companies	maturity
4	[18]	Puchan J., Zeifang A., Leu J.-D.	2019	Industry 4.0 in Practice-Identification of Industry 4.0 Success Patterns	maturity
5	[19]	Trotta D., Garengo P.	2019	Assessing Industry 4.0 Maturity: An Essential Scale for SMEs	maturity
6	[20]	Pirola F., Cimini C., Pinto R.	2019	Digital readiness assessment of Italian SMEs: a case-study research	maturity
7	[21]	Hamidi S.R., Aziz A.A., Shuhidan S.M., Aziz A.A., Mokhsin M.	2018	SMEs maturity model assessment of IR4.0 digital transformation	readiness
8	[22]	Sheen D.-P., Yang Y.	2018	Assessment of Readiness for Smart Manufacturing and Innovation in Korea	readiness
9	[23]	Chonsawat N., Sopadang A.	2019	The development of the maturity model to evaluate the smart SMEs 4.0 readiness	readiness
10	[24]	Brozzi R., D'Amico R.D., Pasetti Monizza G., Marcher C., Riedl M., Matt D.	2018	Design of self-assessment tools to measure industry 4.0 readiness. A methodological approach for craftsmanship SMEs	readiness
11	[25]	Schiffer M., Wiendahl H.-H., Saretz B.	2019	Self-assessment of Industry 4.0 Technologies in Intralogistics for SME's	specific
12	[26]	Krowas K., Riedel R.	2019	Planning Guideline and Maturity Model for Intra-logistics 4.0 in SME	specific
13	[27]	Kolla S., Minufekr M., Plapper P.	2019	Deriving essential components of lean and industry 4.0 assessment model for manufacturing SMEs	specific

Most of the authors used a Likert scale from 1 to 5 for measuring the maturity of Industry 4.0, which was applied also in our work. All works did not go into detail about the single Industry 4.0 concepts and technologies and evaluated only dimensions or fields of application. As one of the most difficult tasks for SMEs is to understand what kind of concepts belong to Industry 4.0, we maintained it was important to develop a much more detailed assessment model. The evaluation might be difficult for SMEs if only a Likert scale with a vague description is shown (e.g., from beginner to expert). We used here the procedure as shown in [26], where each maturity level is well described with examples to simplify the use of the assessment model for SMEs. Further, in all the shown approaches, an evaluation to assess also a target maturity level for SMEs was missing. Not all SMEs need the maximum level of Industry 4.0 for their purposes, thus the maximum maturity level should not be seen as something that must be attained at all costs. In addition, the analyzed assessment models do not show any procedure to rank the potential implementation steps or technologies for a single company. All these points should be included in the new proposed assessment model.

3.2. Identification of Industry 4.0 Concepts and Technologies

To design an assessment model for maturity in the application of Industry 4.0 technologies and concepts, the assessment units must first be identified and defined. As mentioned before in Section 3.1, the aim is to provide a comprehensive method catalog with Industry 4.0 concepts and technologies to support SMEs in the selection of measures to be implemented. Therefore, the research team carried out a systematic literature analysis with the keyword "Industry 4.0" in the database SCOPUS in the first year of the research project. From initially 733 works a large number of publications were excluded based on exclusion criteria like type of paper (only journal articles, reviews, and articles in press were considered), the cover period (from 2011 to 2017), the language (only works in English), the subject area (only papers belonging to engineering, computer science, business, materials science, social sciences, decision sciences, management and accounting, energy, econometrics and finance, multidisciplinary, economics and psychology were considered in the search) were analyzed. The remaining 102 works were screened by three independent researchers to guarantee objectivity in the analysis. The screening process resulted in a total of 27 publications that were defined as firmly pertinent for further analysis, which means that they provided clear information about Industry 4.0 technologies and concepts.

These identified works formed the basis for a subsequent content analysis, from which Industry 4.0 concepts and technologies could be extracted. The content analysis resulted in a first step in 75 Industry 4.0 topics that were condensed by the research team into 42 meaningful Industry 4.0 concepts and technologies used as a basis for the proposed assessment model. The reason for the adopted consolidation lay in the fact that many of the identified Industry 4.0 topics were characterized by a subject or industry-specific nature with the need for generalization.

The broad spectrum of the identified Industry 4.0 concepts and technologies required structuring into several levels of Industry 4.0 dimensions (see also Table 2). In the first dimension level the concepts were classified as follows:

- Operations (concepts/technologies for production and operational processes)
- Organization (concepts/technologies for organizational and management-oriented processes)
- Socio-Culture (concepts/technologies related to corporate culture and employee-related topics)
- Technology (data and process-driven technologies).

In the second dimension level, a total of 21 subdimensions were identified. Table 2 summarizes the identified Industry 4.0 concepts and technologies, including dimension level I and II in alphabetical order.

Table 2. Identified and generalized Industry 4.0 concepts and technologies.

No	Dimension Level I	Dimension Level II	Industry 4.0 Concepts and Technologies
1	Operations	Agile Manufacturing Systems	Agile Manufacturing System
2			Self-Adapting Manufacturing Systems
3			Continuous and Uninterrupted Material Flow Models
4			Plug and Produce
5		Monitoring & Decision Systems	Decision Support Systems
6			Integrated and Digital Real-Time Monitoring Systems
7			Remote Monitoring of Products
8		Big Data	Big Data Analytics
9		Production Planning and Control	Enterprise Resource Planning / Manufacturing Execution System
10	Organization	Business Model 4.0	Digital Product-Service Systems
11			Servitization and Sharing Economy
12			Digital Add-on or Upgrade
13			Digital Lock-In
14			Freemium
15			Digital Point of Sales
16		Innovation strategy	Open Innovation
17		Strategy 4.0	Industry 4.0 Roadmap
18		Supply Chain Management 4.0	Sustainable Supply Chain Design
19			Collaboration Network Models
20	Socio-Culture	Human Resource 4.0	Training 4.0
21		Work 4.0	Role of the Operator
22		Culture 4.0	Cultural Transformation
23		Big Data	Cloud Computing
24		Communication & Connectivity	Digital and Connected Workstations
25			E-Kanban
26			IoT and Cyber-Physical Systems
27	Technology	Cyber Security	Cyber Security
28		Deep Learning, Machine Learning, Artificial Intelligence	Artificial Intelligence
29			Object Self Service
30		Identification and Tracking Technology	Identification and Tracking Technology
31		Additive Manufacturing	Additive Manufacturing (3D Printing)
32		Maintenance	Predictive Maintenance
33			Telemaintenance
34		Robotics & Automation	Automated Storage Systems
35			Automated Transport Systems
36			Automated Manufacturing/Assembly
37			Collaborative Robotics
38			Smart Assistance Systems
39		Product Design and Development	Product Data Management and Product Lifecycle Management
40		Standards 4.0	Cyber-Physical System Standards
41		Virtual Reality, Augmented Reality, and Simulation	Virtual and Augmented Reality
42		Virtual Reality, Augmented Reality, and Simulation	Simulation

4. Maturity Level-Based Assessment Tool of Industry 4.0 for SMEs

4.1. Maturity Levels Used in the Assessment Model

First of all, SMEs have to insert in the assessment model their general data, like operating sector, number of employees, annual revenue, and balance sheet total. These data are used in a later stage to compare their own results with the average of other SMEs of the same size and the same sector. In the next step, SMEs have to define the maturity level of the identified Industry 4.0 concepts and technologies (see Section 3.2) in their own company. As already seen in Section 3.1, many assessment models use such maturity levels to express the progress of implementation of Industry 4.0 in their own company. We wanted to apply the same approach also in our assessment model for Industry 4.0 in SMEs. Each level should specify the peculiarities and the progress of the evaluated enterprises or desires to achieve to become a smart enterprise.

As already mentioned in Section 3.1, we wanted to adopt also a Likert scale with five maturity levels. According to the results of the review of maturity level-based assessment models, it can be confirmed that five stages are most common and suitable to map the implementation or maturity of Industry 4.0 concepts. To facilitate the application of the assessment model in SMEs, the maturity levels are expressed and described not only by numbers from 1 to 5 but with a combination of a single term and a brief statement/example. First, tests with an SME have shown that persons asked to do the assessment based only on a Likert scale from 1 to 5 (therefore described only by numbers) have difficulties deciding which is the right value for the stage of implementation in their company as for many Industry 4.0 concepts they have no experience with what could be the lowest stage or the highest stage of implementation. Next, tests that added a term to the numbers facilitated the assessment, but still caused uncertainty (and therefore a lower quality of the assessment data) due to many new and innovative Industry 4.0 concepts where people needed some more explanation and, if possible, also an example. The explained maturity levels are illustrated in Figure 2 as follows.

INDUSTRY 4.0 CONCEPT	Maturity Level 1	Maturity Level 2	Maturity Level 3	Maturity Level 4	Maturity Level 5
Remote Monitoring of Products	**Products Are Not Monitored** */products are not monitored after delivery*	**Spotwise Product Checks** */products are monitored spotwise by the customer or a sales agent or a technician*	**Periodic Product Checks** */products are monitored by the manufacturer through periodic condition checks*	***Remote Product Monitoring*** */products are digitally monitored by the manufacturer through remote access*	**Remote Product Control** */products are monitored and controlled through remote access*
Big Data Analytics	**No Data Analytics** */no use of existing data*	**Manual Data Analytics of Existing Data** */minimal use of existing data based on Excel or similar*	**Big Data Projects** */collect data in a structured way and perform big data analytics projects through external experts*	**Big Data Analytics Tools** */collect and analyze production and logistics data for process optimization with big data analytics tools*	**Internal Big Data Analytics Experts** */professional application of big data analytics through skilled internal experts (production data analysts)*
ERP/MES	**No ERP system**	**ERP System** */ERP system implemented*	**ERP and PPC System** */production Planning and Control system used for material requirement planning*	**MES Implementation** */MES system or similar implemented but not integrated with ERP*	**ERP/MES Integration** */ERP and MES are integrated and communicate with each other*
Digital Product-Service Systems	**Only Physical Product**	**Maintenance Business Models** */maintenance services sold together with product*	**Product–Service System** */extended services sold together with product*	**Digital Product–Service Systems** */digital services sold together with the product*	**Web/Cloud-based Product–Service Architectures** */digital services available on via web, app or cloud*
Servitization and Sharing Economy	**Ownership Based Business Model** */customer buys physical product (Ownership)*	**Leasing Based Business Model** */customer pays the leasing rate to get ownership*	**Rental Based Business Model** */customer pays a rental rate (no ownership intended)*	**Servitization Model** */customer pays for the service*	**Sharing Economy Platform Model** */customers share access to products or services with other customers*

Figure 2. Exemplary representation of maturity levels of Industry 4.0 concepts.

4.2. Analysis of the Current Status of Implementation and the Target Status

The company completes the assessment of the current status regarding the implementation of Industry 4.0 concepts by selecting one of the maturity levels shown above. In our assessment model, this maturity level is called the 'firm's I4.0 score'. In addition, the company also indicates a degree of maturity that it would like to achieve in the medium term, and where the firm also realistically sees the feasibility of implementation. This objective of a future maturity level is called the 'target level'.

These two values are the basis for the calculation of the 'I4.0 gap' that describes the difference between target level and the firm's I4.0 score. This gap should act as a measure of the difficulty for the SME to achieve the future target level of Industry 4.0 for each single Industry 4.0 concept.

Figure 3 shows a screenshot of the MS Excel-based maturity level-based assessment tool with the fields to be filled in by the SME to determine the firm's I4.0 score and the target level. For the development of a prototype of the assessment model, MS Excel was chosen as it is widely spread in SMEs and easy to use also for nonexpert users of MS Excel. In the final stage of the project a web-based self-assessment model is developed to maximize dissemination.

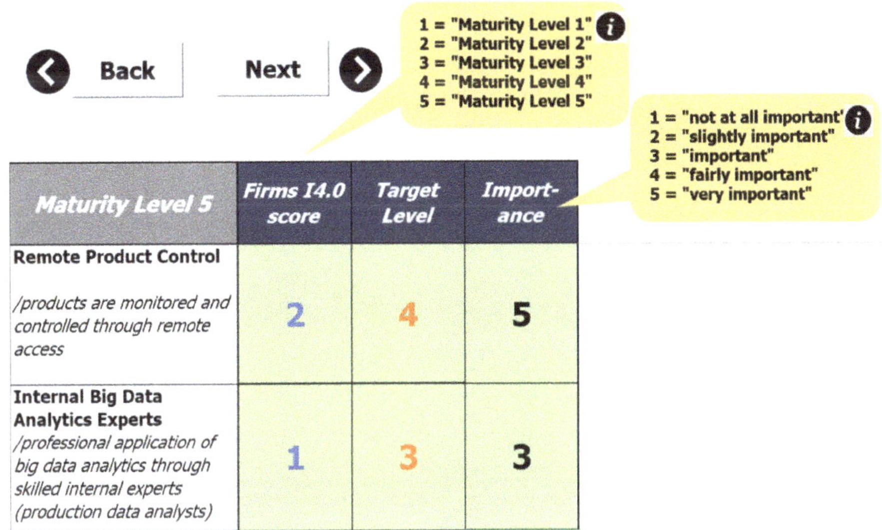

Maturity Level 5	Firms I4.0 score	Target Level	Import- ance
Remote Product Control */products are monitored and controlled through remote access*	2	4	5
Internal Big Data Analytics Experts */professional application of big data analytics through skilled internal experts (production data analysts)*	1	3	3

Figure 3. Fields to be filled in for firm's I4.0 score, target level, and the importance of each Industry 4.0 concept.

Figure 4 shows an example of the visualization of the firm's I4.0 score and the target level automatically generated by the MS Excel prototype of the assessment model. The representation in the radar chart is intended to simplify the evaluation of the assessment for SMEs, as the gap between today and tomorrow is directly visible.

The radar charts illustrate the firm's Industry 4.0 score as well as the firm's target score for each Industry 4.0 concept categorized in an operational I4.0 level, an organizational I4.0 level, a social–cultural I4.0 level, as well as a technological I4.0 level (further subdivided into data-driven and process-driven technological level).

In the visual representation of the results, the technology-related Industry 4.0 concepts are subdivided into a group of data-driven technologies and process-driven technologies (see Table 3).

Table 3. Data-driven and process-driven technologies.

Data-driven Technologies	Process-driven Technologies
Cloud Computing	Additive Manufacturing (3D-Printing)
Digital and Connected Workstations	Predictive Maintenance
E-Kanban	Telemaintenance
Internet of Things and Cyber-Physical Systems	Automated Storage Systems
Cyber Security	Automated Transport Systems
Artificial Intelligence	Automated Manufacturing & Assembly
Object Self Service	Collaborative Robotics
Identification and Tracking Technology	Smart Assistance Systems
	Product Data Management and Product Lifecycle Management
	Cyber-Physical Systems Standards
	Virtual and Augmented Reality
	Simulation

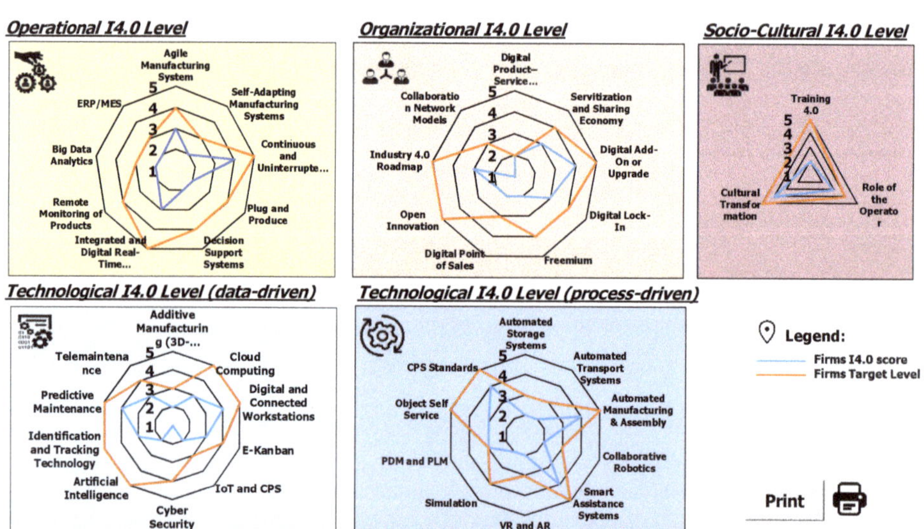

Figure 4. Exemplary visualization of the gap between firm's I4.0 score and the target level.

4.3. Identification and Ranking of Industry 4.0 Concepts and Technologies with the Highest Potential in Implementation

In the next step, the company is asked to provide an assessment of the potential ('importance') of the respective Industry 4.0 concept (see also Figure 3). Also, this evaluation is based on a 5-step Likert scale and is meaningful according to the authors, because not every Industry 4.0 concept might be important for the individual company. While for example, the traceability of products is essential and of the highest importance for certain companies and industries (e.g., automotive sector, food sector), this concept will play a much less important role in many other SMEs.

In this respect, the estimation of the importance is used to express the potential of each Industry 4.0 concept for the individual SME. To prioritize the Industry 4.0 concepts and thus to define the basis for implementation measures, the assessment tool automatically generates a ranking based on the identified potential (see Figure 5).

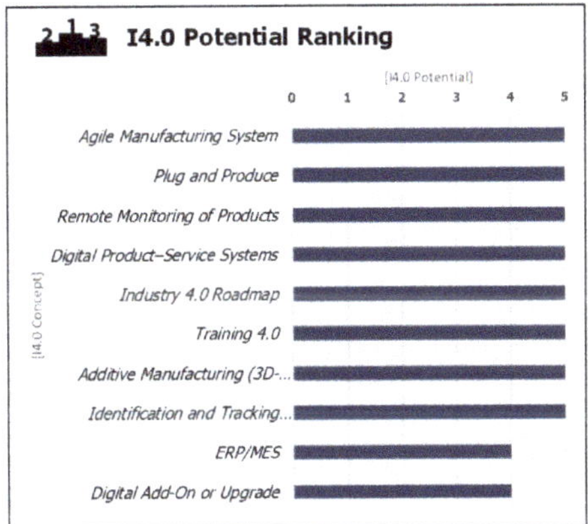

Rank	I4.0 Concept	I4.0 Potential
1	Agile Manufacturing System	5
2	Plug and Produce	5
3	Remote Monitoring of Products	5
4	Digital Product–Service Systems	5
5	Industry 4.0 Roadmap	5
6	Training 4.0	5
7	Additive Manufacturing (3D-Printing)	5
8	Identification and Tracking Technology	5
9	ERP/MES	4
10	Digital Add-On or Upgrade	4

Figure 5. Exemplary visualization of the potential for each Industry 4.0 concept.

4.4. Norm Strategy Matrix for Defining Implementation Measures

Not all of the 42 Industry 4.0 concepts listed will be of high potential and not all of them require the same amount of effort and time for implementation. Therefore it makes sense, especially for SMEs with financial and personnel limits, to strive for a gradual implementation of the Industry 4.0 concepts. Both criteria, namely the gap and the identified potential, are therefore combined in a norm strategy matrix as shown in Figure 6, to facilitate the selection and scheduling of implementation measures for Industry 4.0 in terms of time (short-term vs. medium-term implementable I4.0 concepts). Figure 6 shows the matrix schematically with the following standardization strategies depending on the respective quadrant in the diagram:

- Quick Wins (high potential–low gap)
- Must Haves (high potential–high gap)
- Low Hanging Fruits (low potential–low gap)
- Money Pits (low potential–high gap).

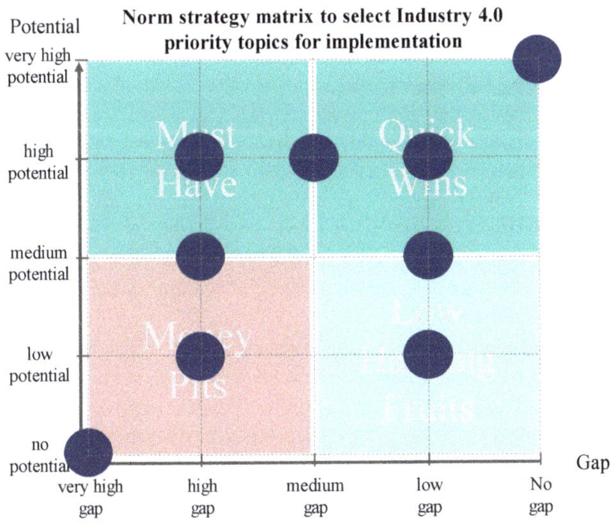

Figure 6. Norm strategy matrix to facilitate the selection and definition of Industry 4.0 measures/projects.

Depending on the position of the Industry 4.0 concepts, the SMEs can then define whether, and to what extent and in what time frame, implementation measures have to be defined.

5. Validation of the Industry 4.0 Assessment in a Field Study with SMEs

5.1. Structure and Procedure of the Field Study

A number of 17 SME companies operating in different industrial sectors from Italy (10), Austria (3), Slovakia (1), and the USA (3) participated in the field study. The participating companies in the field study were mainly operating in industrial goods manufacturing (6), industrial supplies and materials (4), construction (3), food (2), and other industries (2). From these companies, seven were small enterprises with up to 49 employees and 10 were medium-sized enterprises with 50–249 employees. The companies had been collaborating with the research team in the H2020 research project "SME 4.0—Industry 4.0 for SMEs". In the case study companies, the assessment was completed by the company owner/manager or the head of production. In most of the cases, they called for specific concepts experts from product development, IT department, or other operational departments. To identify difficulties in completing the assessment, a representative of the research team was always present. This was done to monitor the time spent on completing the assessment, to record difficulties in completing the assessment, and then to ask the companies for their feedback on the user-friendliness of the assessment. At no time did the researcher present intervene in the evaluation or influence the participant in the assessment.

All participants assessed the firm's I4.0 score, the target level and the importance of the respective Industry 4.0 concepts for their company. As a result and for their benefit, they learned possible Industry 4.0 concepts and technologies, they got an overview of their status and the gap for each concept, and they received a visual ranking of the most potential concepts for their firm.

5.2. Results of the Field Study

In general, the assessed importance/potential of each single Industry 4.0 concept was analyzed in the form of the average value, the respective standard deviation, and the coefficient of variation as a measure of relative variability of the answers as well as the number of answers. This analysis was used by the research team to identify suitable/potential Industry 4.0 concepts from the viewpoint of SMEs. These results are summarized and presented in [28].

In this work, we were more interested in the overall results of the field study and the feedback from the participating persons using the assessment model. Therefore we analyzed the data of the field study to investigate in the first step the maturity level and target level for Industry 4.0 concepts according to the participating SMEs (see Figure 7).

The overall results showed that the current maturity level of Industry 4.0 concepts in SMEs is generally very low. Comparing the single categories and I4.0 concepts depicted in Figure 7 we can see that the 10 concepts with the highest Firms I4.0 score were (from highest to lowest value): (1) agile manufacturing, (2) cloud computing, (3) product data management (PDM) and product lifecycle management (PLM), (4) enterprise resource planning (ERP) and manufacturing execution system (MES), (5) role of the operator, (6) simulation, (7) automation in manufacturing and assembly, (8) integrated and digital real-time monitoring systems, (9) collaboration network models, and (10) self-adapting manufacturing models.

The 10 Industry 4.0 concepts with the highest target level were (from highest to lowest value): (1) integrated and digital real-time monitoring systems, (2) agile manufacturing systems, (3) ERP/MES, (4) Industry 4.0 roadmap, (5) cloud computing, (6) digital and connected workstations, (7) cultural transformation, (8) PDM and PLM, (9) big data analytics, and (10) training 4.0.

Further, it is interesting to know those Industry 4.0 concepts that showed the greatest gap between the current level and the target level (from highest to lowest value): (1) integrated and digital real-time monitoring systems, (2) Industry 4.0 roadmap, (3) Training 4.0, (4) cultural transformation, (5) digital and connected workstations, (6) ERP/MES, (7) internet of things and cyber-physical systems, (8) decision support systems, (9) predictive maintenance, and (10) big data analytics.

In the next step, we analyzed the data to determine the average potential of the Industry 4.0 concepts according to the participating SMEs in the field study (see Figure 8).

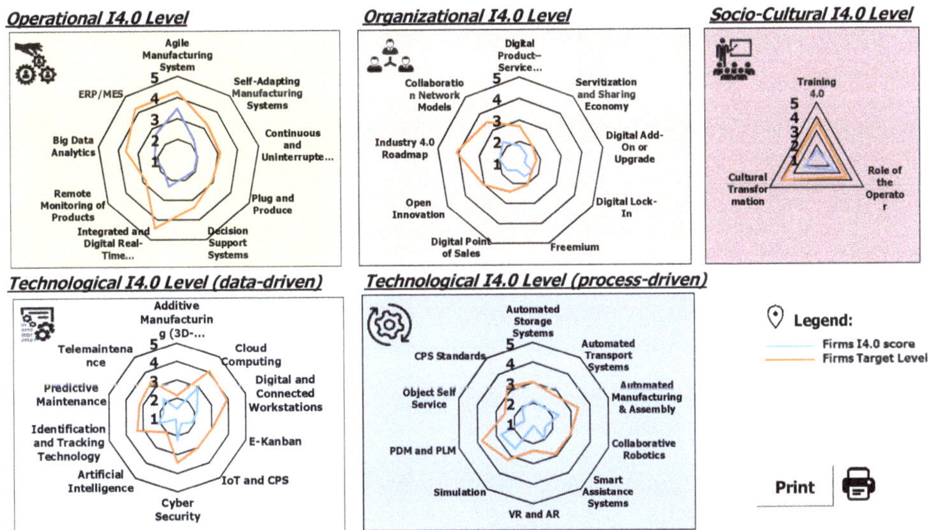

Figure 7. Visualization of average Firms I4.0 Score and Target level of SMEs in the field study.

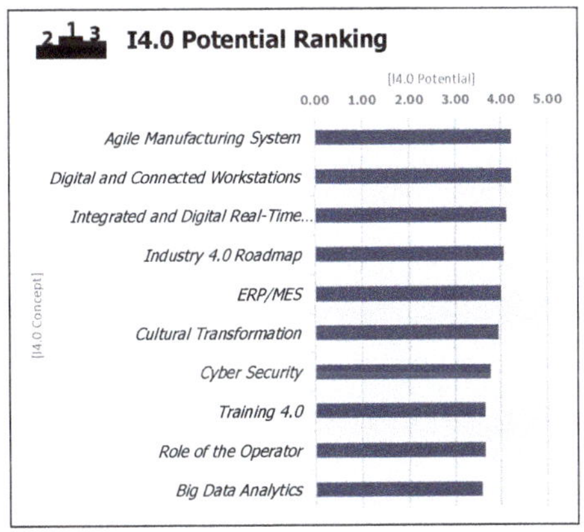

Rank	I4.0 Concept	I4.0 Potential
1	Agile Manufacturing System	4.24
2	Digital and Connected Workstations	4.24
3	Integrated and Digital Real-Time Monitoring Systems	4.12
4	Industry 4.0 Roadmap	4.06
5	ERP/MES	4.00
6	Cultural Transformation	3.94
7	Cyber Security	3.76
8	Training 4.0	3.65
9	Role of the Operator	3.65
10	Big Data Analytics	3.59

Figure 8. Ranking of the average potential of Industry 4.0 concepts according to SMEs in the field study.

Based on the evaluations of the companies in the field study the most promising Industry 4.0 concepts are the ones illustrated in Figure 8. Interestingly, most of these concepts were also evaluated very highly when companies had to determine the current maturity level and the target level. This led to the conclusion that the most promising concepts are not completely new to them and that they are already targeting to increase their competences in these fields. The only concept that was not mentioned before in Figure 7 was cybersecurity. This led to the hypothesis that SMEs are already aware of the future importance of cybersecurity, but they do not feel they are at a good level and also fail to raise their level in this area.

In the third step of the field study, we combined the average data of the Industry 4.0 gap and the identified potential to generate the norm strategy matrix (see Figure 9). In principle, it can be seen that none of the Industry 4.0 concepts mentioned had a low potential and a large gap in implementation. This means that it makes sense to introduce all of the 42 concepts in the long term. There are a few so-called "low hanging fruits" that have a moderate potential with little effort for introducing the target level (which might be high or moderate according to the needs of SMEs). Examples of such concepts are additive manufacturing, servitization and sharing economy, digital upgrade, freemium, or artificial intelligence. It might sound surprising that a technology like artificial intelligence is clustered as a low hanging fruit technology. Here, it must be specified that SMEs in this field study defined a relatively low target level for artificial intelligence. The matrix also showed some concepts that can only be implemented with great effort, but which require a long-term "must have" strategy due to their high potential. Examples for this norm strategy are big data analytics, predictive maintenance, collaborative robotics, real-time monitoring systems, digitalized workplaces, improvement of their ERP, the introduction of MES systems, as well as a cultural change of the company towards developing an Industry 4.0 roadmap and qualifying their people with Training 4.0. The large number of Industry 4.0 concepts mentioned fall within the norm strategy "quick wins", which means that the effort for introduction is relatively moderate, but the concepts show a high potential. Examples of concepts where it makes sense to apply this strategy are agile manufacturing systems, automation, e-kanban, operator 4.0, simulation, cloud computing, smart assistance systems, as well as appropriate cybersecurity solutions for SMEs.

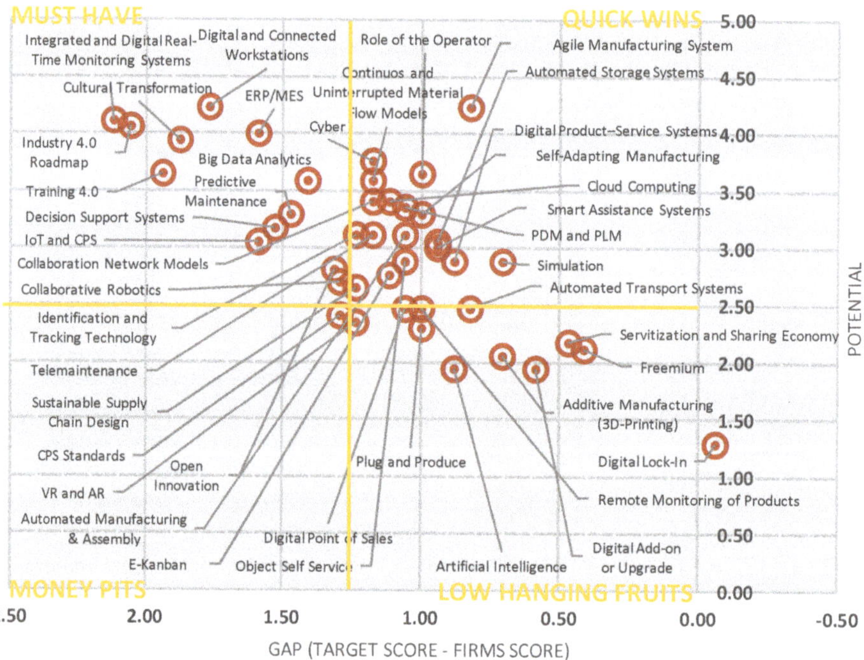

Figure 9. Norm strategy matrix with average values of the Industry 4.0 gap and the potential in the field study.

The norm strategy matrix illustrated in Figure 9 supports SMEs to plan short-term, medium-term, as well as long-term-oriented measures for introducing Industry 4.0 in their company. Thus, the application of the assessment model with a subsequent definition of norm strategies is recommended.

Finally, we asked the participating SMEs about the usability of the assessment model and feedback for improvement. All participants found the assessment to be very instructive and a good exercise for thinking about the status of Industry 4.0 in their own company. Concerning the number of items to be evaluated, the participants were not all of the same opinion. Most felt it very positive that an Industry 4.0 concept catalog was provided to get a better overview of existing Industry 4.0 concepts. However, the catalog of 42 Industry 4.0 concepts requires a lot of time to fill out the assessment, which was considered quite exhausting by the participants. According to the participants, there were no major difficulties in assigning the level of maturity due to the detailed description and examples provided in the assessment model. The companies found it very useful to have an overview in the form of an individual norm strategy matrix at the end of the assessment, as this supported the definition of future Industry 4.0 measures and projects. All in all, the participants found the assessment to be effortful to conduct, but extremely useful for paving the way to implement Industry 4.0 in a systematic manner in the company.

6. Discussion

6.1. Novel Aspects of the Proposed Assessment Model

As described in the introduction and review in Section 3.1, our assessment model differs from existing models. While general assessment models [13–15] (developed for both SMEs and large companies) are often limited to a general Likert scale assessment, our model offers a significant advantage for smaller companies. The additional detailed description of each maturity level for each Industry 4.0 concept (including practical examples) makes it easier for SMEs to determine their own maturity level. This assistance is particularly important as many SMEs do not have highly qualified employees who have experience with all concepts. Another advantage over all previous models is the depth and number of Industry 4.0 concepts that can be evaluated, which means that many companies doing the assessment even start to think about certain perhaps less well-known concepts. Finally, we saw the subsequent derivation of the norm strategy matrix as a decisive difference and novelty of our model, since it is a practical aid especially for SMEs in the selection and introduction of important Industry 4.0 concepts for each individual firm.

6.2. Implications for Academia and Practitioners

For research purposes, the presented assessment model represents an extension and enrichment of already existing assessment models for Industry 4.0. On the one hand, the proposed model is designed for use in SMEs and has also been developed and tested with the involvement of SMEs. On the other hand, the implementation of Industry 4.0 in the whole company or parts of the company is not just evaluated superficially, but existing Industry 4.0 concepts are specifically evaluated on a very detailed level. On the one hand, this type of evaluation goes much deeper than all other currently existing models and at the same time offers a catalog of 42 Industry 4.0 concepts, which have been determined based on scientific contributions. For future research, the assessment model offers the possibility to examine the implementation of the Industry 4.0 concepts in detail. If the assessment is carried out in the future with a sufficient number of SMEs from different industrial sectors, these data can be used for many purposes. For example, it will be possible to determine which Industry 4.0 concepts are most important for different company sizes. In addition, this analysis can also be reduced to individual industrial sectors to understand where and why different dimensions of SMEs or different industrial sectors gave a different assessment. The same analysis can also be extended to the current and target maturity level. Finally, it is a new and original approach, as the assessment of an Industry 4.0 gap and its potential is combined in a norm strategy matrix.

For practitioners from SMEs, the assessment model is an ideal tool to evaluate the current situation of the company and to systematically plan future projects and initiatives. One of the biggest advantages for SMEs is the Industry 4.0 catalog with different concepts and technologies, which also helps small companies to get an overview of Industry 4.0 methods. The validation in practice has shown that the model can be applied in small companies, although the evaluation itself takes some time. The subsequent systematic analysis of the Industry 4.0 gaps and the potential in the norm strategy matrix especially serves SMEs as an orientation guide to better plan future initiatives and to transfer them into an implementation plan for Industry 4.0. The application of the assessment model provides companies practical recommendations for the application of selected Industry 4.0 concepts and how and with which priority they should be implemented.

6.3. Limitations of the Proposed Assessment Model

Despite the many advantages mentioned, the model has certain limitations. One limitation is certainly the already mentioned effort for the evaluation. The validation phase has shown that most companies spend about one to two hours on the evaluation and usually also bring in experts from different areas of the company. Therefore, the user must invest the necessary time to be able to draw the right conclusions from the results. Furthermore, it can be limited if only one person carries out the evaluation because a certain subjective opinion cannot be excluded. Therefore, it is recommended that a team completes the evaluation to improve not only the quality of the answers but also the objectivity of the result. A third limitation to be mentioned is the catalog of 42 Industry 4.0 concepts used as a basis. In the course of the four-year research project from 2017–2020, a literature analysis was conducted to identify existing Industry 4.0 concepts until 2017. A first check by the research team after the development of the model in 2018 and its validation in 2019 showed that the Industry 4.0 concepts are still up-to-date and can be used in this way. However, it cannot be excluded that a few Industry 4.0 concepts might be added in the meantime. The proposed assessment model can be applied to every kind of industry sector. In the sample of SMEs we used for validation there were many different industrial sectors and the model could be used successfully in each of the industries. Of course the proposed model provides a catalog of 42 Industry 4.0 concepts, which not all might be of interest in each sector. This can easily be handled as the companies can assess the importance for their individual company as very low.

7. Conclusions

Industry 4.0 is a chance for SMEs to achieve a new level of competitiveness. Many SMEs are already trying to implement Industry 4.0 [28], but there is still a lack of specific instruments for introducing them [29,30]. In this paper, a maturity level-based assessment model for SMEs is presented. The model is based on 42 Industry 4.0 concepts identified by literature analysis, which are rated on a Likert scale of 1 to 5. In addition to the current maturity level, a target level and the importance and potential of the Industry 4.0 concepts are evaluated by the user. The collected results show in a developed norm strategy matrix, which of the Industry 4.0 concepts should be approached immediately and which may need more time or are not immediately implemented due to a lower potential.

The proposed assessment model shows several advantages compared to already existing models. It provides a detailed overview of existing Industry 4.0 concepts and is very easy for SMEs to adopt. The evaluation of the maturity level is facilitated through a brief description of the five maturity levels for each of the 42 Industry 4.0 concepts. Further, the evaluation and combination of a target level and the potential of each of the Industry 4.0 concepts in the norm strategy matrix allow SMEs to plan and schedule the implementation of Industry 4.0 in a very systematic way.

Further research is planned in the development of a web-based version of the assessment model and its dissemination in SMEs (ongoing work). As soon as enough SMEs complete the assessment model online and allow the use and processing of the anonymized data the research team plans to use these data for further analysis and a benchmarking functionality. In its final version, the assessment model will also show SMEs their position for each Industry 4.0 concept compared to those of the average of companies of a similar size and the same industrial sector. This will allow SMEs to better identify their own competitive situation and researchers to better understand the implementation of Industry 4.0 for each company size and industrial sector as well as its development over the time.

Author Contributions: E.R. led the research team, developed the structure of the maturity level-based assessment model, and conducted the field study for testing and validation purposes; M.U. worked on the literature review and the development of the maturity level-based assessment model; R.A.R. and L.G. contributed to the literature review and screening process as well as in the content analysis; M.W. conducted the analysis of field study data; D.T.M. supervised the work as scientific coordinator of the research project. All authors have read and agreed to the published version of the manuscript.

Funding: This project has received funding from the European Union's Horizon 2020 R&I programme under the Marie Skłodowska-Curie grant agreement No 734713.

Acknowledgments: This work belongs to the project "SME 4.0 – Industry 4.0 for SMEs" (funded in the European Union's Horizon 2020 R&I program under the Marie Skłodowska-Curie grant agreement No 734713). The authors would like to thank all industrial companies that contributed to the field study.

Conflicts of Interest: The authors declare no conflict of interest.

References

1. Matt, D.T. Design of lean manufacturing support systems in make-to-order production. *Key Eng. Mater.* **2009**, *410*, 151–158. [CrossRef]
2. Dallasega, P.; Rauch, E.; Matt, D.T. Sustainability in the supply chain through synchronization of demand and supply in ETO-companies. *Procedia CIRP* **2015**, *29*, 215–220. [CrossRef]
3. Kagermann, H.; Helbig, J.; Hellinger, A.; Wahlster, W. *Recommendations for Implementing the Strategic Initiative INDUSTRIE 4.0: Securing the Future of German Manufacturing Industry*; Final report of the Industrie 4.0 Working Group; Forschungsunion: Frankfurt am Main, Germany, 2013.
4. Zhou, K.; Liu, T.; Zhou, L. Industry 4.0: Towards future industrial opportunities and challenges. In Proceedings of the 12th International Conference on Fuzzy Systems and Knowledge Discovery (FSKD), Zhangjiajie, China, 15–17 August 2015; IEEE: Piscataway, NJ, USA, 2015; pp. 2147–2152.
5. Lee, J.; Bagheri, B.; Kao, H.A. A cyber-physical systems architecture for industry 4.0-based manufacturing systems. *Manuf. Lett.* **2015**, *3*, 18–23. [CrossRef]
6. Lee, J. Industry 4.0 in big data environment. *Ger. Harting Mag.* **2013**, *1*, 8–10.
7. Kang, H.S.; Lee, J.Y.; Choi, S.; Kim, H.; Park, J.H.; Son, J.Y.; Kim, B.H.; Do Noh, S. Smart manufacturing: Past research, present findings, and future directions. *Int. J. Precis. Eng. Manuf. Green Technol.* **2016**, *3*, 111–128. [CrossRef]
8. Thoben, K.D.; Wiesner, S.; Wuest, T. Industrie 4.0" and smart manufacturing-a review of research issues and application examples. *Int. J. Autom. Technol.* **2017**, *11*, 4–16. [CrossRef]
9. Rojas, R.A.; Rauch, E. From a literature review to a conceptual framework of enablers for smart manufacturing control. *Int. J. Adv. Manuf. Technol.* **2019**, *104*, 517–533. [CrossRef]
10. Hübner, M.; Liebrecht, C.; Malessa, N.; Kuhnle, A.; Nyhuis, P.; Lanza, G. A process model for implementing Industry 4.0-Introduction of a process model for the individual implementation of Industry 4.0 methods. *Wt Werkstattstech.* **2017**, *107*, 266–272.
11. Matt, D.T.; Rauch, E.; Riedl, M. Knowledge Transfer and Introduction of Industry 4.0 in SMEs: A Five-Step Methodology to Introduce Industry 4.0. In *Analyzing the Impacts of Industry 4.0 in Modern Business Environments*; Brunet-Thornton, R., Martinez, F., Eds.; IGI Global: Hershey, PA, USA, 2018; pp. 256–282.
12. Anderl, R.; Picard, A.; Wang, Y.; Fleischer, J.; Dosch, S.; Klee, B.; Bauer, J. Guideline Industrie 4.0—Guiding Principles for the Implementation of Industrie 4.0 in Small and Medium Sized Businesses. VDMA. Available online: https://industrie40.vdma.org/documents/4214230/0/Guideline%20Industrie%204.0.pdf/70abd403-cb04-418a-b20f-76d6d3490c05 (accessed on 12 March 2018).
13. Schumacher, A.; Erol, S.; Sihn, W. A maturity model for assessing Industry 4.0 readiness and maturity of manufacturing enterprises. *Procedia CIRP* **2016**, *52*, 161–166. [CrossRef]
14. Jentsch, D.; Riedel, R.; Jäntsch, A.; Müller, E. Factory audit for industry 4.0—Strategic approach for capability assessment and gradual introduction of a smart factory (in German: Fabrikaudit Industrie 4.0: Strategischer Ansatz zur Potenzialermittlung und schrittweisen Einführung einer Smart Factory). *Zwf Z. Für Wirtsch. Fabr.* **2013**, *108*, 678–681. [CrossRef]

15. Ganzarain, J.; Errasti, N. Three stage maturity model in SME's toward industry 4.0. *J. Ind. Eng. Manag.* **2016**, *9*, 1119–1128. [CrossRef]

16. Mittal, S.; Romero, D.; Wuest, T. Towards a Smart Manufacturing Maturity Model for SMEs. In *Advances in Production Management Systems*; Moon, I., Lee, G., Park, J., Kiritsis, D., von Cieminski, G., Eds.; Springer: Cham, Switzerland, 2018; Volume 536, pp. 155–163.

17. Bittighofer, D.; Dust, M.; Irslinger, A.; Liebich, M.; Martin, L. State of Industry 4.0 Across German Companies. In Proceedings of the IEEE International Conference on Engineering, Technology and Innovation 2018, Stuttgart, Germany, 17–20 June 2018; IEEE: Piscataway, NJ, USA, 2018; pp. 1–8.

18. Puchan, J.; Zeifang, A.; Leu, J.D. Industry 4.0 in practice-identification of industry 4.0 success patterns. In Proceedings of the IEEE International Conference on Industrial Engineering and Engineering Management (IEEM 2018), Bangkok, Thauiland, 16–19 December 2018; IEEE: Piscataway, NJ, USA, 2018; pp. 1091–1095.

19. Trotta, D.; Garengo, P. Assessing Industry 4.0 Maturity: An Essential Scale for SMEs. In Proceedings of the 8th International Conference on Industrial Technology and Management (ICITM 2019), Cambridge, UK, 2–4 March 2019; IEEE: Piscataway, NJ, USA, 2019; pp. 69–74.

20. Pirola, F.; Cimini, C.; Pinto, R. Digital readiness assessment of Italian SMEs: A case-study research. *J. Manuf. Technol. Manag.* **2019**. [CrossRef]

21. Hamidi, S.R.; Aziz, A.A.; Shuhidan, S.M.; Aziz, A.A.; Mokhsin, M. SMEs maturity model assessment of IR4. 0 digital transformation. In *Advances in Intelligent Systems and Computing, Proceedings of the 7th International Conference on Kansei Engineering and Emotion Research 2018, KEER 2018, Kuching, Malaysia, 19–22 March 2018*; Lokman, A., Yamanaka, T., Lévy, P., Chen, K., Koyama, S., Eds.; Springer: Singapore, 2018; Volume 739, pp. 721–732.

22. Sheen, D.P.; Yang, Y. Assessment of Readiness for Smart Manufacturing and Innovation in Korea. In Proceedings of the 2018 IEEE Technology and Engineering Management Conference (TEMSCON), Evanston, IL, USA, 28 June–1 July 2018; IEEE: Piscataway, NJ, USA, 2018; pp. 1–5.

23. Chonsawat, N.; Sopadang, A. The Development of the Maturity Model to evaluatethe Smart SMEs 4.0 Readiness. In Proceedings of the International Conference on Industrial Engineering and Operations Management, Bangkok, Thailand, 5–7 March 2019; IEOM: Bangkok, Thailand, 2019; pp. 354–363.

24. Brozzi, R.; D'Amico, R.D.; Monizza, G.P.; Marcher, C.; Riedl, M.; Matt, D. Design of Self-assessment Tools to measure industry 4.0 readiness. A methodological approach for craftsmanship SMEs. In Proceedings of the IFIP International Conference on Product Lifecycle Management 2018, Turin, Italy, 2–4 July 2018; Springer: Cham, Switzerland, 2018; pp. 566–578.

25. Schiffer, M.; Wiendahl, H.H.; Saretz, B. Self-assessment of Industry 4.0 Technologies in Intralogistics for SME's. In Proceedings of the IFIP International Conference on Advances in Production Management Systems 2019, Austin, TX, USA, 1–5 September 2019; Springer: Cham, Switzerland, 2019; pp. 339–346.

26. Krowas, K.; Riedel, R. Planning Guideline and Maturity Model for Intra-logistics 4.0 in SME. In Proceedings of the IFIP International Conference on Advances in Production Management Systems 2019, Austin, TX, USA, 1–5 September 2019; Springer: Cham, Switzerland, 2019; pp. 331–338.

27. Kolla, S.S.V.K.; Minoufekr, M.; Plapper, P. Deriving essential components of lean and industry 4.0 assessment model for manufacturing SMEs. *Procedia CIRP* **2019**, *81*, 753–758. [CrossRef]

28. Rauch, E.; Stecher, T.; Unterhofer, M.; Dallasega, P.; Matt, D.T. Suitability of Industry 4.0 Concepts for Small and Medium Sized Enterprises: Comparison between an Expert Survey and a User Survey. In Proceedings of the International Conference on Industrial Engineering and Operations Management, Bangkok, Thailand, 5–7 March 2019.

29. Saxena, P.; Papanikolaou, M.; Pagone, E.; Salonitis, K.; Jolly, M.R. Digital Manufacturing for Foundries 4.0. In *Light Metals 2020*; Springer: Cham, Switzerland, 2020; pp. 1019–1025.

30. Matt, D.T.; Rauch, E. SME 4.0: The Role of Small-and Medium-Sized Enterprises in the Digital Transformation. In *Industry 4.0 for SMEs*; Palgrave Macmillan: Cham, Switzerland, 2020; pp. 3–36.

 *sustainability*

Article

Industry 4.0—Awareness in South India

Leos Safar [1,*], Jakub Sopko [1], Darya Dancakova [1] and Manuel Woschank [2]

[1] Department of Banking and Investment, Faculty of Economics, Technical University of Košice, Nemcovej 32, 04001 Košice, Slovakia; jakub.sopko@tuke.sk (J.S.); darya.dancakova@tuke.sk (D.D.)

[2] Montanuniversitaet Leoben, Erzherzog-Johann-Strasse 3/1, 8700 Leoben, Austria; manuel.woschank@unileoben.ac.at

* Correspondence: leos.safar@tuke.sk

Received: 28 February 2020; Accepted: 3 April 2020; Published: 15 April 2020

Abstract: Industry 4.0 (I4.0) approaches, frameworks, and technologies have gained an increasing relevance in order to gain sustainable and competitive advantages for industrial enterprises and for small and medium enterprises (SMEs), as well. Contrary to previous studies, which are mainly focused on companies, we conducted a questionnaire-based survey on inhabitants, in an attempt to examine general awareness about I4.0 concepts, in the region of South India. Our findings revealed a rather poor informational level of I4.0 concept and its components, which consequently leads to inadequate future actions and expectations. Moreover, respondents with prior information about I4.0 framework tend to have rather positive opinions and expectations of possible future trends. We emphasize that insufficient knowledge of the potential workforce regarding I4.0 concepts, especially in a region with ascending demographic development, can be considered as one of the main barriers for a successful and sustainable future development towards the 4th industrial revolution.

Keywords: Industry 4.0; Internet of Things; India; SMEs; awareness

---

1. Introduction

In recent years, as a consequence of business and social evolution, a multitude of modern topics emerged, and therefore gained tremendous attention, in the areas of manufacturing, logistics, and organizational development of small and medium enterprises (SMEs), as a consequence of an ongoing business and social evolution. Megatrends, such as globalization, demographics' dynamics, mass customization, technological progress, or climate change, lead to tremendous challenges for both society and the business sector. As a reaction to this very complex and volatile business environment, various strategic initiatives have taken place all over the world, in order to keep pace with the exponential technological development and to achieve sustainable future growth, for example, the "Made in China 2025", Germany's "High Tech Strategy 2020", or the US's "Industrial Internet Consortium" [1]. The aim of these projects is to develop and implement future concepts, frameworks, and technologies (e.g., Internet of Things or Industry 4.0) in order to make industries more effective, competitive, sustainable, and to produce higher value added [2] while minimizing negative impact on environment. Considering the costs, Industry 4.0 initiatives could also improve enterprises' costs management [3–5].

In this study, the authors focus on Industry 4.0 (I4.0) as the most promising concept from both social and manufacturing–orientated point of view. Over the past years, we faced a strong advance of technology among almost all industrial sectors. New business propositions and applications within the business systems were enabled given the new technologies. As Thestrup et al. [6] stated, the systematic gathering of physical and virtual data from users, sensors, or devices, emerged. The so called "Internet of Things" (IoT) [7] was defined as a world-wide network of such objects communicating and operating through standardized communication protocols. However, IoT was recognized after the ITU1 report [8], describing IoT as ability of connecting everyday objects, meaning that people will be able to communicate with objects, the same as objects will be able to communicate among themselves. Prerequisite to such communication is advanced wireless technology (identification technologies and sensors). Logically, IoT can be diversified to Industrial IoT and Commercial IoT, while I4.0 expects all those parts to be interconnected and communicating.

To simplify, the goal of IoT infrastructure is to enable participants (people and objects) to be more flexible, to react appropriately and autonomously, thanks to an information sharing network. Harbor Research [9] suggests that two major strands of technological development emerged in the beginning of the 21st century; first is the mentioned IoT, and secondly, "Internet of People" (IoP, or social networking). These interconnected devices, processes, machines, products, etc., will have significant impact on enterprise's life cycle, efficiency, functioning, and consequently, on the broader economy [10].

To conclude, Sundmaeker et al. [11] defines the IoT as an integrated part of 'Future Internet', or a "dynamic global network infrastructure with self-configuring capabilities based on standard and interoperable communication protocols where physical and virtual 'things' have identities, physical attributes, and virtual personalities and use intelligent interfaces, and are seamlessly integrated into the information network". Internet of Things, as an essential part of I4.0 concept, is already partially adopted by households, with aim of creating a "smart house", despite the fact that not every gadget is appropriately connectable yet [12]. The same problem can be observed among enterprises, especially in the area of SMEs. It assumed that the main barriers for becoming "smart" are insufficient education and little knowledge [10].

Moreover, interconnected objects and subjects are just one prerequisite for the so called "4th industrial revolution", where cyber and physical levels should merge [3]. The term Industry 4.0 points to the 4th industrial revolution, and was firstly presented on Hannover-Messe (one of the biggest international trade fairs, oriented on new and smart technologies) in 2011, while it also indicates the initiative of the German government to improve the environment in the manufacturing sector using new technologies (information regarding the concept was brought up in 2014 at the World Economic Forum in Davos [13]). According to BITKOM (Germany's digital association, founded in 1999 as a merger of individual industry associations in Berlin, representing more than 2500 companies in the digital economy, among them 1000 SMEs, all 400 start-ups), the 4th industrial revolution should allow control over the entire life cycle of the product and value stream, therefore redefine organization entirely. With respect to efficiency oriented on cost-savings and complexity reduction, Modrak and Bednar [14,15] conclude that I4.0 environment will initiate mass customization, mainly because of the ability of each entity throughout the value stream to communicate and identify itself. All these visions and concepts are meant to be environmentally, economically, but mainly socially sustainable. Thestrup et al. [6] also report to the importance of value creation and value constellations for developing the sustainable business models in industrial services. Sustainable development towards I4.0 is however a very complex issue. Leaving a now technical standpoint, we emphasize non-technical aspects of proposed changes within the industries.

As Slusarczyk [16] suggests, the 4th revolution differs from previous revolutions, because it will apply to all aspects of everyday lives, as a consequence of environment, where information will be exchanged between objects, between people, and between people and objects. In other words, real-time data exchange and horizontal and vertical integration of production systems are the main pillars of I4.0 [17], along with cyber security, autonomous systems, capability of analyzing large data sets, virtual reality, and cloud computing. Undoubtedly, such changes would require managerial decisions firstly, due to inevitable initial costs linked to such new technological equipment. Schröder et al. [18] leaves an open question, whether it is even worth implementing I4.0, especially for SMEs, despite the consensus we find among authors describing reduced costs and more efficient processes and environment as a consequence of I4.0. We argue that such dynamics within the industries should be examined deeply, and various elements of sustainable development, not only the economic point of view, should be evaluated [19,20]. The opposite to mentioned cost-saving and cost-reducing is initial need of significant financial expenditures, that are in many occasions out of reach for companies, especially SMEs.

Another very important social aspect of such smart environment is how the intelligent machines will affect the labor market [20–24]. This topic has basically been examined from two perspectives; firstly, by describing requirements towards workers in I4.0 [20–24]; secondly, by examining the standpoint of workers and their outlook or current state of mind [20,25]. Regarding the I4.0 environment, the authors can conclude that there is a growing trend in the literature highlighting the importance of sustainable development and management [26]. We argue that unless reasonable level of awareness and basic knowledge of Industry 4.0 related concepts, parts, and inevitable parts is reached, it will be hard to successfully move towards smart environment, especially in case of less developed regions. Insufficient information base of the eligible work force represents an obstacle for potential employers oriented towards I4.0. Inadequate information and knowledge could also lead potential employees towards wrong or misjudged conclusions or attitudes. Probably the most crowded thought is that bringing in the intelligent machines would steal jobs, again, especially in less developed regions with less qualified, manually involved workforce. Consequently, a lack of awareness and affinity towards any modernization steps could hold the required transformation processes. However, according to Statista [27], countries without any problems with unemployment (e.g., Germany, USA, Japan) report the highest numbers of installed industrial robots per 10,000 employees.

The aim of conducted research is to investigate and analyze general awareness and knowledge regarding Industry 4.0 within the area of South India. Based on an initial analysis, questionnaire-based surveys in the literature show a lack of covering focus on knowledge and attitudes of potential employees within the examined region and in general, as well. This paper concerns the crucial perspective of the potential workforce in the way to find a sustainable process of implementing Industry 4.0 in less developed regions. The questionnaire was created on the basis of previous industry visits and consultations with students, employers, and entrepreneurs in the investigated region. Issues addressed the most by respondents are highlighted and summarized.

The article is further organized as follows: Section 2 provides a problems description, where issues necessary for further research are stated. Section 3 provides a methodology description, while in Section 4, we present obtained results. Section 5 concludes the findings of this research study.

2. Problems Description

The emerging economies should leverage their advantages, such as huge markets, attractive conditions for manufacturing, fast growing economies, and mainly larger labor force with more favorable demography [28]. Admitting that Industry 4.0 will primarily affect manufacturing sector, we face significant discrepancies among countries and regions. Despite the estimate that India will be the world's fastest growing economy in following years [29] with manufacturing sector that could hit 1 trillion US$ in 2025 [30], we doubt the ability of successful transformation towards Industry 4.0, hence we find India and its regions important to examine with respect to the 4th industrial revolution [31]. For example, having Germany—a technology and manufacturing leader, however, with an aging population and lack of labor force—and on the other hand, an emerging country like India—suffering from technological gaps, which put India to a level of Industry 2.0 as Iyer [28] concludes, on the contrary, with strong demography.

There is also a political will to spur manufacturing sector, translated into initiatives such as "Digital India" [32], "Skill India" [33], or "Make in India", with aims to (among others) create sufficient skill sets within the urban poor and rural migrants for inclusive growth, or to increase technological depth in manufacturing in order to increase domestic value addition. In addition, there is the mentioned demographic factor—India has the best demographic dynamics, with approximately 60% of population in age between 15 and 59 [34]. The open question remains, are citizens and workers ready for such development in the foreseeable future?

We accept that within such a huge country, significant disparities among particular regions exist, hence we applied our research only in southern part of India (authors were physically present in the state of Tamil Nadu during data collection). South Indian region includes several states and union territories (Kerala, Andaman and Nicobar Islands, Tamil Nadu, Pondicherry, Andra Pradesh, Karnataka, Telangana, Lakshadweep Islands), which combined accounts for 19.31% of the geographical area of the whole of India. With over 250 million people, South India represents around 20% of the country's population [35]. As of 2016, economic growth of South India was around 17%, compared to 8% growth of the whole of India, while GDP of South India accounted for 30% of total Indian's GDP. Some specific industries are even more important from overall perspectives, such as cotton production (48% of India's entire cotton production comes from South India) or agricultural production (36% of whole state's production comes from South India).

Same as for other countries and regions, the main employers are SMEs. Unfortunately, as Iyer [28] states for India's industrial policy in general, it is old, and lacking critical technology. Many enterprises in this area are old and have long lasting traditions. Despite the established reputation and customers created, they are equipped with insufficient and old devices or machines. Internet access and computer equipment within industries in this region is also rather poor. Since the majority of the research has been conducted in the field of needed modernization, especially with respect to the SMEs in order to successfully transform towards Industry 4.0, we would rather point at the necessity of having a potential labor force ready for such a transformation. It is therefore considered that awareness of I4.0 needs to be continuously expanded and promoted, as confirmed by several authors [10,36–38]. Even if obtaining new machines and gadgets would be economically viable, will there be enough sufficiently educated workers? Throughout the literature, we find papers addressing similar problems within different regions, e.g., concluding that qualified specialists are often not satisfied with the salary, which causes their outflow in favor of richer economic regions, leaving almost no people able to operate such modern machines [39]. We argue that unless some basic level of knowledge regarding the addressed issues is reached within the population, the ability of forming a sustainable path to become competitive in an Industry 4.0 environment is rather limited.

3. Methodology

Research was conducted in the area of South India, where different industries operate in several segments, with a majority representation of SMEs. The aim of this survey-based study is to examine level of awareness and general consciousness of Industry 4.0 among South Indian students, workers, entrepreneurs, in other words, a broad spectrum of citizens. We expect that proper analysis of the gathered responses could provide us with unique and valuable knowledge of the current state of mind of local citizens, along with their current level of internet/connection requiring gadgets/platforms, and further serve as guidance for finding a suitable implementing strategy for new technologies in such areas.

Results presented in this paper concern opinions and knowledge of inhabitants living, studying, working, or doing business in the previously described area. For obtaining responses, a questionnaire was used, and data collection took place from December 2019 to February 2020. As advised by several authors [40,41], we used fixed-choice questions, in order to maintain time efficiency and difficulty of evaluation. Questionnaire was distributed within several traceable ways during the stay of authors in Tamil Nadu (see Table 1). Sample contains 564 unique responses (after removing incomplete and inappropriately filled responses—respondents' answers were checked to confirm all required questions had been answered in a prescribed manner). Respondents were notified in advance that providing answers to this questionnaire is anonymous. All answers provided will serve only for research purposes, and no personal details will be required or stored.

Table 1. Profile of respondents.

Demographics	N	%	Demographics	N	%
Gender			Age		
Female	147	26.1	25 or below	466	82.6
Male	417	73.9	26–35	57	10.1
			36–45	24	4.3
Status			46 and more	17	3.0
Student	438	77.7			
Employed & Entrepreneur	105	18.6	Residential place		
Houseperson & Retired	7	1.2	Andaman and Nicobar Islands *	12	2.1
Unemployed	14	2.5	Andra Pradesh	6	1.1
			Karnataka	10	1.8
Education			Kerala	14	2.5
Higher Secondary and below	181	32.1	Lakshadweep Islands *	4	0.7
Bachelor	256	45.4	Pondicherry *	9	1.6
Master	67	11.9	Tamil Nadu	507	89.9
Doctorate, Medical, Law degree, or higher	60	10.6	Telangana	2	0.4

Note: *Union territory; Source: Prepared by authors.

We divided the questionnaire into four main parts (Figure 1). In the first part, we focused on the social status of the respondent, education, and the place where respondent currently works, studies, or stays. In the second part, we were interested in respondents' basic internet communication and usage of social communication applications. In the third and main part, we looked at the awareness of Industry 4.0 in general among respondents. We asked about key terms such as cloud solutions, mass customization, Internet of Things, Industry 4.0, smart manufacturing, smart cities, etc. In the fourth part, we intended to examine what the I4.0 could bring to the South Indian region from the responders' perspective.

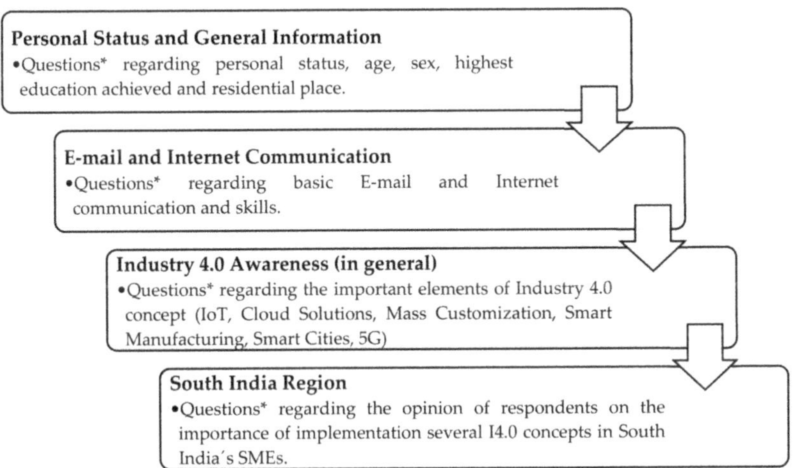

Figure 1. Stages of survey. Note: * For all types of questions, please see Supplementary Materials.

Several scales were used due to substance of the question (full text of the questionnaire and scales of answers is provided in Supplementary Materials). Questions addressing previous experience and general awareness about key terms were scaled binomially (yes/no). Other questions addressing South India region were scaled as 5 levels: Not at all important/no—1; slightly important/rather no—2; no opinion (due to lack of information/knowledge, referring also to "I do not know")—3; fairly important/rather yes—4; very important/yes—5. Supplementary questions regarding usage of social media, email, or e-commerce had specific scales examining frequency of usage.

To analyze responses, we used tables of counts and percentages for the joint distribution of two (severe combinations) categorical variables. We used custom and contingency tables, statistical testing, and generated bar graphs for easier data presentation. Pearson's chi-square test was performed to test the independence between the row and column variables. Pearson's chi-square test requires a large sample. The main rule regarding the sample size is that not more than 20% of expected cells should be less than 5, and none of the expected cells should be less than 1 [42,43]. If the relationship was significant, consequently we used z-test to compare the proportion of column pairs to each other (adjusted by Bonferroni correction) according to the social variables and variables reported by Industry 4.0 areas. For 2×2 tables, we used Fisher's Exact test. The column proportions test shows whether the ratio in one column is significantly different from the ratio in the other column. The test assigns a letter key (A, B, C) to each category reported in column variables. The definition of each comparison of column proportions is discussed in the following section. All statistical outputs were processed in the IBM SPSS Statistics v25.0.

In this study we try to identify the main drivers and barriers of Industry 4.0 readiness in South India. Among the literature, we follow several research questions to investigate their importance on Industry 4.0 readiness in this region. Accordingly, these research issues can be concluded:

- The higher education achieved and higher personal status play an important role on Industry 4.0 readiness in South India's region [28].
- The proper education and requalification are necessary, especially regarding current dynamics (4th industrial revolution), throughout the industries [44–48].
- The importance of the digital behavior of the young people, whose relationship with the digital world and services are very important for their further social and economic development, supported by the integrating social and economic dimensions of sustainable development in the context of I4.0 [38,49].

Further, we will concentrate on presenting the most attention grabbing outcomes and dependences from responses that were statistically proven as significant.

4. Results & Discussion

In order not to confuse respondents and avoid misinterpretations, we provided short descriptions of possibly unknown terms related to our scope (presented in Supplementary Materials). The questionnaire, in its actual form, is composed by twenty-seven questions divided into four main areas, mentioned above.

Firstly, we asked our respondents if they ever heard about key terms related to 4th industrial revolution. As presented in Figure 2, the term "mass customization" is not known by almost 60% of respondents, while, more importantly, the term "Industry 4.0" is unknown to 49.6% of respondents. Rather than focusing only on simple percentage points-presentations of answers observed, we examined and focused mainly on dependences between key answers on a statistically significant basis, as presented further in this chapter.

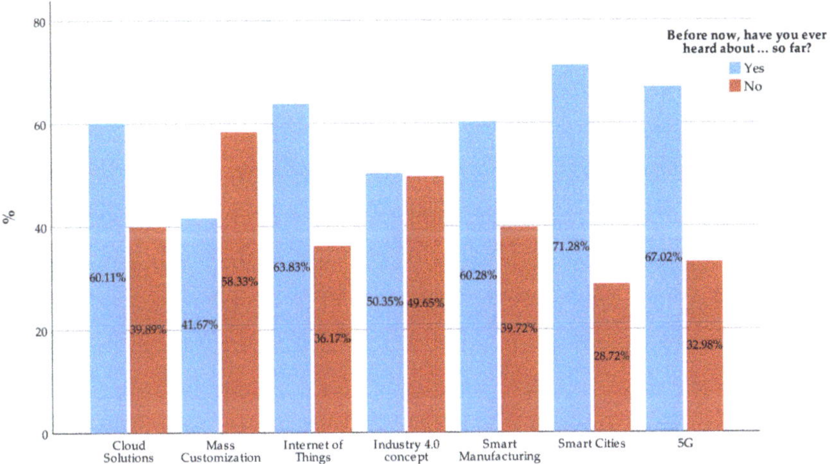

Figure 2. Awareness in general on Industry 4.0.

Before examining key aspects of this survey, we took the first step by examining dependence between age, education, and such awareness. In all tables below, Chi-square statistic (X^2) and *p*-value (Sig.) are presented for each row question, as inevitable assumptions for further column proportions comparison. X^2 refers to Pearson Chi-Square statistic value, obtained by Chi-square test in SPSS, which tests the hypothesis that two variables (row and column) are independent. Sig. refers to significance value, which has the information we are looking for. The lower the Sig., the less likely it is that two variables are unrelated. When the significance value is less than 0.05, we can conclude that there is a relationship between two variables. To understand the relationship between row and column variables, we examine the crosstabulation tables with results of the column proportions tests. As we mentioned in previous section, the column proportions test shows whether the proportion in one column is significantly different from the proportion in the other column. The test assigns a letter key (A, B, C) to each category reported in column variables. We used three significance levels: 0.05 *; 0.01 **; 0.001 ***. Column proportion tests are performed by z-test, and tests are adjusted for all pairwise comparisons within a row of each innermost sub-table using the Bonferroni correction (see Sedgwick [50]). Below we provide Table 2, where statistically significant relationship between answers "No" to above mentioned general awareness questions and education "Higher Secondary and below" can be observed. We find this in line with basic logic that ongoing and deeper education opens possibilities and provides information about new approaches and cutting-edge trends. Similarly, we find a logical relationship within our answers, that higher education (Doctorate, Medical, or Law degree or higher) goes with higher age of the respondent. However, we consider the fact that 46.4% (45.3%) of the group "Higher Secondary and below" answered "No" when asked about "Cloud solutions" ("Internet of Things"), as a result of teaching plans that are not updated sufficiently, not the respondents' inability to learn about possibilities linked to I4.0.

Table 2. Awareness vs. Age and Education.

		Q10 Cloud Solutions So Far? — No (A)	Yes (B)	Q11 Mass Customization so Far? — No (A)	Yes (B)	Q12 Internet of Things So Far? — No (A)	Yes (B)	Q13 Industry 4.0 Concept So Far? (so Called 4th Industrial Revolution) — No (A)	Yes (B)	Q14 Smart Manufacturing So Far? — No (A)	Yes (B)	Q15 Smart Cities So Far? — No (A)	Yes (B)	Q16 5G So Far? — No (A)	Yes (B)	Q17 Do You Have Any Previous Experience with IoT or I4.0 Concept? — No (A)	Yes (B)
X^2	Sig.	31.512	0.000 ***	6.251	0.100	33.499	0.000 ***	44.270	0.000 ***	24.811	0.000 ***	27.827	0.000 ***	18.893	0.000 ***	15.197	0.002 **
Age	25 or below	B				B		B		B		B		B		B	
	26-35		A				A		A		A		A		A		
	36-45						A		A		A		A		A		A
	46 and more		A								A						A
X^2	Sig.	11.945	0.008 **	8.559	0.036 *	28.072	0.000 ***	35.004	0.000 ***	19.037	0.000 ***	53.918	0.000 ***	20.402	0.000 ***	8.560	0.036 *
Education	Higher Secondary and below	B		B		B		B		B		B		B		B	
	Bachelor				A										A		
	Master		A				A		A		A		A		A		
	Doctorate, Medical or Law degree, or higher						B			B				A	B		

Results are based on two-sided tests. For each significant pair, the key of the category with the smaller column proportion appears in the category with the larger column proportion. X^2 refers to Chi-square statistic. Sig. refers to the two-sided asymptotic significance of the chi-square statistic. Significance level for upper case letters (A, B, C): 0.05 *; 0.01 **; 0.001 ***. Tests are adjusted for all pairwise comparisons within a row of each innermost sub-table using the Bonferroni correction; Source: Prepared by authors.

In Table 2, the column proportions test assigns a letter key, (A) or (B), to each category of question Q10–Q17. (A) refers to the answers "No" and (B) to the answers "Yes". The row variables are "Age" and "Education", which have four categories of answers. The two-sided asymptotic significance of chi-square statistics adjusted by Bonferroni correction is less than 0.05 * in all comparisons, except comparison between "Age" and "Mass Customization" (Sig. 0.100). The Sig. value (0.000 ***) is less than 0.001, therefore statistically significant. For the column proportions test associated with the age group "25 or below" and the answers to question Q10, the B key appears in the column "No".

Thus, we can conclude that the proportion of respondents aged "25 or below", who answered the question Q10 about cloud solutions negatively, is greater than the proportion of respondents that answered the question Q10 positively (aged "25 or below"). The same results are listed between the respondents aged "25 or below" and other questions, except the Q11 about mass customization. For the tests associated with "Education", the results indicate the same in the case of "Higher Secondary and below" education for all questions Q10–Q17.

We would like to highlight the relationship between the age group "25 or below" and answers "No" to general questions. In absolute terms, 56.0%, and 41.4%, respectively, of group "25 or below" answered "No" to questions addressing Industry 4.0 and IoT as a crucial stage of 4th industrial revolution, respectively. We consider this as a very poor informational level, especially within the young and flexible group of workers entering labor market. On the contrary, 79.8% (46.6%) of this group is using WhatsApp (Facebook) almost daily, therefore we cannot explain this level of awareness as a result of insufficient conditions for obtaining information or being digitally isolated. Motyl et al. [49] conducted a survey of more than 460 students at three different universities in Italy about the Industry 4.0 concept. The authors point out the importance of the digital behavior of the young people, whose relationship with digital world and services are very important for their further social, but also economic development, ultimately for the development of the region or country. We agree with the authors that in today's environment, it is important to empower a broader knowledge of the general I4.0 concepts and bring well-structured action plans into the educational process. These conclusions should emphasize, on the one hand, the role of education, and the SMEs on the other, which are dependent on an educated workforce in the terms of I4.0 and related IoT.

The second part contains information about the importance of several aspects of doing business from the perspective of respondents, considering SMEs. We present Figure 3 with questions Q18, Q21, Q22, and Q23. We observed a relatively high proportion of responds without any clear opinion regarding each question, while almost one quarter of respondents consider investing in training of workers as "Not at all important". Such attitudes we consider as negative, while on the other hand, almost 40% of respondents consider approaching smart manufacturing as very important.

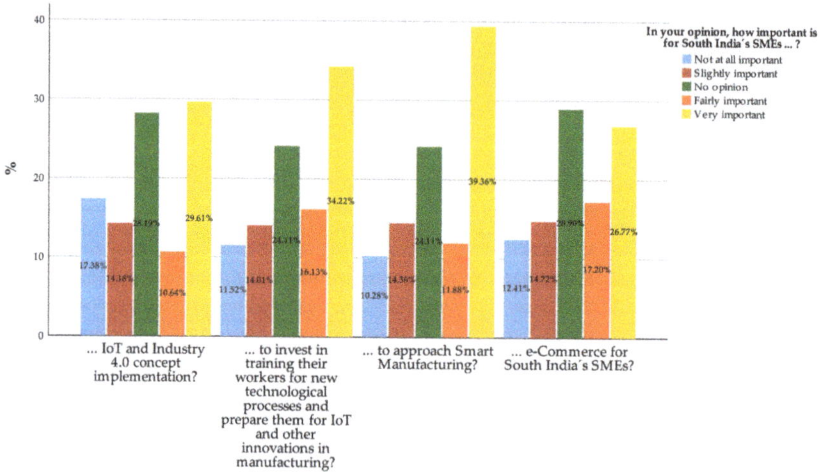

Figure 3. The answers to the questions Q18, Q21–Q23.

In Table 3 below, each column refers to awareness questions mentioned above, and each row refers to questions regarding IoT, I4.0, smart manufacturing, e-commerce, and investing in workers' education.

We then expected the row questions (Q18, Q21–Q23) and column variables (Q10–Q17) would suggests some proportional relations. The fact is, that almost in all situations (where the questions Q18, Q21–Q23 were answered "No opinion", respectively "I don't know"), the proportion of respondents who answered the questions Q10–Q17 negatively is greater compared to the proportion of respondents who answered these questions positively. We argue that such statistical evidence of inability to form an opinion or express expectation stems from obvious lack of information. Stentoft et al. [45] also conclude that the lack of information and knowledge about Industry 4.0 and the lack of employee's readiness are among the main barriers for Industry 4.0. On the contrary, proportion of respondents who answered the questions Q10–Q17 positively is greater compared to the proportion of respondents with negative answers, if we are considering answers "very important" or "fairly important" regarding questions Q18 and Q21–Q23. A possible and logical explanation could be that respondents realize importance of successful transformation of the industries due to previous, at least basic, knowledge about questioned aspects. The most attention grabbing at this stage was, however, finding relations between answers "not at all important" to questions Q18, Q21–23, and questions Q10–Q17 that were answered as "No". The proportion of respondents answering questions Q12, Q14, and Q15 as "No" that also answered Q22 (regarding approaching smart manufacturing from SMEs perspective) as "Not at all important" was significantly higher than the proportion of respondents answering Q12, Q14, and Q15 as "Yes". In total, 10.3% of respondents answered Q22 as "Not at all important", 14.4% answered "Slightly important", and 24.1% answered "No opinion" (or "I don't know"), which makes together 48.8%. We can observe a similar relationship between answers "No" to Q12 and Q14 and answers "Not at all important" to question Q18, addressing importance of implementation of I4.0 from SMEs perspective. Additionally, the relationship between respondents answering Q13 regarding I4.0 as "No" and Q21 addressing investing in training workers answering as "Not at all important" is alarming. This could be seen as a lack of information about inevitable changes in the coming future translating into unclear thoughts about crucial education and training for current and potential employees. This is in contrast with consensus among the authors [46–48] that proper education and requalification is necessary, especially regarding current dynamics throughout the industries.

Table 3. Awareness vs. importance of the modern environment of doing business regarding SMEs.

		Q10		Q11		Q12		Q13		Q14		Q15		Q16		Q17	
						Before now, Have You Ever Heard about the…										Do You Have Any Previous Experience with IoT or I4.0 Concept?	
		Cloud Solutions So Far?		Mass Customization So Far?		Internet of Things So Far?		Industry 4.0 concept So Far? (So Called 4th Industrial Revolution)		Smart Manufacturing So Far?		Smart Cities So Far?		5G So Far?			
		No	Yes	No	Yes	No	Yes	No	Yes	No	Yes	No	Yes	No	Yes	No	Yes
X^2	Sig.	125.083	0.000 ***	61.415	0.000 ***	42.886	0.000 ***	71.854	0.000 ***	59.146	0.000 ***	53.016	0.000 ***	50.778	0.000 ***	47.945	0.000 ***
Q18. In your opinion, is IoT and Industry 4.0 concept implementation important for South India's SMEs?	Not at all important					B				B							
	Slightly important																
	No opinion	B		B		B		B		B		B		B		B	
	Fairly important		A				A						A		A		
	Very important		A		A		A		A		A		A		A		A
X^2	Sig.	71.830	0.000 ***	16.292	0.003 **	13.077	0.011 *	28.767	0.000 ***	31.763	0.000 ***	70.570	0.000 ***	46.957	0.000 ***	24.682	0.000 ***
Q21. In your opinion, how important is it for South India's SMEs to invest in training their workers for new…?	Not at all important							B									
	Slightly important		A												A		
	No opinion	B		B		B		B		B		B		B		B	
	Fairly important		A								A		A				
	Very important		A				A		A		A		A				
X^2	Sig.	36.366	0.000 ***	15.174	0.004 **	17.773	0.001 ***	18.905	0.001 ***	26.573	0.000 ***	55.685	0.000 ***	21.711	0.000 ***	17.629	0.001 ***
Q22. In your opinion, how important is it for SMEs in South India to approach Smart Manufacturing?	Not at all important					B				B		B					
	Slightly important		A								A				A		
	No opinion	B		B				B		B		B		B		B	
	Fairly important						A				A		A				
	Very important		A		A				A				A				A

Table 3. *Cont.*

		Q10		Q11		Q12		Q13		Q14		Q15		Q16		Q17	
		Before now, Have You Ever Heard about the...														Do You Have Any Previous Experience with IoT or I4.0 Concept?	
		Cloud Solutions So Far?		Mass Customization So Far?		Internet of Things So Far?		Industry 4.0 concept So Far? (So Called 4th Industrial Revolution)		Smart Manufacturing So Far?		Smart Cities So Far?		5G So Far?			
		No	Yes	No	Yes	No	Yes	No	Yes	No	Yes	No	Yes	No	Yes	No	Yes
X^2	Sig.	104.602	0.000 ***	40.564	0.000 ***	45.056	0.000 ***	61.571	0.000 ***	78.993	0.000 ***	52.549	0.000 ***	63.905	0.000 ***	46.675	0.000 ***
Q23. In your opinion, how important is e-Commerce for South India's SMEs?	Not at all important											B					
	Slightly important				A		A										
	No opinion	B		B		B		B		B		B		B		B	
	Fairly important		A						A		A		A				
	Very important		A		A		A		A		A		A		A		A

Note: "No opinion" refers to "I don't know". Results are based on two-sided tests. For each significant pair, the key of the category with the smaller column proportion appears in the category with the larger column proportion. X^2 refers to Chi-square statistic. Sig. refers to the two-sided asymptotic significance of the chi-square statistic. Significance level for upper case letters (A, B, C): 0.05 *; 0.01 **; 0.001 ***. Tests are adjusted for all pairwise comparisons within a row of each innermost sub-table using the Bonferroni correction.1 Source: Prepared by authors.

Responds to the questions Q19, Q20, and Q25 are presented further in the Figure 4. The respondents were able to choose one of five options: "No"; "Rather no"; "No opinion" (referring also to "I don't know"); "Rather yes"; "Yes". In each of these questions, we can see a high proportion of respondents who replied to all questions with the "No opinion" ("I don't know"). In Q19, it was more than 32% of respondents, in Q20 more than 33%, and in Q25 more than 21%. This again points towards a lack of information resulting in an inability to form an opinion regarding the issue. On the other hand, the answers "No" and "Rather no" opened further questions that we attempted to examine. Within the age group "25 or below", more than 16% of respondents think that IoT concept will be ineffective for South India's SMEs. Almost 34% of the respondents within this group reported "No opinion". Examining performance of this group also on other questions, we observed nearly 17% of the respondents claiming the SMEs in South India are not ready to implement IoT and I4.0 concept, and as many as 36% of the respondents were unable to make a judgement. For more than 27% of the respondents aged "25 or below", the I4.0 concept is personally unimportant. More than 24% of respondents from the whole sample do not consider the IoT and I4.0 concept as important from a personal point of view.

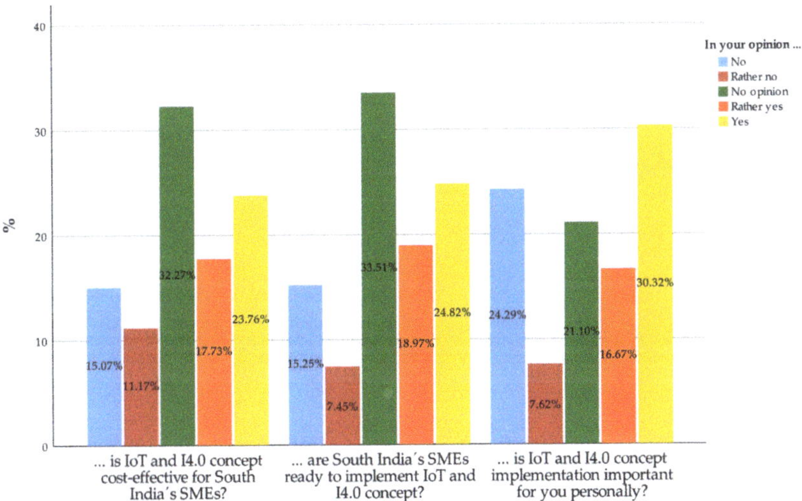

Figure 4. The answers to the questions Q19, Q20, and Q25.

These results are further examined against general awareness in the Table 4. Similarly, we applied column proportions test. For each combination of testing, we also point to the value of the asymptotic significance statistic (Sig.), which in all cases is less than 0.05 * level, and thus variables are related. This table includes also a comparison of the answers to question Q27, which is focused on whether respondents expect any Smart City in South India within the next 10 years. In proportional testing, we found that in three cases (Q12, Q14, and Q17), the Sig. value is higher than the confidence level 0.05 * (0.100; 0.091; 0.234). In such cases, we consider these variables as independent.

Table 4. Awareness vs. questions Q19, Q20, Q25, and Q27.

Columns Q10–Q16 fall under the heading "Before Now, Have You Ever Heard about the…". For each question the two sub-columns are **No (A)** and **Yes (B)**.

		Q10 Cloud Solutions So Far? No (A)	Yes (B)	Q11 Mass Customization So Far? No (A)	Yes (B)	Q12 Internet of Things So Far? No (A)	Yes (B)	Q13 Industry 4.0 Concept So Far? (So Called 4th Industrial Revolution) No (A)	Yes (B)	Q14 Smart Manufacturing So Far? No (A)	Yes (B)	Q15 Smart Cities So Far? No (A)	Yes (B)	Q16 5G So Far? No (A)	Yes (B)	Q17 Do You Have Any Previous Experience with IoT or I4.0 Concept? No (A)	Yes (B)
X^2	Sig.	52.087	0.000 ***	55.046	0.000 ***	28.232	0.000 ***	41.303	0.000 ***	45.401	0.000 ***	17.383	0.002 **	55.819	0.000 ***	72.579	0.000 ***
Q19. In your opinion, is IoT and I4.0 concept cost-effective for South India's SMEs?	No																
	Rather no				A												
	No opinion	B		B		B		B		B		B		B		B	
	Rather yes		A								A		A		A		
	Yes		A				A		A		A		A		A		A
X^2	Sig.	52.379	0.000 ***	16.323	0.003 **	48.810	0.000 ***	37.977	0.000 ***	46.488	0.000 ***	35.185	0.000 ***	59.091	0.000 ***	31.707	0.000 ***
Q20. In your opinion, are South India's SMEs ready to implement IoT and I4.0 concept?	No							B									
	Rather no																
	No opinion	B		B		B		B		B		B		B		B	
	Rather yes						A				A				A		
	Yes		A		A		A		A		A		A		A		A
X^2	Sig.	52.355	0.000 ***	26.645	0.000 ***	44.466	0.000 ***	45.476	0.000 ***	44.201	0.000 ***	73.733	0.000 ***	15.657	0.004 **	26.082	0.000 ***
Q25. In your opinion, is IoT and I4.0 concept implementation important for you personally?	No					B		B				B					
	Rather no		A														
	No opinion	B		B		B		B		B		B		B		B	
	Rather yes		A				A		A		A						
	Yes		A		A		A		A		A		A				A
X^2	Sig.	17.696	0.000 ***	8.219	0.016 *	4.596	0.100	8.943	0.011 *	4.793	0.091	40.112	0.000 ***	15.174	0.001 ***	2.905	0.234
Q27. In your opinion, do you see any Smart City in South India in next 10 years?	No	B						B				B		B			
	No opinion				A							B					
	Yes		A						A				A		A		

Note: "No opinion" refers to "I don't know". Results are based on two-sided tests. For each significant pair, the key of the category with the smaller column proportion appears in the category with the larger column proportion. X^2 refers to Chi-square statistic. Sig. refers to the two-sided asymptotic significance of the chi-square statistic. Significance level for upper case letters (A, B, C): 0.05 *; 0.01 **; 0.001 *** a. Tests are adjusted for all pairwise comparisons within a row of each innermost sub-table using the Bonferroni correction; Source: Prepared by authors.

We highlight the high portion of "No opinion" ("I don't know") answers observed within the set of questions Q19, Q20, and Q25, related to answers "No" (questions Q10–Q17). A similar pattern was observed and described in Table 2. One concern could be potential complexity or difficulty of questions Q19 and Q20, therefore forming a substantiated opinion could be harder for respondent. On the contrary, inability to take a personal stance towards I4.0 and related IoT, we explain as lack of sufficient information, as described previously. Additionally, on a personal level, implementation of I4.0 and IoT (Q25) is not important for respondents answering Q12 and Q13. In total, 24.3% (7.6%) of respondents answered "No" ("Rather no") to a question Q25.

Regarding question Q27, where respondents were asked if they see any Smart City within the region in the next 10 years, in total almost 74% answers were positive, which is rather overconfident, especially in contrast to other studies [28] examining current state of art in India. Putting this question in the context of questions Q10–Q17 results in similar outcomes as for previous sets of questions, where negative answer to Q10–Q17 are related to negative answer addressing Smart City. On the other hand, proportion of respondents answering Q27 positively, that answered also Q10–Q17 positively, is higher on the statistically significant basis than the proportion of those who answered Q10–Q17 negatively.

Thus, we find implementation of any I4.0 related features difficult from non-technical point of view, if respondents' expectations are negative towards companies. On the other hand, throughout each set of questions, we observe a significantly higher proportion of respondents expecting rather positive impacts within questioned aspects, that have previous knowledge or information (which should be also considered as a part of intellectual capital [51]) about key terms addressed in the first part compared to those without such information. To add on, respondents with previous experience with IoT and I4.0 expressed positive expectations with higher frequency compared to those without such experience. On the contrary, we consider some responds to questions addressing Smart cities in South India (Q27), or readiness of SMEs for implementing IoT and I4.0 (Q20), as rather over-confident, considering the current state of art not only in South India [28]. Such observations could however stem from possible drawbacks as sampling error. Other possible limitations could be that only main results are presented. The survey was prepared only in South India's region with high representation of people of state Tamil Nadu, which could lead to possible biasedness. Additionally, authors realize that a higher proportion of employees of SMEs in the survey could lead to more valuable outcomes. Thus, we recommend further examination of the mentioned region because of its huge demographic potential. Possible improvement of the conducted research should be expanding the sample, or expert surveys with representatives of employee associations and other social parties.

5. Conclusions

In this paper, we attempted to examine general awareness, opinions, and attitudes of South India's inhabitants towards Industry 4.0 and its features. Contrary to the majority of studies examining I4.0 readiness and implications, we oriented our research towards citizens rather than SMEs. Conducting a survey using a questionnaire, we gathered unique answers containing crucial information about current state of art regarding addressed issues, the same as future expectations. Besides simple counts of answers, we provided also testing of interdependences between general awareness questions, and several sets of questions addressing various issues.

The main findings suggest that general awareness is quite low (almost 50% of respondents have no prior information of Industry 4.0), which consequently leads to inability to form any opinion regarding effects of such new trends on working and personal life, the same as on living and business environment. Respondents with insufficient knowledge of I4.0 and related IoT then tend to answer negatively regarding questions about possible transformation of SMEs towards I4.0, or they are unable to form any opinion regarding addressed aspects. On the contrary, respondents possessing prior information or knowledge regarding I4.0 and related IoT as its crucial part expressed positive expectations in general. We argue that this approach provides us with important information described in detail in previous section, that should serve as steppingstone in forming a sustainable strategy of implementing I4.0 features for both large enterprises and SMEs. From the perspective of sustainability of the transformation process towards competitive and functional I4.0 environment, we therefore highlight the importance of proper skills, education, working conditions, and informational level, which we consider, to some extent, in line with other studies examining sustainable management and development as a part of I4.0 linked research [6,26,39]. To conclude, improving education and providing proper information is crucial from the authors' perspective.

Based on examined interdependences, we argue that proper education and relevant information dissemination is non-technical, however crucial, in order to achieve a sustainable transformation process of the current environment in South India towards Industry 4.0 and its features. Since we find this approach rather extraordinary, similar survey-based studies in other regions should provide comparable results, and consequently form a broader picture of inhabitants' preparedness and awareness of I4.0.

Supplementary Materials: The following are available online at http://www.mdpi.com/2071-1050/12/8/3207/s1.

Author Contributions: Conceptualization, L.S. and J.S.; methodology, L.S. and J.S.; validation, Darya Dancaková; formal analysis, J.S.; investigation, L.S. and J.S.; resources, L.S., J.S.; writing—original draft preparation, L.S., J.S. and D.D.; writing—review and editing, M.W.; editing and visualization, J.S. and D.D.; supervision, M.W.; project administration, M.W., L.S., and J.S. All authors have read and agreed to the published version of the manuscript.

Funding: This research was funded by the Scientific Grant Agency of the Slovak Republic VEGA, grant number 1/0430/19. This research work is part of actual research activities in the project with acronym SME 4.0, and titled as "SME 4.0—Smart Manufacturing and Logistics for SMEs in an X-to-order and Mass Customization Environment", from the European Union's Horizon 2020 research and innovation program under the Marie Sklodowska-Curie grant agreement No 734713.

Acknowledgments: Authors express gratitude for tremendous support and help to Prof. Naavendra Krishnan, Prof. Sudhakara Pandian, and all SACS MAVMM staff during authors' stay in Madurai, Tamil Nadu.

Conflicts of Interest: The authors declare no conflict of interest.

References

1. Ramsauer, C. Industrie 4.0–Die Produktion der Zukunft. *WINGbusiness* **2013**, *3*, 6–12.

2. Kiel, D.; Arnold, C.; Collisi, M.; Voigt, K.I. The impact of the industrial internet of things on established business models. In Proceedings of the 25th International Association for Management of Technology (IAMOT) Conference, Orlando, FL, USA, 15–19 May 2016; pp. 673–695.

3. Lasi, H.; Fettke, P.; Kemper, H.G.; Feld, T.; Hoffmann, M. Industry 4.0. *Bus. Inf. Syst. Eng.* **2014**, *6*, 239–242. [CrossRef]

4. Posada, J.; Toro, C.; Barandiaran, I.; Oyarzun, D.; Stricker, D.; de Amicis, R.; Vallarino, I. Visual computing as a key enabling technology for industrie 4.0 and industrial internet. *IEEE Comput. Graph. Appl.* **2015**, *35*, 26–40. [CrossRef] [PubMed]

5. Valdez, A.C.; Brauner, P.; Schaar, A.K.; Holzinger, A.; Ziefle, M. Reducing complexity with simplicity-usability methods for industry 4.0. In Proceedings of the 19th triennial congress of the IEA, Melbourne, Australia, 9–14 August 2015; p. 14.

6. Thestrup, J.; Sorensen, T.F.; De Bona, M. Using Conceptual Modeling and Value Analysis to Identify Sustainable m> Business Models in Industrial Services. In Proceedings of the 2006 International Conference on Mobile Business, Copenhagen, Denmark, 26–27 June 2006; p. 7.

7. Brock, D. The Compact Electronic Product Code—a 64-Bit Representation of the Electronic Product Code. Auto-ID White Paper MIT-AUTOID-WH-008. 2001.

8. ITU Internet Reports. *The internet of things*; International Telecommunication Union (ITU): Geneva, Switzerland, 2005.

9. Harbor Research. *Machine-To-Machine (M2M) and Smart Systems Market Opportunity 2010–2014*; Harbor Research, Inc.: Denver, CO, USA, 2011. Available online: http://www.windriver.com/m2m/edk/Harbor_Research-M2M_and_Smart_Sys_Report.pdf (accessed on 12 January 2020).

10. Safar, L.; Sopko, J.; Bednar, S.; Poklemba, R. Concept of SME business model for industry 4.0 environment. *TEM J.* **2018**, *7*, 626–637.

11. Sundmaeker, H.; Guillemin, P.; Friess, P.; Woelfflé, S. Vision and challenges for realising the Internet of Things. *Clust. Eur. Res. Proj. Internet Things Eur. Commision* **2010**, *3*, 34–36.

12. Cui, X. The internet of things. In *Ethical Ripples of Creativity and Innovation*; Palgrave Macmillan: London, UK, 2016; pp. 61–68.

13. Strandhagen, J.W.; Alfnes, E.; Strandhagen, J.O.; Vallandingham, L.R. The fit of Industry 4.0 applications in manufacturing logistics: A multiple case study. *Adv. Manuf.* **2017**, *5*, 344–358. [CrossRef]

14. Modrak, V.; Bednar, S. Using Axiomatic Design and Entropy to Measure Complexity in Mass Customization. *Procedia CIRP* **2015**, *34*, 87–92. [CrossRef]

15. Modrak, V.; Bednar, S. Entropy Based versus Combinatorial Product Configuration Complexity in Mass Customized Manufacturing. *Procedia CIRP* **2016**, *41*, 183–188. [CrossRef]

16. Ślusarczyk, B. INDUSTRY 4.0 – ARE WE READY? *Pol. J. Manag. Stud.* **2018**, *17*, 232–248. [CrossRef]

17. Thoben, K.-D.; Wiesner, S.; Wuest, T. BIBA – Bremer Institut für Produktion und Logistik GmbH, the University of Bremen; Faculty of Production Engineering, University of Bremen, Bremen, Germany; Industrial and Management Systems Engineering, "Industrie 4.0" and Smart Manufacturing – A Review of Research Issues and Application Examples. *Int. J. Autom. Technol.* **2017**, *11*, 4–16.

18. Schröder, C.; Schlepphorst, S.; Kay, R. *Bedeutung der Digitalisierung im Mittelstand (No. 244) IfM-Materialien*; Institut für Mittelstandsforschung (IfM): Bonn, Germany, 2015.

19. Kovacs, O. The dark corners of industry 4.0–Grounding economic governance 2.0. *Technol. Soc.* **2018**, *55*, 140–145. [CrossRef]

20. Eberhard, B.; Podio, M.; Alonso, A.P.; Radovica, E.; Avotina, L.; Peiseniece, L.; Caamaño Sendon, M.; Gonzales Lozano, A.; Solé-Pla, J. Smart work: The transformation of the labour market due to the fourth industrial revolution (I4. 0). *Int. J. Bus. Econ. Sci. Appl. Res.* **2017**, *10*, 47–66.

21. Dallasega, P.; Woschank, M.; Zsifkovits, H.; Tippayawong, K.; Brown, C.A. Requirement Analysis for the Design of Smart Logistics in SMEs. In *Industry 4.0 for SMEs*; Springer: Berlin/Heidelberg, Germany, 2020; pp. 147–162.

22. Zsifkovits, H.E.; Woschank, M. Smart Logistics–Technologiekonzepte und Potentiale. *BHM Berg- Hüttenmännische Monatshefte* **2019**, *164*, 42–45. [CrossRef]

23. Dallasega, P.; Woschank, M.; Ramingwong, S.; Tippayawong, K.; Chonsawat, N. Field study to identify requirements for smart logistics of European, US and Asian SMEs. In Proceedings of the International Conference on Industrial Engineering and Operations Management Bangkok, Bangkok, Thailand, 5–7 March 2019; pp. 844–855.

24. Chaopaisarn, P.; Woschank, M. Requirement Analysis for SMART Supply Chain Management for SMEs. In Proceedings of the International Conference on Industrial Engineering and Operations Management Bangkok, Bangkok, Thailand, 5–7 March 2019; pp. 3715–3725.

25. Wolter, M.I.; Mönnig, A.; Hummel, M.; Schneemann, C.; Weber, E.; Zika, G.; Neuber-Pohl, C. *Industry 4.0 and the Consequences for Labour Market and Economy: Scenario Calculations in Line with the BIBB-IAB Qualifications and Occupational Field Projections (Industrie 4.0 und die Folgen für Arbeitsmarkt und Wirtschaft: Szenario-Rechnungen im Rahmen der BIBB-IAB-Qualifikations-und Berufsfeldprojektionen) (No. 201508_en). Institut für Arbeitsmarkt-und Berufsforschung (IAB)*; Institute for Employment Research: Nuremberg, Germany, 2015.

26. Bonilla, S.H.; Silva, H.R.O.; da Silva, M.T.; Goncalves, R.F.; Sacomano, J.B. Industry 4.0 and Sustainability Implications: A Scenario-Based Analysis of the Impacts and Challenges. *Sustainability* **2018**, *10*, 3740. [CrossRef]

27. The Countries with the Highest Density of Robot Workers. Available online: https://www.statista.com/chart/13645/the-countries-with-the-highest-density-of-robot-workers/ (accessed on 10 January 2019).

28. Iyer, A. Moving from Industry 2.0 to Industry 4.0: A case study from India on leapfrogging in smart manufacturing. *Procedia Manuf.* **2018**, *21*, 663–670. [CrossRef]

29. World Bank. Global Economic Prospects, June 2018: The Turning of the Tide. In *Global Economic Prospects*; World Bank: Washington, DC, USA, 2018.

30. India Manufacturing Report by Indian Brand Equity Forum. Available online: www.ibef.org (accessed on 12 January 2020).

31. Chandran, S.; Poklemba, R.; Sopko, J.; Šafár, L. Organizational Innovation and Cost Reduction Analysis of Manufacturing Process–Case Study. *Manag. Syst. Prod. Eng.* **2019**, *27*, 183–188. [CrossRef]

32. Goswami, H. Opportunities and challenges of digital India programme. *Int. Educ. Res. J.* **2016**, *2*, 78–79.

33. Skill India Portal. Available online: https://www.skillindia.gov.in (accessed on 12 January 2020).

34. Directorate of Intelligence. "CIA–World Factbook". Available online: https://www.cia.gov/library/publications/the-world-factbook/index.html (accessed on 12 January 2019).

35. Census 2011 (Final Data)-Demographic Details, Literate Population (Total, Rural Urban). Planning Commission, Government of India. Available online: http://www.planningcommission.gov.in (accessed on 12 January 2020).

36. Matt, D.T.; Rauch, E. SME 4.0: The Role of Small-and Medium-Sized Enterprises in the Digital Transformation. In *Industry 4.0 for SMEs*; Springer: Berlin/Heidelberg, Germany, 2020; pp. 3–36.

37. Burgess, S. (Ed.) *Managing Information Technology in Small Business: Challenges and Solutions: Challenges and Solutions*; IGI Global: Hershey, PA, USA, 2001.

38. Kagermann, H. Change through digitization—Value creation in the age of Industry 4.0. In *Management of Permanent Change*; Springer: Berlin/Heidelberg, Germany, 2015; pp. 23–45.

39. Ingaldi, M.; Ulewicz, R. Problems with the Implementation of Industry 4.0 in Enterprises from the SME Sector. *Sustainability* **2020**, *12*, 217. [CrossRef]

40. Schwarz, N.; Hippler, H.J. What response scales may tell your respondents: Informative functions of response alternatives. In *Social Information Processing and Survey Methodology*; Springer: Berlin/Heidelberg, Germany, 1987; pp. 163–178.

41. Schuman, H.; Presser, S.; Ludwig, J. Context effects on survey responses to questions about abortion. *Public Opin. Q.* **1981**, *45*, 216–223. [CrossRef]

42. Agresti, A.; Kateri, M. *Categorical Data Analysis*; Springer: Berlin/Heidelberg, Germany, 2011.

43. Armitage, P.; Berry, G.; Matthews, J.N.S. *Statistical Methods in Medical Research*; John Wiley Sons: Hoboken, NJ, USA, 2008.

44. Lin, D.; Lee, C.K.M.; Lau, H.; Yang, Y. Strategic response to Industry 4.0: An empirical investigation on the Chinese automotive industry. *Ind. Manag. Data Syst.* **2018**, *118*, 589–605. [CrossRef]

45. Stentoft, J.; Jensen, K.W.; Philipsen, K.; Haug, A. Drivers and Barriers for Industry 4.0 Readiness and Practice: A SME Perspective with Empirical Evidence. In Proceedings of the 52nd Hawaii International Conference on System Sciences, Maui, HI, USA, 8–11 January 2019.

46. Coskun, S.; Kayikci, Y.; Gencay, E. Adapting Engineering Education to Industry 4.0 Vision. *Technologies* **2019**, *7*, 10. [CrossRef]

47. Benešová, A.; Tupa, J. Requirements for Education and Qualification of People in Industry 4.0. *Procedia Manuf.* **2017**, *11*, 2195–2202. [CrossRef]

48. Schuster, K.; Groß, K.; Vossen, R.; Richert, A.; Jeschke, S. Preparing for Industry 4.0 – Collaborative Virtual Learning Environments in Engineering Education. *Engineering Education 4.0* **2016**, *4*, 477–487.

49. Moty, B.; Baronio, G.; Uberti, S.; Speranza, D.; Filippi, S. How will Change the Future Engineers' Skills in the Industry 4.0 Framework? A Questionnaire Survey. *Procedia Manuf.* **2017**, *11*, 1501–1509. [CrossRef]

50. Sedgwick, P. Multiple significance tests: The Bonferroni correction. *BMJ* **2012**, *344*, e509. [CrossRef]

51. Pastor, D.; Glova, J.; Lipták, F.; Kovac, V. Intangibles and methods for their valuation in financial terms: Literature review. *Intang. Cap.* **2017**, *13*, 387. [CrossRef]

Article

On LSP Lifecycle Model to Re-design Logistics Service: Case Studies of Thai LSPs

Sunida Tiwong [1], Sakgasem Ramingwong [2,3,*] and Korrakot Yaibuathet Tippayawong [2,3]

[1] Ph.D.'s Degree Program in Industrial Engineering, Department of Industrial Engineering, Faculty of Engineering, Chiang Mai University, Chiang Mai 50200, Thailand; sunida.tiwong@gmail.com

[2] Center of Excellence in Logistics and Supply Chain Management, Chiang Mai University, Chiang Mai 50200, Thailand; ktippayawong@gmail.com

[3] Department of Industrial Engineering, Faculty of Engineering, Chiang Mai University, Chiang Mai 50200, Thailand

* Correspondence: sakgasem@gmail.com; Tel.: +66-089-700-9083

Received: 13 February 2020; Accepted: 13 March 2020; Published: 19 March 2020

Abstract: Improving service logistics is crucial in order to reciprocate customer needs. The paper aims to validate the Logistics Service Provider (LSP) Lifecycle Model for re-designing logistics service in three LSP case studies in Thailand. The lifecycle-stage evaluation was adapted to identify the current status in its lifecycle. Afterward, logistics service strategies were implemented according to the voice of the customer by Quality Function Deployment (QFD). The study combined the Logistics Service Provider (LSP) Lifecycle Model with the application of Industry 4.0 (I4.0) to improve service logistics. Case studies showed the implementation of the service logistics strategies with the feasibility solution of Industry 4.0.

Keywords: LSP Lifecycle Model; Industry 4.0; Quality Function Deployment; Best-Worst Method

---

1. Introduction

Global enterprises fulfill customer satisfaction with trust and royalty [1,2]. The service sector is extremely significant in this competition [3,4]. In addition, the Logistics Service Provider (LSP) has become a role for smoothing cooperation, progressing association, and improving the value-added throughout the entire supply chain [5,6]. After the Industrial Revolution, LSPs have been essential to managing the supply chain in the global economy. It is a task for LSPs to offer what customer needs to compete in the service outsourcing that has arisen around the world [7]. Service innovation and efficiency improvement are needed to respond to the complexity of customer requirements [8]. The LSP Lifecycle Model was created to improve customer satisfaction throughout the whole lifecycle of the logistics services, comprising such states of the lifecycle as design, test, logistics operation, after-sales service, service evaluation, and decomposition. There are three phases [9]: the Beginning of Life (BOL), the Middle of Life (MOL), and the End of Life (EOL) [10]. The evaluation is needed to classify the phase of logistics service, which can be recognized to improve logistics services. The development and re-designing are important parts for examining service innovation in response to customer needs [11]. Quality Function Deployment (QFD) is a well-known methodology for design and re-design to acknowledge customer requirements [12–14]. Yet, implementing QFD with LSPs for re-designing appropriate services in order to extend the service lifecycle is questionable and, indeed, challenging. Customer satisfaction can be recorded as Customer Requirements (CRs), which is translated into Engineering Characteristics (ERs) to offer what customer needs [15,16]. Logistics service can be improved or develop new service from the Voice of Customer (VOC) [14,17–20].

Whilst the LSPs' end-customer lifestyles have changed due to technological advancement, their requirements have become more complicated and sophisticated than ever before. On-demand delivery, real-time information, and Circular Economy (CE) [21] are now among the basic customer requirements. Furthermore, customer behavior has changed too. Quick and immediate responses are expected. Information Technology (IT) has been increasingly used in the global business to satisfy the customer in terms of supply [22,23]. Today, Industry 4.0 is being used to accomplish the customer target, operation, and improvement in manufacturing [24,25]. Some examples of the advancement of Industry 4.0 are applying to self-learning automation, big data analytics, real-time information, Internet of Things (IoT), and smart sensors [26,27]. This improvement expands to the logistics industry in the form of, for example, on-demand delivery which offers an abrupt response to the customer [28,29]. The trend has changed rapidly around the world and the logistics service is demanded to support it. Therefore, LSPs need to innovate and create new service to address these complicated customer trends [30]. This paper demonstrates the use of the LSP Lifecycle Model to re-design the logistics service using three case-study LSPs in Thailand. It aims to address the research questions of whether the developed LSP Lifecycle Model is valid and if the Industry 4.0 concept is applicable and can be beneficial for LSPs. Firstly, the study identified the stage of the service. Secondly, services were re-designed using QFD. Thirdly, customer requirements were prioritized using the Multiple-Criteria Decision-Making (MCDM) tool. Finally, the Industry 4.0 concept was mapped to address these demands.

2. Literature Review

The goal of this literature review is to review if the LSP Lifecycle Model can be integrated with the Industry 4.0 concept. Whilst the Industry 4.0 is common and recognized to be beneficial to the industry and manufacturing sectors, the solid approach in logistics sectors is otherwise. In addition to maintaining service quality [31,32] and a standardized management system [33], re-designing the logistics service is required to extend its lifecycle.

2.1. LSP Lifecycle Model

LSPs play an important role in smooth operation and increasing the value-added in supply-chain management [34]. LSPs must satisfy customer requirements by collecting and analyzing data and managing the operations to resolve problems and to develop services that can support the whole lifecycle. The complexity of the logistics service is separated into activities which are based on what LSPs attempt to offer to match the customer needs. The level of party LSP is determined based on a set of characteristics that correspond to five levels, namely, 1PL, 2PL, 3PL, 4PL, and 5PL [10]. The level of service and commitment are incremental to the level of cooperation and adaptation of IT [35,36]. Agility is a highly important determinant of impressing the target group. The LSP Lifecycle Model cooperates with the physical flow (material flow) and information flow throughout each phase from forward and backward. The LSP Lifecycle Model is created by combing the logistics service characteristics of the entire lifecycle to fulfill customer requirements. The model comprises of three lifecycle phases: the Beginning of Life (BOL), the Middle of Life (MOL), and the End of Life (EOL). Creating and designing the logistics service is the first stage of implementing a new service to satisfy the customers [37]. Collecting and analyzing information data can assist in accommodating all of the requirements and lead to suggestions for service innovation. Evaluation and decomposition are the last phases of improving the service to satisfy customer needs throughout the lifecycle (see Figure 1).

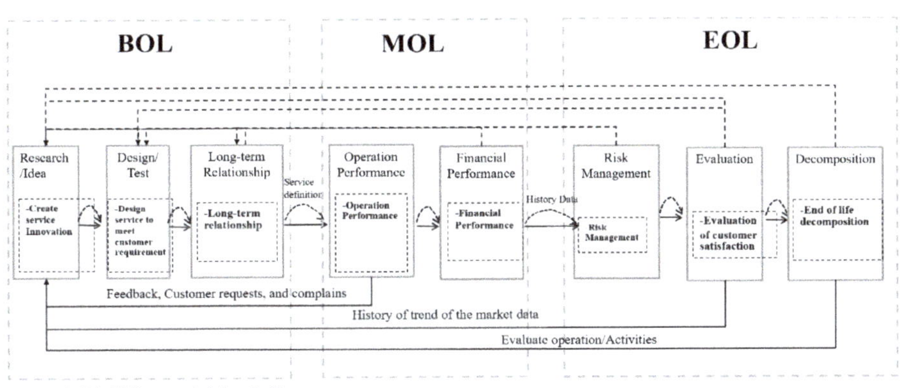

Figure 1. Interactions and relations in the Logistics Service Provider (LSP) Lifecycle Model [10]. Three lifecycle phases: BOL, Beginning of Life; MOL, Middle of Life; and EOL, End of Life.

2.2. Industry 4.0

Information Technology (IT) has been widely applied in manufacturing and the supply chain [22,23]. Industry 4.0 (I4.0) was first recognized in 2011 for progressing high-technology to advance the competitiveness in quality, cost, operation, and risk [38–41]. Supporting quick response, the phenomenon to offer customer satisfaction through "the right product at the right time and suitable cost" is now compulsory. The attractiveness of I4.0 is influenced by self-automation, Artificial Intelligence (AI), virtual technology, and real-time information [42]. Demand prediction and sales forecasting are examples of how to address customer expectations [43]. The main segments of I4.0 consist of Cyber-Physical Systems (CPSs), Internet of Things (IoT), Internet of Service (IoS), and smart factories or smart manufacturing. Real-time information, track, and trace technology are imperative parts of the logistics service, leading to trust and loyalty in the business market. Axiomatic Design and I4.0 were adapted to create service innovation using the LSP Lifecycle Model which creates new activities [44]. Figure 2 shows the strategies created based on I4.0 through the entire lifecycle of the logistics service. Figure 2 presents the Design Parameters (DP) in the Axiomatic Design, responding to customer need based on the LSP Lifecycle Model [44].

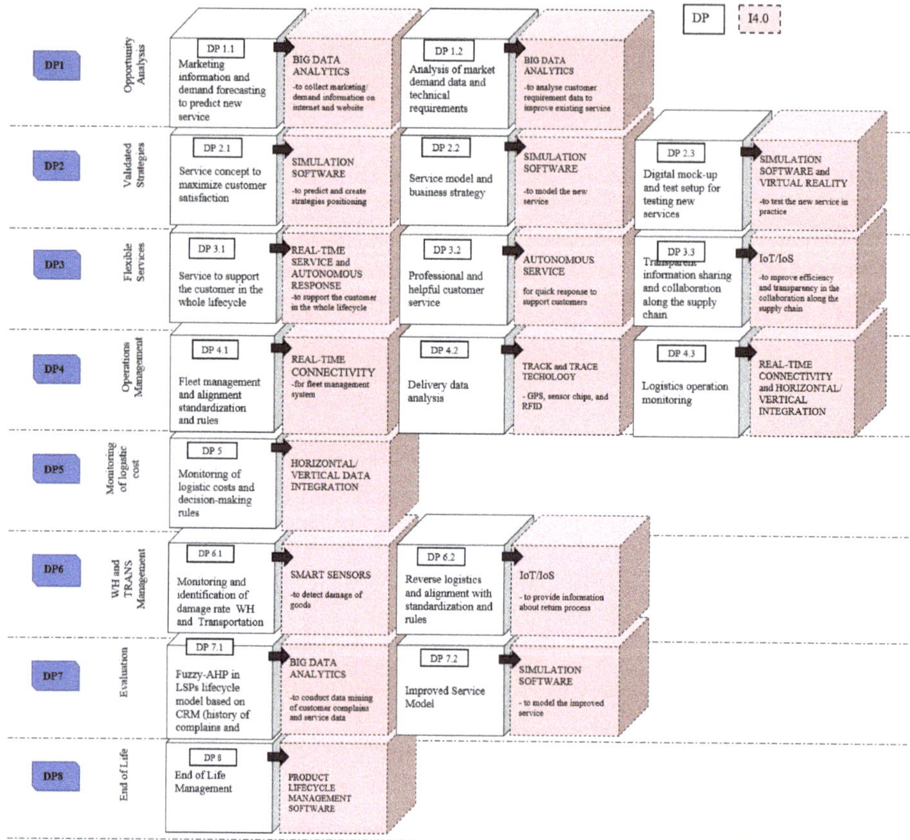

Figure 2. The implementation of Industry 4.0 (I4.0) in the LSP Lifecycle Model [44]. DP, Design Parameter.

3. Methodology

This section examines the two phases of the research methodology (see Figure 3). The first phase identified and evaluated the stage of the logistics service lifecycle. The second phase was the validation of the LSP Lifecycle Model. The study selected three case-study LSPs in Thailand.

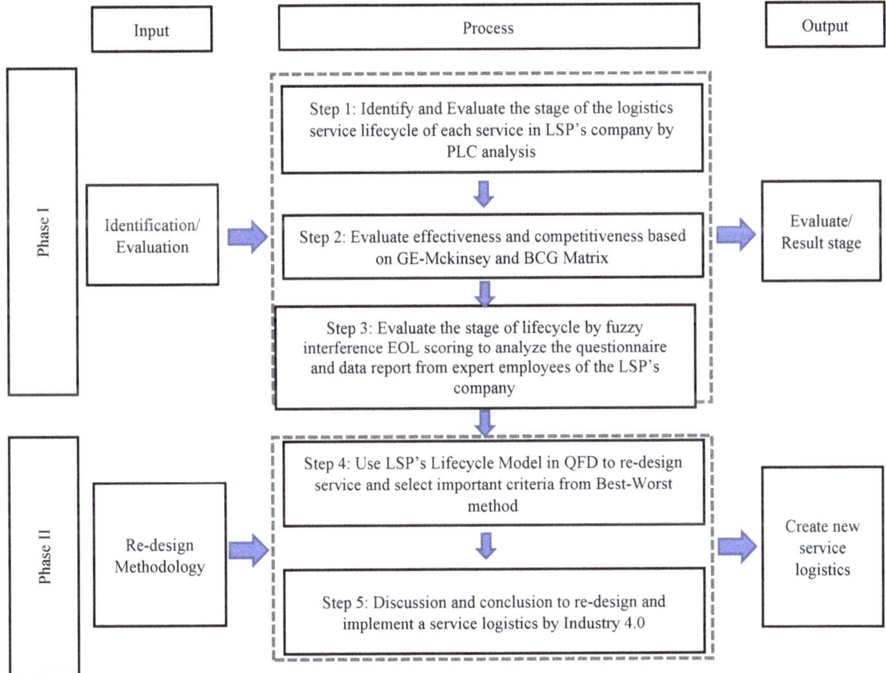

Figure 3. Phases of the research methodology. PLC, Production Lifecycle; QFD, Quality Function Deployment.

3.1. Phase I: Identification and Evaluation of the Lifecycle

The identification and evaluation of logistics service lifecycles involve recognizing the "As-Is" status of the product or service from the current status. "As-Is" is imperative in finding the correct way to be competitive in the global market [45,46]. A stage of the lifecycle can be evaluated using Production Lifecycle Analysis (PLC Analysis), Pareto analysis, and matrix analysis to estimate the current status of the lifecycle.

Step 1. PLC Analysis: The analysis consists of four stages of the lifecycle: introduction, growth, maturity, and decline. These four stages are analyzed in the market size, prime cost, profits, target customer, competitor, marketing strategy, product variety, and selling price.

Step 2. Matrix Analysis for qualitative evaluation: The effectiveness and competitiveness of the service are evaluated. GE-Mckinsey and BCG matrix can be used to estimate the target-market in business [47]. GE-Mckinsey matrix can evaluate the performance of product transition point by industry attractiveness (x-axis) and business strength (y-axis). It shows if any are weak, medium, or strong. BCG matrix has four fields in the market: question marks, dogs, stars, and cash cows. Question marks (problem child) products are those of high growth in the market but low in the market share. This field appears for new products to launch in the market. Dogs represent a low-growth product and low market share. This field is low-profit but normally is highly produced with high inventory costs. The company should get rid of this product. Cash-cow products are the products with a high market share in the slow-growing environment. This field can be a high-profit product but the company must determine if their product is mature in the market. The company should either improve or re-design these products to respond to customer requirements. Star products are those with high market demand in the rapid-growth industry. The company needs these products.

Step 3. Fuzzy Inference model: Qualitative information can be extracted from previous steps to quantify numbers using fuzzy logic. Fuzzy logic consists of three steps: fuzzification, rule evaluation, and defuzzification. The Fuzzy Inference model processes three main inputs: fuzzy of the BCG matrix, fuzzy of the GE matrix, and fuzzy of the EOL scoring. At first, the fuzzification determines the quantitative input as a value of 0 or 1. The membership function is to define the graph of the fuzzy set. For example, a fuzzy graph for the BCG matrix shows brand awareness on X-axis and sale volumes on Y-axis. Figure 4 shows the membership function on the BCG matrix, which separates 2 levels on X-Axis and Y-axis; low and high (Figure 4a,b). The second step, "If … Then rules," shows "If … Then rules" to find the relationship of the service. The last step, defuzzification, is used to evaluate and find the numerical values of the membership, using center of gravity techniques to compute defuzzification, namely, dogs, question marks, cash cows, and stars (Figure 4c). The evaluation of EOL scoring combines the analytics of the BCG matrix (X-axis) and the GE matrix (Y-axis) to classify the stage of logistics service.

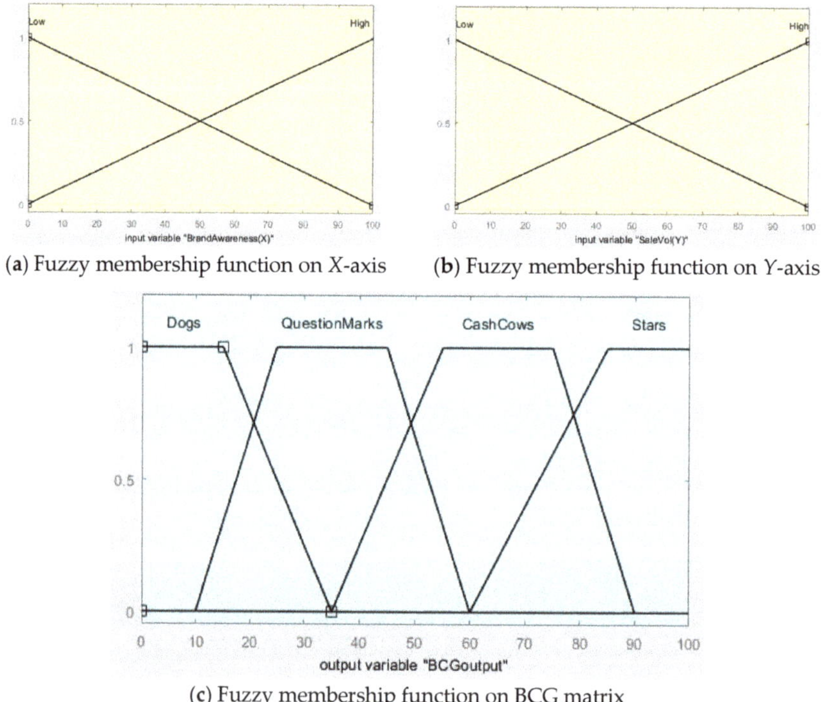

(**a**) Fuzzy membership function on X-axis   (**b**) Fuzzy membership function on Y-axis

(**c**) Fuzzy membership function on BCG matrix

Figure 4. Fuzzy interference model in membership function for the BCG matrix.

The identification and evaluation of logistics service is a key part to detecting the current status and positioning in the global market. This information is vital to improving and developing service that meets customer requirements and strengthens loyalty.

3.2. Phase II: Re-Designing Logistics Service

After identifying the stages of the logistics service, the selected logistics service in maturity and decline stages is then used as the case of re-design. The adaptation of the Best-Worst Method is for reflecting customer requirements and interest (Step 4). In Step 5, service is re-designed based on customer requirements using QFD and the application I4.0 in technical requirements.

3.2.1. Step 4: Best-Worst Method

Best-Worst method (BWM) is a recent Multi-Criteria Decision-Making (MCDM) method, developed to reduce the complexity of pairwise criteria in the AHP (Analytic Hierarchy Process) method [48–53]. Compared to AHP, BWM uses less comparison (BWM= 2n − 3, AHP= n(n − 1)/2), yet the consistency is better [48,54]. BWM is used to solve various problems, namely, supplier selection [55], location selection [50,53,56], service quality improvement [57], product design selection [49,58,59], supply-chain management [60], and performance evaluation [52,61]. BWM allows the decision-maker to select the best and the worst criteria to be transformed into the weight of each criterion using linear programming.

3.2.2. Step 5: Quality Function Deployment (QFD)

QFD is a famous product design technique that considers whether to address customer requirement and improve product quality or reduce costs and time [62–65]. The method translates "Voice of Customer" (VOC) or Customer requirement (CRs) into Engineering Characteristics (ECs) to create Design Requirements (DRs) for each phase of product development and production [64,65]. QFD adapts the CRs within the product to improve the quality of products, reduce time, costs, resources, and control the capability of production. Complete QFD consists of four phases: the product planning matrix, the part planning matrix, the process planning matrix, and the production/operation planning matrix. Information is required for both forward and backward data to develop a traceability method in all of the various phases [66]. VOC (WHATs) is combined with quality characteristics (HOWs) to create new products or to estimate costs and quality of production. This paper uses QFD to improve the matured or declined service by re-designing an appropriate lifecycle transition. VOC is defined from internal and external requirements and the service lifecycle requirement of what the customer needs [16].

4. LSP Companies and Service Re-Design: Case Study

In 2019, LSPs in Thailand have grown by 7–9% YoY according to the report from the Kasikorn Research Center [67]. This paper summarizes the results of three case studies of small, medium, and large LSP companies in Thailand. The companies provide several logistics services from local and national-wide transportation to moving services in the city area. The results of the identification and evaluation stages of the lifecycle (Step 1) are shown in Table 1. The results will be discussed in the following section. Then, the developing service logistics (Step 2) is described.

Table 1. The results of the identification and evaluation stage of the lifecycle of the case study LSP.

Service	PLC	GE (X)	GE (Y)	GE	GE Score	BCG (X)	BCG (Y)	BCG	BCG Score	EOL	EOL Score
Case Study 1	Growth	75	80	Investment/ Growth	77.31	70	65	Stars	57.82	Medium	61.15
Case Study 2	Growth	70	55	Selectivity/ Growth	53.15	15	70	Question Mark	39.44	Medium	50.00
Case Study 3	Maturity	65	75	Selectivity/ Growth	62.45	70	15	Cash cows	39.44	Medium	52.24

4.1. Case Study 1: Small Moving Service LSP

Case study 1 is a small LSP in the North of Thailand. The company offers a moving service which can be for office, house, or exhibition. Target groups are local customers, for example, students, schools, and offices in the Chiang Mai area. The questionnaire was answered by a senior manager in the company.

4.1.1. Identification and Evaluation Stage of the Lifecycle: Case Study 1

The results of the PLC analysis shows that the service is at the growth stage due to positive customer feedback and constant sales growth. The service is highly profitable and there are few competitors in the market. In addition, the company has a strong market strategy that guarantees package delivery. The company also offers a variety of services to fulfill customer requirements. Figure 5 shows the company is positive in the GE-Mckinsey matrix. The service is in the Investment/Growth stage due to the business strength stemming from the fully equipped moving service compared to other competitors. The service attractiveness is also high due to the demand in the exhibition market. Home and office movings are also growing as a result of the urbanization trend. The GE score is 77.31, which shows the attractiveness of this moving business. The service is in the stars stage in the BCG matrix due to high sales volume and brand awareness based on the history of the business. This service and business unit are spun off from the LSP mother company, providing service since 1988. The BCG score is 57.82. From these evaluations, it is found that the stage of the lifecycle of this service is in the medium stage, with an EOL score of 61.15. The results show the company responds well to customer needs in the industry (see Table 1).

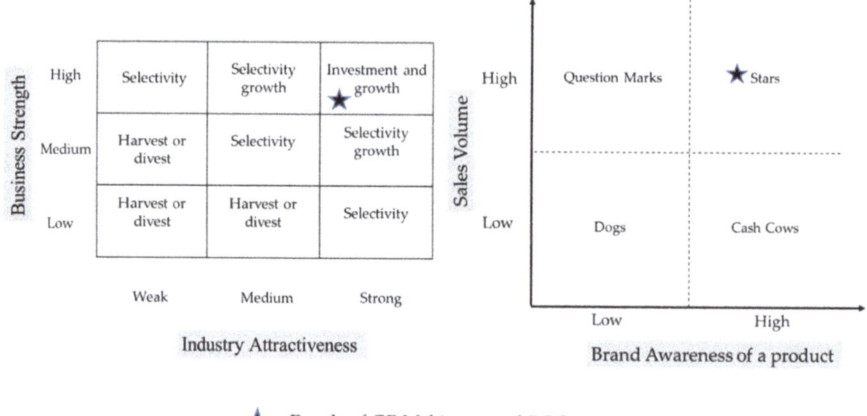

Figure 5. The analysis of GE Mckinsey and BCG matrix for Case Study 1.

4.1.2. Service Re-Design: Case Study 1

To improve the service by fulfilling customer requirements, the evaluation of the factor is conducted using the BWM (weight scale 1–9). The factors are prioritized from most to least important in Table 3. Design service to meet customer requirement criteria is the most significant criterion. The least important criterion is the evaluation of customer satisfaction. It leads to optimized weight for main attributes, shown in Tables 2 and 3.

Table 4 illustrates the scores on the Important Rate (IMP), analyzed by the BWM, which are allocated in QFD. The concern is in creating and re-designing the service to respond to customer needs. The aim is to build up loyalty by offering tracking service after sales, which will support fulfilling and delivery service. The high competition in the moving market is the main part that needs innovation. I4.0 suggests the following tools: text response, real-time tracking, and real-time monitoring of moving operation. This will help the customer to manage their plans easily. Moreover, IT is adapted to replace paper and to increase customer satisfaction. Engineering Characteristics and target setting are shown in Table 5.

Table 2. Comparison vectors for the best and worst criteria: Case Study 1.

Best to Others	Create Service Innovation	Design Service to Meet Customer Requirement	Long-Term Relationship	Operations Performance	Financial Performance	Risk Management	Evaluation of Customer Satisfaction	End of Life Decomposition
Best Criterion: Long-Term Relationship	4	1	3	5	2	8	9	7
Others to Worst								Worst criterion: End of life decomposition
Create Service Innovation								6
Design Service to Meet Customer Requirements								9
Long-Term Relationship								7
Operations Performance								4
Financial Performance								8
Risk Management								2
Evaluation of Customer Satisfaction								1
End of Life Decomposition								3

Table 3. Optimized weight for the main criterion: Case Study 1.

Criterion	Weights	ξ^{L}**
Create Service Innovation	0.1023	0.0731
Design Service to Meet Customer Requirements	0.3361	
Long-Term Relationship	0.1364	
Operations Performance	0.0818	
Financial Performance	0.2046	
Risk Management	0.0511	
Evaluation of Customer Satisfaction	0.0292	
End of Life Decomposition	0.0584	

** ξ^{L} is used as an indicator for the comparison.

Table 4. House of Quality (HoQ): Case Study 1. IMP, Important Rate.

Technical Requirement		IMP	BOL			MOL		EOL		
Customer Need			Collect and Analyse Data from Customer Suggestions and Marketing data	Design and Test Service to Satisfy Customer Requirements	Implement Customer Relationship Management Initiation	Improve Efficiency of Logistics Operations	Reduce Logistics Cost	Analyse and Evaluate the Risk of Warehouse and Transportation	Evaluate and Improve Service Life Cycle	Evaluate the End of Life Cycle
BOL	Create Service Innovation	0.1023	9							
	Design Service to Meet Customer Requirements	0.3361		9						
	Long-term Relationship	0.1364			3					
MOL	Operations Performance	0.0818				9				
	Financial Performance	0.2046					9			
EOL	Risk Management	0.0511						1		
	Evaluation of Customer Satisfaction	0.0292							3	
	End of Life Decomposition	0.0584								3
Importance of Technical Indicators			0.92	3.02	0.41	0.74	1.84	0.05	0.09	0.18
Relative Importance of Technical Indicators			12.71	41.74	5.65	10.16	25.41	0.71	1.21	2.42

Table 5. Details of service re-design and improvement: Case Study 1.

	Engineering Characteristics	**Target Setting**
FR2.	Design and test service to satisfy customer requirements	Simulation and virtual technologies to create and test company strategies
	FR2.1 Understand the added value for the customer in service innovation	Using simulation software to predict and create strategic positioning
	FR2.2 Design the new/adapted service offer	Using simulation software and other tools to model the new service
	FR2.3 Test the new service for practical applicability	Using simulation software and virtual reality to test the new service in practice
FR4.	Reduce logistics cost	Horizontal/vertical data integration to enable sophisticated business intelligence for cost controlling, use paperless strategies to reduce cost
FR1.	Create service innovation	Big data analytics to collect and analyze customer requirement data and marketing demand
	FR1.1 Identify customer requirements and opportunities to generate new service	Using Big Data Analytics to collect marketing/ demand information on internet and website
	FR1.2 Innovate existing service to increase customer satisfaction	Using Big Data Analytics to analyze customer requirement data and evaluation data from the customer to improve existing service

4.2. Case Study 2: Courier Service Provider

Case study 2 is a medium-sized LSP offering courier services. The company uses passenger buses to move passengers and parcels within the routes mostly in the North of Thailand. The prominent point of the company is the specific route and time, which the customer can use to specify the departure and arrival times. The service of interest is parcel service. Targeted customers are local businesses, people, students, and hospitals. The questionnaire was answered by a senior manager of the company.

4.2.1. Identification and Evaluation Stage of the Lifecycle: Case Study 2

The service is identified and evaluated as a growth stage due to the continuous growth of sales and profit. The service is strong with the strategy of on-time delivery and flexibility of service, from agriculture products to motorcycles. The GE-Mckinsey matrix shows that the service is at the selective-growth stage. Business strength is medium due to the limitation of vehicles and its fixed route. The customer can pick the delivery up at the station only. Industry attractiveness is at the high stage due to the popularity of e-commerce. The service area is specific to its northern route. Thus, the parcel can be delivered quickly. For the BCG matrix, the service is in the question mark stage. Sale volume is high but the brand awareness of the product is low because of several competitors in the market, both domestic and overseas. The price is the key factor. From these evaluations, the service is considered to be at the medium stage of its lifecycle. Details are shown in Table 1.

4.2.2. Service Re-Design: Case Study 2

To improve service, fulfillment is suggested. The focus is to enhance the delivery service (see Table 5). A long-term relationship is the most important factor. Evaluation of customer satisfaction and financial performance are among the most important factors. The decision-maker decides to improve on these issues to gain trust and loyalty from the customer. At the start, the company will survey for suggestions and use this information to improve the service. I4.0 is applied by developing the website. Evaluation and improving service is highly recommended to increase customer satisfaction. The data can be collected from several available platforms such as social media. The feedback data is a critical part of re-designing the service. For example, if tracking is not online and updated on the platform, the customer cannot check the delivery status. GPS and RFID can be used to address the issue. Then the customers will be able to access real-time information. On-time delivery will be possible to satisfy customer expectations; the customer should be able to access real-time information. In addition, parcel tracking and tracking information can be sent to the customer (via SMS), informing them when their package is due to arrive. Table 6 shows an example of the measurements to improve the service life cycle for the case-study company.

Table 6. Details of service re-design and improvement: Case Study 2.

	Engineering Characteristics	**Target Setting**
FR7	Evaluate and improve the service life cycle	Big data analytics and simulation
FR7.1	Evaluate the service life cycle	Collecting Data from the branch, application online (Facebook, line) and feedback data
FR7.2	Re-design and improve service after evaluation	Simulation software to model the improved service

4.3. Case Study 3: Large LSP

Case study 3 is a large LSP in the North of Thailand. The company has more than 500 trucks and more than 1000 employees. The service is the delivery of mass products between Chiang Mai and Bangkok. Most of the customers in Chiang Mai are local small and medium Enterprises (SMEs) and farmers who sent their goods to Bangkok. On the contrary, most of the products from Bangkok are industry products, shipped to wholesalers and retailers in Chiang Mai. From the surveying of the senior manager who is responsible for the company strategy, the results are as follows.

4.3.1. Identification and Evaluation Stage of the Lifecycle: Case Study 3

The service is at the maturity stage due to the market size and low prime costs. Profit is high but is declining. Selling price is competitive-edged but it cannot be reduced because there is not much profit. The GE-Mckinsey matrix suggests that the service is at the selectivity growth stage. Industry attractiveness is at the middle stage. As a result, this business may face a lot of competition. Business strength is at a high level due to the variety of vehicles such as trucks and trailers, as well as a refrigerated container which can cater to a larger variety of customer requirements. The BCG matrix identifies the service at the cash cow stage. BCG(X) shows the middle volume of sales but BCG(Y) shows high brand awareness because the company has been in business for a long time and has gained loyalty from their customers. Identification and evaluation showed that the stage of the lifecycle is medium. Details are shown in Table 1.

4.3.2. Service Re-Design: Case Study 3

The improvement to satisfy the customer is significant and the data is suggesting that the company should re-design its delivery service. From BWM, long-term relationship criterion is the most important one. The least important criterion is end of life decomposition. Table 7 shows an example of CRM implementation for the case-study company.

Table 7. Details of service re-design and improvement: Case Study 3.

	Engineering Characteristics	Target Setting
FR3	Implement CRM initiation	Internet of Things/service and real-time autonomous service to support customer
FR3.1	Focus long term partnership	Real-time service, track and trace technology for support product delivery
FR3.2	Respond quickly to the customer	Autonomous service for quick response to support customers, Application Technology to offer the client
FR3.3	Increase supplier relationship	Internet of Things/Service to improve efficiency and transparency in the collaboration along the supply chain (e.g., responsive information platform)

5. Discussion

The study confirms the significance of technology advancement in business competitiveness. In the case of logistics management, Industry 4.0 can fulfill the needs of Smart Logistics [42,68–70]. However, in the case of LSPs, limited research has clearly demonstrated the benefits of implementing the Industry 4.0 concept [71,72], especially for re-designing service to accommodate customer needs depending on their lifecycle stages.

It can be seen from the case studies that the Industry 4.0 concept can be applicable for re-designing logistics services. With different growth stages and business strengths, the lifecycle stage of the service can be identified. As a result, a suitable measurement can be suggested based on customer requirements and technology readiness. Industry 4.0 tools, such as IoT, Big Data Analytics, and Virtual Technology, are found to be supportive if the LSP wishes to accommodate their customers. This will increase their competitiveness in the highly ambitious market. The implementation can be costly and complicated [73], however further study must be conducted to determine if the measures are feasible.

6. Conclusions

In the Fourth Industrial Revolution, the customers require high-quality service from LSPs. Service innovation is an important tool to compete in the world economy. This paper identified and investigated each stage of the LSP service re-design. In the first step, the status of a lifecycle state was evaluated using PLC analysis, Pareto analysis, matrix analysis (GE, BCG matrix), and fuzzy interference. The second step re-designed the matured services for attaining customer satisfaction. QFD was used to create and re-design logistics services. In the significant evaluation of criteria, BWM was applied to estimate the important criteria in LSP's Lifecycle Model that had eight criteria. The paper showed the results from three logistics services which were at the maturity stage and that each company needed Industry 4.0 to improve their service quality. For case study 1, the most important factor was to design service that meets customer requirements. To create new service and increase information, responsiveness is an important factor to advance their operations. There are available tools of Industry 4.0 to address these issues such as simulation, virtual technology, horizontal/vertical data integration, and Big Data Analytics. As a result, the new service can be modeled and tested. The costs can be reduced. Data can be analyzed to improve the existing service to meet customer requirements. For case study 2, on-time delivery was the main point of competitiveness in the business. A long-term relationship was also the main interest. Therefore, it was suggested that the service data should be used and analyzed in order to improve service. The data should also be shared with the customers to improve their satisfaction. For case study 3, the main focus was concerned with CRM initiation. Fulfilling customer satisfaction to sustain their clients and to generate loyalty and trust was recommended. Here, IoT, IoS, and autonomous service are a few examples of what can be used for that purpose.

The paper shows that Industry 4.0 can be applied to re-design logistics service, but with caution. Although the developed LSP Lifecycle Model and the presented methodology can provide suggestive measures for LSP per their service requirements, there are some limitations. Whilst the first phase of identification and evaluation can give contemplative and multi-dimensional observation of the service of interest, the second phase of new service re-design can only be suggestive and the result is rather dynamic. The implementation of the measures to solidly validate the model is required, which can be even more challenging.

Author Contributions: Conceptualization, S.T., S.R., and K.Y.T.; methodology, S.T. and S.R.; validation, S.T., S.R., and K.Y.T.; formal analysis, S.T. and S.R.; investigation, S.R. and K.Y.T.; resources, S.T.; writing—original draft preparation, S.T.; writing—review and editing, S.R. and K.Y.T.; visualization, S.T.; supervision, S.R. and K.Y.T.; project administration, K.Y.T. All authors have read and agreed to the published version of the manuscript.

Funding: Horizon 2020: 734713.

Acknowledgments: This research is part of the project "Industry 4.0 for SMEs" from the European Union's Horizon 2020 research and innovation program under the Marie Skłodowska-Curie grant agreement No 734713. This research work was partially supported by Chiang Mai University, Thailand.

Conflicts of Interest: The authors declare no conflict of interest.

References

1. Nyadzayo, M.W.; Khajehzadeh, S. The antecedents of customer loyalty: A moderated mediation model of customer relationship management quality and brand image. *J. Retail. Consum. Serv.* **2016**, *30*, 262–270. [CrossRef]
2. Zawadzki, P.; Żywicki, K. Smart product design and production control for effective mass customization in the industry 4.0 concept. *Manag. Prod. Eng. Rev.* **2016**, *7*, 105–112. [CrossRef]
3. Christiane, H.; Hariolf, G. Innovation in the service sector: The demand for service-specific innovation measurement concepts and typologies. *Res. Policy* **2005**, *34*, 517–535.
4. Kowalkowski, C.; Gebauer, H.; Kamp, B.; Parry, G. Servitization and deservitization: Overview, concepts, and definitions. *Ind. Mark. Manag.* **2017**, *60*, 4–10. [CrossRef]
5. Lam, J.S.L.; Dai, J. Environmental sustainability of logistics service provider: An ANP-QFD approach. *Int. J. Logist. Manag.* **2015**, *26*, 313–333. [CrossRef]
6. Costes, N.F.; Roussat, C. Supply Chain Integration: Views from a Logistics Service Provider. *Supply Chain Forum* **2011**, *12*, 20–30. [CrossRef]
7. Goldsby, T.J.; Zinn, W.; Closs, D.J.; Daugherty, P.J.; Stock, J.R.; Fewcett, S.E.; Waller, M. Editional Reflections on 40 Years of the Journal of Business Logistics: From the Editors. *J. Bus. Logist.* **2019**, *40*, 4–29. [CrossRef]
8. Alkhatib, S.F.; Darlington, R.; Yang, Z.; Nguyen, T.T. A novel technique for evaluating and selecting logistics service providers based on the logistics resource view. *Expert Syst. Appl.* **2015**, *42*, 6976–6989. [CrossRef]
9. Pedrosa, A.D.M.; Blazevic, V.; Jasmand, C. Logistics innovation development: A micro-level perspective. *Int. J. Phys. Distrib. Logist. Manag.* **2015**, *45*, 313–332. [CrossRef]
10. Tiwong, S.; Ramingwong, S. Lifecycle Management Model Review and Design for LSP. In Proceedings of the 2nd International Conference on High Performance Compilation Computing and Communications, Hong Kong, China, 15–17 March 2018; Volume 19, pp. 88–92.
11. Ginting, R.; Ali, A.Y. Triz or DFMA combined with QFD as product design methodology: A review. *Pertanika J. Sci. Technol.* **2016**, *24*, 1–25.
12. Chowdhury, M.M.H.; Quaddus, M.A. A multi-phased QFD based optimization approach to sustainable service design. *Int. J. Prod. Econ.* **2016**, *171*, 165–178. [CrossRef]
13. Lam, J.S.L.; Bai, X. A quality function deployment approach to improve maritime supply chain resilience. *Transp. Res. E Logist. Transp. Rev.* **2016**, *92*, 16–27. [CrossRef]
14. Harano, M.; Santoso, A.; Prayogo, D.N. How kansei engineering, kano and QFD improve logistics services. *Int. J. Technol.* **2017**, *8*, 1070–1081.
15. Bolar, A.A.; Tesfamariam, S.; Sadiq, R. Framework for prioritizing infrastructure user expectations using Quality Function Deployment (QFD). *Int. J. Sustain. Built Environ.* **2017**, *6*, 16–29. [CrossRef]

16. Ji, J.J.; Ying, P.; Ying, L. Prioritizing engineering characteristics based on customer online reviews for quality function deployment. *J. Eng. Des.* **2014**, *25*, 303–324.

17. Kurtulmuşoğlu, F.B.; Pakdil, F.; Atalay, K.D. Quality improvement strategies of highway bus service based on a fuzzy quality function deployment approach. *Transportmetrica* **2016**, *12*, 175–202. [CrossRef]

18. Yang, S.; Liu, J.; Wang, K.; Miao, Y. An Uncertain QFD Approach for the Strategic Management of Logistics Services. *Math. Probl. Eng.* **2016**, *2016*, 1486189. [CrossRef]

19. Ho, W.; He, T.; Lee, C.K.M. Strategic logistics outsourcing: An integrated QFD and fuzzy AHP approach. *Expert Syst. Appl.* **2012**, *39*, 10841–10850. [CrossRef]

20. Lin, Y.; Pekkarinen, S. QFD-based modular logistics service design. *J. Bus. Ind. Mark.* **2011**, *26*, 344–356. [CrossRef]

21. Fonseca, L.M.; Domingues, J.P.; Pereira, M.T.; Martins, F.F.; Zimon, D. Assessment of circular economy within Portuguese organizations. *Sustainability* **2018**, *10*, 2521. [CrossRef]

22. Rüßmann, M.; Lorenz, M.; Gerbert, P.; Waldner, M.; Justus, J.; Engel, P.; Harnisch, M. *Industry 4.0: The Future of Productivity and Growth in Manufacturing Industries*; Boston Consulting Group: Boston, MA, USA, 2015.

23. Choi, D.; Song, B. Exploring Technological Trends in Logistics: Topic Modeling-Based Patent Analysis. *Sustainability* **2018**, *10*, 2810. [CrossRef]

24. Wollschlaeger, M.; Sauter, T.; Jasperneite, J. The future of industrial communication automation networks in the era of the internet of things and industry 4.0. *IEEE Ind. Electron. Mag.* **2017**, *11*, 17–27. [CrossRef]

25. Schumacher, A.; Erol, S.; Sihn, W. A maturity for assessing Industry 4.0 readiness and maturity of manufacturing enterprises. *Procedia CIRP* **2016**, *52*, 161–166. [CrossRef]

26. Govindan, K.; Cheng, T.C.E.; Mishra, N.; Shuka, N. Big data analytic and application for logistics and apply chain management. *Transport Res. E-Logist.* **2018**, *114*, 343–349. [CrossRef]

27. Ivanov, D.; Dolgui, A.; Sokolov, B. The impact of digital technology and Industry 4.0 on the ripple effect and supply chain risk analytics. *Int. J. Prod. Res.* **2019**, *57*, 829–846. [CrossRef]

28. Zhang, Y.; Liu, S.; Liu, Y.; Li, R. Smart box-enabled product-service system for cloud logistics. *Int. J. Prod. Res.* **2016**, *54*, 6693–6706. [CrossRef]

29. Silva, V.L.D.S.; Kovaleski, J.L.; Pagani, R.N. Technology transfer in the supply chain oriented to industry 4.0: A literature review. *Technol. Anal. Strateg.* **2019**, *31*, 546–562. [CrossRef]

30. Tao, F.; Cheng, J.; Qi, Q.; Zhang, M.; Zhang, H.; Sui, F. Digital twin-driven product design, manufacturing and service with big data. *Int. J. Adv. Manuf. Technol.* **2017**, *94*, 3563–3576. [CrossRef]

31. Clegg, B.; Kersten, W.; Koch, J. The effect of quality management on the service quality and business success of logistics service providers. *Int. J. Qual. Reliab. Manag.* **2010**, *27*, 185–200.

32. Kilibarda, M.; Zečević, S.; Vidović, M. Measuring the quality of logistic service as an element of the logistics provider offering. *Total Qual. Manag. Bus. Excell.* **2012**, *23*, 1345–1361. [CrossRef]

33. Zimon, D.; Madzik, P.; Sroufe, R. Management systems and improving supply chain processes: Perspectives of focal companies and logistics service providers. *Int. J. Retail Distrib. Manag.* **2020**, in press. [CrossRef]

34. Liu, C.L.; Lee, M.Y. Integration, supply chain resilience, and service performance in third-party logistics providers. *Int. J. Logist. Manag.* **2018**, *29*, 5–21. [CrossRef]

35. Drakovic, M. Organization of outsourcing in logistics partnership between the seaports of Montenegro and Slovenia. *MJSS* **2013**, *9*, 93–113.

36. Talty, M.; Moutmihi, M. From the logistics function to the logistics service: A literature Review. *GJMBR* **2015**, *15*, 1–7.

37. Matt, D.T.; Rauch, E.; Fraccaroli, D. A Three level model for the design, planning and operation of changeable production systems in distributed manufacturing. In Proceedings of the 5th International Conference on Changeable, Agile, Reconfigurable and Virtual Production (CARV 2013), Munich, Germany, 6–9 October 2013; pp. 23–28.

38. Ford, M. *Rise of the Robots: Technology and the Threat of a Jobless Future*, 1st ed.; Basic Books: New York, NY, USA, 2015.

39. Hermann, M.; Pentek, T.; Otto, B. Design Principles for Industrie 4.0 Scenarios. In Proceedings of the 2016 49th Hawaii International Conference on System Sciences (HICSS), Koloa, HI, USA, 5–8 January 2016; pp. 3928–3937.

40. Jabbour, A.B.L.D.S.; Jabbour, C.J.C.; Filho, M.G.; Roubaud, D. Industry 4.0 and the circular economy: A proposed research agenda and original roadmap for sustainable operations. *Ann. Oper. Res.* **2018**, *270*, 273–286.

41. Kaipia, R.; Turkulainen, V. Managing integration in outsourcing relationships—The influence of cost and quality priorities. *Ind. Mark. Manag.* **2017**, *61*, 114–129. [CrossRef]

42. Hofmann, E.; Rüsch, M. Research paper Industry 4.0 and the current status as well as future prospects on logistics. *Comput. Ind.* **2017**, *89*, 23–34. [CrossRef]

43. See-To, E.W.K.; Ngai, E.W.T. Customer reviews for demand distribution and sales nowcasting: A big data approach. *Ann. Oper. Res.* **2018**, *270*, 415–431. [CrossRef]

44. Tiwong, S.; Rauch, E.; Šoltysová, Z.; Ramingwong, S. Industry 4.0 for Managing Logistic Service Providers Lifecycle. In Proceedings of the 13th International Conference on Axiomatic Design (ICAD 2019), Sydney, Australia, 18–20 October 2019.

45. Shen, L.; Zhou, J.; Skitmore, M.; Xia, B. Application of a hybrid Entropy–McKinsey Matrix method in evaluating sustainable urbanization: A China case study. *Cities* **2015**, *42*, 186–194. [CrossRef]

46. Oh, J.; Han, J.; Yang, J. A fuzzy-based decision-making method for evaluating product discontinuity at the product transition point. *Comput. Ind.* **2014**, *65*, 746–760. [CrossRef]

47. Oh, J.; Yang, J.; Lee, S. Managing uncertainty to improve decision-making in NPD portfolio management with a fuzzy expert system. *Expert Syst. Appl.* **2012**, *39*, 9868–9885. [CrossRef]

48. Brunelli, M.; Razaei, J. A multiplicative best-worst method for multi-criteria decision making. *Oper. Res. Lett.* **2019**, *47*, 12–15. [CrossRef]

49. Gupta, H.; Barua, M.K. Identifying of technological innovation for Indian MSMEs using best-worst multi-criteria decision making method. *Technol. Forecast. Soc.* **2016**, *107*, 69–79. [CrossRef]

50. Tian, Z.P.; Zhang, H.Y.; Wang, J.Q.; Wang, T.L. Green supplier selection using improved TOPSIS and Best-Worst method under intuitionistic fuzzy environment. *Informatica* **2018**, *29*, 773–800. [CrossRef]

51. Velasquez, M.; Hester, P.T. An Analysis of Multi-Criteria Decision Making Methods. *Int. J. Oper. Res.* **2013**, *10*, 56–66.

52. Razaei, J. Best-Worst multi-criteria decision making method. *Omega* **2015**, *53*, 49–57. [CrossRef]

53. Safarzadeh, S.; Khansefid, S.; Barzoki, M.R. A group multi-criteria decision-making based on best-worst method. *Comput. Ind. Eng.* **2018**, *126*, 111–121. [CrossRef]

54. Sadjadi, S.; Karimi, M. Best-worst multi-criteria decision-making method: A robust approach. *Decis. Sci. Lett.* **2018**, *7*, 323–340. [CrossRef]

55. Aboutorab, H.; Saberi, M.; Asadabadi, M.R.; Hussain, O. ZBWM: The Z-number extension of Best Worst Method and its application for supplier development. *Expert Syst. Appl.* **2018**, *107*, 115–125. [CrossRef]

56. Kheybari, S.; Kazemi, M.; Razaei, J. Bioethanol facility location selection using best-worst method. *Appl. Energy* **2019**, *242*, 612–623. [CrossRef]

57. Gupta, H. Evaluating service quality of airline industry using hybrid best worst method and VIKOR. *J. Transp. Manag.* **2018**, *68*, 35–47. [CrossRef]

58. Omrani, H.; Alizadeh, A.; Emrouznejad, A. Finding the optimal combination of power plants alternatives: A multi response Taguchi-neural network using TOPSIS and fuzzy best-worst method. *J. Clean. Prod.* **2018**, *203*, 210–223. [CrossRef]

59. Maghsoodi, A.; Mosavat, M.; Hafezalkotob, A.; Hafezalkotob, A. Hybrid hierarchical fuzzy group decision-making based on information axioms and BWM: Prototype design selection. *Comput. Ind. Eng.* **2019**, *127*, 788–804. [CrossRef]

60. Ahmadi, H.B.; Sarpong, S.; Razaei, J. Assessing the social sustainability of supply chains using Best Worst Method. *Resour. Conserv. Recycl.* **2017**, *126*, 99–106. [CrossRef]

61. Salimi, N.; Razaei, J. Evaluating firms' R&D performance using best worst method. *Eval. Program Plan.* **2018**, *66*, 147–155.

62. Yang, L.; Xing, K. Innovative Conceptual Design Approach for Product Service System based on TRIZ. In Proceedings of the 10th International Conference on Service Systems and Service Management (ICSSSM), Hong Kong, China, 17–19 July 2013.

63. Liao, C.N.; Kao, H.P. An evaluation approach to logistics service using fuzzy theory quality function deployment and goal programming. *Comput. Ind. Eng.* **2014**, *68*, 54–64. [CrossRef]

64. Delice, E.K.; Güngör, Z. A new mixed integer linear programming model for product development using quality function deployment. *Comput. Ind. Eng.* **2011**, *57*, 906–912. [CrossRef]
65. Franceschini, F.; Galetto, M.; Maisano, D.; Mastrogiacomo, L. Prioritization of engineering characteristics in QFD in the case of customer requirements orderings. *Int. J. Prod. Res.* **2015**, *53*, 3975–3988. [CrossRef]
66. Sharma, N.; Singhi, R. Logistics and supply chain management quality improvement of supply chain process Through Vendor Managed Inventory: A QFD Approach. *J. Supply Chain Manag.* **2018**, *7*, 23–33.
67. Growth of E-commerce and Logistics (in Thai). Available online: https://kasikornbank.com/th/business/sme/ KSMEKnowledge/article/KSMEAnalysis/Pages/E-Commerce_Logistic.aspx (accessed on 28 December 2017).
68. Barreto, L.; Amaral, A.; Pereira, T. Industry 4.0 implications in logistics: An overview. *Procedia Manuf.* **2017**, *13*, 1245–1252. [CrossRef]
69. Witkowski, K. Internet of things, big data, industry 4.0—Innovative solutions in logistics and supply chains management. *Procedia Eng.* **2017**, *182*, 763–769. [CrossRef]
70. Maslarić, M.; Nikoličić, S.; Mirčetić, D. Logistics response to the industry 4.0: The physical internet. *Open Eng.* **2016**, *6*, 2391–5439. [CrossRef]
71. Zsifkovits, H.; Woschank, M.; Ramingwong, S.; Wisittipanich, W. State-of-the-Art Analysis of the Usage and Potential of Automation in Logistics. In *Industry 4.0 for SMEs*; Palgrave Macmillan: Cham, Switzerland, 2020; pp. 193–212.
72. Dallasega, P.; Woschank, M.; Zsifkovits, H.; Tippayawong, K.; Brown, C.A. Requirement Analysis for the Design of Smart Logistics in SMEs. In *Industry 4.0 for SMEs*; Palgrave Macmillan: Cham, Switzerland, 2020; pp. 147–162.
73. Ramingwong, S.; Manopiniwes, W.; Jangkrajarng, V. Human factors of thailand toward industry 4.0. *Manag. Res. Pract.* **2019**, *11*, 15–25.

Article

Problems with the Implementation of Industry 4.0 in Enterprises from the SME Sector

Manuela Ingaldi * and Robert Ulewicz

Faculty of Management, Czestochowa University of Technology, al. Armii Krajowej 19b, 42-200 Czestochowa, Poland; robert.ulewicz@wz.pcz.pl
* Correspondence: manuela.ingaldi@wz.pcz.pl; Tel.: +48-34-3250-426

Received: 3 November 2019; Accepted: 24 December 2019; Published: 26 December 2019

Abstract: Industry is currently undergoing a revolution (called Revolution 4.0) related to the far-reaching integration of all production areas through the digitization and the creation of new communication channels. The Polish economy generated a GDP of USD 524.5 billion in 2017, of which small and medium enterprises generated about 50% of revenue and in which microenterprises accounted for the largest share in generating GDP; i.e., around 30.2%. The aim of the research is to determine the adaptation possibilities of the small and medium-sized enterprises (SME) sector to Industry 4.0 solutions. Pilot research was carried out in the Czestochowa Industrial District. Enterprises from the SME sector were asked to provide an information about the used technologies and to determine the level of their organization's readiness for sustainable development through technological transformation. Financial resources as well as a lack of specialized support in obtaining new technologies were defined as problems as far as transformation is concerned. The solution of the diagnosed problem is the development of a platform aimed at integrating the potential of enterprises from the SME sector in order to undertake joint actions for sustainable development.

Keywords: sustainability; Industry 4.0; SMEs; e-business modelling

1. Introduction

Each enterprise works in a turbulent environment. Such conditions apply to different spheres of business activity. It is important for an enterprise to observe such changes and, if necessary, to react accordingly. Only an appropriate reaction allows an enterprise to adapt to this situation. The global industrial landscape has changed dramatically over the past few years due to the rapid technological development and innovation in production processes [1]. In recent years, many initiatives have been launched for intelligent production in the world in order for a significant share of industry in the economy to be stored and regained [2,3].

In recent years, a new, large industrial revolution can be observed, affecting the functioning of enterprises in the market and resulting in different changes. This is the fourth wave of socio-economic evolution, called Industry 4.0 [4,5]. The development towards Industry 4.0 has a significant impact on the production industry. It presents the latest trends in automation technologies that are gaining popularity in the manufacturing industry. Additionally, Industry 4.0 has become a new topic for scientists dealing in disciplines such as management and business economics; thus, many publications on this subject have appeared [6]. Industry 4.0 seems to be one of the most promising ideas in terms of boosting industrialization and industrial competitiveness in EU countries [7].

The fourth industrial revolution differs from the previous ones because it applies to all aspects of our lives. Within its framework, industry processes and commercializes the exchange of information between people and between people and objects, as well as between the objects themselves [8]. Industry 4.0 gives the industry a new perspective that allows it to work with new technologies to achieve maximum efficiency with the minimal use of resources in the manufacturing industry. It affects the strategies and operations of enterprises, as well as the relationships between enterprises, customers and suppliers. In addition, decentralized production processes play an important role in Industry 4.0 through intelligent products that know their destinations, history, etc. [9].

The beginnings of Industry 4.0 stemmed from the strategy the German government started in 2013, focusing on achieving greater operational efficiency and productivity through automation. The first information about the Industry 4.0 concept appeared in 2014 at the World Economic Forum conference in Davos [10–12].

The horizontal and vertical integration of production systems are the foundations of Industry 4.0. These systems are based on real-time data exchange and flexible production to enable customized production [13,14]. The pillars of this evolution are, among others, the Internet of Things (allowing global access to data and machines coupled into the network), autonomous robots/systems (autonomy of production processes that allows for the full adjustment of production to the needs of the individualized market), cyber security, huge amounts of data fpr processing (analysis of big data sets, Big Data and Cloud Computing), additive printing (3D), advanced simulations, and virtual and extended reality [8,15–18].

One of the main elements of Industry 4.0 is the Internet of Things. Since the end of the 20th century, it has been possible to observe the progressive digitization of everything that surrounds us, as well as the rapid development of the Internet. Currently, most European enterprises have access to the Internet, and many of them offer their products via the Internet. Due to global communication, products and components will be able to exchange data with any IoT devices, thus significantly expanding their functions [19]. Digital transformation has not only changed the way the organization works, but, as a result, the market is also transforming into a full value chain. Industry 4.0 causes essential changes to issues such as the economy, work environment and skill development [20].

Virtual and extended reality is yet another significant element of Industry 4.0. It allows the collection of large amounts of data and the creation of various types of platforms by a larger group of enterprises. It also facilitates the exchange of a variety of information by employees within one enterprise, but also within a group of enterprises. This exchange does not take place in the real world, but a virtual one, which can be accessed from different places in the world.

Such changes take into account certain elements related to the concept of sustainable development. From the environmental point of view, they allow for a smaller impact on the natural environment and better use of materials. From the economic point of view, however, they are associated with large financial expenditures, in particular the purchase of modern, highly developed machines, which for many enterprises is unfortunately impossible. Intelligent machines can reduce the demand for employees in the labor market, which indicates a social aspect. The growing trend in the literature towards sustainable development within the Industry 4.0 model highlights the importance of this topic, as well as various aspects that can be reliably resolved [21].

End-to-end engineering, which covers the entire product life cycle, should include an intelligent network and digitization relating to individual stages of the product life cycle: it should start from the raw material to the production system, then be applied to the use of the product, and only end with decommissioning [22].

Many researchers have contributed to the concept of Industry 4.0 by describing the theoretical aspects of this issue and then supporting them with practical solutions [12,22–28]. Nevertheless, there are many topics related to Industry 4.0 that require deeper analysis. One of these issues is linking Industry 4.0 with the concept of sustainable development, especially in the SME sector.

The development of enterprises using the assumptions of Industry 4.0 results in significant opportunities for implementing sustainable production, all due to the use of the omnipresent infrastructure of information and communication technologies (ICT). An important aspect in the time of the fourth industrial revolution is that the SME sector has not been destroyed; therefore, the doctrine of sustainable development is bound to work here. In current literature, it is an important determinant of the environmental dimension of sustainable development. Main resources such as products, materials, energy and water and their allocation can be implemented by smart-linked value creation modules [22]. Scientists still have to wait to determine how the implications of implementing Industry 4.0 will affect local and global economies. It should be pointed out that no more than 7% of research on Industry 4.0 focuses on the issue of sustainable development [1].

The industry has contributed to social welfare, which means that very high-quality products that meet virtually all customers' needs are produced, but at the same time, employees are also provided with the proper working conditions. Simultaneously, it can be observed that the current production structure still lacks adequate environmental sustainability. However, it is important that the industrial sector produces in an economically, environmentally and socially sustainable manner to achieve sustainable development [29].

Owing to Industry 4.0, new models of industrial management have been developed for small and medium-sized enterprises (SMEs). This concept, supported by, among others, numerous new technologies, is more flexible and cheaper compared to traditional IT systems used in enterprises; e.g., ERP (enterprise resource planning) and MES (manufacturing execution system). Unfortunately, small and medium-sized enterprises are insufficiently equipped to make use of the new opportunities that can aid production planning and control [30].

Sustainable development in the aspect of Industry 4.0 for the SME sector is, however, a very complex issue, not only from the technical point of view, but also from a non-technical standpoint; i.e., considering organizational factors and human resources. The generational change resulting from the natural state of affairs results in the appearance of employees with new competences and skills in the market, while, simultaneously, such abilities are often ahead of the adaptation possibilities of the technological infrastructure of the SME sector [7,31,32].

Such a situation can be observed in the research region. In total, 80% of the analyzed enterprises have problems with human resources. There are no specialists such as operators of universal devices available on the market. The available staff are able to operate advanced processing equipment which is unavailable for economic reasons for SMEs. This is the cause of the closure of economic activity by almost 80% of business entities that have opened their economic activity in the SME sector over the last three years [33].

Some of the production enterprises have changed into redistributive enterprises, outsourcing production—mainly to China, where they can deal with the real revolution of robotization and production automation resulting from the established priorities of the 5th Five Year Plan, as well as the objectives of the Made in China 2025 program. A considerable problem related to sustainable development in the SME sector, which is extremely noticeable due to economic factors and human resources, has been detected. It is possible to observe the liquidation of jobs and the transfer of production to China, where there is a large concentration of production and the exchange of the structure of business entities [34,35].

However, it should be emphasized that the Chinese market is also subject to dynamic changes and a significant increase in production robots per employee. In Poland, there are 20–30 robots per 10,000 employees. In the Czech Republic and Slovakia, this ratio is almost three times higher, and in Germany, this number is 10 times higher. In China, where there is a very rapid increase in wages, it is estimated that, by 2020, there will have been 100 robots per 10,000 employees. In 2017, the Digital Economy and Society Index (DESI) reported that Poland was 23rd in the group of 28 EU Member States. In 2016, 8.2% of Polish enterprises used Cloud Computing services.

In order to maintain sustainable development in the SME sector of the metal industry, taking into account the economic factor, external funds should be obtained to change the technological equipment. There is, however, a problem related to the product portfolio, which may be unsatisfactory in relation to the new machinery. The production load may be insufficient to obtain a positive financial result.

The main aim of the conducted research is to analyze the level of automation and robotization and to assess the current state of enterprises from the SME sector and to check whether these enterprises would be interested in using modern communication channels based on the platform structure, enabling the optimization of the product portfolio of producers based on a consolidation and optimization of the investment strategy in terms of obtaining modern flexible technologies compliant with the Industry 4.0 standard.

In this article, the results of pilot research related to the impact of changes in Industry 4.0 on small and medium-sized enterprises in the Czestochowa Industrial District were presented.

This paper concerns a very important issue related to the implementation of Industry 4.0 in small and medium enterprises. The survey that was used for the study was created on the basis of the initial analysis (preliminary questionnaire and expert interview). The major problems related to the implementation of Industry 4.0 in enterprises of the SME sector which were mentioned most often by respondents have been summarized. The survey was created as a result of initial analysis and was sent to the research enterprises. The set of questions that were used in the research and which are presented in the paper can be used by other scientists for similar research. The results of the pilot research themselves, presented in this paper, indicate the level of encountered problems and may form the basis for further research carried out by other research centers.

The remaining part of the article is organized as follows: Section 2 presents the problem description, which was the basis for further research. Section 3 includes the presentation of the research materials and method. In Section 4, the main results with their discussion are presented, while in Section 5, the main conclusions of the paper are presented.

2. Problems Description

The Czestochowa Industrial District is the third largest industrial district of the Silesian Voivodship. The basis of this district is the metallurgical and metal industries. It should also be emphasized that, in this district, enterprises from the SME sector are the most numerous. In many cases, they are family enterprises which have been passed from generation to generation.

Many enterprises from this district are old and have a lasting tradition. They have established a reputation on the market and created a group of their own customers. Unfortunately, they are equipped with old devices, which often break down. Due to the low profit, low productivity and low production level, they cannot afford to buy new, modern machines which are already fully automated. Often, it may also turn out that, even if it was possible to buy a machine, it would not be economically viable, because the efficiency of such a machine would be too high for a given enterprise. Enterprises from the SME sector in Poland are poorly equipped in terms of both computers and Internet access.

Generally, all over the world, small and medium-sized enterprises often do not have sufficient funds to invest in the latest technologies and must allocate capital very effectively and carefully. [17,36]. Other problems are difficulties in hiring staff, the high cost of staff and big competition.

Old equipment is primarily a problem from the production and economic point of view for a given enterprise. Old machines often break down, require a great deal of renovation, are not very efficient and often cause incompatibilities from the point of view of employees and are also dangerous to use. This affects the costs of the enterprise's operation, especially the cost of repairs and thus result in high prices of the sold products. Moreover, old equipment also has a greater impact on the natural environment surrounding such enterprises, increasing waste and emissions to the atmosphere. It should be mentioned that Czestochowa and the surrounding area are part of the Cracow-Czestochowa Upland; i.e., green and tourist areas. However, there is a continuous increase in investment outlays for new and used fixed assets [37].

Another problem that is common in the whole country as well as in the region is the lack of employees. Currently, the employees more often dictate conditions and the employers try to convince them to work for them. This is due to the fact that many people, after Poland's accession to the European Union, left the country to look for a job abroad. Many of these people are qualified employees—specialists who were not satisfied with their salary in Poland. Consequently, there are almost no people who can handle specialized machines. The outflow of employees also affects the economics of the research region.

More and more employees from other countries are hired: employees from Ukraine, Turkey or Serbia, however, do not have all the necessary skills. Workers from China, despite having such skills, are, however, not very flexible, and the cost of their transport and stay in Poland is very high. Also, communicating with such employees poses a problem as most of them do not speak Polish, and they often do not even know any of the congress languages, which would facilitate communication. It should also be emphasized that the Polish language is one of the most difficult to learn; furthermore, the forecasts indicate that many employees—for example, from Ukraine—may soon leave Poland to work in German enterprises due to the liberalization of the German labor market for that country.

One of the ideas that would help solve the above problems would be a platform in which several enterprises would buy machines together in order to reduce the purchase costs per enterprise and make better use of their production capacity. A group of employees should be employed to operate the machines, and both would be paid by all the involved enterprises on the basis of the capacity utilization. The platform could also offer sales of production capacities to other interested entrepreneurs, as well as taking advantage of the possibility of co-employing employees or exchanging information.

The idea of creating a platform is not new. It was used by other enterprises from other industrial sectors. In many cities in eastern Europe, stationary versions of such platforms are created; i.e. technology parks, special branches at industry chambers or clusters. They bring together enterprises with similar production and gather good practices and experience focused on specific industries. However, it should be checked whether the enterprises of the research region would be willing to take part in the described platform.

Research on Industry 4.0 in the SME sector has been conducted by scientists in other countries. According to Turkish researchers, most Turkish SMEs currently use sensor systems, automation and preventative maintenance. Nevertheless, at the same time, many enterprises do not have knowledge about the Internet of Things, additive (3D) production, augmented reality and large data. Most enterprises believe that high costs are unfortunately the main barrier to the implementation of Industry 4.0 technology. Research has also pointed out that SMEs need more information on the adoption of Industry 4.0 [38].

The Chinese government wants to build intelligent production systems that will become competitive on the global market and which will be at the level of ubiquitous information systems and communication systems. China's goal is to reduce their dependence on imported products by purchasing technologies for production; for example, in the automotive sector [39,40].

European Member States and their individual regions are involved in adapting their innovation systems to the objectives of Industry 4.0, and Europe as a whole faces the challenge of striking a balance between elements such as promoting excellence in research and innovation, sustainable development and the positioning of less developed regions to make the most of the benefits of the ongoing industrial revolution [41].

The patronage of the European Union is a centralized guideline regarding the implementation of Industry 4.0. Several important programs have been created at the European level, but many European countries have come out with their own initiatives and their own funding objectives and programs. The US is working on technology for future production. The consortium also includes important German companies such as Siemens and Bosch. The main focus is on IoT and the connection of future factories. According to Industry 4.0, production should be more efficient while value-added processes should be optimized. In addition, enterprises want greater machine availability as well as a high level of customized production [39].

Researchers from Estonia emphasize the relationship between networking, organizational development, structural framework conditions and sustainable development in the context of Industry 4.0. Many production facilities never develop a mature organizational culture, so employees often lack a common organizational way of thinking, which may cause organizational inefficiencies in the production plant, as well as impede communication and cooperation [7].

3. Methodology

Pilot research was carried out in the production enterprises of the SME sector in the area of the Silesian Voivodship. The industry, which is located in this voivodship, has been experiencing huge changes in recent years. Previously, this was a region in which mines and enterprises from the metallurgical industry predominated; currently, production companies prevail.

The results presented in this paper concern enterprises from the Czestochowa Industrial District. As a test sample, 200 business entities from the selected area were involved. The aim of the research was to identify areas and problems in the field of Industry 4.0 and to identify opportunities to find potential solutions in the field of Industry 4.0 for manufacturing enterprises from the SME sector. The classification for the research sample was a minimum of 10 years of production activity and the size of employment that classified the enterprise in the SME sector.

Based on the results of the preliminary questionnaire, supplemented by an expert interview, the main areas and problems in the case of enterprises of the SME sector from our region related to the implementation of solutions in the field of Industry 4.0 were defined. The solutions used in Germany have been adopted as a reference point. Activities related to the implementation of Industry 4.0 are supported at a government level, as evidenced by the implementation of the recommendations issued [42]. A special Internet platform has even been created [43]. Therefore, one should expect significant progress in these activities in the coming years. Previously defined problems and solutions used in Germany were the basis for the research presented in the paper.

The research consists of three stages (Figure 1): In the first part, the characteristics of the enterprises participating in the research were made and their current status was assessed. This stage had a form of a survey. The second stage took the form of an expert interview with the enterprises who agreed to share information on the problems in Industry 4.0 they had encountered. In the third stage, a survey was again used—this time, to assess the possibility of creating a platform to facilitate Industry 4.0. Both surveys were completed by the same enterprises (in an electronic way).

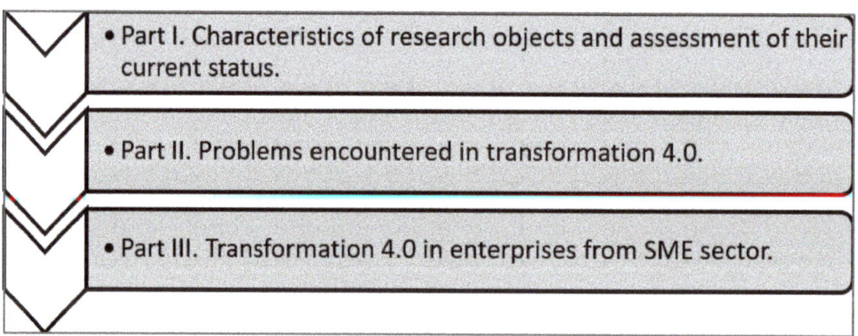

Figure 1. Stages of the research.

Part I. Characteristics of research objects and assessment of their current status.

This part concerned issues related to the type, form of production, automation and the used software. This part took the form of a questionnaire survey. This part of the research was conducted in spring 2018. The question area concerned the following:

1. Type of production organization used in the enterprise: unit, serial, mass–serial, mass;
2. The level of automation and robotization of production, assembly, transport and storage operations;
3. Use of industrial robots;

4.   Use of an autonomous transport system;
5.   Use of the product modeling programs;
6.   Use of the modeling production processes programs;
7.   Use of the CIM system (Computer Integrated Manufacturing);
8.   JIT (Just in Time) and KANBAN integration with suppliers and customer;
9.   VSM (Value Stream Mapping) studies for the current state of the enterprise;
10.   Measurement and monitoring of the value stream and the manner of performing these operations (manually, computer system);
11.   Independence or integrating the operation of a computer system in an enterprise;
12.   Use of the TPM method (Total Productive Maintenance);
13.   Use of OEE indicators (Overall Equipment Effectiveness);
14.   Identification of products, tools and materials in the system; e.g., the use of chips, bar codes or RFID (Radio-frequency identification);
15.   Possibilities for sending commands directly by the system to particular elements;
16.   Types of process database, database servers (own or Cloud) and their security;
17.   Use of 3D printing technology during product prototyping;
18.   Use of 3D printing technology during the production process;
19.   Individualization of the product, mass customization and application of modular structure of products;
20.   Use of AI or artificial intelligence in an enterprise.

Due to the fact that different rating scales were used, it was decided to use the methodology proposed by R. Kolman in quality engineering for the overall assessment of product quality [44]. For this purpose, one of the transformational methods was chosen—i.e., grading gradually (used for both measurable and non-measurable features)—and then the individual groups of responses were transformed to the interval (0.1), where 1 means the full modernity of the enterprise. In this way, it was possible to compare the responses and assess the overall current status of the research enterprises.

In the case of area 2, the respondents were asked to assess the level of automation and robotization of individual operations on a scale of 1–5, where 1 meant manual machine processes and meant 5 full automation and robotization. Then, the assessments were transformed as follows: 1, 0.1; 2, 0.3; 3, 0.5; 4, 0.7, 5: 0.9.

In the case of areas 3–10 and 12–20, respondents could indicate whether such an instrument is used in their enterprise, choosing one of the following responses: I do not know, no, sometimes, yes. In this case, the transformation was the following: I do not know, 0.1; no, 0.3; sometimes, 0.7; yes, 0.9.

Additional information was obtained for areas 10 and 16. In area 10, people were asked whether metering and monitoring was carried out manually (0.3) or by a computer (0.9), while in the case of area 16, they were asked whether the enterprises' databases were their own databases (0.3) or the Cloud (0.9). A separate scale was given to area 11. Enterprises were asked about the operation of a computer system: an independent system (0.3) or an integrated system (0.9).

Part II. Problems encountered in Industry 4.0.

The second part of the research took the form of an expert interview. It concerned the identification of obstacles and problems related to the use of various types of instruments aimed at modernizing the enterprise. This part of the research was conducted in summer and autumn 2018. The following questions were asked:

1.   What are the obstacles associated with increasing the level of automation and robotization of production processes?
2.   What are the obstacles associated with increasing the level of automation and robotization of warehouse processes?

3. What are the obstacles associated with increasing the level of automation and robotization of transport processes?
4. What are the obstacles associated with the use of robots and process simulation programs?
5. Is there a need to monitor processes in real time?
6. What are the obstacles associated with the monitoring of the processes in real time?
7. Does the enterprise see the need to switch to a pull system and use JIT, KANBAN or other systems?
8. What are the problems associated with the implementation of JIT, KANBAN or other systems or on the side of the system itself, people or integration with external entities?
9. What are the problems with monitoring processes, documenting and selling data and their use in the field of process improvement and optimization?
10. What are the problems with the implementation and use of TPM?
11. What are the problems with monitoring and supervision over the material, product and tool during the process?
12. What problems did the enterprise encounter when individualizing the product?

Due to the fact that these were open questions, the grouping methodology proposed in the CIT method was applied [45,46] and used in the assessment of the service quality, but without any division into positive and negative answers. The responses for individual areas were grouped according to their similarity, and so it was possible to indicate which responses (groups of responses) were the most common.

Part III. Industry 4.0 in enterprises from the SME sector.

Based on the expert interview and the most frequently indicated problems, a second questionnaire was developed to obtain the answer to whether creating a platform aimed at initiating activities to facilitate Industry 4.0 is justified in the opinion of entrepreneurs from the SME sector. Then, the survey was sent to companies that agreed to take part in the first research. The research was conducted from January to April 2019. The questions from the second questionnaire concerned the following:

1. The average age of owned production machines;
2. The production capacity in relation to the expectations and the level of its possible increase or the sale of a possible surplus;
3. Planning the purchase of new production equipment;
4. Problems related to planning the purchase of such equipment;
5. Possible problems with the number of employed employees and actions eliminating shortages in employment;
6. Willingness to participate in a platform that would bring together similar enterprises;
7. Willingness to share the purchase and use of machines within the platform;
8. Willing to co-employ employees within the platform;
9. Exchange of information regarding machines and employees within the platform;
10. Willingness to sell/buy production capacity as part of the platform;
11. Willingness to jointly develop a product within the platform.

The research was conducted among the enterprises whose headquarters are located in the Czestochowa Industrial District. The research was addressed to enterprises from the SME sector; however, large enterprises employing over 250 people were also included in the characteristics of the respondents. In this way, surveys completed by enterprises not from the SME sector could be eliminated. The results were obtained from 194 enterprises; however, seven of them were rejected.

In total, 28% of the surveyed objects were enterprises in which there was a unit type of production, 53% of respondents declared serial production, 8% mass–serial production, and 11% mass production.

4. Results and Discussion

The analysis of pilot research in the field of evaluation of production entities in the aspect of preparation for Industry 4.0 allowed us to draw interesting results. The assessment of the current status of surveyed enterprises is shown in Figure 2.

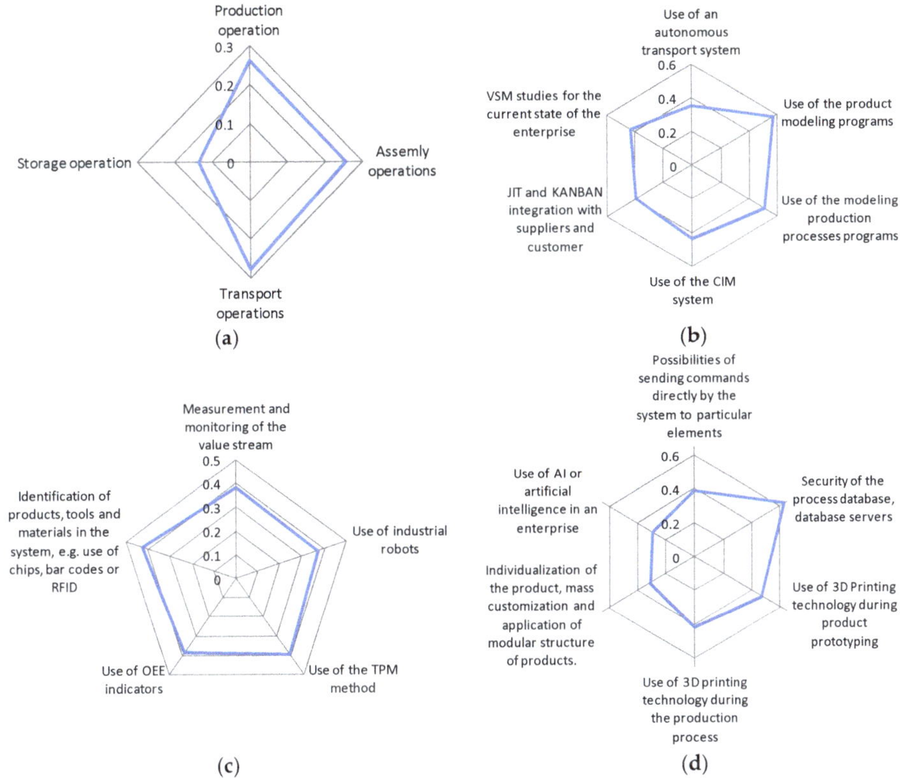

Figure 2. (a) The level of automation and robotization of individual operations; (b), (c), (d) assessment of the current state in the research enterprises.

From the analysis of Figure 2a, it can be concluded that the lowest level of automation and robotization among individual operations was recorded in the case of the storage operations (over 0.1). In the case of the other three types of operations, this level was similar but still low (about 0.3). The research enterprises are equipped with old, non-automated devices, which often break down. In general, the level of automation and robotization of Polish enterprises from the SME sector is lower than similar enterprises from other European Union countries.

Some of the research enterprises indicated that they sometimes use different programs for product modeling and for modeling production processes. They also secure process databases and database servers, but it should be emphasized that this results from the new provisions of the GDPR (General Data Protection Regulation) in force in the European Union.

Most instruments are used very rarely by a small number of enterprises. It was not known whether enterprises do not know these instruments or have problems using them. The expert interview examined the reason for this situation.

Summarizing the first part of the study, it can be pointed out that the average assessment of the current state of the surveyed enterprises of the SME sector in the range (0; 1) is 0.38, which is low.

The research enterprises are equipped with old, non-automated equipment, which often breaks down. Only a few of them use a variety of instruments supporting production processes. Many of them have no idea about the existence of such instruments or do not know their names.

Based on the obtained results, it can be concluded that production systems in the SME sector are characterized by a low level of system integration as well as a low level of robotization and automation. This is mainly caused by the size of these enterprises and the lack of capital for investments.

The next part of the study (expert interview) concerned the identification of obstacles and problems related to the use of various types of instruments aimed at modernizing the enterprise. The most important results are presented in Figures 3–5.

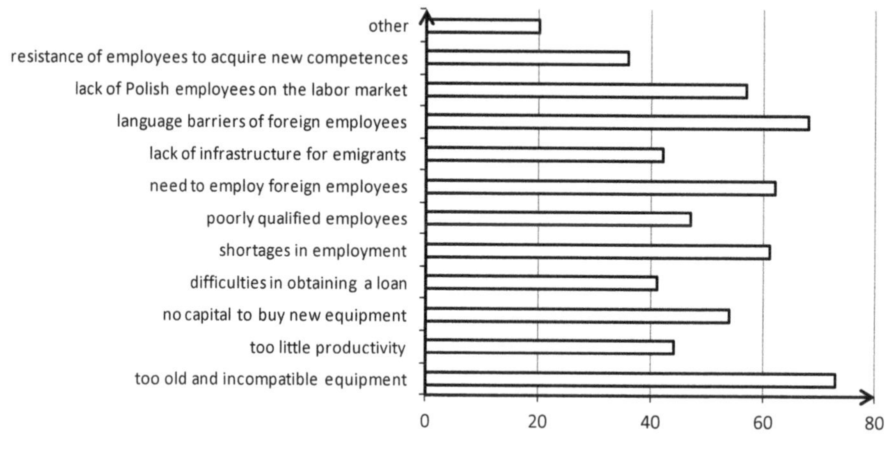

Figure 3. Obstacles associated with increasing the level of automation and robotization of different processes in enterprises.

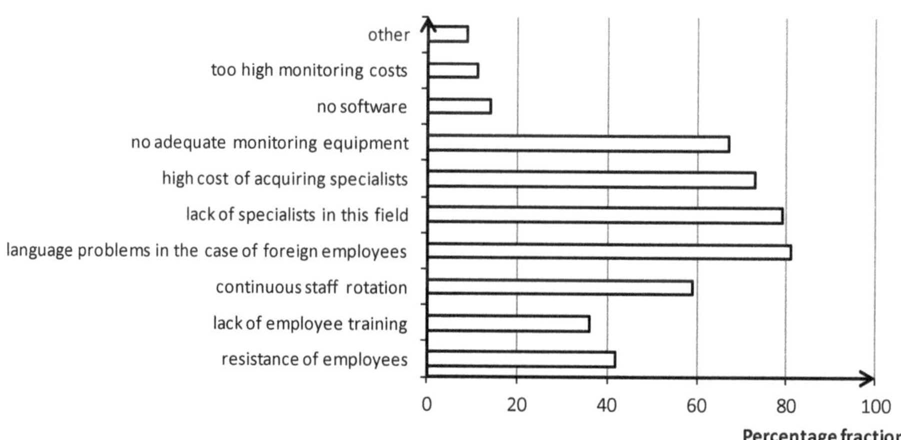

Figure 4. Problems with monitoring processes, documenting and selling data and their use in the field of process improvement and optimization, and monitoring and supervision of the material, product and tools during the process.

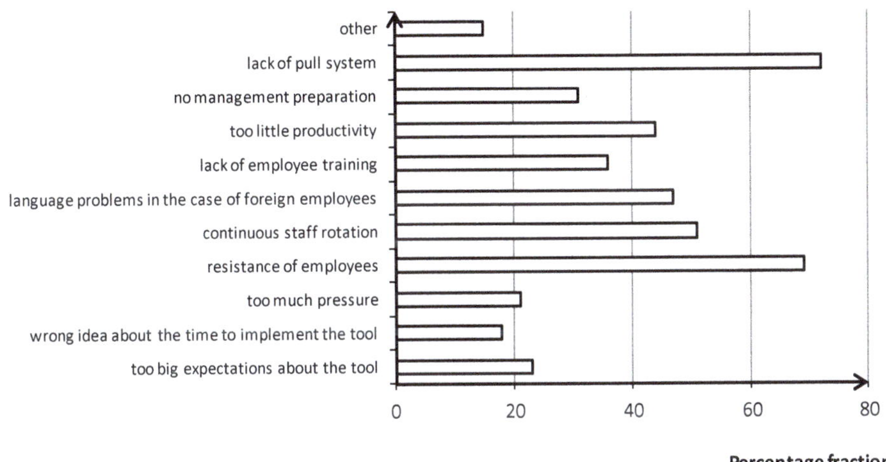

Figure 5. Problems when implementing various systems.

During the expert interview, the research enterprises showed that there is a need to monitor individual processes in real time. This is mainly related to the need to constantly control the course of these processes. These enterprises can also see the need to switch to a pull system and use JIT, KANBAN or other systems. This is related to the type of production that takes place in these facilities, which depends on the customers' orders and the sizes of these enterprises, because most enterprises from the SME sector cannot afford freezing capital.

The enterprises indicated many problems related to the implementation or use of various instruments aimed at improving their business. The resistance of employees is an important problem. Management must convince employees that changes are good and important and explain the need to implement them, but in the beginning, the same management must be convinced of such changes.

The problem with staff is a significant problem in Poland, including among research enterprises. It is an important social aspect of the functioning of enterprises, which determines the success of the entire enterprise and the introduced changes.

During the research, the enterprises pointed to the lack of specialists—operators who are able to handle the so-called analog production systems and the lack of specialists in the field of automation, robotics and IT. The employment of people especially from the second group is associated with high costs of obtaining and paying for this professional group.

Currently, enterprises hire specialists mainly from Turkey, Serbia and Ukraine; however, there are language barriers, sometimes cultural barriers, and a lack of infrastructure for foreign workers.

The results of the expert interview suggest that a certain solution may be the integration of SME sector production entities in the transition to Industry 4.0 through the development of a common product portfolio based on technological similarity. As a result, it is possible to reduce the costs of purchasing technical resources, reduce costs, and increase productivity through the optimal use of a universal machine park and greater individualization of the product.

The most important part of the research concerned the creation of a platform aimed at initiating actions to facilitate Industry 4.0 and especially its success among enterprises from SME sector. These results are shown in Figures 6–9.

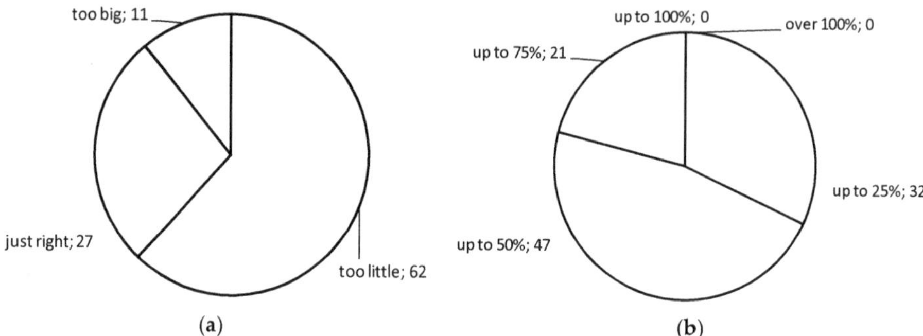

Figure 6. (**a**) Current production capacity in relation to the expectations of the enterprise; (**b**) if the production capacity is too small, then, how much would the enterprise like to increase it?

Only 27% of the research enterprises assessed their production capacity as just right. For 62%, the production capacity was too small. 32% of respondents would willingly increase their production capacity to 25%, and another 47% of respondents would increase their capacity even by up to 50%. However, they are not able to do this at their own expense. All companies with overcapacity are interested in its possible sale, and those with too little production capacity are interested in its possible purchase.

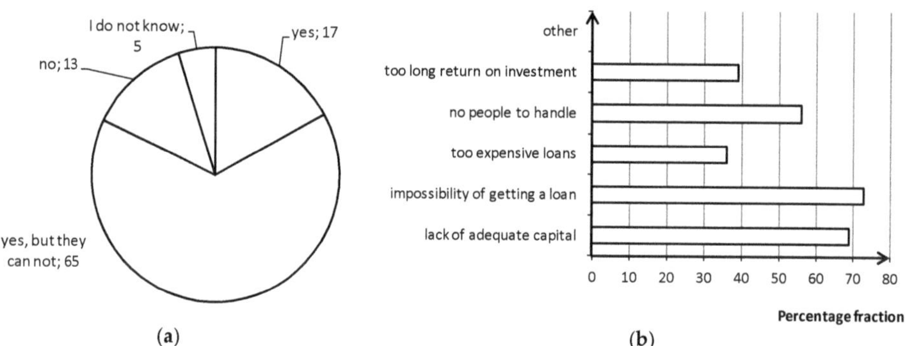

Figure 7. (**a**) Is the company interested in purchasing new equipment? (**b**) If they cannot, why? (possibly more than one answer).

Over 80% of all respondents were interested in purchasing new, more advanced equipment. Unfortunately, 65% of them, despite their willingness, were unable to do it. The most important reasons for this situation are the lack of adequate capital and the impossibility of obtaining a loan. Without the capital, the purchase of new equipment is not possible.

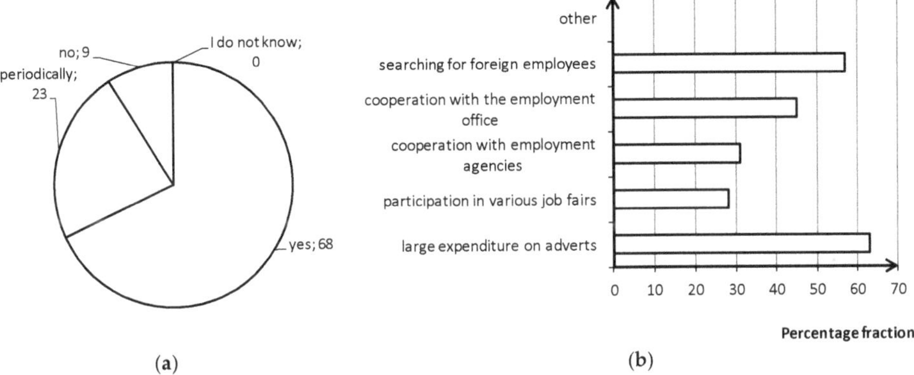

Figure 8. (**a**) Does the enterprise have problems with a lack of employees? (**b**) What is done in case of a lack of employees? (possibly more than one answer).

In total, 68% of respondents have problems with a lack of employees, while another 23% indicated the occurrence of this problem periodically. The enterprises are trying to solve this problem in various ways. Among the most common solutions were large expenditures on advertisements, searching for foreign employees and cooperation with employment offices.

It should be reminded that there is a lack of specialists in Poland. After Poland's accession to the European Union, there was a huge outflow of employees of all types from the domestic market, which resulted in the labor market changing from an employer's market to an employees' market. Perhaps these enterprises should be interested in cooperation with local high schools, universities, which could educate students in the subjects indicated by them.

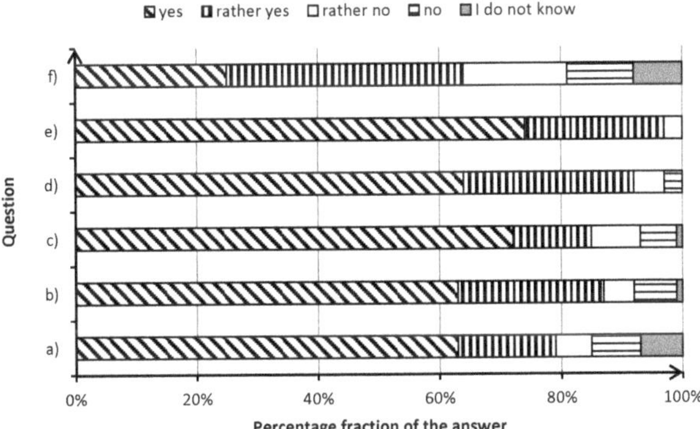

Figure 9. (**a**) Is the enterprise interested in taking part in a platform bringing together similar enterprises with similar problems? (**b**) Would the enterprise within such a platform be interested in the joint purchase and use of machines? (**c**) Would the enterprise within such a platform be interested in co-employing employees e.g. to operate these machines? (**d**) Would the enterprise within such a platform be interested in exchanging information on machines and employees? (**e**) Would the enterprise within such a platform be interested in selling/buying production capacity? (**f**) Would the enterprise within such a platform be interested in jointly development of a product?

In total, 63% of the research enterprises expressed their willingness to take part in a platform bringing similar enterprises with similar problems, while another 16% were also interested, although not completely sure of this decision. In total, 63% of respondents would be interested in the joint purchase and use of machines, while another 24% were not sure about this until the end. According to the respondents, co-employing employees would be a good idea (72% answered yes and 13% inclined towards yes). This would help to solve problems with employment, especially regarding the employment of specialists. The enterprises were also interested in exchanging information on machines and employees within such a platform. An important point was the ability to sell/buy production capacity, which was declared by 74% of the research enterprises (plus 23% who were not entirely certain).

The enterprises were least interested in the joint development of products (only 25% of indications for yes, and 39% were not entirely certain). This may be related to the fact that these enterprises belong to different industries and produce different products, or it is possible that these enterprises may also, despite their willingness to cooperate, want to remain competitive in the market.

From the conducted research, it can be concluded that enterprises from the SME sector from the Czestochowa Industrial District are focused on changes, but they cannot always afford them. However, if it were possible, they would like to cooperate with similar enterprises to invest in new equipment or employees.

5. Conclusions

As Herzog, Buchmeister, Beharic, et al. wrote [47], radical changes and transformations of various types of production systems, which are oriented on the assumptions of Industry 4.0, are possible mainly due to the development of disruptive technologies and the digital era. The fourth industrial revolution is a fact based on digitization; it has a variety of uses, from everyday tasks in the home to the automation of production processes. This industry is based on the systematic use of information technology. This situation is affected by many elements, including the development of automation, data processing and exchange, manufacturing techniques and the management organization of all processes. The key solutions that are used in Industry 4.0 can have a significant impact on the overall situation of the enterprise. The most important factors include increasing productivity and the availability of production resources, increasing equipment and production efficiency, and increasing value per employee [48].

However, it should be remembered that Industry 4.0 is not only about technology, but also about new ways of working and the role of people in industry. The Internet plays an increasingly important role in everyday life—not only of ordinary people, but above all of enterprises. E-enterprises are becoming more and more popular, which is caused by growing social awareness with relation to the role of the Internet. Activities that are undertaken by electronic means may reach a greater number of customers over a much faster time period and at a lower cost [49].

An important element which is often overlooked in the case of research on Industry 4.0 is sustainable development. It should be remembered that the structure of sustainable management can be introduced in different ways [50,51]. Two important elements of sustainable development are economics and society, from which it can be concluded that employees and their proper motivation are the key issue here.

Whether a management team successfully develops its enterprise and increases its competitiveness and income depends on many factors [52]. This is much more difficult for small and medium-sized enterprises, as enterprises in this sector have at their disposal less capital, fewer employees, less productivity, and less of much else besides. They may be supported in their effort by specific-oriented models [53], a deeper analysis of a process [54] and special technologies [55] or approaches [56,57] giving them a competitive advantage.

It is not easy for small and medium-sized enterprises to transform according to Industry 4.0. First of all, this is due to the economic aspect and the human factor. The product and technology portfolio available to small and medium enterprises is not adapted to the following industrial transformation. Therefore, they must look for solutions that, to some extent, may allow them to do this.

In this paper, analyses of the level of automation and robotics, an assessment of the current state of enterprises from the SME sector and a verification of whether these enterprises would be interested in using modern communication channels based on the platform structure, enabling the optimization of the product portfolio of producers based on a consolidation and optimization of the investment strategy in terms of obtaining modern flexible technologies compliant with the Industry 4.0 standard, were presented. In this paper, the results of pilot research were included. This research took the form of a survey and expert interview. Pilot research was carried out in the Czestochowa Industrial District.

As a result of the conducted research, a hypothesis can be put forward, whose confirmation requires additional, further research. The determinants of implementing Industry 4.0 in the SME sector are difficult and demanding. The reason for this situation is the fragmentation of the technological and product portfolio in this sector, as well as the high level of costs of automation and robotization of production processes.

The human factor is another factor determining the effectiveness of the transformation process. Unlike in the case of China, where automation and robotization are perceived through the human–machine or the machine–human perspectives, in the research area presented in this paper, this process equals the loss of a job, hence the resistance of employees to making changes.

The summary of this research results is the determination of the three main barriers to Industry 4.0 in the SME sector. The first barrier is the narrow product portfolio of SMEs, which does not guarantee full use of the efficiency of automated and autonomous production systems. The second barrier is the cost of obtaining money; i.e., the funds for a given investment. An adverse condition is also the turbulence of the environment from the micro and macro aspect.

The results presented in the paper are not without limitations. The research survey was created on the basis of a preliminary survey and expert interview with a small group of enterprises of the SME sector. The authors could not find the main problems in the field of the Industry 4.0; therefore, some factor could have been omitted. Only the most important results are presented, and consequently, some elements might be overlooked, which may affect the understanding of the results. The pilot research will allow minor changes to be made to the survey form before proceeding to the next stage of the study.

The research presented in the paper will be continued in other parts of Poland and later will also be continued in other European countries. Additionally, the results of these research works will be compared with each other.

Author Contributions: Both authors participated equally in the preparation of this article. R.U. created the methodology; M.I. conducted the experiments. All authors have read and agreed to the published version of the manuscript.

Funding: This research received no external funding.

Conflicts of Interest: The authors declare no conflict of interest.

References

1. Maresova, P.; Soukal, I.; Svobodova, L.; Hedvicakova, M.; Javanmardi, E.; Selamat, A.; Krejcar, O. Consequences of Industry 4.0 in Business and Economics. *Economies* **2018**, *6*, 46. [CrossRef]
2. Maszke, A.; Dwornicka, R.; Ulewicz, R. Problems in the Implementation of the Lean Concept at a Steel Works—Case Study. In Proceedings of the 12th International Conference Quality Production Improvement—QPI 2018, Zaborze, Poland, 18–20 June 2018; EDP Science: Les Ulis, France, 2018.
3. Kleszcz, D. Barriers and opportunities in the implementation of lean manufacturing in the ceramic industry. *Prod. Eng. Arch.* **2018**, *19*, 48–52. [CrossRef]
4. Salvador, R.M.; de la Cruz, J.M. Presence of Industry 4.0 in additive manufacturing: Technological trends analysis. *DYNA* **2018**, *93*, 597–601. [CrossRef]
5. Thames, L.; Schaefer, D. Software-defined Cloud Manufacturing for Industry 4.0. *Procedia CIRP* **2016**, *2*, 12–17. [CrossRef]

6. Piccarozzi, M.; Aquilani, B.; Gatti, C. Industry 4.0 in Management Studies: A Systematic Literature Review. *Sustainability* **2018**, *10*, 3821. [CrossRef]

7. Prause, G.; Atari, S. On sustainable production networks for Industry 4.0. *Intern. J. Entrep. Sustain. Issues* **2017**, *4*, 421–431. [CrossRef]

8. Ślusarczyk, B. Industry 4.0—Are We Ready? *Pol. J. Manag. Stud.* **2018**, *17*, 232–248. [CrossRef]

9. Kagermann, H.; Lukas, W.; Wahlster, W. Industrie 4.0: Mit dem Internet der Dinge auf dem Weg zur 4.0 industriellen Revolution. *VDI Nachrichten* **2011**, *13*, 2.

10. Strandhagen, J.W.; Alfnes, E.; Strandhagen, J.O.; Vallandingham, L.R. The fit of Industry 4.0 applications in manufacturing logistics: A multiple case study. *Adv. Manuf.* **2017**, *5*, 344–358. [CrossRef]

11. Liao, Y.; Deschamps, F.; Loures, E.D.F.R.; Ramos, L.F.P. Past, Present and Future of Industry 4.0—A Systematic Literature Review and Research Agenda Proposal. *Int. J. Prod. Res.* **2017**, *55*, 3609–3629. [CrossRef]

12. Lasi, H.; Fettke, P.; Kemper, H.G.; Feld, T.; Hoffmann, M. Industry 4.0. *Bus. Inf. Syst. Eng.* **2014**, *6*, 239–242. [CrossRef]

13. Li, G.; Hou, Y.; Wu, A. Fourth Industrial Revolution: Technological Drivers, Impacts and Coping Methods. *Chin. Geogr. Sci.* **2017**, *27*, 626–637. [CrossRef]

14. Thoben, K.D.; Wiesner, S.; Wuest, T. Industrie 4.0 and Smart Manufacturing—A Review of Research Issues and Application Examples. *Int. J. Autom. Technol.* **2017**, *11*, 4–16. [CrossRef]

15. Pereira, A.C.; Romero, F. A review of the meanings and the implications of the Industry 4.0 concept. *Procedia Manuf.* **2017**, *13*, 1206–1214. [CrossRef]

16. Zhou, K.; Liu, T.; Zhou, L. August. Industry 4.0: Towards Future Industrial Opportunities and Challenges. In Proceedings of the 2015 12th International Conference on Fuzzy Systems and Knowledge Discovery (FSKD), Zhangjiajie, China, 15–17 August 2015; IEEE: Piscataway, NJ, USA, 2015.

17. Erol, S.; Jaeger, A.; Hold, P.; Ott, K.; Sihn, W. Tangible industry 4.0: A scenario-based approach to learning for the future of production. *Procedia Cirp* **2016**, *54*, 13–18. [CrossRef]

18. Wittenberg, C. Cause the trend industry 4.0 in the automated industry to new requirements on user interfaces? In *Human–Computer Interaction: Users and Contexts. Lecture Notes in Computer Science*; Kurosu, M., Ed.; Springer: Cham, Switzerland, 2015; Volume 9171, pp. 238–245.

19. Hitpass, B.; Astudillo, H. Editorial: Industry 4.0 Challenges for Business Process Management and Electronic-Commerce. *J. Theor. Appl. El. Commer. Res.* **2019**, *14*, 1–3.

20. Sony, M. Industry 4.0 and lean management: A proposed integration model and research propositions. *Prod. Manuf. Res. Open Access J.* **2018**, *6*, 416–432. [CrossRef]

21. Bonilla, S.H.; Silva, H.R.O.; da Silva, M.T.; Goncalves, R.F.; Sacomano, J.B. Industry 4.0 and Sustainability Implications: A Scenario-Based Analysis of the Impacts and Challenges. *Sustainability* **2018**, *10*, 3740. [CrossRef]

22. Stock, T.; Seliger, G. Opportunities of Sustainable Manufacturing in Industry 4.0. *Procedia CIRP* **2016**, *40*, 536–541. [CrossRef]

23. Cheng, G.J.; Liu, L.T.; Qiang, X.J.; Liu, Y. Industry 4.0 development and application of intelligent manufacturing. In Proceedings of the 2016 International Conference on Information System and Artificial Intelligence (ISAI), Hong Kong, China, 24–26 June 2016; IEEE: Berrechid, Morocco, 2016.

24. Fonseca, L.M. Industry 4.0 and the digital society: Concepts, dimensions and envisioned benefits. In Proceedings of the International Conference on Business Excellence, Bucharest, Romania, 22–23 March 2018; Sciendo: Bucharest, Romania, 2018.

25. Kagermann, H. Change through digitization—Value creation in the age of Industry 4.0. In *Management of Permanant Change*; Albach, H., Meffert, H., Pinkwart, A., Eds.; Springer Gabler: Wiesbaden, Germany, 2015; pp. 23–45.

26. Rubmann, M.; Lorenz, M.; Gerbert, P.; Waldner, M.; Justus, J.; Engel, P.; Harnisch, M. *Industry 4.0: The Future of Productivity and Growth in Manufacturing Industries*; Boston Consulting Group: Boston, MA, USA, 2015; pp. 1–20.

27. Yen, C.T.; Liu, Y.C.; Lin, C.C.; Kao, C.C.; Wang, W.B.; Hsu, Y.R. Advanced manufacturing solution to Industry 4.0 trend through sensing network and Cloud Computing technologies. In Proceedings of the 2014 IEEE International Conference on Automation Science and Engineering (CASE), Taipei, Taiwan, 18–22 August 2014; IEEE: Piscataway, NJ, USA, 2014.

28. Bertola, P.; Teunissen, J. Fashion 4.0. Innovating Fashion Industry Through Digital Transformation. *RJTA* **2018**, *22*, 352–369. [CrossRef]

29. Ingaldi, M.; Jursova, S. Economy and Possibilities of Waste Utilization in Poland. In Proceedings of the METAL 2013 22nd International Conference on Metallurgy and Materials, Brno, Czech Republic, 15–17 May 2013; TANGER Ltd.: Ostrava, Czech Republic, 2013.

30. Moeuf, A.; Pellerin, R.; Lamouri, S.; Tamayo-Giraldo, S.; Barbaray, R. The industrial management of SMEs in the era of Industry 4.0. *Int. J. Prod. Res.* **2018**, *56*, 1118–1136. [CrossRef]

31. Müller, J.M.; Voigt, K.I. Sustainable Industrial Value Creation in SMEs: A Comparison between Industry 4.0 and Made in China 2025. *IJPEM-GT* **2018**, *5*, 659–670.

32. Müller, J.M.; Buliga, O.; Voigt, K.I. Fortune Favors the Prepared: How SMEs Approach Business Model Innovations in Industry 4.0. *Technol. Forecast. Soc.* **2018**, *132*, 2–17. [CrossRef]

33. Glowny Urzad Statystyczny. Available online: www.stat.gov.pl (accessed on 10 August 2018).

34. Iskierka, S.; Krzemiński, J.; Weżgowiec, Z. Zapotrzebowanie rynku pracy na informatyków a praktyka dydaktyczna. *Dydaktyka Informatyki* **2017**, *12*, 33–42. [CrossRef]

35. Skwarska, A. Pracownicy z Ukrainy Odpowiedzią na Problemy Polskich Pracodawców? Available online: http://media.pracuj.pl/18776-pracownicy-z-ukrainy-odpowiedzia-na-problemy-polskich-pracodawcow (accessed on 9 April 2018).

36. Bar, K.; Herbert-Hansen, Z.N.L.; Khalid, W. Considering Industry 4.0 Aspects in the Supply Chain for an SME. *Prod. Eng. Res. Dev.* **2018**, *12*, 747–758. [CrossRef]

37. Raport o Stanie Sektora Małych i Średnich Przedsiębiorstw w Polsce. Available online: https://www.parp.gov.pl/component/publications/publication/raport-o-stanie-sektora-msp-w-polsce-2017 (accessed on 11 October 2017).

38. Gergin, Z.; Üney-Yüksektepe, F.; Gençyılmaz, M.G.; Aktin, T.; Gülen, K.G.; İlhan, D.A.; Dündar, U.; Cebeci, O.; Çavdarlı, A.I. Industry 4.0 Scorecard of Turkish SMEs. In Proceedings of the International Symposium for Production Research 2018, Vienna, Austria, 28–31 August 2018; Durakbasa, N.M., Gencyilmaz, M.G., Eds.; Springer: Cham, Switzerland, 2018.

39. Ehlers, N.; Loew, R.; Bleimann, U. Industry 4.0 national and international. In Proceedings of the Collaborative European Research Conference (CERC 2016), Cork, Ireland, 23–24 September 2016.

40. Overmaat, B. World Wide Race—Future of Production: Industrie 4.0 in Europa/Asien/USA. Available online: https://engineered.thyssenkrupp.com/world-wide-wettlauf-industrie-4-0-aktivitaeten-in-europe-asien-usa (accessed on 9 April 2016).

41. Ciffolilli, A.; Muscio, A. Industry 4.0: National and Regional Comparative Advantages in Key Enabling Technologies. *Eur. Plann. Stud.* **2018**, *26*, 2323–2343. [CrossRef]

42. Kagermann, H.; Wahlster, W.; Helbig, J. Securing the Future of German Manufacturing Industry. Recommendations for Implementing the Strategic Initiative INDUSTRIE 4.0. Final Report of the Industrie 4.0 Working Group. Available online: https://www.din.de/blob/76902/e8cac883f42bf28536e7e8165993f1fd/recommendations-for-implementing-industry-4-0-data.pdf (accessed on 19 December 2019).

43. Weyer, S.; Schmitt, M.; Ohmer, M.; Goreck, D. Towards Industry 4.0—Standardization as the crucial challenge for highly modular, multi-vendor production systems. *IFAC-PapersOnLine* **2015**, *48*, 579–584. [CrossRef]

44. Kolman, R. *Inżynieria Jakości*, 1st ed.; PWE: Warsaw, Poland, 1992; ISBN 978-83-2080-867-4.

45. Jubenville, T.; Cairns, S. An Introduction to the Enhanced Critical Incident Technique. *Int. J. Qual. Methods* **2016**, *15*, 1.

46. Ingaldi, M. Overview of the main methods of service quality analysis. *Prod. Eng. Arch.* **2018**, *18*, 54–59. [CrossRef]

47. Herzog, N.V.; Buchmeister, B.; Beharic, A.; Gajsek, B. Visual and optometric issues with smart glasses in Industry 4.0 working environment. *Adv. Prod. Eng. Manag.* **2018**, *13*, 417–428.

48. Resman, M.; Pipan, M.; Šimic, M.; Herakovič, N. A new architecture model for smart manufacturing: A performance analysis and comparison with the RAMI 4.0 reference model. *Adv. Prod. Eng. Manag.* **2019**, *14*, 153–165. [CrossRef]

49. Skowron-Grabowska, B.; Sukiennik, K. Innovations in e-enterprises on the Polish market. *Procedia Comput. Sci.* **2015**, *65*, 1046–1051. [CrossRef]

50. Grabara, J. The another point of view on sustainable management. *Qual. Access Success* **2017**, *18*, 344–349.

51. Ingaldi, M.; Ulewicz, R. How to Make E-Commerce More Successful by Use of Kano's Model to Assess Customer Satisfaction in Terms of Sustainable Development. *Sustainability* **2019**, *11*, 4830. [CrossRef]
52. Dobes, K.; Kot, S.; Kramolis, J.; Sopkova, G. The Perception of Governmental Support in The Context of Competitiveness of SMEs in the Czech Republic. *J. Compet.* **2017**, *9*, 34–50. [CrossRef]
53. Pietraszek, J.; Gadek-Moszczak, A.; Torunski, T. Modeling of Errors Counting System for PCB Soldered in the Wave Soldering Technology. *Adv. Mater. Res.* **2014**, *874*, 139–143. [CrossRef]
54. Sygut, P.; Klimecka-Tatar, D.; Borkowski, S. Theoretical analysis of the influence of longitudinal stress changes on band dimensions during continuous rolling process. *Arch. Metall. Mater.* **2016**, *61*, 183–188. [CrossRef]
55. Zorawski, W.; Chalys, R.; Radek, N.; Borowiecka-Jamrozek, J. Plasma-sprayed composite coatings with reduced friction coefficient. *Surf. Coat. Technol.* **2008**, *202*, 4578–4582. [CrossRef]
56. Gadek-Moszczak, A. History of stereology. *Image Anal. Stereol.* **2017**, *36*, 151–152. [CrossRef]
57. Gadek-Moszczak, A.; Radek, N.; Wronski, S.; Tarasiuk, J. Application the 3D Image Analysis Techniques for Assessment the Quality of Material Surface Layer Before and After Laser Treatment. *Adv. Mater. Res.* **2014**, *874*, 133–138. [CrossRef]

Article

Complexity Assessment of Assembly Supply Chains from the Sustainability Viewpoint

Vladimir Modrak [1,*], Zuzana Soltysova [1] and Daniela Onofrejova [2]

[1] Faculty of Manufacturing Technologies, Technical University of Kosice, 080 01 Presov, Slovakia; zuzana.soltysova@tuke.sk

[2] Faculty of Mechanical Engineering, Institute of Management, Industrial and Digital Engineering, Technical University of Kosice, 042 00 Kosice, Slovakia; daniela.onofrejova@tuke.sk

* Correspondence: vladimir.modrak@tuke.sk; Tel.: +421-55-602-6449

Received: 25 October 2019; Accepted: 11 December 2019; Published: 13 December 2019

Abstract: Assembly supply chain systems are becoming increasingly complex and, as a result, there is more and more need to design and manage them in a way that benefits the producers and also satisfies the interests of community stakeholders. The structural (static) complexity of assembly supply chain networks is one of the most important factors influencing overall system complexity. Structures of such networks can be modeled as a graph, with machines as nodes and material flow between the nodes as links. The purpose of this paper is to analyze existing assembly supply chain complexity assessment methods and propose such complexity metric(s) that will be able to accurately reflect not only specific criteria for static complexity measures, but also selected sustainability aspects. The obtained results of this research showed that selected complexity indicators reflect sustainability facets in different ways, but one of them met the mentioned requirements acceptably.

Keywords: assembly supply chain; sustainability; complexity indicators; testing criteria

---

1. Introduction

In the past decades, product and manufacturing process developments have faced rapidly changing market needs. This global trend has resulted in more complex products and manufacturing processes to produce them. As is well known, assembly processes play a key role in production systems. Therefore, any effort for the improvement and optimization of the assembly process is vital to manufacturing competitiveness, since about 50% of the product cost should be ascribed to the assembly phase [1]. One possible way to improve manufacturing competitiveness is to reduce assembly process complexity and the associated cost. Even though the system complexity is inherent and cannot be avoided, it has to be kept affordable. Before defining system complexity for specific purposes, it is important to present some background on complexity as such. According to Flood and Carson [2], "complexity does not solely exist in things, to be observed from their surface and beneath their surface". Flood [3] adds that if systems are tangible things, then complexity and system are synonymous, where system is prime. According to Faulconbridge and Ryan [4], "a system necessarily has a boundary through which it or its elements interact with elements or systems outside the boundary". This system-inherent property allows us to adopt graph complexity measures for the complexity quantification of engineered systems [5]. Reynolds [6] points out that the structural modeling approach is applicable for clearly structured systems, in contrast to the behavioral approach, where structure is not assumed a priori. In the context of the architectural design, system complexity is proportional to the risk that the functional requirements cannot be met [7]. In order to briefly introduce a structural complexity measurement problem, it would be useful to mention the fact that prevalent approaches toward measuring system complexity are based on entropy theory (see, e.g., [8–11]). Entropy can be interpreted in different ways, but in terms of technical systems, it is construed as a measure of information. This entropy measure was introduced by Shannon and Weaver [12], who reduced the concept of entropy to pure probability theory. Their considerations were adopted by Frizelle and Woodcock [13] in order to define static and dynamic system complexity. In regard to the scope of this paper, our primary concern is static system complexity, defined as the amount of information needed to specify the system and its components. System complexity belongs to general systems theory, because it has been applied to different kinds of systems, including technical, social, and biological networks [14–16].

The purpose of this paper is to analyze existing complexity assessment methods to measure assembly supply chain (ASC) complexity and propose accurate complexity metric(s) that will include not only specific criteria for static complexity measures, but also sustainability viewpoints.

2. Related Works

The recent literature offers several different views on the relation between the sustainability and complexity of manufacturing systems, including ASCs. For instance, Peralta et al. [17] argue that it is necessary to establish new approaches that reduce increasing complexity of manufacturing systems in the design and management stage in order to promote their sustainability. For that purpose, they developed a so-called fractal model for sustainable manufacturing. Moldawska [18] proposed a complexity-based model of sustainable manufacturing, assuming that sustainable manufacturing organization tends to maintain its own sustainability, as well as contribute to the sustainability of the world. Relations between supply chains and sustainability were also treated in works [19,20] in which the authors analyzed the effect of distributed manufacturing systems on sustainability. The study [21] explored the influence of mass customization strategy on the configuration complexity of assembly systems, and their authors proposed bounds for configuration complexity scale. ASCs in terms of mass customization can lead to very complex systems due to the high number of product variants. It has been proved that variety itself also has a significant impact on productivity, increased costs, and quality [22,23]. Therefore, such companies need to consider integrating their business practices with sustainability and reducing supply chain costs to achieve a competitive advantage [24–26].

3. Methodology Framework

When exploring the relation between ASC structural complexity and sustainability, one has to consider the fact that lower ASC structural complexity has a positive implication on material costs, energy costs, and organizational costs. Moreover, it is rather clear that the first of the two items are directly related not only to economic burden, but also to environmental quality. Therefore, the two fundamental elements of sustainability practice, i.e., economic prosperity and environmental protection, is further the main subject of our interest in the context of ASCs. Because it is rather complicated to sufficiently estimate the relation between the structural complexity of ASCs and social implications, the social aspect of sustainability in our approach is omitted. Nevertheless, it has to be mentioned that increasing organizational complexity can cause positive implications on social advancement [27]. When comparing the three cost items (material, energy, and organizational), the first two of them can be empirically considered to be more important from a sustainability viewpoint than organizational costs. However, organizational costs have to be perceived as relevant in a given nexus of sustainability practice, since any organizational activity uses materials and energy.

With the aim to fulfill the goal of this research defined in the introduction section with the encompassing above outlined relation between ASC complexity and sustainability, the following methodological framework is proposed and used (see Figure 1).

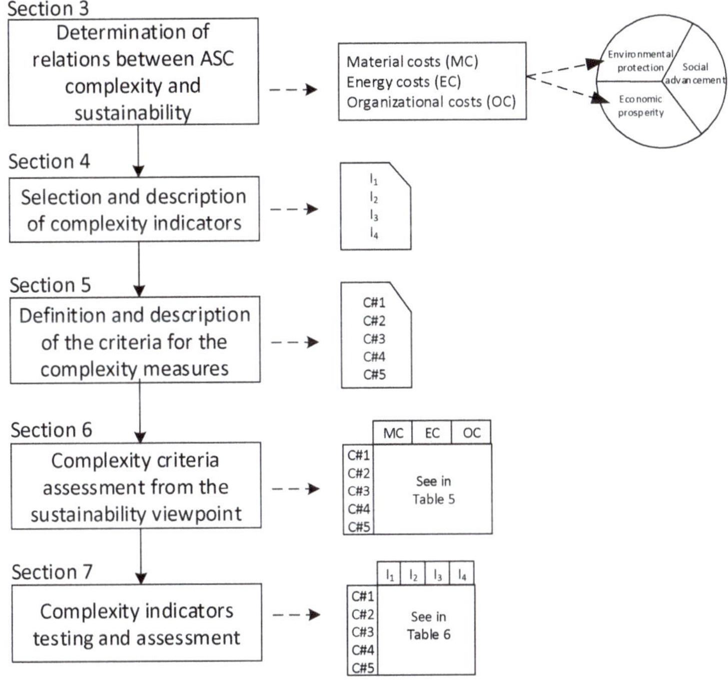

Figure 1. The sequence of the methodological procedures for the selection of the most suitable complexity indicator.

4. Description of Possible ASC Structural Complexity Indicators

In order to identify reliable and consistent complexity indicator(s), four relevant complexity methods were described and mutually compared, namely index of vertex degree, modified flow complexity, system design complexity, and process complexity indicator.

4.1. Index of Vertex Degree

This indicator was originally developed by Bonchev [28] to quantify the structural complexity of general networks. He adopted Shannon's information theory by applying the entropy of information $H(\alpha)$ in describing a message of N symbols. Further it is assumed that such symbols are distributed according to some equivalence criterion α into k groups of $N_1, N_2, \ldots, N_k$ symbols. Then, entropy of information $H(\alpha)$ is calculated by the formula:

$$H(\alpha) = -\sum_{i=1}^{k} p_i log_2 p_i = -\sum_{i=1}^{k} \frac{N_i}{N} log_2 \frac{N_i}{N}, \tag{1}$$

where p_i specifies the probability of the occurrence of the elements of the ith group.

Further, he substituted symbols or system elements for the vertices, and defined the probability for a randomly chosen system element i to have the weight w_i as $p_i = w_i/\sum w_i$ with $\sum w_i = w$, and $\sum p_i = 1$.

Then, the probability for a randomly chosen vertex i in the complete graph of V vertices to have a certain degree $deg\,(v)_i$ can be expressed by the formula:

$$p_i = \frac{\deg(v)_i}{\sum_{i=1}^{V} \deg(v)_i} \tag{2}$$

Assuming that the information can be defined, according Shannon [29], as $I = H_{max} - H$, where H_{max} is the maximum entropy that can exist in a system with the same number of elements, the information entropy of a graph with a total weight W and vertex weights w_i can be expressed in the form of the equation:

$$H(W) = Wlog_2 W - \sum_{i=1}^{V} w_i log_2 w_i \tag{3}$$

Since the maximum entropy is when all w_i = 1, then H_{max} = $Wlog_2 W$, and by substituting $W = \sum deg(v)_i$ and $w_i = deg(v)_i$, the information content of the vertex degree distribution of a network, called the vertex degree index (I_{vd}), is expressed as follows [28]:

$$I_{vd} = \sum_{i=1}^{V} deg(v)_i log_2 deg(v)_i \tag{4}$$

The vertex degree index was subsequently applied to measure the structural complexity of assembly supply chains, and compared with other existing complexity measures [30].

The selected complexity indicator was applied on the following examples of ASCs by Hamta et al. [31] (see Figure 2). All the five ASCs had the same number of input components, but they differed in the number of operations and the number of machines.

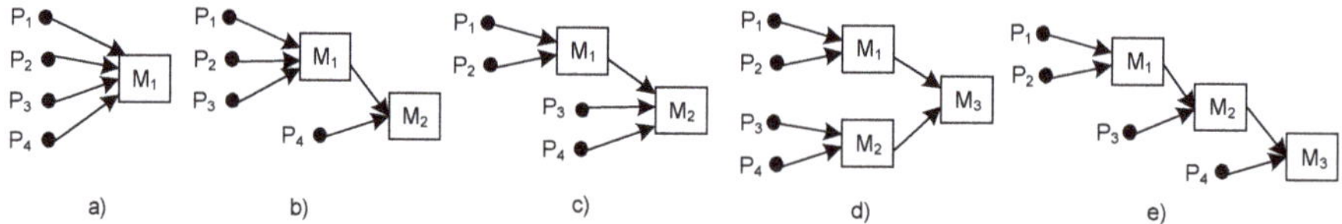

Figure 2. The possible assembly supply chain (ASC) network with four input components and the corresponding structural alternatives.

The complexity values obtained by using Equation (4) are shown in Table 1.

Table 1. Obtained complexity values by I_{vd}.

Graph	(a)	(b)	(c)	(d)	(e)
I_{vd}	8 bits	10 bits	9.51 bits	11.51 bits	11.51 bits

4.2. Process Complexity Indcator

An additional ASC structural complexity measure that was considered was the so-called process complexity indicator (PCI), which was introduced for the purpose of enumerating the operational complexity of manufacturing processes. Its expression is as follows [32]:

$$PCI = -\sum_{i=1}^{M} \sum_{j=1}^{P} \sum_{k=1}^{O} p_{ijk} \cdot log_2 p_{ijk}, \tag{5}$$

where p_{ijk} means the probability that part j is being proceeded by operation k by individual machine i based on the scheduling order; O is the number of operations according to parts production; P is the number of parts produced in the manufacturing process; and M is the number of all machines of all types in the manufacturing process.

It is assumed that machines in a given manufacturing process are organized in a serial and/or parallel manner in Equation (5). The probability that part j is being processed due to operation k on an individual machine i is calculated in the following way. In the case when a part is processed on machines in a serial manner, then p_{ijk} equals $1/M_s$, where M_s presents the number of machines organized in serial. In the case when a part is processed on machines in a parallel manner, then $p_{ijk} = 1/M_p$, where M_p represents the number of machines organized in parallel. In the case where we have a serial/parallel arrangement of machines and a part is processed on one of the parallel machines, then p_{ijk} equals $1/M_s.M_p$.

The complexity values obtained by using Equation (5) are shown in Table 2.

Table 2. Obtained complexity values by process complexity indicator (PCI).

Graph	(a)	(b)	(c)	(d)	(e)
PCI	0 bits	3 bits	2 bits	4 bits	4.17 bits

4.3. System Design Complexity

Guenov [33] proposed three indicators for architectural design complexity measurement based on Boltzmann's entropy theory (see, e.g., [34]) and axiomatic design theory [35]. Those indicators can be principally applied also as ASC complexity measures. However, only one of them has been treated as a potential indicator to measure ASC structural complexity [36]. Its description is as follows. Let us denote N as the number of interactions within a design matrix, and $N_1, N_2, \ldots, N_k$ as the numbers of interactions per each design parameter (DP) of the same matrix. Then, the so-called degree of disorder W can be expressed by the formula:

$$\Omega = C_N^{N_1} * C_{N-N_1}^{N_2} * C_{N-N_1-N_2}^{N_3} \cdots * \cdots C_{N-N_1-\ldots-N_{K-1}}^{N_K} = \frac{N!}{N_1!N_2!\ldots N_K!},\tag{6}$$

where:

$$C_N^{N_1} = \frac{N!}{(N-N_1)!N_1!}\tag{7}$$

The multiplicity Ω in Equation (6) is often called the degree of disorder.

The state of gas body g at a given time t where the gas body consists of N molecules, each characterized by n magnitudes φ_j was considered by Boltzmann. For each magnitude φ_j, its interval of admitted values is divided into small intervals of equal length Δ_j. Then, the n-dimensional space, also known as μ-space or module space, can be divided into a system of cells of equal volume: $v^\mu = \Delta_i, \ldots, \Delta_n$. K is the number of these cells in the total range of admitted values, R^μ; then: $v^\mu = V^\mu/K$, where V^μ is the volume of R^μ. The μ-cells are analogous to the cells Q_j $(j = 1, \ldots, K)$ in the classification system. f_j means the density in Q_j, i.e., the number of molecules per unit of μ-volume: $f_j = N_j/v^\mu$. The function defined by Boltzmann for a statistical description is:

$$\mathrm{H} = \sum_{j=1}^{K}\left[f_j ln f_j\right]v^\mu\tag{8}$$

According to Equation (8) and $f_j = N_j/v^\mu$, where the volume is assumed equal to unity, the following formula for complexity measure can be derived:

$$\sum N_j ln N_j\tag{9}$$

where N_j is interpreted as the number of interactions per single DP.

Its application to measure the structural complexity of ASC was provided by Modrak and Soltysova [37], where the model of assembly process is structured with two groups of objects, while the first group is denoted as input components (ICs) and the second group as workstations. The input components are assembled at these workstations. Then, for the purpose of measuring process complexity, the following transformation is proposed: ICs are substituted by DPs, according to relation DPs = B. PVs (where PVs mean process variables and B is the design matrix that defines the characteristics of the process design) and workstations are considered as PVs. Subsequently, the assembly process structure is transformed into design matrix (DM) with DP–PV relations and finally, the structural complexity of ASC can be enumerated.

The complexity values obtained by using Equation (9) are shown in Table 3.

Table 3. Obtained complexity values by system design complexity (SDC).

Graph	(a)	(b)	(c)	(d)	(e)
SDC	5.55 nats	8.84 nats	6.93 nats	8.32 nats	10.23 nats

4.4. Modified Flow Complexity

The next possible ASC complexity indicators were developed by Crippa [38]. The most suitable indicator of these is considered to be the so-called modified flow complexity (*MFC*) indicator. The *MFC* indicator enumerates all tiers (including Tier 0), nodes, and links and adds all these counts, weighted with determined α, β, and γ and coefficients. Nodes and links are counted only once, even if they are repeated in the graph. Node and link repetition is included in coefficients. The MFC indicator can be enumerated by the following equations [38]:

$$MFC = \alpha.T + \beta.N + \gamma.L \tag{10}$$

$$\alpha = MTI = \frac{TN - N}{(T-1).N} \tag{11}$$

$$\beta = MTR = \frac{TN}{N} \tag{12}$$

$$\gamma = MLR = \frac{LK}{L} \tag{13}$$

where *MTI* is multi-tier index; *MTR* is multi-tier ratio; *MLR* is multi-link ratio; N is the number of nodes; TN is the number of nodes per ith tier level; L is number of links; LK is number of links per ith tier level; T is the number of tiers.

The complexity values obtained by using Equation (10) are shown in Table 4.

Table 4. Obtained values enumerated by modified flow complexity (MFC).

Graph	(a)	(b)	(c)	(d)	(e)
MFC	9	11	11	13	13

5. Definition of Testing Rules for ASC Complexity Indicators

To prove the selected complexity indicators, all of them were verified through defined criteria for complexity measures that might also be taken into account to assess their validity. Such rules were specified and proposed for static complexity metrics by Deshmuk et al. [9] and Olbrich et al. [39]. Deshmuk et al. [9] analyzed different factors influencing the static complexity of manufacturing systems, and defined four conditions that the static complexity metric must satisfy. They are as follows [9]:

Rule#1: Static complexity should increase with the number of parts and the number of machines and operations required to process the part mix.

Rule#2: Static complexity should increase with increases in sequence flexibility for the parts in the production batch.

Rule#3: Static complexity should increase as the sharing of resources by parts increases.

Rule#4: If the original part mix is split into two or more groups, then the complexity of processing should remain constant.

Due to the fact that Rule#3 and Rule#4 are not relevant for assembly supply chains, only two of them were adopted in terms of ASC systems.

Olbrich et al. [39] studied how static complexity measures behave if the system size is increased and explored three special cases of adding an independent element, two independent subsystems, and two identical copies. In this context, the authors proposed three specific requirements or rules for a reliable complexity measure that must be met:

Rule#1: Additional independent element: The element has no structure itself, so it has no complexity of its own. Because it is independent of the rest of the system, the complexity should not change.

Rule#2: Union of two independent systems: Because there are no dependencies between the two systems, the complexity of the union should be simply the sum of the complexities of the subsystems.

Rule#3: Union of two identical copies: Because there is no need for additional information to describe the second system, one could argue that the complexity should be equal to the complexity of one system. One has, however, to include the fact in the description that the second system is a copy of the first one. At least this part should not be extensive with respect to the system size.

However, any of the three rules appear to be impactful for ASC complexity metrics, and therefore were not directly used for the purpose of testing and comparing the complexity indicators.

After deeper consideration, the two rules by Deshmuk et al. [9], namely Rules 1 and 2, were adopted by us into the four testing criteria (C). Moreover, we added one additional criterion (C#5) based on the increasing number of echelons. The criteria are summarily shown as follows:

C#1: Static complexity should increase with the number of parts required to process the part mix.

C#2: Static complexity should increase with the number of machines required to process the part mix.

C#3: Static complexity should increase with the number of operations required to process the part mix.

C#4: Static complexity should increase with increases in sequence flexibility for the parts in the production batch.

C#5: Static complexity should increase with the number of echelons while the number of parts, machines, and operations is constant.

6. Analysis of Testing Criteria from Sustainability Viewpoint

Our description of relations between ASC structural complexity and sustainability seems to be useful to categorize the abovementioned testing criteria based on their importance to the sustainability objectives. In any given case, it was expected that the complexity indicators would reflect the three sustainability facets, i.e.:

- (Direct) energy costs;
- (Direct) material costs;
- (Direct) organizational costs.

Therefore, each of the criteria might validate at least one of the three cost items. By analyzing how these criteria reflect the abovementioned cost items, it can be found that they cover a different number of the cost items. Summarily, these differences are shown in Table 5.

Table 5. Testing criteria and their reflection on cost items.

Testing Criteria	Material Costs	Energy Costs	Organizational Costs
C#2	✔	✔	✔
C#3	-	✔	✔
C#1	-	-	✔
C#4	-	-	✔
C#5	-	-	✔

As can be seen, all criteria include in themselves organizational costs. C#3 moreover covers energy costs and C#2 relates to all of the cost items. This viewpoint was further applied for the purpose of comparing the usability of ASC complexity indicators.

7. Testing and Comparison of ASC Complexity Indicators

This part of the paper is focused on the comparison of two assembly supply chains, $ASC_1(i,j,k)$ and $ASC_2(i,j,k)$. One of them is more complex according to the defined criteria (C#1–C#5). P represents parts, j is the number of parts, O represents operations, k is the number of operations, M represents machines, and i is the number of machines.

These criteria were tested using the theoretical examples shown in Figures 3–7. In C#1, #2, #4, and #5, the number of machines was equal to the number of operations. In C#3, the number of operations was increasing.

7.1. Testing of C#1

C#1 is proposing the following statement:

If i is constant ($i = k$) and j is increasing, then the structural complexity of $ASC(i,j-1,k) < ASC(i,j,k)$.

Let us test $ASC_1(3;4;3)$ and $ASC_2(3;5;3)$, shown in Figure 3. While in Figure 3a, ASC_1 consists of three machines and operations ($i = k = 3$), and the number of parts equals four ($j = 4$), in Figure 3b, ASC_2 contains three machines and operations ($i = k = 3$), and the number of parts equals five ($j = 5$).

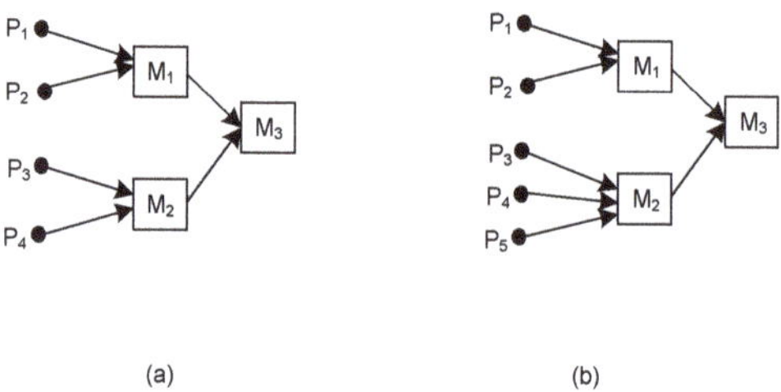

Figure 3. ASC consisting of: (**a**) three machines, three operations, and four parts; (**b**) three machines, three operations, and five parts.

Applying Equation (4), according to complexity indicator I_{vd}, we obtained the following results for static complexity using I_{vd1} (for ASC_1) and I_{vd2} (for ASC_2):

$$I_{vd1} = 3 \times \log_2 3 + 3 \times \log_2 3 + 2 \times \log_2 2 = 11.51 \text{ bits}$$

$$I_{vd2} = 3 \times \log_2 3 + 4 \times \log_2 4 + 2 \times \log_2 2 = 14.76 \text{ bits}$$

Applying Equation (5), according to complexity indicator PCI, we obtained the following results for PCI_1 and PCI_2:

$$PCI_1 = -(0.5 \times \log_2 0.5 + 0.5 \times \log_2 0.5 + 0.5 \times \log_2 0.5 + 0.5 \times \log_2 0.5) = 4 \text{ bits}$$

$$PCI_2 = -(0.5 \times \log_2 0.5 + 0.5 \times \log_2 0.5 + 0.5 \times \log_2 0.5 + 0.5 \times \log_2 0.5 + 0.5 \times \log_2 0.5 + 0.5 \times \log_2 0.5) = 5 \text{ bits}$$

Applying Equation (9), according to complexity indicator system design complexity (SDC), we obtained the following results for SDC_1 and SDC_2:

$$SDC_1 = 4 \times \ln 4 + 2 \times \ln 2 + 2 \times \ln 2 = 8.32 \text{ nats}$$

$$SDC_2 = 5 \times \ln 5 + 2 \times \ln 2 + 3 \times \ln 3 = 12.73 \text{ nats}$$

Applying Equation (10), according to complexity indicator MFC, we obtained the following results for MFC_1 and MFC_2:

$$MFC_1 = 0 \times 3 + 1 \times 7 + 1 \times 6 = 13$$

$$MFC_2 = 0 \times 3 + 1 \times 8 + 1 \times 7 = 15$$

According to these complexity indicators, it was proved that the complexity of $ASC_1(3,4,3) <$ $ASC_2(3,5,3)$, then $ASC(i,j-1,k) < ASC(i,j,k)$.

7.2. Testing of C#2

C#2 is proposing the following statement:

If j is constant and i is increasing $(i = k)$, then the structural complexity of $ASC(i-1,j,k) < ASC(i,j,k)$.

Let us test $ASC_1(3,4,3)$ and $ASC_2(4,4,4)$, shown in Figure 4. While in Figure 4a, ASC_1 consists of three machines and operations $(i = k = 3)$, and the number of parts equals four $(j = 4)$, in Figure 4b, ASC_2 contains four machines and operations $(i = k = 4)$, and the number of parts equals four $(j = 4)$.

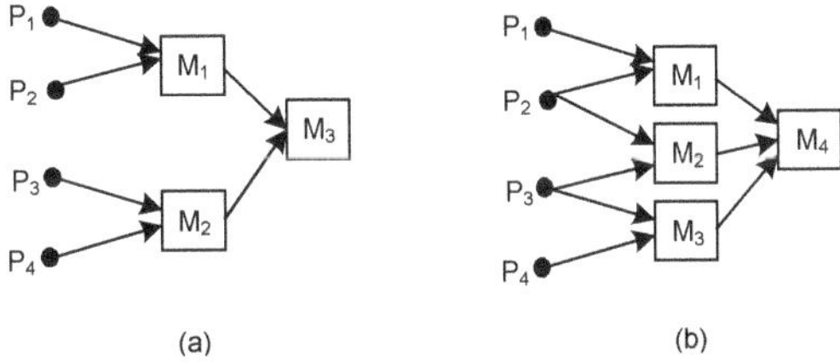

Figure 4. ASC consisting of: (**a**) three machine, three operations, and four parts; (**b**) four machines, four operations, and four parts.

Applying Equation (4), according to complexity indicator I_{vd}, we obtained the following results for I_{vd1} and I_{vd2}:

$$I_{vd1} = 3 \times \log_2 3 + 3 \times \log_2 3 + 2 \times \log_2 2 = 11.51 \text{ bits}$$

$$I_{vd2} = 3 \times \log_2 3 + 3 \times \log_2 3 + 3 \times \log_2 3 + 3 \times \log_2 3 = 19.02 \text{ bits}$$

Applying Equation (5), according to complexity indicator PCI, we obtained the following results for PCI_1 and PCI_2:

$$PCI_1 = -(0.5 \times \log_2 0.5 + 0.5 \times \log_2 0.5 + 0.5 \times \log_2 0.5 + 0.5 \times \log_2 0.5) = 4 \text{ bits}$$

$$PCI_2 = -(0.5 \times \log_2 0.5 + 0.5 \times \log_2 0.5 + 0.25 \times \log_2 0.25 + 0.25 \times \log_2 0.25 + 0.5 \times \log_2 0.5 +$$
$$0.25 \times \log_2 0.25 + 0.25 \times \log_2 0.25 + 0.5 \times \log_2 0.5 + 0.5 \times \log_2 0.5 + 0.5 \times \log_2 0.5) = 5 \text{ bits}$$

Applying Equation (9), according to complexity indicator SDC, we obtained the following results for SDC_1 and SDC_2:

$$SDC_1 = 4 \times \ln 4 + 2 \times \ln 2 + 2 \times \ln 2 = 8.32 \text{ nats}$$

$$SDC2 = 4 \times \ln 4 + 2 \times \ln 2 + 2 \times \ln 2 + 2 \times \ln 2 = 9.7 \text{ nats}$$

Applying Equation (10), according to complexity indicator MFC, we obtained the following results for MFC_1 and MFC_2:

$$MFC_1 = 0 \times 3 + 1 \times 7 + 1 \times 6 = 13$$

$$MFC_2 = 0 \times 3 + 1 \times 8 + 1 \times 9 = 17$$

According to these complexity indicators, it was proved that $ASC_1(3,4,3) < ASC_2(4,4,4)$, then $ASC(i-1,j,k) < ASC(i,j,k)$.

7.3. Testing of C#3

C#3 is proposing the following statement:

If i and j are constant and k is increasing, then the structural complexity of ASC($i,j,k-1$) < ASC(i,j,k).

Let us test $ASC_1(3,4,3)$ and $ASC_2(3,4,4)$, shown in Figure 5. While in Figure 5a, ASC_1 consists of three machines and operations ($i = k = 3$), and the number of parts equals four ($j = 4$), in Figure 5b, ASC_2 contains three machines ($i = 3$), four operations ($k = 4$), and the number of parts equals four ($j = 4$).

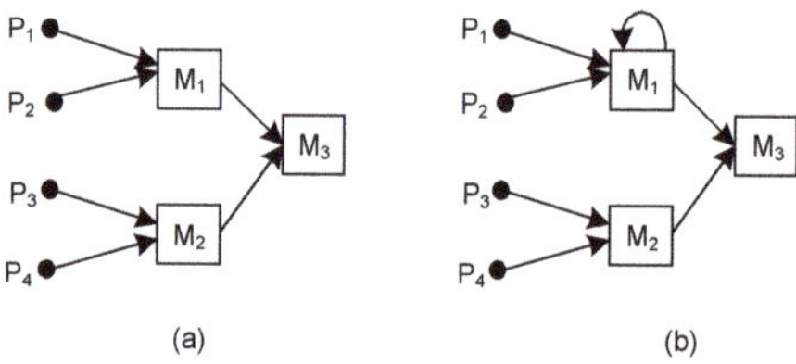

Figure 5. ASC consisting of: (**a**) three machines, three operations, and four parts; (**b**) three machines, four operations, and three parts.

Applying Equation (4), according to complexity indicator I_{vd}, we obtained the following results for I_{vd1} and I_{vd2}:

$$I_{vd1} = 3 \times \log_2 3 + 3 \times \log_2 3 + 2 \times \log_2 2 = 11.51 \text{ bits}$$

$$I_{vd2} = 5 \times \log_2 5 + 3 \times \log_2 3 + 2 \times \log_2 2 = 18.37 \text{ bits}$$

Applying Equation (5), according to complexity indicator PCI, we obtained the following results for PCI_1 and PCI_2:

$$PCI_1 = -(0.5 \times \log_2 0.5 + 0.5 \times \log_2 0.5 + 0.5 \times \log_2 0.5 + 0.5 \times \log_2 0.5) = 4 \text{ bits}$$

$$PCI_2 = -(0.25 \times \log_2 0.25 + 0.25 \times \log_2 0.25 + 0.5 \times \log_2 0.5 + 0.25 \times \log_2 0.25 + 0.25 \times \log_2 0.25 +$$
$$0.5 \times \log_2 0.5 + 0.5 \times \log_2 0.5 + 0.5 \times \log_2 0.5 + 0.5 \times \log_2 0.5 + 0.5 \times \log_2 0.5) = 5 \text{ bits}$$

Applying Equation (9), according to complexity indicator SDC, we obtained the following results for SDC_1 and SDC_2:

$$SDC_1 = 4 \times \ln 4 + 2 \times \ln 2 + 2 \times \ln 2 = 8.32 \text{ nats}$$

$$SDC_2 = 4 \times \ln 4 + 2 \times \ln 2 + 2 \times \ln 2 = 8.32 \text{ nats}$$

Applying Equation (10), according to complexity indicator MFC, we obtained the following results for MFC_1 and MFC_2:

$$MFC_1 = 0 \times 3 + 1 \times 7 + 1 \times 6 = 13$$

$$MFC_2 = 0 \times 3 + 1 \times 7 + 1 \times 7 = 14$$

According to the three complexity indicators I_{vd}, PCI, and MFC, it was proved that the structural complexity of $ASC_1(3,4,3)$ < $ASC_2(3,4,4)$, then ASC($i,j,k-1$) < ASC(i,j,k).

7.4. Testing of C#4

C#4 is proposing the following statement:

If i, j, and k are constant and the number of routings is increasing, then the structural complexity of ASC(i,j,k) < ASC(i,j,k) with an increasing number of routings.

Let us test $ASC_1(3,4,3)$ and $ASC_2(3,4,3)$, shown in Figure 6. While in Figure 6a, ASC_1 consists of three machines and operations ($i = k = 3$), and the number of parts equals four ($j = 4$), in Figure 6b, ASC_2 consists of three machines and operations ($i = k = 3$), and the number of parts equals four ($j = 4$), but P_2 and P_3 can be processed on M_1 or M_2.

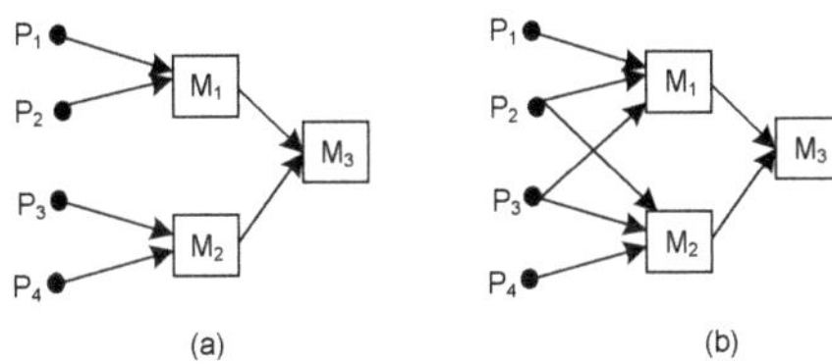

Figure 6. ASC consisting of: (**a**) three machines, three operations, and four parts; (**b**) three machines, three operations, and four parts.

Applying Equation (4), according to complexity indicator I_{vd}, we obtained the following results for I_{vd1} and I_{vd2}:

$$I_{vd1} = 3 \times \log_2 3 + 3 \times \log_2 3 + 2 \times \log_2 2 = 11.51 \text{ bits}$$

$$I_{vd2} = 4 \times \log_2 4 + 4 \times \log_2 4 + 2 \times \log_2 2 = 18 \text{ bits}$$

Applying Equation (5), according to complexity indicator PCI, we obtained the following results for PCI_1 and PCI_2:

$$PCI_1 = -(0.5 \times \log_2 0.5 + 0.5 \times \log_2 0.5 + 0.5 \times \log_2 0.5 + 0.5 \times \log_2 0.5) = 4 \text{ bits}$$

$$PCI_2 = -(0.5 \times \log_2 0.5 + 0.5 \times \log_2 0.5 + 0.25 \times \log_2 0.25 + 0.25 \times \log_2 0.25 + 0.5 \times \log_2 0.5 +$$
$$0.25 \times \log_2 0.25 + 0.25 \times \log_2 0.25 + 0.5 \times \log_2 0.5 + 0.5 \times \log_2 0.5 + 0.5 \times \log_2 0.5) = 5 \text{ bits}$$

Applying Equation (9), according to complexity indicator SDC, we obtained the following results for SDC_1 and SDC_2:

$$SDC_1 = 4 \times \ln 4 + 2 \times \ln 2 + 2 \times \ln 2 = 8.32 \text{ nats}$$

$$SDC_2 = 4 \times \ln 4 + 3 \times \ln 3 + 3 \times \ln 3 = 12.14 \text{ nats}$$

Applying Equation (10), according to complexity indicator MFC, we obtained the following results for MFC_1 and MFC_2:

$$MFC_1 = 0 \times 3 + 1 \times 7 + 1 \times 6 = 13$$

$$MFC_2 = 0 \times 3 + 1 \times 7 + 1 \times 8 = 15$$

According to these complexity indicators, it was proved that the complexity of $ASC_1 < ASC_2$, then $ASC(i,j,k) < ASC(i,j,k)$ with an increasing number of routings.

7.5. Testing of C#5

C#5 is proposing the following statement:

If i, j, and k are constant and the number of echelons is increasing, then the complexity of $ASC(i,j,k) < ASC(i,j,k)$ with an increasing number of echelons.

Let us test $ASC_1(3,4,3)$ and $ASC_2(3,4,3)$, shown in Figure 7. While in Figure 7a, ASC_1 consists of three machines and operations ($i = k = 3$), the number of parts equals four ($j = 4$), and the number of echelons equals two, in Figure 7b, ASC_2 contains three machines ($i = 3$), four operations ($k = 4$), the number of parts equals four ($j = 4$), and the number of echelons equals three.

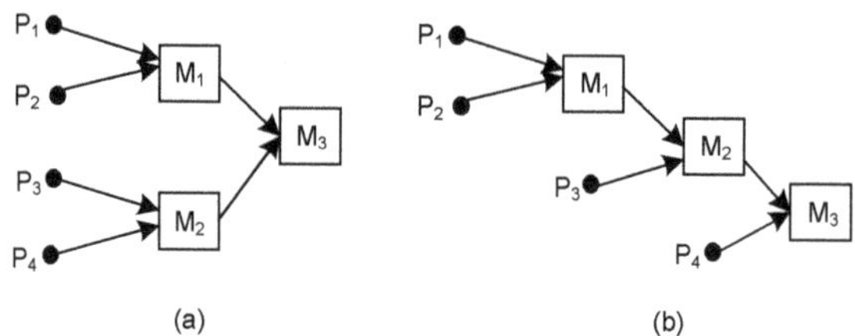

Figure 7. Manufacturing system consisting of: (**a**) three machines, three operations, four parts, and two echelons; (**b**) three machines, three operations, four parts, and three echelons.

Applying Equation (4), according to complexity indicator I_{vd}, we obtained the following results for I_{vd1} and I_{vd2}:

$$I_{vd1} = 3 \times \log_2 3 + 3 \times \log_2 3 + 2 \times \log_2 2 = 11.51 \text{ bits}$$

$$I_{vd2} = 3 \times \log_2 3 + 3 \times \log_2 3 + 2 \times \log_2 2 = 11.51 \text{ bits}$$

Applying Equation (5), according to complexity indicator PCI, we obtained the following results for PCI_1 and PCI_2:

$$PCI_1 = -(0.5 \times \log_2 0.5 + 0.5 \times \log_2 0.5 + 0.5 \times \log_2 0.5 + 0.5 \times \log_2 0.5) = 4 \text{ bits}$$

$$PCI_2 = -(0.33 \times \log_2 0.33 + 0.33 \times \log_2 0.33 + 0.33 \times \log_2 0.33 + 0.33 \times \log_2 0.33 + 0.33 \times \log_2 0.33 + 0.33 \times \log_2 0.33 + 0.5 \times \log_2 0.5 + 0.5 \times \log_2 0.5 + 1 \times \log_2 1) = 4.17 \text{ bits}$$

Applying Equation (9), according to complexity indicator SDC, we obtained the following results for SDC_1 and SDC_2:

$$SDC_1 = 4 \times \ln 4 + 2 \times \ln 2 + 2 \times \ln 2 = 8.32 \text{ nats}$$

$$SDC_2 = 4 \times \ln 4 + 3 \times \ln 3 + 2 \times \ln 2 = 10.23 \text{ nats}$$

Applying Equation (10), according to complexity indicator MFC, we obtained the following results for MFC_1 and MFC_2:

$$MFC_1 = 0 \times 3 + 1 \times 7 + 1 \times 6 = 13$$

$$MFC_2 = 0 \times 4 + 1 \times 7 + 1 \times 6 = 13$$

According to the three complexity indicators PCI, SDC, and MFC, it was proved that the complexity of ASC_1 < the complexity of ASC_2, then $ASC(i,j,k) < ASC(i,j,k)$ with an increasing number of echelons.

7.6. Comparison and Selection of ASC Complexity Measure

A mutual comparison of ASC complexity methods is shown in the following Table 6:

Table 6. Mutual comparison of ASC complexity methods.

Criteria	I_{vd}	SDC	MFC	PCI
C#2	✔	✔	✔	✔
C#3	✔	X	✔	✔
C#1	✔	✔	✔	✔
C#4	✔	✔	✔	✔
C#5	X	✔	X	✔

Based on the obtained results, it can be seen that PCI for ASCs could be considered as the most suitable.

8. Discussion of Results, Implications and Limitations

Summarizing the obtained results from the testing of the ASC complexity indicators, the following statements can be provided:

(i)   All described indicators sufficiently reflect organizational aspects of ASCs.
(ii) Three of the complexity indicators, namely, I_{vd}, MFC, and PCI, can be effectively used to measure ASC complexity in order to identify how ASC structural variants are influencing organizational costs, as well as energy costs.
(iii) The PCI complexity measure reflects all of the three cost items and covers the two crucial dimensions of sustainability, economic and environmental.

In addition, it has to be emphasized that the impact of ASC complexity on social development can only be anticipated based on the theoretical assumption that higher organizational complexity, induced by a company's favorable development, could positively influence social sustainability. In such a case, it would be expected that organizational complexity is growing not only along various dimensions, such as production volume, but also with the increasing number of employees [40,41]. However, the related literature is more or less ambivalent about whether organizational complexity has positive or negative effects on firm performance [42]. Moreover, if we consider that organizational complexity is frequently defined as "the amount of differentiation that exists within different elements constituting the organization" [43], then it is rather clear that any of the structural complexity indicators can identify organizational complexity changes within a company. Thus, it cannot be excluded that changes of the internal structural complexity will not impact the organizational complexity. Based on the above formulations, it can be concluded that, when comparing two or several ASC alternative solutions, the one with the lowest structural complexity can be considered as the most sustainable and cost-effective approach. Therefore, structural complexity measures can be primarily used as potential indicators for the indirect assessment of material costs and related energy consumption. This drawback of the presented approach can be seen as the main limitation of the analyzed indicators to reflect ASC complexity along with the sustainability issues.

This evaluation method can be effectively used, especially in terms of small and medium size enterprises (SMEs), since it has been found that SMEs are particularly sensitive to the internal complexity environment, as well as to the external complexity environment [44]. For this reason, SMEs need to pay attention also to the complexity management approach in order to entirely manage their sustainability in a turbulent business environment.

9. Conclusions

Finally, it is useful to emphasize that the presented research construct and obtained findings are pertinent for ASCs for the completion of standard products, i.e., not for those that are mass customized. Mass customization, as a new paradigm, also brings significant changes into the assembly process principles. For example, ASCs in terms of mass customization have to be as modular as possible, and it means that modularity has to be considered as a primary property of ASCs. Taking this fact into consideration, the structural complexity of ASCs can be seen as only a complementary indicator to optimize ASCs. Accordingly, the featured related research could be oriented specially to study relations between the complexity and modularity of ASCs in terms of mass customization.

Author Contributions: V.M. provided the introduction of this paper, described relations between ASC complexity and sustainability, defined testing rules for ASC complexity indicators, and summarized the findings in the conclusion section. Z.S. worked on the selection and description of possible ASC complexity indicators, and provided the testing and comparison of ASC complexity indicators. D.O. helped with the analysis of relations between testing criteria and cost items.

Funding: This research was funded by the Scientific Grant Agency of the Slovak Republic VEGA, grant number 1/0419/16. Besides, this research was funded by the European Union's Horizon 2020 research and innovation program under the Marie Skłodowska-Curie, grant number 734713.

Conflicts of Interest: The authors declare no conflict of interest.

References

1. Fan, J.; Dong, J. Intelligent virtual assembly planning with integrated assembly model. In Proceedings of the IEEE International Conference on Systems, Man and Cybernetics. Conference Theme-System Security and Assurance, Washington, DC, USA, 8 October 2003; pp. 4803–4808.
2. Flood, R.L.; Carson, E.R. *Dealing with Complexity: An Introduction to the Theory and Application of Systems Science*, 2nd ed.; Springer Science & Business Media: New York, NY, USA, 2013; pp. 1–280.
3. Flood, R.L. Liberating systems theory. In *Liberating Systems Theory*; Springer: Boston, MA, USA, 1990; pp. 11–32.
4. Faulconbridge, R.I.; Ryan, M.J. *Systems Engineering Practice*; Argos Press: Canberra, Australian, 2018; pp. 1–320.
5. Efatmaneshnik, M.; Ryan, M.J. Fundamentals of system complexity measures for systems design. In *INCOSE International Symposium*; Wiley: Hoboken, NJ, USA, 2015; Volume 25, pp. 876–890.
6. Reynolds, P.A. *Introduction to International Relations*, 3rd ed.; Routledge: New York, NY, USA, 2016; pp. 1–358.
7. Suh, N.P. *The Principles of Design*; Oxford University Press: New York, NY, USA, 1990; pp. 1–401.
8. Gell-Mann, M.; Lloyd, S. Information measures, effective complexity, and total information. In *Complexity*; John Wiley & Sons: New York, NY, USA, 1996; Volume 1, pp. 44–52.
9. Deshmukh, A.V.; Talavage, J.J.; Barash, M.M. Complexity in manufacturing systems. Part 1: Analysis of static complexity. *IIE Trans.* **1998**, *30*, 645–655. [CrossRef]
10. Frizelle, G.; Frizelle, G. *The Management of Complexity in Manufacturing*; Business Intelligence: London, UK, 1998; pp. 1–320.
11. Hu, S.J.; Zhu, X.; Wang, H.; Koren, Y. Product variety and manufacturing complexity in assembly systems and supply chains. *CIRP Annals* **2008**, *57*, 45–48. [CrossRef]
12. Shannon, C.E.; Weaver, W. *The Mathematical Theory of Communication*; University of Illinois Press: Urbana, IL, USA, 1949; pp. 1–125.
13. Frizelle, G.; Woodcock, E. Measuring complexity as an aid to developing operational strategy. *Int. J. Oper. Prod. Manag.* **1995**, *15*, 26–39. [CrossRef]
14. Gare, A. *Systems Theory and Complexity: Introduction*; Routledge: Abingdon-on-Thames, UK, 2000; Volume 6, pp. 327–339.
15. Schilling, M.A. Toward a general modular systems theory and its application to interfirm product modularity. *Acad. Manag. Rev.* **2000**, *25*, 312–334. [CrossRef]
16. Baldwin, C.Y.; Clark, K.B. Managing in an Age of Modularity. In *Managing in the Modular Age: Architectures, Networks, and Organizations*; John Wiley & Sons: New York, NY, USA, 2003; Volume 149, pp. 84–93.
17. Peralta, M.E.; Marcos, M.; Aguayo, F.; Lama, J.R.; Córdoba, A. Sustainable Fractal Manufacturing: A new approach to sustainability in machining processes. *Proc. Eng.* **2015**, *132*, 926–933. [CrossRef]
18. Moldavska, A. Model-based sustainability assessment–an enabler for transition to sustainable manufacturing. *Proc. Cirp* **2016**, *48*, 413–418. [CrossRef]
19. Rauch, E.; Dallasega, P.; Matt, D.T. Sustainable production in emerging markets through Distributed Manufacturing Systems (DMS). *J. Clean. Prod.* **2016**, *135*, 127–138. [CrossRef]
20. Rauch, E.; Dallasega, P. Sustainability in manufacturing and supply chains through distributed manufacturing systems and networks. In *Science and Technology*; Elsevier: Amsterdam, The Netherlands, 2017.
21. Modrak, V.; Marton, D.; Bednar, S. The influence of mass customization strategy on configuration complexity of assembly systems. *Proced. CIRP* **2015**, *33*, 538–543. [CrossRef]
22. Fisher, M.L. What is the right supply chain for your product. In *Operations Management: Critical Perspectives on Business and Management*; Routledge: New York, NY, USA, 2003; Volume 4.
23. Fisher, M.L.; Ittner, C.D. The impact of product variety on automobile assembly operations: Empirical evidence and simulation. *Manag. Sci.* **1999**, *45*, 771–786. [CrossRef]
24. Khan, S.A.R.; Dong, Q. Impact of green supply chain management practices on firms' performance: An empirical study from the perspective of Pakistan. *Environ. Sci. Pollut. Res.* **2017**, *24*, 16829–16844. [CrossRef]
25. Gunasekaran, A.; Spalanzani, A. Sustainability of manufacturing and services: Investigations for research and applications. *Int. J. Prod. Econ.* **2012**, *140*, 35–47.
26. Gouda, S.K.; Saranga, H. Sustainable supply chains for supply chain sustainability: Impact of sustainability efforts on supply chain risk. *Int. J. Prod. Res.* **2018**, *56*, 5820–5835. [CrossRef]

27. Cicmil, S.; Marshall, D. Insights into collaboration at the project level: Complexity, social interaction and procurement mechanisms. *Build. Res. Inf.* **2005**, *33*, 523–535. [CrossRef]
28. Bonchev, D.; Buck, G.A. Quantitative measures of network complexity. In *Complexity in Chemistry, Biology and Ecology*; Bonchev, D., Rouvray, D.H., Eds.; Springer: New York, NY, USA, 2005; pp. 191–235.
29. Shannon, C.E. A mathematical theory of communication. *Bell Syst. Technol. J.* **1948**, *27*, 379–423. [CrossRef]
30. Modrak, V.; Marton, D. Modelling and complexity assessment of assembly supply chain systems. *Proc. Eng.* **2012**, *48*, 428–435. [CrossRef]
31. Hamta, N.; Shirazi, M.A.; Behdad, S.; Ghomi, S.F. Modeling and measuring the structural complexity in assembly supply chain networks. *J. Intell. Manuf.* **2018**, *29*, 259–275. [CrossRef]
32. Modrak, V.; Soltysova, Z. Development of operational complexity measure for selection of optimal layout design alternative. *Int. J. Prod. Res.* **2018**, *56*, 7280–7295. [CrossRef]
33. Guenov, M.D. Complexity and cost-effectiveness measures for systems design. In *Manufacturing Complexity Network Conference*; Frizelle, G., Richards, H., Eds.; Mathematics: Cambridge, UK, 2002; pp. 455–465.
34. Perrot, P. *A to Z of Thermodynamics*; Oxford University Press: Oxford, UK, 1998.
35. Suh, N.P. Axiomatic Design. In *Advances and Applications*; Oxford University Press: Oxford, UK, 2001; pp. 1–503.
36. Modrak, V.; Bednar, S.; Soltysova, Z. Application of axiomatic design-based complexity measure in mass customization. *Proced. CIRP* **2016**, *50*, 607–612. [CrossRef]
37. Modrak, V.; Soltysova, Z. Axiomatic design based complexity measures to assess product and process structures. In *MATEC Web of Conferences*; EDP Sciences: Les Ulis, France, 2018; Volume 223, pp. 1–9.
38. Crippa, R.; Bertacci, N.; Larghi, L. Representing and measuring flow complexity in the extended enterprise: The D4G approach. In Proceedings of the RIRL 2006—Sixth International Congress of Logistics Research, Pontremoli, Italy, 3–6 September 2006.
39. Olbrich, E.; Bertschinger, N.; Ay, N.; Jost, J. How should complexity scale with system size? *Eur. Phys. J. B* **2008**, *63*, 407–415. [CrossRef]
40. Scheidt, C.; Li, L.; Caers, J. (Eds.) *Quantifying Uncertainty in Subsurface Systems*; John Wiley & Sons: Hoboken, NJ, USA, 2018; Volume 236, pp. 1–304.
41. Grasso, A.; Convertino, G. Collective intelligence in organizations: Tools and studies. *Comput. Support. Coop. Work (CSCW)* **2012**, *21*, 357–369. [CrossRef]
42. Larsen, M.M.; Manning, S.; Pedersen, T. The ambivalent effect of complexity on firm performance: A study of the global service provider industry. *Long Range Plan.* **2019**, *52*, 221–235. [CrossRef]
43. Dooley, K. Organizational Complexity. In *International Encyclopedia of Business and Management*; Thompson Learning: London, UK, 2002; pp. 5013–5022.
44. Okreglicka, M.; Gorzeń-Mitka, I.; Ogrean, C. Management challenges in the context of a complex view-SMEs perspective. *Proced. Econ. Financ.* **2015**, *34*, 445–452. [CrossRef]

 *sustainability*

Article

Challenges of Industry 4.0 Technology Adoption for SMEs: The Case of Japan

Martin Prause

Institute for Industrial Organization, WHU—Otto Beisheim School of Management, 56179 Vallendar, Germany; martin.prause@whu.edu

Received: 26 August 2019; Accepted: 16 October 2019; Published: 19 October 2019

Abstract: In the light of several national advanced manufacturing strategies such as Industry 4.0 in Germany or the Made in China 2025 initiative in China, this article examines the challenges of Industry 4.0 adoption of Japanese small and medium-sized manufacturing firms. A technology adoption model for Industry 4.0 is developed and empirically tested with 38 manufacturing companies. The results yield that the market uncertainty of the firm's business is a significant driver for adoption in the short, medium, and long-term. Relative competitive advantage matters in the short term and top management support in the long-term. No support has been identified concerning advanced manufacturing complexity and market transparency of Industry 4.0 solutions.

Keywords: advanced manufacturing; industry 4.0; SME; technology adoption model

---

1. Introduction

Manufacturing is the backbone of large economies such as the U.S., Europe, China, or Japan. Changing demographics, globalization, scarcity of resources, the challenges of climate change, and mass customization are the megatrends that challenge the future of manufacturing [1]. These changes imply volatile, uncertain, complex, and ambiguous environments for firms and affect them across their strategic environment [2]. Various initiatives emphasized the urgency for advanced manufacturing strategies to tackle those challenges and support economic growth [3].

Industry 4.0 facilitates the balancing act of internal and external complexity by shifting traditional production systems from a structured centralized control to decentralized control. It is a specific deployment of an advanced manufacturing strategy. The core principles of Industry 4.0 are modularization, self-regulation, and digital integration across business functions and within and beyond the organizational boundaries. Industry 4.0 induces product innovation based on the usage of intelligent sensor and actor systems to facilitate context-sensitive production processes and ICT based process innovation to integrate production processes across the value chain, value network, and product lifecycle.

While the manufacturing industry is the backbone of large economies such as the U.S., Europe, China, or Japan, small and medium-sized enterprises (SMEs) are the foundation for manufacturing industries. Small and medium-sized enterprises are very flexible in adapting new technologies and catering niche markets, while large corporates are better on scale efficiencies but slower in adapting innovations [4]. Therefore, it is essential to study the challenges of information technology adoption concerning the concept of Industry 4.0 for small and medium-sized enterprises to accelerate the diffusion of advanced manufacturing.

This article contributes to the body of knowledge in two ways. First, a technology adoption model is presented that accounts for the particularities of Industry 4.0 information technologies, and second, this model is empirically tested, and the results are discussed.

2. Technology Characteristics of Industry 4.0

Industry 4.0 is defined as the "[...] integration of cyber-physical-systems in production and logistics as well as the application of the Internet of Things in industrial processes. This includes the consequences for the value chain, business models, services, and work environment" [5]. While some advanced manufacturing strategies include the usage of new materials, advanced human-machine interaction, assistive (ambient) systems, and factory virtualization, the common element in all definitions is the linkage of physical systems and virtual systems using information and communication technologies for facilitating manufacturing processes.

It consists of two innovation types: product and process innovation. On the product innovation side, it is the integration of cyber-physical systems (CPS) to integrate context-aware and self-adaptive production processes. Using machine learning techniques on higher ICT levels, the monitored data from sensors can be used to control embedded actors to respond to environmental changes (self-regulatory). This is depicted in the stages of the 5C architecture of a CPS, ranging from smart connections to data conversion, cyber integration, cognition, and self-configuration [6]. Based on technology such as the Internet of Things (machine-to-machine communication), Internet of Service (machine-to-human communication), and Internet of People (virtual human-to-human communication), CPS integrates the physical world of productions vertically. Successful implementation of CPS is based on "[...] horizontal integration through value networks, end-to-end digital integration of engineering across the entire value chain, vertical integration, and networked manufacturing systems" [5]. This implies that product and process innovation has to be implemented in combination and not separately.

From a technical perspective, these two elements of the advanced manufacturing strategies, cyber-physical systems, and digital integration of production and business processes are not new in the industrial context. Notably, both parts are the core principles of a so-called Smart Factory. The Smart Factory idea is based on transferring the ubiquitous (wireless) computing to an industrial context [7].

A Smart Factory is a "[...] Factory that context-aware assists people and machines in execution of their tasks [...] by systems working in background, so-called Calm-systems and context-aware applications" [8]. A Smart Factory masters the complexity challenge by modularization of its components and self-regulating mechanics based on context-aware applications. A self-adaptive system, such as the Smart Factory, addresses the complexity issue by automatically modifying itself in response to changes in its operating environment. A Smart Factory is robust and increases production efficiency by enabling communication flows among humans, machines, and resources, just like communication flow in social networks [5].

The key implication of linking production processes with business functions using context-aware systems is the ubiquitous real-time availability and analysis of data for self-regulatory behavior [9]. Other authors refer to the Smart Factory concept as Industrial Internet, Intelligent Manufacturing, Digital Factory, Ubiquitous Factory, or Factory 4.0 [10,11].

The technological enablers for deploying cyber-physical systems and ICT-integrated production-business processes are advanced technologies in Big Data Analytics, Cloud Computing, Internet of Things, and virtualization/simulation technologies for education and assistance [2,12]. However, not all enablers have reached a certain readiness level [13]. Due to the uncertainty of economic rents involved with each technological readiness level, SMEs will adopt mature technologies, likely adopt technologies that reached the evaluation stages and hesitate to adopt technologies in its infancy. Specifically, maturity models for small-and-medium-sized enterprises should be different from multinational-enterprises due to their specific requirements [13].

3. Technology Adoption Models

A list of models, tailored to the particular product and process innovation characteristics and the SME context, is added to the annex of this article. General concepts differentiate between the adoption perspective of an organization and an individual, where this study takes the firm's view.

Roger's innovation diffusion theory (IDT) states that potential adopters evaluate the following technical characteristics to assess the adoption of an innovation: first, the relative advantage of the innovation compared to the competition and its compatibility with the existing infrastructure are the most significant adoption determinants [14]. Second, in terms of complexity and trialability, a firm also evaluates the external organizational context to suppliers, customers, or governmental authorities in terms of efficiency gains or possible constraints from the adopted innovation [15].

The Technology, Organization and Environment framework (TOE) [16] is a widely accepted framework [17] which identified three constructs that are vital to adopt an innovation: (1) availability, best practices, and equipment, (2) firm size, communication processes and managerial structure, and (3) industry characteristics, market characteristics, and technology support.

Following the arguments by Scott [18], institutional environments such as business partners or competitors shape organizational structures and actions, which leads to the inter-organizational model (IO) based on the constructs of (1) perceived benefits, (2) organizational readiness, and (3) external pressure. Raymond [19] underlines this approach and postulates that IDT and TOE frameworks "[...] need[s] to be enriched when the innovation relates to complex technologies with an inter-organizational locus of impact, for which adoption decisions are linked (e.g., when imposed by business partners) and when the innovation is adopted by organizations" [19].

This has been implemented by Chau and Tam [15], where the authors propose a model based on the TOE framework and added product-specific characteristics to study the adoption of open (software) system standards: (1) external environment e.g., market uncertainty, (2) organizational technology such as complexity of IT infrastructure, satisfaction with existing system, formalization on system development & management, and (3) open system characteristics like perceived benefits, perceived barriers, interoperability, and interconnectivity.

While Iacovou et al. [20] incorporated inter-organizational aspects, Chatterjee et al. [21] argued that the adoption of specific information and communication technologies are guided by intra-organizational cooperation. Following the argumentation by Swanson [22], the adoption decision must be coordinated. This is reflected in their model, which incorporates top management factors such as beliefs towards innovation, participation in the adoption process, and strategic investment rationale.

4. Method

For the development of the technology adoption model, Industry 4.0 is defined as the digital integration of the production system with the company's business functions using self-regulatory sensor-actor networks (CPS) in combination with information and communication technologies. The constituting constructs are: (1) the usage of technologies which continuously monitor processes from inbound logistics, production and outbound logistics to regulate the respective processes autonomously depending on changes in the environment, (2) the real-time availability and analysis of the monitored data to other business functions such as administration, research and development, service, marketing and sales and, (3) the usage of software system for automatic data exchange between business functions such as administration, research and development, service, marketing and sales. For the sake of ease use in a survey, these elements are summarized as self-adaptive technologies and digitalized processes in the individual survey items.

On the one hand, a CPS is considered as product innovation due to its technological foundation. On the other hand, the aligned software-oriented information technologies are considered process innovation because they combine, align, and integrate the CPS components with the existing business processes. This study follows the approach of Fichman and Kemerer [23] and Chau and Tam [15], where the authors highlight that a complicated technological innovation, such as Industry 4.0, should add innovation specific adoption characteristics in addition to the traditional ones. Therefore, the research model builds on the TOE framework and incorporates independent variables from other models to match the context of small and medium-sized enterprises and the characteristics of the innovations. The dependent variable is a construct of adoption intention and the early stages of technology assimilation as defined by the Guttman scale.

Specific characteristics, which stifle or foster the adoption of advanced manufacturing by SMEs, are highlighted in the study by Wischman et al. [12]. Although the study looks at the broader context of Industry 4.0 in Germany, the factors are assumed to be transferable to the defined advanced manufacturing context in this article because the core elements are identical. According to Wischmann et al. [12], the main hurdles for SMEs to adopt Industry 4.0 are based on (1) financial constraints, (2) an immature IT to support a smooth integration of Industry 4.0 technologies, (3) lack of top management support to lead and digital vision and provide the necessary budget, (4) lack of skills of the workforce, or (5) resistance due to technological uncertainty and immature standards and protocols. The institutional theory is incorporated in this model because, based on the determinants of Swanson's framework [22], Industry 4.0 is a core business function innovation [21]. This study includes only intra-organizational factors and excludes inter-organizational factors because, contrary to EDI, the interconnectivity of firms is not a necessity for cyber-physical systems and data exchange within the organization (Figure 1).

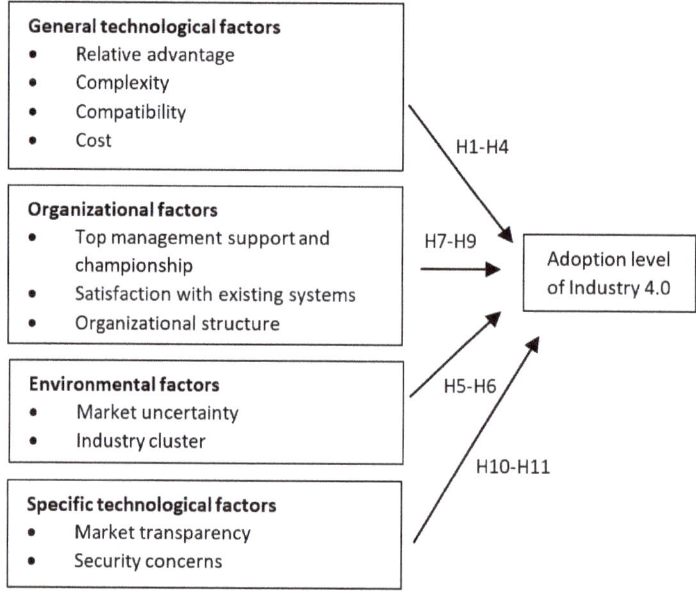

Figure 1. The Industry 4.0 Adoption Model proposing the hypotheses from H1 to H11.

4.1. General Technological Factors

4.1.1. Complexity

Complexity is the degree to which the technologies are perceived as challenging to understand and use [24,25], and it is typically negatively correlated with adoption [25]. The lack of knowledge is the second major obstacle for Industry 4.0 adoption for SMEs [12]. Complexity can be measured in terms of number and diversity of relationships and number and diversity of elements within a given. Cyber-physical systems consist of multiple specialized heterogeneous devices and operate in a changing environment. In combination with data analysis and ICT integration with business functions, product innovation is considered to be complex. Technological communication standards and protocols must be established to guarantee a smooth integration of data exchange across business functions. The orchestration of multiplicity of hardware and software is a complex task [25]. Based on established measures by Premkumar and Roberts [25] and Jeon et al. [26] the indicator is operationalized as:

- The skills required to use self-adaptive technologies and digitalized processes are too complex for our employees.
- Integrating self-adaptive technologies and digitalized processes in our current work practices will be very difficult.

Hypothesis H1. *The lower the perceived complexity of the advanced manufacturing technologies, the more likely they will be adopted.*

4.1.2. Compatibility

Compatibility is the degree to which technological innovation can be easily integrated with the existing infrastructure and processes [24]. If the innovation is compatible with current practices and technological infrastructure, adoption is more likely, according to Cooper and Zmud [27]. The shift from centralized controlled production to decentralized (self-adaptive) controlled production is a substantial change for the organization. According to Premkumar and Roberts [25], it is essential for small businesses that the changes are compatible with the organizational culture. Otherwise, it may result in resistance. Chatterjee et al. [21] highlight, in the context of e-business, that higher technological compatibility of the innovation with the existing system implies higher adoption capability of the organization. Old production systems are strong inhibitors to adopt Industry 4.0. Based on these measures, this indicator is operationalized as:

- The changes introduced by self-adaptive technologies and digitalized processes are consistent with the firm's existing beliefs/values.
- The self-adaptive technologies and digitalized processes are compatible with existing IT infrastructure.
- The self-adaptive technologies and digitalized processes are compatible with the firm's existing experiences with similar systems.

Hypothesis 2 (H2). *The greater the perceived compatibility of the advanced manufacturing technologies with current infrastructure, values, and beliefs, the more likely they will be adopted.*

4.1.3. Relative Advantage

Relative advantage is the degree of additional benefit for the organization from adopting the innovation compared to the status quo, according to Roberts [24]. A recent study by Commerzbank [28] on the impact of the digital trend on manufacturing firms revealed that firms tend to accelerate ICT integration to establish new business models and innovations to counter the competitive pressure and short product life cycles. Relative advantages are referred to in recent models as a perceived benefit [20]. Roger's and Iacovou's model indicate both that understanding the relative advantage increases the likelihood of adoption. Therefore, advanced manufacturing is expected to be able to give organizations a better competitive advantage. Based on established measures by Iacovou et al. [20], Premkumar and Roberts [25], Hsu et al. [29], Gibbs and Kraemer [30] and Oliveira and Martins [31], the indicator is operationalized as:

- Self-adaptive technologies and digitalized processes will allow us to better respond to customer needs.
- Self-adaptive technologies and digitalized processes will allow us to cut costs in our operations.
- Implementing self-adaptive technologies and digitalized processes will increase the profitability of our business.
- Self-adaptive technologies and digitalized processes provide timely information for decision making.

Hypothesis 3 (H3). *The greater the perceived relative advantage of the advanced manufacturing technologies, the more likely they will be adopted.*

4.1.4. Cost

Tornatzky and Klein [14] state that technologies that are perceived to be low in cost are more likely to be adopted. Also, Premkumar and Roberts [25] highlight that for small businesses, the cost of hardware/software is a sever inhibitor for adoption. The study by Wischmann et al. [12] revealed that lack of resources, both financial and human resources, is the major obstacle for small and medium-sized companies. Based on established measures by Premkumar and Roberts [25] and Jeon et al. [26], the indicator is operationalized as:

- The costs of adoption of these self-adaptive technologies and digitalized processes are far greater than the benefits.
- The cost of maintenance and support of these self-adaptive technologies and digitalized processes are very high for our business.
- The amount of money and time invested in training employees to use these self-adaptive technologies and digitalized processes is very high.

Hypothesis 4 (H4). *The smaller the perceived cost/benefit ratio of the advanced manufacturing technologies, the more likely they will be adopted.*

4.2. Environmental Factors

4.2.1. Market Uncertainty

External organizational factors like megatrends or economic fluctuations affect the way business is conducted. Market uncertainty stems from all technological, social, ecological, and political areas such as technological disruptions, degree of competition, volatility in demand, uncertain supply, and political [15]. This market uncertainty induces tension to the outer and inner complexity of a firm, which as to be balanced [2]. Self-adaptive systems such as the Smart Factory facilitate the balancing act [10]. The authors explain how cyber-physical systems support mass customization and handle market uncertainties. Therefore, it is assumed that a higher degree of market uncertainty is positively correlated with the likelihood of adopting advanced manufacturing technologies. In other studies, this construct is represented by competitive pressure [25,26], perceived industry pressure [32], or competition intensity [29]. Based on established measures by Chau and Tam [15], this indicator is operationalized as:

- The market for firms' products/services is stable.
- The competition of firms' products/services is stable.
- The demand for firms' products/services is stable.
- The degree of loyalty of their major customers is stable.
- The frequency of price-cutting in the industry is stable.

Hypothesis 5 (H5). *The higher the market uncertainty for the company's business, the more likely the advanced manufacturing technologies will be adopted.*

4.2.2. Industry Cluster

Since the first industrial revolution, it has been observed that firms tend to cluster in specific geographic regions. While this has been driven by the availability of resources in the first place and the need to secure them, this changed with the second industrial revolution when the infrastructure (power and transportation) allowed firms to settle independently of large cities. With the rise of automation and machinery-based manufacturing, access to the labor market, and close proximity to suppliers became more critical [33]. This clustering of firms creates externalities such as reduced labor cost or knowledge spillover because, in a reciprocal way, it attracts more people to the region, which in turn reduces the cost of labor [34]. However, the degree of those positive externalities depends on the affinity of the firm to the existing cluster. A cluster of similar manufacturing SMEs (similar/same industry) will benefit from a large specialized workforce to a greater extent than SMEs from different industries due to labor pooling and sharing of inputs. In the Marshall-Arrow-Romer model [35], this is described as externalities due to specialization, where this leads to knowledge spillovers, and if similar firms collocate in one region (cluster), it fosters innovation due to competition.

On the other hand, in Jacob's model [36], SMEs from different industries strive for innovation due to knowledge spillovers based on different labor skills in high-tech industries [34]. In addition to industry affinity externalities, whether based on specialization or diversification, Alacer and Chung [37] highlight the importance of firm size on the agglomeration effects. Large firms typically experience a higher knowledge drain than SMEs, which benefit from it (knowledge gain). Based on this argumentation, the indicator for this construct is defined as:

- The company resides in a geographic, industrial area.
- The companies within the geographic, industrial area stem from many diversified industries.

Hypothesis 6 (H6). *The higher diversity in an industry cluster, the more likely the advanced manufacturing technologies will be adopted.*

4.3. Organizational Factors

4.3.1. Top Management Support and Championship

According to Chatterjee et al. [21], top management championship is an important factor for assimilating information technologies, such as Web technologies, because it defines institutional norms and values how the company perceives the innovation. If the top management believes in the innovation and actively shapes the vision and strategy, this serves as a strong signal to the organization to overcome obstacles in the assimilation process. The significance of top management beliefs and active participation has been demonstrated in various studies [21]. Thus, top management support is positively related to adoption [25]. This is supported by Wischmann et al. [12], which highlight that the lack of a digital vision by the top management is an inhibitor to implement Industry 4.0. Based on established measures by Chatterjee et al. [21], this indicator is operationalized as:

- The top management is likely to invest funds for self-adaptive technologies and digitalized processes.
- The top management is willing to take risks in the adoption process of self-adaptive technologies and digitalized processes.
- The top management is likely to be interested in adopting the self-adaptive technologies and digitalized processes.
- The top management is likely to consider the adoption of the self-adaptive technologies and digitalized processes as strategically important.
- The top management has articulated a vision or strategy for the organizational use of self-adaptive technologies and digitalized processes.

Hypothesis 7 (H7). *The higher the top management support and championship for advanced manufacturing technologies, the more likely they will be adopted.*

4.3.2. Satisfaction with the Existing System

The shift from centralized production to decentralized (self-adaptive) production is a substantial change for the organization. As highlighted by McGaughey and Snyder [38], human resistance to change was one major factor that hampered the success of computer integrated manufacturing in the 1990s. Resistance to change or motivation for change stems from two sources: lack of skills or satisfaction with the existing system. Because a lack of skills is already covered by the complexity construct, this construct follows the argument by Rogers [24] that a low satisfaction level with the existing system is perceived as a performance gap and will motivate the organization to improve performance. Based on established measures by Chau and Tam [15], this indicator is operationalized as:

- The existing production system serves the needs of the company.
- The existing IT system serves the needs of the company.
- The cost/performance of the production system satisfies the top management.
- The cost/performance of the current IT system satisfies the top management.

Hypothesis 8 (H8). *The higher the satisfaction with the existing system, the less likely the advanced manufacturing technologies will be adopted.*

4.3.3. Organizational Structure

The effect of organizational structures such as centralization (degree of organizational hierarchy) and decision-making processes is debatable according to Kimberly and Evanisko [39]. A centralized organizational structure can be positively or negatively correlated with technology adoption behavior. In the context of advanced manufacturing technologies, this study follows the argument by Chatterjee et al. [21]: To successfully implement a complex technology, which affects multiple organizational entities, the needed intensity of interactions and collaborations among the managers of these entities is a critical factor in the adoption of advanced manufacturing technologies. Thus, it is expected that a decentralized organizational structure with flatter decision structures will be associated with the adoption of new technologies. Based on these established measures this indicator is operationalized as:

- The company decision making is highly concentrated at top management levels.
- The company extensively utilizes cross-functional work teams for managing the day to day operations.
- The company has reduced the formal organizational structure to integrate operations more fully.

Hypothesis 9 (H9). *The more decentralized the organization, the more likely the advanced manufacturing technologies will be adopted.*

4.4. Specific Technological Factors

4.4.1. Market Transparency for Advanced Manufacturing Technologies

Market transparency refers to the availability of information and solutions to implement the technologies. It is reflected in governmental efforts to promote information, establish public-partnerships, and establish measures to compare technologies solutions of different provides. A highly fragmented market, with various isolated solution providers, with every one having its standard, implies an immature industry.

Thus, market transparency is a proxy for technological maturity. Small and medium-sized companies are more likely to adopt mature Industry 4.0 technologies than an immature one. The study by Wischmann et al. [12] also highlights that a lack of user-transparency, technological standards, and availability of solutions are severed obstacles for small and medium-sized enterprises. Based on this argument, the following indicators are proposed:

- Information about products and services on self-adaptive technologies and digitalized processes are widely available.
- The market for self-adaptive technologies and digitalized processes is transparent concerning product and service features.
- The market for self-adaptive technologies and digitalized processes is transparent concerning product and service costs.

- Information on standards and protocols for self-adaptive technologies and digitalized processes is widely available.

Hypothesis 10 (H10). *The higher the market transparency, the more likely advanced manufacturing technologies will be adopted.*

4.4.2. Security Concerns of Advanced Manufacturing Technologies

Cyber-physical systems and its digital integration along the business function share are a unique feature with e-business security concerns. Firms conducting Internet-based e-business have less control over data as in legacy systems. Also, the integration along the value chain implies the diffusion of core corporate data, and even if its access restricted, its several physical access points for fraud usage as increased.

These security concerns are even more prevalent in the context of self-regulating factories. According to Bauer et al. [2], information security based on confidentiality, integrity, and availability is a prerequisite or even the decisive factor in adopting Industry 4.0 technologies. Based on established measures by Zhu et al. [40], this indicator is operationalized as:

- The company is very concerned about the security and privacy of data and transactions using self-adaptive technologies and digitalized processes.
- Our trading partners are very concerned about the security of data and privacy using self-adaptive technologies and digitalized processes.

Hypothesis 11 (H11). *The higher the security concerns, the less likely advanced manufacturing technologies will be adopted.*

4.5. Dependent Variable

Advanced manufacturing technologies are currently in its evaluation stage for scale based industrial applications, and many public-private initiatives have just recently been planned or have just been set up to promote prototypes and technology champions [41]. Thus, it is way too early to investigate general adoption stages.

Therefore, this study follows first the approach by Tsai et al. [42] and investigates the general intention to adopt advanced manufacturing strategies and subsequently includes only the early stages of technology assimilation based on the Guttman scale in Fichman and Kemerer [23]. On a five-point Likert scale ranging from strongly agree to disagree the intention to adopt is measured according to Tsai et al. [42]:

- The company is seriously contemplating to adopt self-adaptive technologies and digitalized processes within one to three years from now.
- It is critical for the company to adopt self-adaptive technologies and digitalized processes within one to three years from now.
- My firm is likely to adopt self-adaptive technologies and digitalized processes within one to three years from now.

5. Results

A survey has been sent out to 38 selected manufacturing SMEs in Japan according to the relative representation. The descriptives of the firms are presented in Table 1. The company sizes range between 10 (12 firms), 50 (13 firms), 100 (eight firms), and more than 200 employees (five firms).

Table 1. Manufacturing Industries.

Industry	Number of Firms
Automated Equipment and Robots	2
Basic Metals	2
Computer, Electronics and Optical Products	4
Control Equipment	2
Electrical Equipment	4
Fabricated Metal Products	1
Machinery and Equipment	9
Motor vehicles, trailers, and semi-trailers	1
Non-metallic mineral products	1
Transport Equipment	1
Paper Products	1
Pipes and Pumps	1
Mold Manufacturing	1
Precision Manufacturing	5
Other Manufacturing	3

The survey has been translated to Japanese and back to English to ensure semantic consistency. The survey consists of the statements described in the previous section, and the interviewees indicated to which extent, on a five-point Likert scale, they agree or disagree with each statement.

The model is evaluated for reliability, convergent validity, and discriminant validity. The Construct reliability is assessed by computing Cronbach's alpha. Cronbach's alpha is a measure to check the internal consistency of a construct based on the average correlations between the construct items (Table 2). Cronbach's alpha should be above 0.7. If the Cronbach's alpha value is below 0.7, this is an indicator that the items measure different concepts, or the items are not understandable/are ambiguous. Therefore, only the constructs Complexity, Compatibility, Relative Advantage, Market Uncertainty, Top Management Support, Market Transparency, and the dependent variables are considered. Based on the remaining constructs a principal component analysis (correlation matrix, varimax) is used to assess the convergent and discriminant validity. The Compatibility construct had to be dropped because of too many over-loadings and insignificancy (loading > 0.5).

The remaining five components explain 72% of the variance and have significant loadings in their respective parts and are uncorrelated across the components. Heteroskedasticity, the variance of the error term, has been tested with the Breusch-Pagan/Cook-Weisberg, yielding a $chi^2 = 0.18$. This means that the variance of the error terms does not change with the observations.

After the determinants have been validated, the regression analysis is performed on the adjusted model. A linear regression analysis is performed for the three scenarios of short-term (one year), medium-term (two years), and long-term (three years) adoption intention.

Table 2. Cronbach's Alpha.

Technology Adoption Construct	Cronbach's Alpha
Relative Advantage	0.86
Complexity	0.74
Compatibility	0.79
Cost	0.64
Top Management Support and Championship	0.79
Satisfaction with Existing Systems	0.5
Organizational Structure	0.82
Market Uncertainty	0.69
Industry Cluster	0.34
Market Transparency	0.88
Security Concerns	0.67

The final regression model is shown in Figure 2. The regression shows that Market Uncertainty (of the company's business) is a significant driver for adoption intention in the short, medium, and long-term. Also, while the Relative Advantage matters in the short-term, strong indicators for adoption intention in the long-term are Top Management Support of a company. No support has been identified for the other constructs. This indicates that the adoption of advanced or smart manufacturing for SMEs is market-driven, whether there is a strong need to change and build a strong relative competitive advantage or not. This is facilitated in the long-term by the top management. Therefore, this study highlights that the adoption of Industry 4.0 technologies and processes is primarily driven by external factors and less by internal drivers. This is aligned with the results obtained in [43], where the authors highlight that Industry 4.0 in SMEs is typically related to Cloud Computing to improve operational efficiency and only to a smaller degree used in real business cases for digital transformation.

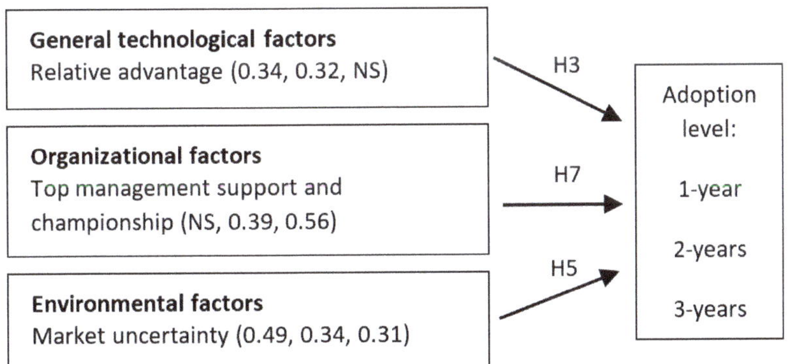

Figure 2. The resulting Industry 4.0 Adoption Model. NS= No support. The number in brackets shows the beta coefficients for one, two, and three years adoption levels.

The consequences of external pressure as the main driver has advantages and disadvantages. Especially for small and medium-sized enterprises, it could lead to late adoption of Industry 4.0, because compared to mass markets catered by large enterprises, niche markets are less competitive. Thus, external pressure signals could be misinterpreted or necessary technology adoption could be deferred. However Industry 4.0 standards and protocols have not reached a unified maturity level, thus being a late adopter could refrain SMEs to tap into additional costly investments. Especially in the light of interoperability challenges [44] and unsolved cyber-security concerns [45], late adoption for SMEs might be beneficial.

Funding: This research was partially funded by Japan Society for the Promotion of Science (PE15003).

Acknowledgments: The author would like to express his gratitude to Motohashi Kazuyuki for advising and hosting the author at his institute. The author would like to express his appreciation and thanks to the two anonymous referees who have been a tremendous help to improve the article.

Conflicts of Interest: The author declares no conflict of interest.

References

1. Abele, E.; Reinhart, G. *Zukunft der Produktion: Herausforderungen, Forschungsfelder, Chancen*; Hanser: Munich, Germany, 2011.
2. Bauer, W.; Schlund, S.; Marrenbach, D.; Ganschar, O. Industrie 4.0—Volkswirtschaftliches Potenzial für Deutschland. Available online: www.produktionsarbeit.de (accessed on 15 October 2019).
3. Ezell, S.; Atkinson, R. The Case for a National Manufacturing Strategy. Available online: http://www2.itif.org/2011-national-manufacturing-strategy.pdf (accessed on 15 October 2019).

4. Nooteboom, B. Innovation and Diffusion in Small Firms: Theory and Evidence. *Small Bus. Econ.* **1994**, *6*, 327–347. [CrossRef]

5. Kagermann, H.; Wahlster, W. Umsetzungsempfehlung für das Zukunftsproject Industrie 4.0. Available online: https://www.bmbf.de/files/Umsetzungsempfehlungen_Industrie4_0.pdf (accessed on 15 October 2019).

6. Lee, J.; Bagheri, B.; Kao, H. A Cyber-Physical Systems architecture for Industry 4.0-based manufacturing systems. *Manuf. Lett.* **2015**, *3*, 18–23. [CrossRef]

7. Zuehlke, D. SmartFactory—Towards a factory of things. *Annu. Rev. Control* **2010**, *34*, 129–138. [CrossRef]

8. Lucke, D.; Constantinescu, C.; Westkämper, E. Smart factory—A step towards the next generation of manufacturing. In *Manufacturing Systems and Technologies for the New Frontier*; Springer: London, UK, 2008; pp. 115–118.

9. Bauernhansl, T.; Hompel, M.T.; Vogel-Heuser, B. *Industrie 4.0 in Produktion, Automatisierung und Logistik*; Springer Fachmedien Wiesbaden: Wiesbaden, Germany, 2014.

10. Radziwon, A.; Bilberg, A.; Bogers, M.; Madsen, E.S. The Smart Factory: Exploring Adaptive and Flexible Manufacturing Solutions. *Procedia Eng.* **2014**, *69*, 1184–1190. [CrossRef]

11. Wang, S.; Wan, J.; Li, D.; Zhang, C. Implementing Smart Factory of Industrie 4.0: An Outlook. *Int. J. Distrib. Sens. Netw.* **2016**, *12*, 1–10. [CrossRef]

12. Wischmann, S.; Wangler, L.; Botthof, A. Studie Industrie 4.0—Volkswirtschaftliche Faktoren für den Standort Deutschland. Available online: https://vdivde-it.de/system/files/pdfs/industrie-4.0-volks-und-betriebswirtschaftliche-faktoren-fuer-den-standort-deutschland.pdf (accessed on 15 October 2019).

13. Mittal, S.; Khan, M.; Muztoba, A.; Romero, D.; Wuest, T. A critical review of smart manufacturing & Industry 4.0 maturity models: Implications for small and medium-sized enterprises (SMEs). *J. Manuf. Syst.* **2018**, *49*, 194–214.

14. Tornatzky, L.; Klein, K. Innovation Characteristics and Innovations Adoption-Implementation: A Meta-Analysis of Findings. *IEEE Trans. Eng. Manag.* **1982**, *29*, 28–45. [CrossRef]

15. Chau, P.; Tam, K.Y. Factors Affecting the Adoption of Open Systems: An Exploratory Study. *MIS Q.* **1997**, *21*, 1–24. [CrossRef]

16. DePietro, R.; Wiarda, E.; Fleischer, M. The context for change: Organization, technology and environment. In *The Processes of Technological Innovation*; Tornatzky, L., Fleischer, M., Eds.; Lexington Books: Lexington, KY, USA, 1990; pp. 151–175.

17. Oliveira, T.; Maria, M. Literature Review of Information Technology Adoption Models at Firm Level. *Electron. J. Inf. Syst. Eval.* **2011**, *14*, 110–121.

18. Scott, W.R. *Institutions and Organizations*; Sage Publications: Thousands Oaks, CA, USA, 2001.

19. Raymond, L. Determinants of Web Site Implementation in Small Businesses. *Int. Res.* **2001**, *11*, 411–422. [CrossRef]

20. Iacovou, C.; Benbasat, I.; Dexter, A. Electronic Data Interchange and Small Organizations: Adoption and Impact of Technology. *MIS Q.* **1995**, *19*, 465–485. [CrossRef]

21. Chatterjee, D.; Grewal, R.; Sambamurthy, V. Shaping up for E-Commerce: Institutional Enablers of the Organizational Assimilation of Web Technologies. *MIS Q.* **2002**, *26*, 65–89. [CrossRef]

22. Swanson, E.B. Information systems innovation among organizations. *Manag. Sci.* **1994**, *40*, 1069–1092. [CrossRef]

23. Fichman, R.G.; Kemerer, C.F. The assimilation of software process innovations: An organizational learning perspective. *Manag. Sci.* **1997**, *43*, 1345–1363. [CrossRef]

24. Rogers, E. *Diffusion of Innovations*; Free Press: New York, NY, USA, 1995.

25. Premkumar, G.; Roberts, M. Adoption of new information technologies in rural small businesses. *Omega* **1999**, *27*, 467–484. [CrossRef]

26. Jeon, B.N.; Han, K.S.; Lee, M.J. Determining factors for the adoption of e-business: The case of SMEs in Korea. *Appl. Econ.* **2006**, *38*, 1905–1916. [CrossRef]

27. Cooper, R.B.; Zmud, R.W. Information technology implementation research: A technological diffusion approach. *Manag. Sci.* **1990**, *36*, 123–139. [CrossRef]

28. Commerzbank, A.G. Management im Wandel: Digitaler, Effizienter, Flexibler! Available online: www.unternehmerperspektiven.de/portal/media/unternehmerperspektiven/up-studien/up-studien-einzelseiten/up-pdf/Studie15-Mai-2015-Management-im-Wandel.pdf (accessed on 15 October 2019).

29. Hsu, P.-F.; Kraemer, K.; Dunkle, D. Determinants of E-Business Use in U.S. Firms. *Int. J. Electron. Commer.* **2006**, *10*, 9–45. [CrossRef]

30. Gibbs, J.; Kraemer, K. A cross-country investigation of the determinants of scope of e-commerce use: An institutional approach. *Electron. Mark.* **2004**, *14*, 124–137. [CrossRef]

31. Oliveira, T.; Martins, M.F. Understanding e-business adoption across industries in European countries. *Ind. Manag. Data Syst.* **2010**, *110*, 1337–1354. [CrossRef]

32. Kuan, K.; Chau, P. A perception-based model for EDI adoption in small business using a technology—Organization—Environment framework. *Inf. Manag.* **2001**, *38*, 507–521. [CrossRef]

33. Marschall, A. *Principles of Economics: An Introductory Volume*; Macmillan and Co.: London, UK, 1920.

34. Beaudry, C.; Schiffauerova, A. Who's right, Marschall or Jacobs? The localization versus urbanization debate. *Res. Policy* **2009**, *38*, 318–337. [CrossRef]

35. Glaeser, E.; Kallal, H.; Scheinkman, J.; Shleifer, A. Growth in cities. *J. Political Econ.* **1992**, *100*, 1126–1152. [CrossRef]

36. Jacobs, J. *The Economies of Cities*; Random House: New York, NY, USA, 1969.

37. Alcacer, J.; Chung, W. Location Strategy for agglomeration economies. *Strat. Manag. J.* **2014**, *35*, 1749–1761. [CrossRef]

38. McGaughey, R.E.; Snyder, C.A. The obstacles to successful CIM. *Int. J. Prod. Econ.* **1994**, *37*, 247–258. [CrossRef]

39. Kimberly, J.; Evanisko, M. Organizational Innovation: The Influence of Individual, Organizational, and Contextual Factors on Hospital Adoption of Technological and Administrative Innovations. *Acad. Manag. J.* **1981**, *24*, 689–713.

40. Zhu, K.; Dong, S.; Xu, S.X.; Hally, M. Innovation diffusion in global contexts: Determinants of post-adoption digital transformation of European companies. *Eur. J. Inf. Syst.* **2006**, *15*, 601–616. [CrossRef]

41. BMBF. Industrie 4.0—Innovationen für die Produktion von Morgen. Available online: https://www.bmbf.de/upload_filestore/pub/Industrie_4.0.pdf (accessed on 15 October 2019).

42. Tsai, M.-C.; Lee, W.; Wu, H.-C. Determinants of RFID adoption intention: Evidence from Taiwanese retail chains. *Inf. Manag.* **2010**, *47*, 255–261. [CrossRef]

43. Moeuf, A.; Pellerin, R.; Lamouri, S.; Tamayo-Giraldo, S.; Barbaray, R. The industrial management of SMEs in the era of Industry 4.0. *Int. J. Prod. Res.* **2018**, *56*, 1118–1136. [CrossRef]

44. Lu, Y. Industry 4.0: A survey on technologies, applications and open research issues. *J. Ind. Inf. Integr.* **2017**, *6*, 1–10. [CrossRef]

45. Babiceanu, R.; Radu, F.; Seker, R. Big Data and virtualization for manufacturing cyber-physical systems: A survey of the current status and future outlook. *Comput. Ind.* **2016**, *81*, 128–137. [CrossRef]

Article

Event-Driven Online Machine State Decision for Energy-Efficient Manufacturing System Based on Digital Twin Using Max-Plus Algebra

Junfeng Wang [1,*], **Yaqin Huang** [1], **Qing Chang** [2] **and Shiqi Li** [1]

[1] Department of Industrial and Manufacturing System Engineering, Huazhong University of Science and Technology, Wuhan 430074, China; hyqyx2016@163.com (Y.H.); sqli@hust.edu.cn (S.L.)

[2] Department of Mechanical and Aerospace Engineering, University of Virginia, Charlottesville, VA 22904, USA; qc9nq@virginia.edu

* Correspondence: wangjf@hust.edu.cn; Tel.: +86-(0)27-8754-1034

Received: 17 August 2019; Accepted: 12 September 2019; Published: 14 September 2019

Abstract: Energy-efficient manufacturing is an important aspect of sustainable development in current society. The rapid development of sensing technologies can collect real-time production data from shop floors, which provides more opportunities for making energy saving decisions about manufacturing systems. In this paper, a digital twin-based bidirectional operation framework is proposed to realize energy-efficient manufacturing systems. The data view, model view, and service view of a digital twin manufacturing system are formulated to describe the physical systems in virtual space, to perform simulation analysis, to make decisions, and to control the physical systems for various energy-saving purposes. For online energy-saving decisions about machines in serial manufacturing systems, an event-driven estimation method of an energy-saving window based on Max-plus Algebra is presented to put the target machine to sleep, considering real-time production data of a system segment. A practical, simplified automotive production line is used to illustrate the effectiveness of the proposed method by simulation experiments. Our method has no restriction on machine failure mode and predefined parameters for energy-saving decision of machines. The proposed approach has potential use in synchronous and asynchronous manufacturing systems.

Keywords: energy efficient operation; manufacturing system; stochastic event; digital twin; Max-plus Algebra; MATLAB-Simulink

---

1. Introduction

Industrial activities are the most important factor that leads to carbon dioxide (CO_2) emissions in the world. The manufacturing industry in China accounts for 58.27% of the total CO_2 emissions, which increased by approximately 220.77% in the past twenty years [1]. With the increased sustainable consensus, many emission reduction policies were issued to undertake environmental responsibility and duty in China. More and more manufacturers take effort to reduce energy costs. Energy-efficient manufacturing has been paid more attention considering productivity, flexibility, efficiency and environmental effect in production management [2].

A discrete manufacturing system is a kind of complex, stochastic, and dynamic system with interactions among machines and buffers. It is very common that many machines stay in an idle state, resulting from starvation or blockage during a working day. Statistically, the machine at an idle state consumes about 70% energy of their processing state [3]. Many modern Computerized Numerical Control (CNC) machines and robots have multiple idle states with different energy consumptions. An important strategy for energy-saving operations is to put the idle machine to sleep and to wake it up again at an appropriate time point without sacrificing the system productivity. The machine control requires real-time production data. New, emerging Internet of Things (IoT) technology, e.g., smart sensors and meters, provide multilevel monitoring of manufacturing system (almost) in real time [4]. The collected data provide more opportunities for energy-saving decisions at the machine, process, and system levels.

Usually, discrete event simulation can be used to evaluate system performances including energy cost [5]. However, discrete event simulation is computationally expensive and time consuming, which is not suitable for online decision-making. As a new, emerging technology for modeling, simulation, and optimization, digital twin can reflect real-time states of physical systems [6]. Based on various models, digital twin not only can carry out simulations in virtual space before production but also can deliver decisions from virtual space to physical space during production, which realizes closed-loop feedback and interactions between physical systems and virtual systems. In this paper, a framework of digital-twin-based energy-efficient operation of manufacturing systems is proposed to support bidirectional interactions between virtual and physical systems. An event-driven Max-plus Algebra method is presented and validated online to predict the dynamic sleep duration of a machine for energy-saving operations at the system level.

The rest of this paper is organized as follows. The literature review is given in Section 2. Section 3 describes an operation framework for energy-efficient manufacturing systems based on digital twin technology. An event-driven energy-saving window approach for online machine state decision is proposed using Max-plus Algebra in Section 4. In Section 5, a practical serial manufacturing system is used to demonstrate the effectiveness of our method and discussions are given. Conclusions and future works are drawn in Section 6.

2. Literature Review

2.1. Modeling and Simulation of Energy-Efficient Manufacturing

Modeling and simulation are important methods for performance evaluation and optimization of energy-efficient manufacturing [7]. The simulation models of machine tools, processes, systems, or products can be used to carry out various what-if analyses. The production manager can understand the impacts of different manufacturing parameters on energy consumption and, thus, make decisions to optimize the parameters for energy-efficient manufacturing.

From the machine tool and process view, Lee et al. [8] proposed a simulation method based on a virtual machine tool to estimate the machining condition considering the energy consumption function or component model. Based on dynamic and kinematic behaviors of machine tools, Bi et al. [9] explored relations of machining parameters (cutting force, depth, and feed rate) and energy consumption. Kim et al. [10] proposed an additive regression algorithm to formulate the relationships between CNC machine energy consumption and machining parameters. From the product aspect, Seow et al. [11] built a simulation model to evaluate the total embodied product energy based on direct and indirect consumed energy.

At the production or manufacturing system levels, simulation-based prediction of energy consumption can support a manager in making decisions considering dynamic changes of production plans, energy prices, and environmental impact. Process chain simulation was an energy-oriented manufacturing system simulation approach [12]. The EnergyBlocks simulation method integrated energy-efficiency criteria with evaluation and decision processes during production system planning and scheduling [13]. Prabhu et al. [14] presented a simulation model that continuously integrated variable power consumption at the machine level and discrete nature at the production level to characterize energy dynamics in discrete manufacturing systems. By integrating a sustainability-enabled product lifecycle management information model with energy-consumption metrics, Zhao et al. [15] estimated the impact of energy control on actual energy consumption of a heating, ventilation, and air conditioning system; a ventilation system; and a production system.

The current modeling and simulation approaches focused on offline evaluation, prediction, and optimization for energy-efficient manufacturing. In order to incorporate real-time production data of shop floors, a data-driven method was always adopted for alleviating the burden of modeling and simulation [16].

2.2. Energy-Aware Manufacturing System Scheduling and Control

Taking energy consumption as one objective in scheduling; a multi-objective optimization problem can be formalized, and near-optimal job sequences can be searched by heuristic algorithms or exact methods [17]. Shrouf et al. [18] presented a mathematical model to minimize energy costs for a single machine system, and genetic algorithm technology has been used to obtain optimal solutions. For a two-machine system, Fang et al. [19] proposed a mathematical programming model that considered peak power load, energy consumption, and associated carbon footprint in addition to cycle time. In a flow shop with a reference schedule, Bruzzone et al. [20] built a mixed integer programming model where the reference schedule was modified to account for energy consumption without changing the jobs' assignment and sequencing. An energy-efficient flexible job shop scheduling model considering total production energy consumption and makespan was formulated in Reference [21], and a non-dominated sorting genetic algorithm-II was adopted to solve the model. Luo et al. [22] developed an ant colony optimization meta-heuristic to optimize both makespan and electric power cost with the presence of time-of-use electricity prices in a hybrid flow shop.

Quantitative policies and fuzzy rules can describe the operation knowledge of machine/system for energy-efficient manufacturing. Mouzon et al. [23] described several switch-off dispatching policies to minimize total energy consumption in a one-buffer, one-machine system. For a single machine system with stochastic inter-arrival times, the N-policy, Upstream Policy, and Downstream Policy were defined to switch the machine off when the conditions of the policies were satisfied [24]. By formulating four ON–OFF control rules in serial production lines with Bernoulli machines, Jia et al. [25] analyzed system performance measures in order to achieve high energy efficiency while maintaining high productivity. Wang et al [26] designed a fuzzy controller considering real-time data of buffers/machines in order to switch machine states and realize energy-efficient production.

Some works used an analytical method to study energy-saving opportunities of manufacturing systems. Chang et al. [27] proposed an energy opportunity window of a machine, which was predicted based on an approximated analytical model considering random downtime events. Sun and Li [28] presented an analytical estimation of energy control opportunities considering buffer utilization. Zou et al. [29] developed a stochastic analytical model which could predict shutdown time and recovery time of machines based on discrete-time Markov Chains in a stochastic production system. Li et al. [30] developed an analytical approach to quantitatively predict the system level production loss resulting from an energy-saving control event.

The multi-objective scheduling methods of energy-aware manufacturing are usually used to optimize operation plans before actual production. The complex and stochastic events of real shop floors are simplified or ignored in scheduling algorithms. It is noted that the control policies or rules for real-time energy-saving operation of production systems have many parameters to be determined and adjusted with time-consuming work. The advantage of the analytical models is its mathematical theory. Most of the analytical models did not consider the associated events during energy-saving operation, and only a constant energy opportunity window was predicted.

2.3. Digital Twin of Cyber-Physical Manufacturing and Production System

Digital twin [31], a near-real-time virtual representation of a physical component, product or system enriched with sensing data, is a digital and dynamic model in the cyber-space that is Digital twin [31], a near-real-time virtual representation of a physical component, product, or system enriched with sensing data, is a digital and dynamic model in cyberspace that is completely consistent with its physical ones in real space. It is considered a key technology in realizing cyber–physical production systems (CPPS) for Industry 4.0 [32]. Digital twin can simulate the characteristics, behavior, and performance of its physical counterpart in a timely fashion, which shows gigantic application potential in manufacturing system analysis and control [33,34].

In the aspects of production management, optimization, and control, Zhuang et al. [35] proposed a framework of digital-twin-based smart production management and control approach for complex product-assembly shop-floors. By using Plant simulation software, Vachálek et al. [36] built a digital twin for the optimization of production plans, which was designed in a way that it triggered physical and virtual productions as well. Coronado et al. [37] described the development and implementation of a new manufacturing execution system that combines MTConnect data with production data collected from operators. The work was integral to realizing a complete digital model of the shop floor for production control and optimization.

Considering digital-twin-based energy consumption evaluation and optimization of a production plan, Karanjkar et al. [38] presented an IoT-driven digital twin for energy optimization in an automated printed circuit-board assembly line. A digital twin of the line was built, and a buffering-based solution for improving energy efficiency was evaluated using simulations. Lu et al. [39] proposed an open system architecture for energy-efficient cyber–physical production network for facilitating distributed manufacturing. A cyber twin of a physical manufacturing asset was built, and the energy consumption of feature-based machining process models were evaluated in connected virtual factories. Zhang et al. [40] proposed a framework of equipment energy consumption management for a digital-twin shop floor. The data and models were discussed for monitoring, analysis, and optimization of equipment energy consumption.

It can be concluded that the most current works on digital twin of manufacturing or production system are used to monitor, evaluate, and visualize the real system based on unidirectional mapping from physical space to digital space. There is little research on models of online real-time production control from digital space to physical space, which is the main contribution in this paper considering energy consumption and productivity.

3. Digital Twin Based Operation Framework of Energy-Efficient Manufacturing System

Energy-efficient operation decisions of production systems require data from shop floors, carries out evaluations in virtual space, and makes control actions on the shop floor. Digital twin is the digital representation of a physical object. The main purpose of a production system digital twin is to facilitate decision-making activities based on simulation or computation before, during, and after production. The real-time monitoring, simulation, prediction, and control of production systems are vital to improving productivity, efficiency, and flexibility. At present, the interaction between physical and digital production systems is mainly offline interactions, lacking continuous and online interaction [33]. In this section, an operation framework with a three-view representation (Figure 1), i.e., data view, model view, and service view, is proposed to digitalize a manufacturing system, which enables us to perform simulations, to make decisions, and to control the physical system for different energy-saving purposes.

In the physical space of Figure 1, the real manufacturing/production system is a complex, diverse, and dynamic system with sensible ability in the context of Industry 4.0. The multisource and heterogeneous data (e.g., behavior, event, and state) of physical objects (e.g., machines, materials, and parts) are monitored and transmitted to databases based on IoT sensors (e.g., smart meters, radio frequency identification reader, and tags), which enhance the perception ability for physical manufacturing systems. The energy-efficient decisions made in virtual space based on digital twin are implemented through actuators of machines on the shop floor. A three-view digital twin of manufacturing systems for energy-saving operations is described as follows, which enables two-way direct or indirect mapping and interaction between the physical and the virtual manufacturing systems.

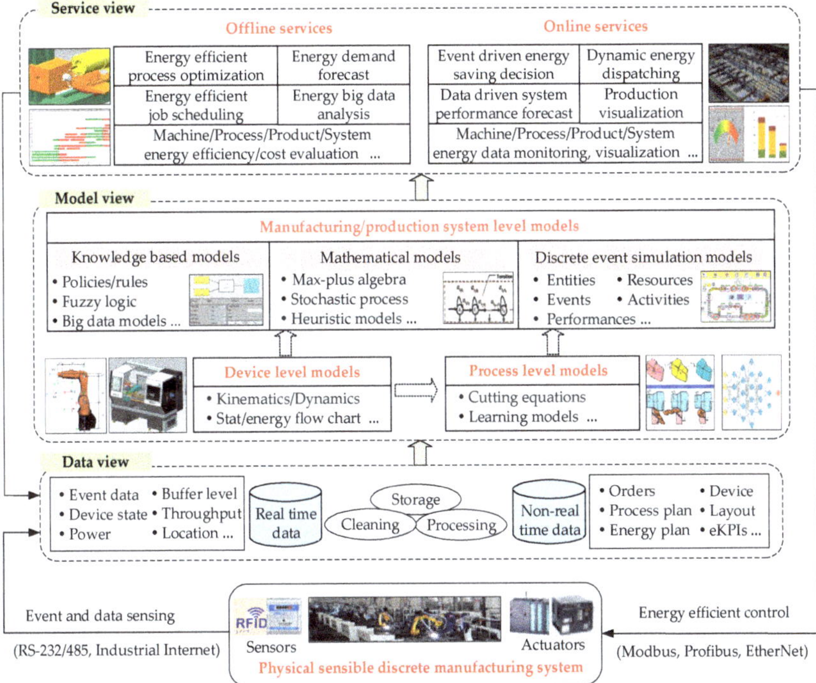

Figure 1. Framework of digital-twin based energy efficient operation of manufacture systems. eKPIs = energy-related key performance indicators.

3.1. Data View

The data of energy-efficient manufacturing collected from different sources has the "4V" characteristics (i.e., volume, velocity, variety, and value) of big data considering the spatial and time dimensions. The big data for energy-efficient operation of manufacturing system includes real-time and non-real-time data. The real-time data, such as electrical power of device, location of parts, buffer level, and machine throughput, is monitored and collected during production, which is the most important data source for online energy-saving decisions. The layout data, process plan, maintenance plan, and job schedule are the non-real-time data of the system collected or generated from the physical and cyber space before production. The energy-related key performance indicators (eKPIs), such as energy demand, energy cost, and energy efficiency, are also non-real-time data mined or acquired based on simulation after production. The energy big data can be structured or semi-structured. For example, the data of a machine, such as rated power, spindle speed, and cutting rate, is structured data. The process plan of a part is semi-structured data which has several flexible aspects (e.g., feature sequence, operation type, and machine selection).

The methods for cleaning, reduction, integration, and storage of energy-efficient manufacturing data can be referenced in Reference [41] and is not the focus of the paper. The high real-time processing ability of the storm real-time computing framework [42] can be adopted to process the time-series energy data. The Hadoop computing framework [43] can be used to process the non-real-time data. In order to efficiently store the big data of energy-efficient manufacturing, the Not only Structured Query Language [44] can be chosen for these large-scale structured and semi-structured datasets.

3.2. Model View

In digital space, the parameter correlations of the objects/systems are represented in simulation or computation models, which are core components of digital twin and the base of the application services. The models describe the working principle of machines with power supply, the energy consumption resources of the machining process, and the state space evolution of production systems. The models at three different levels are built for energy-efficient decisions in this framework.

At the device level, the physics-based energy model is required to formulate power consumption and dynamic behavior of industrial robot and CNC machine [40,45], and thus the energy-efficient trajectory planning, operating parameters, and power state of machine components can be completed. At the process level, the machining parameters, e.g., spindle speed, feed rate, and depth of cut, have a notable influence of energy consumption for the same job features [9]. The mathematical cutting equations or learning algorithms should be constructed for energy-consumption calculation and prediction of machining process.

The models at the system level depict the interaction among the objects in production systems. Various models such as knowledge-based models, mathematical models, and discrete event simulation models are established for different decisions of energy-efficient manufacturing. The policies/rules [23–25] and fuzzy logic [26] are expert knowledge for the energy-saving operation of production systems. The big data analysis models [46], e.g., neural networks, rough set theory, and deep learning, can mine implicit relationships of production parameters considering energy cost and productivity. The stochastic process [28] and algebra mathematical models [47] usually are used to evaluate the system eKPIs. The heuristic models (e.g., simulated annealing, genetic algorithm, and particle swarm optimization algorithm) for multi-objective optimization [17] are important for production scheduling. The discrete event simulation models are very popular for system evaluation and optimization as stated in the review section. The complex relations among entities, events, and activities of a production system are scheduled based on queuing theory in simulation models.

3.3. Service View

At the top of Figure 1, many kinds of application services are provided for online and offline scheduling, evaluating, forecasting, visualizing, monitoring, and controlling of energy-efficient manufacturing based on simulation or computation models.

The offline services are usually initiated on-demand by managers to determine the appropriate energy-efficient process parameters or job sequences in advance before production. The simulation services allow managers to perform what-if analyses on various production plans and to optimize schedules for minimizing total energy cost while maintaining higher productivity. After production, the future energy management [48,49], such as energy policy, energy benchmark, energy recycle, and manufacturing process/production schedule, can also be analyzed based on the collected data. The data-mining application services [46] can be formulated to discover hidden valuable information and knowledge from the enormous production datasets with energy indicators.

From the online services view, a near-real-time dynamic visualization application of production- and energy-related information can be generated based on discrete event simulation models with collected data, which provide more direct perceptions of the physical shop floor status. By creating the latest snapshot of the physical manufacturing system status based on the data-driven simulation method [16], the forecasting application is adaptive and flexible to any change in terms of actual production settings and can provide prediction services of system performances and eKPIs using online forward simulation experiments. Feedback control based on decisions in virtual space is the distinct characteristic of digital twin. The energy-saving opportunities during production come from the transient states of manufacturing system [27]. An online, event-driven energy-saving decision service for machine state control is proposed in this paper to execute short-term analysis by mapping the exact state of a physical system into the digital world at run time. The decision results will be sent to physical machines for realizing energy-efficient production.

4. Event-Driven Online Decision Model of Energy Saving Window Using Max-Plus Algebra

4.1. Physical Manufacturing System and Modeling Assumptions

In this paper, a typical serial production system (Figure 2) is used to illustrate energy efficient operation at the system level based on the above described digital twin framework. In order to build the mathematical model of the physical system for energy saving decision and control, the following assumptions are made.

1. A physical serial production system is composed of m machines and $m-1$ buffers. There is one buffer between each adjacent machine.
2. Each buffer B_i ($i = 1, 2, \ldots, m-1$) has a finite capacity C_i and its real-time level is I_i.
3. Each machine M_i ($i = 1, 2, \ldots, m$) has a constant cycle time P_i.
4. The reliability model of machines can be any probability distribution.
5. The bottleneck machine (BN) of the system is denoted as M_b ($1 \leq b \leq m$).
6. The first machine M_1 is never starved, and the last machine M_m is never blocked.
7. The transportation time between machines and buffers is not considered.

Figure 2. Typical serial production system.

4.2. Methodology of Event-Driven Online Energy Saving Decision and Control of Machine States

Usually, time-based scheme is adopted at shop floor to collected production data with predefined sampling intervals and data-driven method is used to make operation decisions [41]. Generally speaking, the idle activities during production bring energy-saving opportunities. The idle activities generate from the events of blockage or starvation. Thus, event-based scheme can be more effective and economical considering the big data processing burden of time-based scheme [50]. Event-driven decision has shown many potential applications in manufacturing, such as manufacturing information system architecture [51], make-to-order production [52], and manufacturing execution system [53]. The most related research of event-based energy efficient production control is conducted by Li et al. [30,54], where a sequence of disruption events was defined to estimate the production loss for energy saving operation. In this paper, four traditional events of a machine, i.e., blockage event (BE), starvation event (SE), failure event (FE) and recover event from failure (RE), are monitored at the physical shop floor, which trigger energy-saving decision in digital space and control machines in physical system.

Based on previous works [27,55–57], an energy-saving window (ESW$_i$) of a target machine M_i is defined as a time length of M_i being purposely put to sleep without bringing production loss to the system in this study. A system segment (SG$_i$) of a target machine M_i is defined as the partial system including all machines and buffers between the target machine and bottleneck machine. The estimation of ESW in the next section will consider the bottleneck machine of the system, of which the production interruption will most likely result in system production loss [57]. The main flow of event-driven online energy-saving decision and control is shown in Figure 3 and explained as follows.

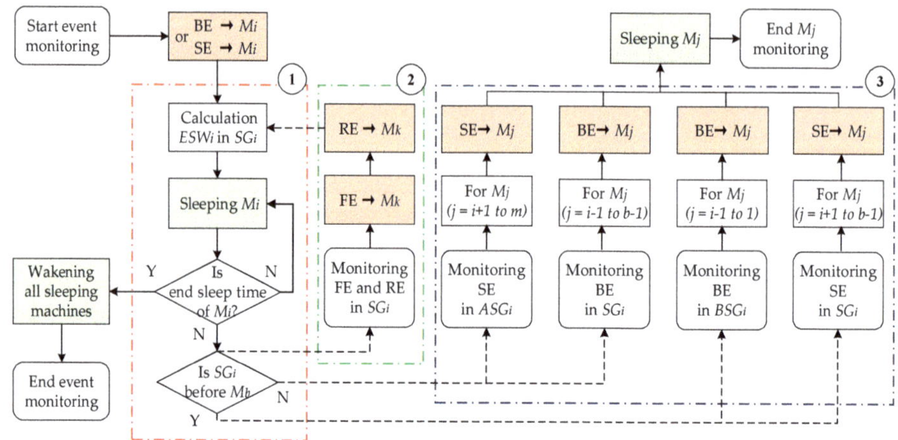

Figure 3. Event-driven online machine energy saving decision and control flow. Note: Abbreviations defined before Figure 2.

As shown in Figure 3, the main decision flow (rectangle 1) will be triggered if there is a BE or SE of a target machine M_i during production. The ESW of this target machine is estimated based on the current states of SG_i. Then, the physical machine is switched into sleep mode. During the sleep duration of a target machine, different events of machines will be monitored and machine states will be controlled for energy-saving operation. In the flow of rectangle 2, the FE and then the RE of a machine in SG_i are monitored. If an RE occurs on a machine, the main decision flow is retriggered to estimate a new ESW of the target machine. The underlying principle is that the failure of a machine in SG_i can provide more energy-saving time for the target machine. In the flow of rectangle 3, different machines of the system are monitored according to the place of SG_i, considering the bottleneck M_b. If the target machine M_i is before M_b, a new system segment before SG_i is named as BSG_i. The BE of machines in BSG_i and the SE of machines in SG_i are monitored. For each machine, from near to distant, related to the target machine, the machine in BSG_i or SG_i will be put to sleep if it is blocked or starved. If the target machine M_i is after M_b, a new system segment after SG_i is named ASG_i and similar decisions are made on each machine before the target machine ends its sleep. All the machines that were put to sleep will be woken when the ESW of the target machine is exhausted and a decision round ends.

Compared with time-based data sampling in the fuzzy logic method [26], our event-driven method greatly eases the burden of data collection and conforms to the stochastic of the system. The dynamic updating of the ESW of the target machine during one round decision is more reasonable compared with the constant opportunity windows [27,29,55–57] and will be illustrated in Section 5.

From the flow in Figure 3, our event-driven energy-saving decision method is also different from the event-based energy-efficient production control in References [30,54], which used predefined energy-saving control events to analyze the system performances.

4.3. Estimation of Energy Saving Window Based on System Segment using Max-Plus Algebra

Discrete event simulation is usually time consuming and cannot be used as the online tool for real-time decisions. Max-plus Algebra is a mathematical tool that can model discrete event systems in linear equations analogous to traditional state space dynamic equations [58] and has higher computation efficiency. Some works used Max-plus Algebra to model and analyse manufacturing systems, such as parametric analysis of mixed-model assembly lines [59], makespan calculation of a job-shop [60], and flow shop scheduling problems [61]. In this section, Max-plus Algebra is used to estimate the ESW of a target machine for energy saving operation based on previous work [62].

As aforementioned, ESW represents the sleep time of the target machine without affecting the production of M_b. If the target machine M_i is located before the bottleneck M_b ($1 \leq i < b$), the sleep of the target machine will result in the starvation of the M_b. We define TE_i as the time length that all work-in-processes (WIP) in buffers of SG_i are cleared out. In order to recover the production of M_b immediately after waking the target machine, the time length from entering the target machine to arriving at buffer B_{b-1} must be eliminated and donated as TR_i. If the target machine M_i is located after the bottleneck M_b ($b < i \leq m$), putting the target machine to sleep will cause a blockage of the M_b if all buffers in SG_i are full. We define TF_i as the time duration that all buffers in SG_i are filled from their current levels. The ESW of the target machine M_i is estimated as follows:

$$ESW_i = \begin{cases} TE_i - TR_i & (1 \leq i < b) \\ TF_i & (b < i \leq m) \end{cases} \tag{1}$$

The time length TE_i, TR_i, and TF_i can be estimated using the Max-plus Algebra method, which can calculate the time instant of a system state. Here, the estimation of TE_i is illustrated as an example. The following variables are defined to describe the system states for target machine M_i ($1 \leq i < b$).

- The total number of WIP in all buffers of SG_i is denoted as n.
- u_j (k) is the time instant when the kth ($1 \leq k \leq n$) part in SG_i is available to M_j ($i + 1 \leq j \leq b$).
- u_j is an column vector of u_j (k).
- x_j (k) is the time instant when M_j starts processing the kth part.

Apparently, we can get the following formula if the current time clock is zero.

$$TE_i = x_b(n) \tag{2}$$

where x_b (n) means the time instant that bottleneck machine M_b starts processing the last part in SG_i.

At a decision time point, there are I_{j-1} parts in buffer B_{j-1} that are available to machine M_j. After the processing activity of M_j, a new part will be sent to B_j and will be available to machine M_{j+1}. The vector u_j is formulated as follows:

$$u_{i+1} = \underbrace{[0,0,\ldots,0]}_{I_i}{}^T \tag{3}$$

$$u_{i+2} = \underbrace{[0,0,\ldots,0,}_{I_{i+1}} x_{i+1}(1) + P_{i+1}, x_{i+1}(2) + P_{i+1}, \ldots, x_{i+1}(I_i) + P_{i+1}]^T \tag{4}$$

$$u_b = \underbrace{[0,0,\ldots,0,}_{I_{b-1}} x_{b-1}(1) + P_{b-1}, x_{b-1}(2) + P_{b-1}, \ldots, x_{b-1}(I_i + I_{i+1} + \ldots + I_{b-2}) + P_{b-1}]^T \tag{5}$$

For each part in buffers of SG_i, the time instant that M_j starts processing the part is as follows:

$$x_{i+1}(k) = \max(u_{i+1}(k), x_{i+1}(k-1) + P_{i+1}, x_{i+2}(k-1+I_{i+1}-C_{i+1})) \tag{6}$$

$$x_{i+2}(k) = \max(u_{i+2}(k), x_{i+2}(k-1) + P_{i+2}, x_{i+3}(k-1+I_{i+2}-C_{i+2})) \tag{7}$$

$$x_{b-1}(k) = \max(u_{b-1}(k), x_{b-1}(k-1) + P_{b-1}, x_b(k-1+I_{b-1}-C_{b-1})) \tag{8}$$

$$x_b(k) = \max(u_b(k), x_b(k-1) + P_b) \tag{9}$$

In state space form, it can be described as follows:

$$\begin{aligned} x(k) = \quad & A \otimes u(k) \oplus D \otimes x(k-1) \oplus F^1 \otimes x_{i+2}(k-1+I_{1+1}-C_{i+1}) \\ & \oplus F^2 \otimes x_{i+3}(k-1+I_{i+2}-C_{i+2}) \oplus \ldots \oplus F^{b-i} \otimes x_b(k-1+I_{b-1}-C_{b-1}) \end{aligned} \tag{10}$$

The coefficient matrices A, D and F are formulated as follows:

$$\mathbf{A} \in R_{\varepsilon}^{(b-i+1)\times(b-i+1)}, \ \mathbf{D} \in R_{\varepsilon}^{(b-i+1)\times n}, \ \mathbf{F}^l \in R_{\varepsilon}^{(b-i+1)\times(b-i+1)}$$

$$\mathbf{A} = (a_{ij}) = \begin{cases} e & i=j; \\ \varepsilon & otherwise \end{cases} \tag{11}$$

$$\mathbf{D} = (d_{ij}) = \begin{cases} P_i & i=j; \\ \varepsilon & otherwise \end{cases} \tag{12}$$

$$\mathbf{F}^l = (f_{ij}) = \begin{cases} e & i=l, j=l+1 \ (l=1,2,\ldots,b-i) \\ \varepsilon & otherwise \end{cases} \tag{13}$$

where e is the null element and ε is the identity element in Max-plus Algebra.

Based on the above method, the state equations of the system segment will be quickly initialized using real time production data collected from shop floor and the ESW of a machine can be estimated for online energy saving decision and control.

5. Simulation Experiments and Discussion

5.1. Simulation Case of a Serial Manufacturing System

The event-driven energy saving decision model in our digital twin of an energy efficient manufacturing system can be used to control the physical system online under Industry 4.0 environment in the future, which realizes the communication from virtual space to physical space. In order to imitate the operation procedure of the physical production system, the model of the real production line is built in MATLAB-Simulink [63]. The running process of the modeled system generates real-time production data and events, which are monitored, collected and sent to digital twin. The energy-saving decision based on Max-plus Algebra is built in digital space as a service. The decision is driven based on defined events and the decision results are sent to the mimetic physical production system in MATLAB-Simulink, where the machines are controlled for energy-saving operation. The whole process simulates the bidirectional communication between physical space and virtual space.

A typical serial manufacturing system including six machines and five buffers (6M5B) in References [26,29] is taken to illustrate the effectiveness of our method. This system is a simplified version of a real automotive production line. The parameters are presented in Tables 1 and 2, where mean time between failure (MTBF) and mean time to repair (MTTR) are assumed as exponential distribution for this case. The simulation duration is three weeks, i.e., 30,240 min. The electricity rate is assumed to be $0.2/kWh. It is assumed that no power is consumed in machine sleep mode for comparing performances with the methods in References [26,29].

Table 1. Machine parameters of the 6M5B serial automotive production line.

Machine	M_1	M_2	M_3	M_4	M_5	M_6
MTBF (min)	5422	6301.2	11,872.2	5440.2	6412.8	6250.8
MTTR (min)	130.8	208.2	409.8	279.6	205.2	250.8
Cycle time (min)	3.5	4.3	2.7	9.4	1.1	5.9
Power rate (kW)	450	300	240	288	660	360

Table 2. Buffer parameters of the 6M5B serial automotive production line.

Buffer	B_1	B_2	B_3	B_4	B_5
Capacity	120	150	160	50	150
Initial Level	70	30	50	40	50

Three scenarios are simulated as follows: baseline scenario with no energy saving decision and control (S1), only one machine being controlled for energy saving (S2), and multi-machine except the bottleneck machine being controlled for energy saving (S3). These three scenarios are simulated for 20 trials-each. After simulation of baseline scenario S1, machine M_4 is recognized as the system bottleneck.

5.2. Experiment Results and Analysis

The simulation results of system performances, i.e., System throughput (STP), System throughput Loss (STPL), Total system energy cost (TSEE), Energy cost per part (ECPP), and Energy cost saving per part (ECSPP), are shown in Table 3 with 95% confidential interval (CI). It can be seen that the energy cost saving per part is remarkable based on an event-driven digital twin model of the machine control method. When multiple-machines are controlled in S3, there is almost 57.24% energy-cost saving and only 2.20% system throughput is lost compared with baseline scenario S1.

Table 3. Simulation results of the three scenarios.

	STP with 95% CI	STPL (%)	TSEE ($) with 95% CI	ECPP ($) with 95% CI	ECSPP (%)
S1	3168.45 [3150.06, 3186.84]	-	225,732.57 [225,070.14, 226,395.00]	71.25 [70.81, 71.70]	-
S2 (M_3)	3162.40 [3144.77, 3180.03]	0.19%	183,762.24 [175,341.88, 192,182.59]	58.08 [55.60, 60.55]	18.48%
S2 (M_5)	3114.80 [3096.59, 3133.01]	1.96%	155,183.59 [154,795.41, 155,571.78]	49.83 [49.61, 50.05]	30.06%
S3	3098.70 [3078.35, 3119.05]	2.20%	94,424.29 [93,875.72, 94,972.87]	30.47 [30.31, 30.64]	57.24%

Figure 4 shows the machine states time in three scenarios, where the machine states include processing (PR), blockage (BL), starvation (ST), failure (FL), and sleep (SL). For machine M_1, M_2, and M_3 before the BN (M_4), the blockage time gradually decreases and the sleep time (energy-saving time) gradually increases in S2 and S3 compared with S1. The processing time of the machines almost remain unchanged, which means the energy-saving control does not affect processing work and converts the blockage time to sleep time. For machines M_5 and M_6 after the BN (M_4), the energy-saving decision and control method can effectively change the starvation time into sleep time. For example, the total starvation time of M_5 in S1 is 86.37% of simulation time. In S3, the starvation time of M_5 is entirely converted to sleep time. The total processing time of M_5 in S1 (3438.2 min) and S3 (3376.6 min) almost remain the same, while the sleep time of M_5 is 86.57% of simulation time in S3. The control of machine states results in the energy-cost reduction as shown in Table 3.

We recorded the data of an event-driven energy-saving decision process during simulation experiments. Table 4 shows some data from S3, and the meanings are explained as follows.

An SE of M_3 is monitored at 19,772.1 min, and M_3 is taken as the target machine. An energy-saving decision is triggered based on the system segment SG_3 (i.e., M_3-B_3-M_4), and the ESW of M_3 is estimated as 110.1 min. At the same time, the M_3 is put to sleep. No energy-saving event is monitored following 110.1 min, and then M_3 is woken at 19,882.2 min. For M_5, an SE is monitored at 20,679.6 min and the target machine is set as M_5. The ESW_5 is calculated as 470.0 min, considering system segment SG_5 (i.e., M_4-B_4-M_5) in record No.4. Before the end of sleep time, there is a failure event of M_4 at time 20,757.8 (No.5) and then a recover event of M_4 at time 20,813.3 (No.6). Therefore, the energy-saving decision process of M_5 is retriggered at time 20,813.3 and a new ESW_5 is estimated as 385.4 min. The ESW_5 is updated dynamically based on the event-driven method. Later, M_6 in ASG_5 (i.e., B_5-M_6) has an SE and is put to sleep at time 20,944.8. The sleep states of M_5 and M_6 are ended at time 21,198.7 (No.9). M_5 and M_6 have another decision round shown in record No.10, 12, and 14. For M_3, a BL is monitored at 21,065.4 min and taken as the target machine. The ESW_3 is calculated as 1501.3 min (No.8). At the later procedure, M_2 and M_1 are put to sleep in succession because of the BE in BSG_3 (i.e., M_1-B_1-M_2-B_2). At the end of the sleep time of M_3 (i.e., 22,566.7 min), M_1, M_2, and M_3 are woken up together. It can be seen that multiple machines can be controlled for energy-saving operation based on the event-driven method considering online production data.

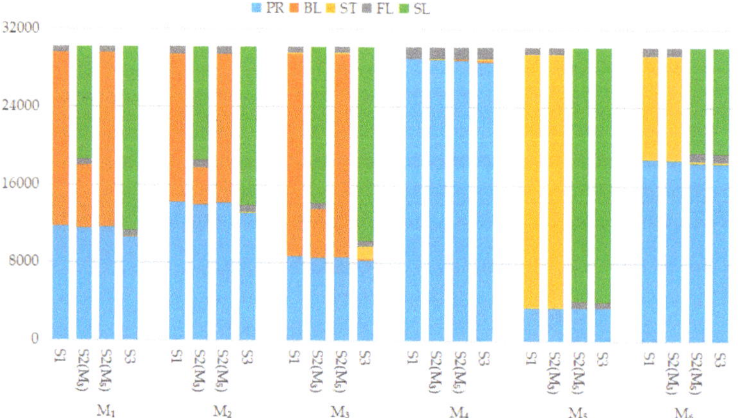

Figure 4. The state time distribution of machines in three scenarios.

Table 4. Event-driven energy-saving decision process in scenario S3.

No.	Time (min.)	Event-Machine	Control Action	ESW_i (min.)	Segment
1	19,772.1	SE - M_3	Sleeping M_3	110.1 (ESW_3)	SG_3
2	19,882.2	-	Wakening M_3	-	SG_3
3	From 19,882.2 to 20,679.6	-	-	-	-
4	20,679.6	SE-M_5	Sleeping M_5	470.0 (ESW_5)	SG_5
5	20,757.8	FE-M_4	-	-	-
6	20,813.3	RE-M_4	Sleeping M_5	385.4 (ESW_5)	SG_5
7	20,944.8	SE-M_6	Sleeping M_6	-	ASG_5
8	21,065.4	BE-M_3	Sleeping M_3	1501.3 (ESW_3)	SG_3
9	21,198.7	-	Wakening M_5, M_6	-	-
10	21,259.2	SE-M_5	Sleeping M_5	470.0 (ESW_5)	SG_5
11	21,517.3	BE-M_2	Sleeping M_2	-	BSG_3
12	21,524.3	SE-M_6	Sleeping M_6	-	ASG_5
13	21,648.8	BE-M_1	Sleeping M_1	-	BSG_3
14	21,729.2	-	Wakening M_5, M_6	-	-
15	22,566.7	-	Wakening M_1, M_2, M_3	-	-

Figure 5 shows the changing process of machine states (M_5) in S1 and S3, which is randomly selected from the experiments between 20,000 to 24,000 min. In baseline S1 with no energy-saving decision and control, there are many interval starvations and processing times of M_5 due to the empty state of buffer B_4. For example, there is one part in B_4 at time 20,158.6 (S1) and in M_5 that begins to process the part for 1.1 min. After time 20,159.7, the frequent starvation and processing state occurs alternately. Therefore, many normal starvation states of M_5 with short duration consume a lot of non-value-added energy. In controlled scenario S3, the M_5 is put to sleep until time 20,619.2. At time 20,619.2, M_5 is woken up and processes parts continuously for 61.6 min until time 20,680.7. Then, M_5 is put to sleep again in a new event-driven decision round. Comparing the states of M_5 in S1 and S3, it can be concluded that control of machine energy-saving states brings energy-cost reduction of manufacturing systems.

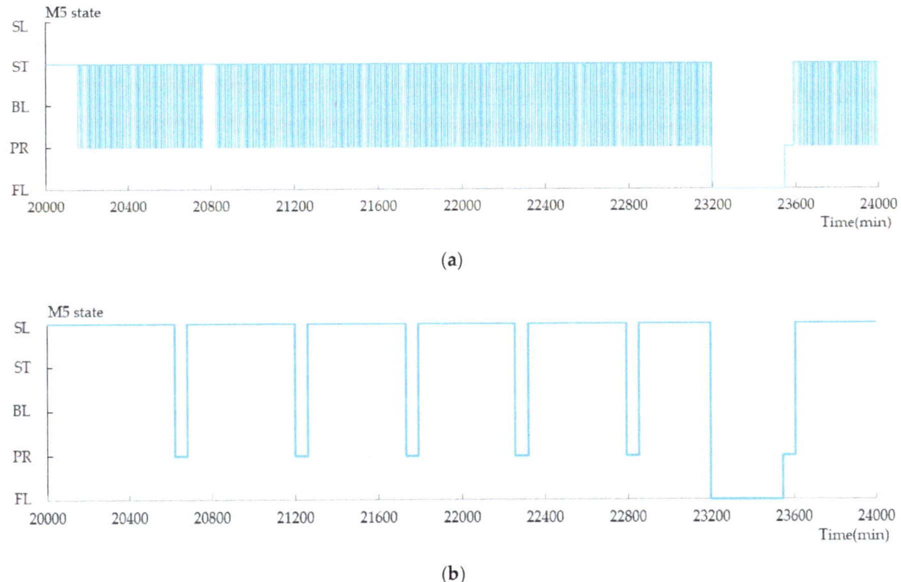

Figure 5. The machine state changing process of M5: (a) S1 and (b) S3.

The eKPIs of the system under multi-machine control scenario in this paper are compared with fuzzy logic method [26] and mathematical model [29] in Table 5. The three methods of energy-saving control all result in reduction of system throughput less than 3%, which is acceptable in practical manufacturing systems. Our method achieves the most energy-cost reduction per part (57.23%) compared with the other two methods, and the effectiveness of this paper is proved.

Table 5. Comparision of the system eKPIs in different methods.

Method	Reduction of System Throughput (%)	Reduction of System Energy Cost (%)	Reduction of Energy Cost per Part (%)
This paper	2.20%	58.17%	57.23%
[26]	0.23%	51.76%	51.66%
[29]	2.70%	27.00%	24.98%

Except for the better performances, our method has some advantages for practical use in shop floors. The first advantage is that there is no restriction on the failure mode of machines in our method. The mathematical models [24,25,29] for energy-saving decision are all based on specific failure modes of machines. When the machine recovers from failure, the event-driven method will re-estimate the ESW of the machine. Non-predefined control parameter is the second advantage of our method. In fuzzy logic method [26], the threshold values and decision intervals of the machines should be defined in advance for energy-saving control. The method in Reference [29] should assume the subjective value of discount ratio and collaboration ratio. The third advantage is that our method can be used in synchronous manufacturing systems and asynchronous manufacturing systems. The method in Reference [29] can only be used in an asynchronous line.

6. Conclusions

With the development of IoTs, energy-efficient manufacturing is provided more opportunities in an Industry 4.0 environment, where the digital twin plays an important role for real-time production monitoring, simulation analysis, and feedback control of physical systems. Considering most current works focus on the uniderection mapping from physical space to virtual space, a digital-twin-based bidirectional operation framework of energy-efficient manufacturing system is proposed to digitalize the physical manufacturing systems, to perform simulations and decisions for different energy-saving purposes in virtual space, and to control the power states of physical machines. The three views, i.e., data view, model view, and service view, are described to map the physical system to digital system. An online event-driven energy-saving decision method based on Max-plus Algebra is presented to switch machines to sleep mode with consideration of maximum energy saving and minimum effect on system throughput. The Max-plus Algebra model is used to snapshot the states of system segment and to estimate the ESW of the machines. A simplified real automotive production line is taken as an example to carry out simulation experiments, and the results illustrate the effectiveness of the proposed method compared with other methods in literatures. Not relying on machine failure mode and non-predefined parameters in our method makes it simple to be used in synchronous and asynchronous systems. The potential application of our method in more complex line, e.g., parallel line, assembly line, or disassembly line, will be experimented in the future.

Author Contributions: The authors contribute equally to this study.

Funding: This research was funded by National Natural Science Foundation of China grant number No. 71571075.

Acknowledgments: Authors thank the anonymous reviewers for their valuable comments.

Abbreviations

IoT	Internet of Things
CNC	Computerized Numerical Control
CPPS	cyber-physical production systems
eKPIs	energy-related key performance indicators
BN	bottleneck machine
BE	Blockage event of a machine
SE	Starvation event of a machine
FE	Failure event of a machine
RE	Recover event from failure of a machine
ESW	Energy saving window of a machine
SG_i	System segment between target machine M_i and bottleneck machine
BSG_i	System segment before SG_i
ASG_i	System segment after SG_i
WIP	Work-in-process
TE_i	The time length that all the WIP in buffers of SG_i is cleared out
TR_i	The time length of a part from entering target machine to arriving buffer B_{b-1}
TF_i	The time duration that all buffers in SG_i are filled from their current levels
STP	System throughput
STPL	System throughput loss
TSEE	Total system energy cost
ECPP	Energy cost per part
ECSPP	Energy cost saving per part
PR	Processing state
BL	Blockage state
ST	Starvation state
FL	Failure state
SL	Sleep state
MTBF	Mean time between failure
MTTR	Mean time to repair

References

1. Sun, W.; Hou, Y.; Guo, L. Analyzing and forecasting energy consumption in china's manufacturing industry and its subindustries. *Sustainability* **2019**, *11*, 99. [CrossRef]
2. Salonitis, K.; Ball, P. Energy Efficient Manufacturing from Machine Tools to Manufacturing Systems. *Procedia CIRP* **2013**, *7*, 634–639. [CrossRef]
3. Gutowski, T.; Dahmus, J.; Thiriez, A. Electrical energy requirements for manufacturing processes. In Proceedings of the 13th CIRP International Conference on Life Cycle Engineering, Leuven, Belgium, 31 May–2 June 2006; pp. 623–628.
4. Shrouf, F.; Miragliotta, G. Energy management based on Internet of Things: Practices and framework for adoption in production management. *J. Clean. Prod.* **2015**, *100*, 235–246. [CrossRef]
5. Thiede, S.; Seow, Y.; Andersson, J.; Johansson, B. Environmental aspects in manufacturing system modeling and simulation: State of the art and research perspectives. *CIRP J. Manuf. Sci. Tec.* **2013**, *6*, 78–87. [CrossRef]
6. Kritzinger, W.; Karner, M.; Traar, G.; Henjes, J.; Sihn, W. Digital Twin in manufacturing: A categorical literature review and classification. *IFAC-PapersOnLine* **2018**, *51*, 1016–1022. [CrossRef]
7. Garwood, T.L.; Hughes, B.R.; Oates, M.R.; Connor, D.O.; Hughes, R. A review of energy simulation tools for the manufacturing sector. *Renew. Sust. Energ. Rev.* **2018**, *81*, 895–911. [CrossRef]
8. Lee, W.; Kim, S.H.; Park, J.; Min, B.K. Simulation-based machining condition optimization for machine tool energy consumption reduction. *J. Clean. Prod.* **2017**, *150*, 352–360. [CrossRef]
9. Bi, Z.M.; Wang, L. Optimization of machining processes from the perspective of energy consumption: A case study. *J. Manuf. Syst.* **2012**, *31*, 420–428. [CrossRef]
10. Kim, S.; Meng, C.; Son, Y.J. Simulation-based machine shop operations scheduling system for energy cost reduction. *Simul. Model. Pract. Theory* **2017**, *77*, 68–83. [CrossRef]
11. Seow, Y.; Rahimifard, S.; Woolley, E. Simulation of energy consumption in the manufacture of a product. *Int. J. Comput. Integ. Manuf.* **2013**, *26*, 663–680. [CrossRef]
12. Herrmann, C.; Thiede, S.; Kara, S. Energy oriented simulation of manufacturing systems: Concept and application. *CIRP Ann.* **2011**, *60*, 45–48. [CrossRef]
13. Weinert, N.; Chiotellis, S.; Seliger, G. Methodology for planning and operating energy-efficient production systems. *CIRP Ann.* **2011**, *60*, 41–44. [CrossRef]
14. Prabhu, V.; Taisch, M. Simulation Modeling of Energy Dynamics in Discrete Manufacturing Systems. In Proceedings of the 14th IFAC Symposium on Information Control Problems in Manufacturing, Bucharest, Romania, 23–25 May 2012; pp. 740–745.
15. Zhao, W.B.; Jeong, J.W.; Noh, S.D.; Yee, J.T. Energy simulation framework integrated with green manufacturing-enabled PLM information model. *Int. J. Pr Eng. Manuf. Green Technol.* **2015**, *2*, 217–224. [CrossRef]
16. Wang, J.; Chang, Q.; Xiao, G.; Wang, N.; Li, S. Data driven production modeling and simulation of complex automobile general assembly plant. *Comput. Ind.* **2011**, *62*, 765–775. [CrossRef]
17. Gahm, C.; Denz, F.; Dirr, M.; Tuma, A. Energy-efficient scheduling in manufacturing companies: A review and research framework. *Eur. J. Oper. Res.* **2016**, *248*, 744–757. [CrossRef]
18. Shrouf, F.; Ordieres-Meré, J.; García-Sánchez, A.; Ortega-Mier, M. Optimizing the production scheduling of a single machine to minimize total energy consumption costs. *J. Clean. Prod.* **2014**, *67*, 97–207. [CrossRef]
19. Fang, K.; Uhan, N.; Zhao, F.; Sutherland, J.W. A new approach to scheduling in manufacturing for power consumption and carbon footprint reduction. *J. Manuf. Syst.* **2011**, *30*, 234–240. [CrossRef]
20. Bruzzone, A.A.G.; Anghinolfi, D.; Paolucci, M.; Tonelli, F. Energy-aware scheduling for improving manufacturing process sustainability: A mathematical model for flexible flow shops. *CIRP Ann.* **2012**, *61*, 459–462. [CrossRef]
21. Zhang, Z.W.; Wu, L.H.; Peng, T.; Jia, S. An improved scheduling approach for minimizing total energy consumption and makespan in a flexible job shop environment. *Sustainability* **2019**, *11*, 179. [CrossRef]
22. Luo, H.; Du, B.; Huang, G.Q.; Chen, H.P.; Li, X.L. Hybrid flow shop scheduling considering machine electricity consumption cost. *Int. J. Prod. Econ.* **2013**, *146*, 423–439. [CrossRef]
23. Mouzon, G.; Yildirim, M.B.; Twomey, J. Operational methods for minimization of energy consumption of manufacturing equipment. *Int. J. Prod. Res.* **2007**, *45*, 4247–4271. [CrossRef]

24. Frigerio, N.; Matta, A. Energy-efficient control strategies for machine tools with stochastic arrivals. *IEEE Trans. Autom. Sci. Eng.* **2015**, *12*, 50–61. [CrossRef]

25. Jia, Z.; Zhang, L.; Arinez, J.; Xiao, G. Performance analysis for serial production lines with Bernoulli machines and real-time WIP-based machine switch-on/off control. *Int. J. Prod. Res.* **2016**, *54*, 6285–6301. [CrossRef]

26. Wang, J.F.; Fei, Z.C.; Chang, Q.; Fu, Y.; Li, S.Q. Energy-saving operation of multistage stochastic manufacturing systems based on fuzzy logic. *Int. J. Simul. Model.* **2019**, *18*, 138–149. [CrossRef]

27. Chang, Q.; Xiao, G.; Biller, S.; Li, L. Energy saving opportunity analysis of automotive serial production systems. *IEEE Trans. Autom. Sci. Eng.* **2013**, *10*, 334–342. [CrossRef]

28. Sun, Z.; Li, L. Opportunity estimation for real time energy control of sustainable manufacturing systems. *IEEE Trans. Autom. Sci. Eng.* **2013**, *10*, 38–44. [CrossRef]

29. Zou, J.; Chang, Q.; Arinez, J.; Xiao, G. Data-driven modeling and real-time distributed control for energy efficient manufacturing systems. *Energy* **2017**, *127*, 247–257. [CrossRef]

30. Li, Y.; Wang, J.; Chang, Q. Event-based production control for energy efficiency improvement in sustainable multistage manufacturing systems. *ASME J. Manuf. Sci. Eng.* **2018**, *141*, 021006. [CrossRef]

31. Negri, E.; Fumagalli, L.; Macchi, M. A review of the roles of digital twin in CPS-based production systems. *Procedia Manuf.* **2017**, *11*, 939–948. [CrossRef]

32. Uhlemann, T.H.J.; Lehmann, C.; Steinhilper, R. The Digital twin: Realizing the cyber-physical production system for industry 4.0. *Procedia CIRP* **2017**, *61*, 335–340. [CrossRef]

33. Bao, J.S.; Guo, D.S.; Li, J.; Zhang, J. The modelling and operations for the digital twin in the context of manufacturing. *Enterpol. Inf. Syst.* **2019**, *13*, 534–556. [CrossRef]

34. Fei, T.; Cheng, J.; Qi, Q.; Meng, Z.; He, Z.; Sui, F. Digital twin-driven product design, manufacturing and service with big data. *Int. J. Adv. Manuf. Technol.* **2017**, *94*, 3563–3576.

35. Zhuang, C.; Liu, J.; Xiong, H. Digital twin-based smart production management and control framework for the complex product assembly shop-floor. *Int. J. Adv. Manuf. Technol.* **2018**, *96*, 1149–1163. [CrossRef]

36. Vachálek, J.; Bartalský, L.; Rovný, O.; Šišmišová, D.; Morháč, M.; Lokšík, M. The digital twin of an industrial production line within the industry 4.0 concept. In Proceedings of the 2017 International Conference on Process Control, Strbske Pleso, Slovakia, 6–9 June 2017; pp. 258–262.

37. Coronado, P.D.U.; Lynn, R.; Wafa, L.; Parto, M.; Wescoat, E.; Kurfess, T. Part data integration in the Shop Floor Digital Twin: Mobile and cloud technologies to enable a manufacturing execution system. *J. Manuf. Syst.* **2018**, *48*, 25–33. [CrossRef]

38. Karanjkar, N.; Joglekar, A.; Mohanty, S.; Prabhu, V.; Raghunath, D.; Sundaresan, R. Digital twin for energy optimization in an SMT-PCB assembly line. In Proceedings of the 2018 IEEE International Conference on Internet of Things and Intelligence System, Bali, Indonesia, 1–3 November 2018; pp. 85–89.

39. Lu, Y.Q.; Peng, T.; Xun, X. Energy-efficient cyber-physical production network: Architecture and technologies. *Comput. Ind. Eng.* **2019**, *129*, 56–66. [CrossRef]

40. Zhang, M.; Zuo, Y.; Tao, F. Equipment energy consumption management in digital twin shop-floor: A framework and potential applications. In Proceedings of the 2018 IEEE International Conference on Networking, Sensing and Control, Zhuhai, China, 27–29 March 2018; pp. 1–5.

41. Zhou, K.; Fu, C.; Yang, S. Big data driven smart energy management: From big data to big insights. *Renew. Sustain. Energy Rev.* **2016**, *56*, 215–225. [CrossRef]

42. Yang, W.; Liu, X.; Zhang, L.; Yang, L.T. Big data real-time processing based on storm. In Proceedings of the 12th IEEE International Conference on Trust, Security and Privacy in Computing and Communications, Melbourne, Australia, 16–18 July 2013; pp. 1784–1787.

43. Shafer, J.; Rixner, S.; Cox, A.L. The Hadoop distributed file system: Balancing portability and performance. In Proceedings of the 2010 IEEE International Symposium on Performance Analysis of Systems & Software, White Plains, NY, USA, 28–30 March 2010; pp. 122–133.

44. Cattell, R. Scalable SQL and NoSQL data stores. *ACM Sigmod. Rec.* **2011**, *39*, 12–27. [CrossRef]

45. Paryanto, P.; Brossog, M.; Bornschlegl, M.; Franke, J. Reducing the energy consumption of industrial robots in manufacturing systems. *Int. J. Adv. Manuf. Technol.* **2015**, *78*, 1315–1328. [CrossRef]

46. Wu, X.; Zhu, X.; Wu, G.Q.; Ding, W. Data mining with big data. *IEEE Trans. Knowl. Data Eng.* **2014**, *26*, 97–107.

47. Seleim, A.; Elmaraghy, H. Generating max-plus equations for efficient analysis of manufacturing flow lines. *J. Manuf. Syst.* **2015**, *37*, 426–436. [CrossRef]

48. Zhang, Y.F.; Ma, S.Y.; Yang, H.D.; Lv, J.X.; Liu, Y. A big data driven analytical framework for energy-intensive manufacturing industries. *J. Clean. Prod.* **2018**, *197*, 57–72. [CrossRef]

49. May, G.; Stahl, B.; Taisch, M.; Kiritsis, D. Energy management in manufacturing: From literature review to a conceptual framework. *J. Clean. Prod.* **2017**, *167*, 1464–1489. [CrossRef]

50. Liu, Q.; Wang, Z.; He, X.; Zhou, D.H. A Survey of Event-Based Strategies on Control and Estimation. *Syst. Sci. Control Engrg.* **2014**, *2*, 90–97. [CrossRef]

51. Theorin, A.; Bengtsson, K.; Provost, J.; Lieder, M.; Johnsson, C.; Lundholm, T.; Lennartson, B. An event-driven manufacturing information system architecture for industry 4.0. *Int. J. Prod. Res.* **2017**, *55*, 1297–1311. [CrossRef]

52. Yao, X.; Zhang, J.; Li, Y.; Zhang, C. Towards flexible RFID event-driven integrated manufacturing for make-to-order production. *Int. J. Comput. Integr. Manuf.* **2018**, *31*, 228–242. [CrossRef]

53. Fang, J.; Huang, G.Q.; Li, Z. Event-driven multi-agent ubiquitous manufacturing execution platform for shop floor work-in-progress management. *Int. J. Prod. Res.* **2013**, *51*, 1168–1185. [CrossRef]

54. Li, Y.; Chang, Q.; Ni, J.; Brundage, M.P. Event-Based Supervisory Control for Energy Efficient Manufacturing Systems. *IEEE Trans. Autom. Sci. Eng.* **2018**, *15*, 92–103. [CrossRef]

55. Chang, Q.; Biller, S.; Xiao, G. Transient analysis of downtimes and bottleneck dynamics in serial manufacturing systems. *J. Manuf. Sci. Eng.* **2010**, *132*, 051015. [CrossRef]

56. Chang, Q.; Ni, J.; Bandyopadhyay, P.; Biller, S.; Xiao, G. Maintenance opportunity planning system. *J. Manuf. Sci. Eng.* **2007**, *129*, 661–668. [CrossRef]

57. Gu, X.; Lee, S.; Liang, X.; Garcellano, M.; Diederichs, M.; Ni, J. Hidden maintenance opportunities in discrete and complex production lines. *Expert. Syst. Appl.* **2013**, *40*, 4353–4361. [CrossRef]

58. De Schutter, B.; Boom, T. Max-plus algebra and max-plus linear discrete event systems: An introduction. In Proceedings of the 2008 9th International Workshop on Discrete Event Systems, Goteborg, Sweden, 28–30 May 2008; pp. 36–42.

59. Seleim, A.; El Maraghy, H. Parametric analysis of mixed-model assembly lines using max-plus algebra. *CIRP J. Manuf. Sci. Tec.* **2014**, *7*, 305–314. [CrossRef]

60. Singh, M.; Judd, R. Efficient calculation of the make-span for job-shop systems without re-circulation using max-plus algebra. *Int. J. Prod. Res.* **2014**, *52*, 5880–5894. [CrossRef]

61. Kubo, S.; Nishinari, K. Applications of max-plus algebra to flow shop scheduling problems. *Discret. Appl. Math.* **2018**, *247*, 278–293. [CrossRef]

62. Huang, Y.; Wang, J.; Li, S. Max-plus algebra based machine sleep decision for energy efficient manufacturing. In Proceedings of the 2018 Chinese Automation Congress, Xian, China, 30 November–2 December 2018; pp. 3986–3991.

63. Simulink. Simulation and Model-Based Design—MATLAB & Simulink. Available online: https://www.mathworks.com/products/simulink.html (accessed on 1 May 2019).

Article

Concept and Evaluation of a Method for the Integration of Human Factors into Human-Oriented Work Design in Cyber-Physical Production Systems

Hendrik Stern [1,2,*,†,‡] **and Till Becker** [3,‡]

[1] BIBA—Bremer Institut für Produktion und Logistik GmbH at the University of Bremen, 28359 Bremen, Germany

[2] Faculty of Production Engineering, University of Bremen, 28359 Bremen, Germany

[3] Faculty of Business Studies, University of Applied Sciences Emden/Leer, 26723 Emden, Germany

* Correspondence: ste@biba.uni-bremen.de

† Current address: Hochschulring 20, 28359 Bremen, Germany.

‡ These authors contributed equally to this work.

Received: 21 July 2019; Accepted: 16 August 2019; Published: 20 August 2019

Abstract: Due to the shift from mainly manual labor to an increased portion of cognitive tasks in manufacturing caused by the introduction of cyber-physical systems, there is a need for an updated collection of adequate design principles for user interfaces between humans and machines. Thus, we developed a method for the determination and evaluation of such design principles. It is based on human factors methods and facilitates the assessment of specific work design elements which are supposed to have a significant effect on work performance and the perception of work in cyber-physical production systems (CPPS). Within the application of the developed method, we derived an overview of key design elements in CPPS, developed an experimental platform, and conducted two empirical studies with a total of $n = 68$ participants. This way, three design elements were investigated, and the findings transferred into preliminary design principles. We can state that the method can be used both for a better understanding of the mechanisms between human factors and work in CPPS. Besides, it helps to provide a catalogue of design principles applicable to SMEs to promote more efficient and successful integration of workers into CPPS.

Keywords: human factors; Industry 4.0; cyber-physical systems; cyber-physical production systems; anthropocentric design; Operator 4.0; human–machine interaction

---

1. Introduction

Cyber-physical systems (CPS) describe a new type of technological systems in which physical elements are equipped with computers (embedded systems). These are networked with each other and can exchange information. Thus, real objects and processes are associated with virtual objects and processes [1]. This results in an increasing intelligence of products and systems [2]. These smart products are characterized by the fact that they are equipped with computers for storing product-specific information which can be used for data exchange and localization. The intelligent products can thus influence their own manufacture, use, or disposal [3].

In the field of manufacturing environments, intelligent systems include intelligent machine tools, test equipment, transport vehicles, and products. They continuously collect data about their condition and share it with other cyber-physical systems. The analysis and use of these data [1] reveal great potential for process improvements [4–6]. An example is the Smart Production, whose functionalities are made possible as a result of data exchange [7,8]. In detail, there are various applications for cyber-physical systems in production. For example, autonomous cooperating production processes, use of a collaborative robot, preventive maintenance or virtual reality/augmented reality systems can be used and implemented [3,5,6]. Many of these ideas are not new and therefore not entirely attributable to Industry 4.0, but can now unfold their potential as a result of the greater amount of data available [6]. The reorganization of manufacturing as a result of automation or changes in work systems will have an impact on existing jobs. It can be assumed that people in the Industry 4.0 environment both perform different activities and require different qualifications for these tasks [5].

How should the Work 4.0 associated with Industry 4.0 be designed in a both socially and economically sustainable way? This question is of main interest in the present paper. Section 1 provides background information and existing fundamentals on the change of work systems, human-oriented work design, human–machine interaction, and human factors both in general and in cyber-physical production systems (CPPS). In Section 2, we outline the underlying method of carrying out a human factors study and analyze available human-centered research in CPPS to determine their level of integrating human factors principles and methods. Section 3 is separated into four subsections and presents the results of our research. Firstly, work design elements for Human–machine interfaces in CPPS are proposed (Section 3.1). Further, a method for the integration of human factors into human-oriented work design in CPPS is shown. It facilitates the assessment of the proposed work design elements and quantifies their effect on work performance and the perception of work in CPPS and thus leads to design principles (Section 3.2). Then, an experimental platform designed to carry out these CPPS-related human factors studies easily is presented (Section 3.3). Finally, we apply and validate the developed method. This includes the hypothesis of key design elements out of the proposed work design elements, the design and implementation of representative work tasks as a human factors study, the execution, and evaluation of the human factors study, and the final development of design principles for interfaces in CPPS (Section 3.4). Conclusively, in Section 4, the followed approach is summarized and interpreted. We also point out subsequent future research ideas.

1.1. Change of Human Work in Manufacturing

Work in future manufacturing systems will be transformed by cyber-physical systems due to the technical and organizational shift described above. Despite the increased computerization and automation, studies say that human work is considered to remain very important for the future [9]. As a result, these studies expect the development of hybrid production systems, which involve collaborative work between people and machines. The specific application of the physical and mental abilities of humans can generate further added value [9,10]. Hirsch-Kreinsen [10] described the design of the human–machine interface as a central challenge. Only if the operators, which are part of the CPPS, can control and comprehensively understand them, decisions can be made that are consistent with the automated system (cf. Ironies of Automation [11]).

Further, CPPS lead to rationalization effects through digitization, networking, and automation [12]. Therefore, experts expect less simple, repetitive work and more qualified work in relation with the cyber-physical systems [9]. As a result, the type of work that will be required in the future is likely to be different. According to a study by the Prognos Institute, increasing demand for highly qualified workers and decreasing demand for low-skilled work and auxiliary activities can be assumed [13]. This effect is typical for technical and organizational change processes [12]. Additionally, Brossardt [14] highlighted changes in particular task areas: production-related occupations are declining. Here, technological development has an impact on the form of rationalization and automation. An increase can be observed for the knowledge-oriented field of occupation. This reflects the tendency towards a knowledge society with a lower need for manual but a higher need for knowledge-based activities [14].

1.2. Anthropocentric Work Design in CPPS

Work organization deals with actions that contribute to changing existing work systems or creating new ones. These actions include the design of work organization, links to other work systems, work tasks, equipment, workplaces, or working environment. They can be distinguished by different perspectives:

- From an economic or engineering perspective, the work design aims at improving the performance and efficiency of the value creation process.
- From a human factors perspective, work design deals with the creation of work systems that enable safe work that is neither physically nor mentally exhaustive [15].

Significant contributions to human-oriented work design were made by Hackman and Oldham [16], Luczak [17], Rohmert [18], and Ulich [19]. Their criteria include both physical and mental human needs and attributes. For example, the employees' work should be designed in a way that feasibility, non-harmfulness, and appropriateness of the tasks is ensured, along with satisfaction and personal development. Various standards (e.g., DIN 6385 [20] and DIN EN 894 [21]) also include principles and approaches in the ergonomic design of (interactive) work systems. The work systems must be designed to be compatible with the performance of the person, the expectations of the operator, and the type of work task.

1.2.1. Human–Machine Interaction

Human–machine interaction is a part of every socio-technical work system in which people and technology collaborate. This refers both to systems in which the machines are merely human work equipment and to systems in which humans and machines act as collaborating, independent actors. It enables two-way dependent communication between humans and machines, so called dialogues. Each dialogue process consists of several interfaces: Input and output, dialogue, and tools. DIN EN ISO 9241 provides a set of rules in this regard, which serves to design and evaluate these dialogues [22]. They should, therefore, be appropriate, self-explanatory, faithful to expectations, adaptable, robust, and conducive to learning. Further design principles were proposed by Butz and Krüger [23]. They provided basic design rules that focus on the accessibility of human–machine interfaces and intuitive operation. Regarding human–machine interaction in CPPS, Peissner and Hipp [24] identified further challenges and preliminary solutions. They considered human-oriented design, the role of the humans, multimodal interaction, the design of automation, and individualization as essential success factors. Table 1 provides an overview of these measures and gives examples for each.

Table 1. Approaches to human–machine interaction in CPPS according to Peissner and Hipp [24].

Approach	Exemplary Measures
Human-oriented design	Attractive design of user interfaces to increase user acceptance and work performance Human-oriented development of work systems through early involvement of the work force Iterative design processes of the work areas Improving the user experience and work motivation through the use of gamification or innovative tools
Role of human actors	Clear Presentation and Context-Sensitive Display of Information for Complexity Reduction and Error Avoidance Mobile workstations for fast reactions to events Availability of manifold information through real-time data, assistance systems, and flexible tools Use of physical analogies and consistency in interface design Active error avoidance through use of constraints and plausibility checking

Table 1. *Cont.*

Multi-modal interaction	Use of different input and output methods for usability improvement and error avoidance
Design of automation	Use of comprehensible automation for improved acceptance and error avoidance
Individualization	User-dependent design of interfaces

1.2.2. Human Factors

In addition to the disciplines of work design and human–machine interaction, the area of human factors also covers issues raised by the design of CPPS. Human factors can be understood as all physical, psychological, and social characteristics of the humans, which influence the action in socio-technical systems [25]. The associated scientific discipline is also referred to as human factors, and it deals with the role of humans in complex systems, the design of work equipment, and the human-oriented adaptation of the working environment [26]. On the one hand, it aims to carry out basic research in the field of the interaction between people and technology and, on the other hand, to produce solutions for practical problems. The aim is to avoid the negative consequences that can arise from this interaction. As a result, human well-being can be increased, and the safety and function of the human–machine system improved [25]. Human-oriented design of the CPPS is thereby of high importance.

In the area of manufacturing systems, the human factors address the workers who operate or monitor production machines. The gain of knowledge is mainly achieved through social scientific research methods such as behavioral observations, interviews, or questionnaires. A combination of these methods enables the recording of externally observable actions and utterances as well as the acquisition of "inner" information, such as action strategies, satisfaction, or expert knowledge [25].

1.2.3. Human Factors in CPPS

For CPPS in particular, cognitive ergonomics play an essential role. In contrast to conventional ergonomics, it is less concerned with physical and anthropometric characteristics than with the cognitive, emotional, and motivational needs and abilities of people. It does not only affect those informational processes that occur during task processing. Rather, through the design of the work equipment or the task, it also has an impact on how the work is organized by the worker. It affects the entire work process, starting with the task preparation and ending with the completion of the task [27]. Part of cognitive ergonomics with regard to human–computer interaction is also the consideration of emotional and motivational aspects. For example, researchers investigate how assistance systems should give hints or recommendations to workers in order to trigger the desired consideration and response from users, but avoid defense reactions or ignorance [25].

CPPS show a high process and work complexity as well as an increased level of automation. These aspects need to be considered more closely about human factors. Complex situations have specific characteristics. First, they are extensive in terms of recording and processing the work task. They contain a large number of variables that have to be included in the consideration. These variables are not isolated from each other but interlinked. As a result, they influence each other in direction, type, and intensity. Measures on one variable may, therefore, have (undesirable) side-effects on other variables. Furthermore, complex situations are characterized by opacity, which can cause parts of the overall situation to remain hidden for the acting person. For example, variables may be unknown, their interlinking unclear, or their current state not determinable. Consequently, actions must be taken under uncertainty. Finally, complex situations are dynamic. The variables and their interaction change over time. Thus, the states of the variables can change even without the influence of the acting person or show a different state at the time of implementation of a measure than at the time of the first observation [28]. While in everyday life, many tasks can be met by routines, complex conditions lead to the necessity to actively organize tasks. There are, therefore, various ways of dealing with complexity:

behavioral prevention and structural prevention. Behavioral prevention aims to provide strategies to deal with new and complex situations by training. Thus, it increases the flexibility of the workers. Structural prevention tries to reduce complexity by design in the best possible way. The reduced complexity then shows a higher similarity with standard situations and can be handled better [28].

The change of work as a result of the introduction of CPPS leads to the computerization of workplaces and further automation of processes (see also Section 1.1). Therefore, many of the future workplaces present themselves as human–machine systems. Consequently, in these cases, tasks are not carried out by only one person but by the interaction of (several) workers and machines. At this point, the term machine includes both production machines and cyber-physical assistance systems. The design of these systems in terms of a human-oriented work is in the area of interest of the human factors. Automation is the process of transferring activities from humans to machines. The automated activities can be manual or cognitive: Manzey [29] mentioned manual work, manual control, and control activities on the one hand, and judgment and decision-making activities on the other hand. An important design issue of automated systems lies in the distribution of tasks among the various actors of the human–machine system. This applies both to the allocation of tasks to different people and the assignment of tasks to workers or machines. The assignment of tasks to machines serves different purposes: utilization of economic advantages, error reduction, quality improvement, or improvement of work for humans.

The quality of the division of tasks between people and machines in practice depends crucially on the interaction between them [29]. There are various approaches to answering these allocation questions. The first approach is an automation strategy with a focus on technology and costs: according to this strategy, all tasks which can be easily and cost-effectively automated are transferred to machines. Humans are left with those tasks where automation is impossible or costly. The goals of this strategy are primarily to increase the efficiency and reliability of processes. There is little consideration given to the impact of this strategy on workers [26,29]. A second approach is an automation strategy that focuses on the capabilities of the actors. They should be relevant to the tasks assigned to them. Often, however, an isolated view of people on the one hand and machines, on the other hand, cannot meet the idea of a system-wide perspective. As a result of this criticism, a third approach regards people and automation as a complementary, hybrid work system. According to this idea, humans always bear the responsibility for the systems and thus take on the role of a so-called "leading control" [30]. Their tasks include:

- the planning of the tasks of the machines;
- the transfer of tasks to the machines;
- the monitoring of the execution of the activities;
- the intervention in case of unintended outcomes; and
- the learning from experience in handling the machines.

Examples of implementing this human role in the design include the creation of active involvement of humans and the access to complete information about the activities of the automation. Humans and machines should be considered as equal and independent actors of the system [29,30]. The usability and performance of a human–machine system which contains automated components depend to a large extent on humans having a reasonable degree of confidence in automation. An appropriate situational awareness is present when the operator is informed about the current system status and can foresee its further development [29,31]. Various reasons can lead to a lack of situational awareness. Examples are a lack of monitoring of the automation, changed feedback channels (for example, haptic or visual impressions of the production machine may no longer exist when the task is changed to leading control), or lack of transparency of the automation. Often a lack of situational awareness goes hand in hand with a loss of manual skills. This can be critical if operators need to regain control of the system or have to identify errors (see again [11]).

2. Materials and Methods

The application of methods is an elementary part of human factors research. They are utilized to conduct evaluations and predictions on the interaction among humans and machines concerning productivity, safety, and work satisfaction [32]. In this way, the performance of a human–machine system can be evaluated, and the requirements and effects of the work on the working persons can be assessed. The findings can be used to develop new and improved systems [32,33]. In many cases, the starting point for a human factors research project lies in the need to improve a human–machine system in terms of work conditions and efficiency. In addition, the objective may be to describe fundamental relationships in the interaction of humans and machines that are detached from a concrete problem [32]. Thus, research can be fundamental or applied. Fundamental studies are used to understand the factors that influence the function of the human–machine system. The results of these studies are basic principles of action and theories [32,34]. Applied research aims to transfer the results of fundamental research into the development and evaluation of human–machine systems [32].

2.1. Design and Analysis of Experiments

A standard method of human factors are experimental studies [32]. They can be used to show whether there are causal relationships between the design of a human–machine system and effects on the humans involved, the system, or the system performance. Therefore, in an experimental study, one or more variables are modified in order to induce observable effects [32,33]. The goal of a subsequent statistical evaluation of the experiments is to answer the question whether the selected independent variables (factors) have an influence on the dependent variables. Both descriptive and closing statistics methods are used. While the former summarizes and presents the collected data, the latter concludes whether the identified differences are significant or only the result of a random variation [33]. Descriptive statistics in the field of human factors often include the representation of mean values and standard deviations. Typically, the values of the dependent variables are determined for the respective levels of the independent variables and statements are made about their variance [33]. If there is a factorial design with several independent variables, the main effect of an independent variable and interaction effects between several independent variables can be determined [35]. For experimental designs that follow a factorial design with several independent variables in different levels, an analysis of variance (ANOVA) is often performed (if different conditions are met, see [35]). ANOVA provides a statistical test that takes into account the observed mean values of the different groups. It allows a decision to be made whether these differences are large enough to conclude that there are similar differences in the underlying populations [36].

2.2. Available Human Factors Research in CPPS

In the past, several research projects have dealt with the effects of human-oriented working conditions on the physical and mental condition of employees on the one hand and the economical parameters of companies on the other hand. These works revealed connections between human-oriented measures (e.g., ergonomic design or cognitive simplification) and their positive consequences (e.g., increased job satisfaction and improved product quality) [37–41]. In addition, a cost–benefit assessment of ergonomic measures in companies showed mostly positive effects of these measures by their ergonomic intentions as well as short amortization phases [42]. While the majority of these studies were carried out based on a classical understanding of ergonomics, in the context of Industry 4.0, the human factor is currently once again the subject of research. Various papers deal with the analysis, potential, and impact of human factors [43–46]. In addition, an interdisciplinary approach to the implementation of the human factors, the consideration of social changes in product development, and the role of technology acceptance by users has been discussed.

According to the previous explanations in Section 1.1, the work systems in semi-automated production systems are characterized by intensive use of the interface between humans and machines.

Here, cognitive assistance systems are used to establish this connection between the automated production parts and the workers. In a study by Rauch et al. [47], existing research and contributions were identified using a systematic literature review, which pursued a human-oriented approach in the development of cognitive assistance systems. These research contributions are examined in more detail below in order to evaluate them in terms of considering human factors in the design process of the cognitive assistance systems.

Chen et al. [48] developed a framework for early consideration of human factors during the planning process of an assembly system. By integrating ergonomics analyses into a virtual assembly system, the expected stress on the employees can be evaluated. The result of the work of Yang et al. [49] is a concept for a human-oriented virtual factory in which work-relevant issues, such as ergonomics, collaboration, or training, are integrated into the planning process at an early stage. These factors are taken into account both in the product design process and in factory and process planning. Santochi and Failli [50] considered possible approaches for sustainable work in production, which is the result of the interplay of physical and mental working conditions and motivational influencing factors. They also suggested possible approaches for improving a typical assembly workplace using cognitive assistance systems. Romero et al. [51] proposed a human-centered reference architecture for the conception of semi-automated systems. They intended to develop production systems which are characterized by adaptable automation and thus support employees meeting their needs, for example, the requirements of elderly employees or trainees. Hold et al. [52] described a procedure for integrating physical and cognitive assistance systems into CPPS. They aimed to uncover suitable application possibilities and potentials.

Ohtsuka et al. [53] described a cognitive assistance system for collaboration with robots. The system recognizes the desired operation of the employee and provides support. This results in faster task processing. The work of Weiss et al. [54] contains the idea of a collaborative work system between humans and robots. For the developed system, studies were carried out to determine the interaction between the actors and to derive improvement potentials. The authors used human factors methods. They proposed an assistance system integrated into the worker's clothing for robot control and monitoring. Soffker et al. [55] presented a method for mapping human–process interaction in manufacturing work. It can be used to develop intelligent assistance systems, which contribute to an in-depth process of awareness. The authors proposed an assessment of suitable channels to be used to interact with workers (e.g., auditory or visual channels). Pacaux-Lemoine et al. [56] proposed starting points for cognitive assistance systems to be used by workers with control tasks in autonomous production systems. A prototypical implementation led to both a reduction of the energy consumption and of the throughput times.

The work of Gorecky et al. [57] deals with the development of virtual training and knowledge transfer systems for worker qualification and for improving communication in production systems. The authors described the prototypical implementation of two exemplary systems. Paelke [58] considered the development of an augmented reality assistance system to support employees in part selection and assembly during an assembly process. Different visualization concepts are developed for the information to be displayed. In addition, a test of the system with a large number of study participants is described. Another augmented reality system was presented by Rauh et al. [59]. Here, an augmented reality assistance system is developed to support employees during maintenance activities in the production process. This work includes a human-oriented development and evaluation of the prototype system by involving the users in different phases of the design process.

Level of Integration of Human Factors

The presented research works deal with cognitive assistance systems, which can be used either in the planning process of production systems or during the manufacturing process of the products. They aim to improve working conditions for employees and to optimize process quality. A human-oriented approach was followed in each case. These ideas were evaluated concerning the type and extent of consideration of human-oriented design principles and methods of human factors. They were checked qualitatively for the fulfillment of the following criteria:

- Explicit consideration of designing a human-oriented assistance system (use of design principles of human-oriented work design and human–machine interaction in general and for automated systems in particular)
- Explicit consideration of the interface between worker and cognitive assistance system
- Methodical integration of human factors into the design process of assistance systems (use of an approach which integrates human factors concepts and standard methods)

Table 2 shows the results. A checked research paper indicates the fulfillment of a corresponding criterion. It can be noted that only a low level of criteria compliance could be found. Only two out of twelve research papers fulfill one of the criteria. Altogether, we can conclude that the focus of the contributions is mainly on the technical and organizational development of cognitive assistance systems, but only to a limited extent on the methodological integration of the human factors.

Table 2. Analysis of research papers taken out of a review by Rauch et al. [47] on available human-oriented approaches for the development of cognitive assistance systems: Design, explicit consideration of designing a human-oriented assistance system; Interface, explicit consideration of the interface between worker and cognitive assistance system; and Methods, methodical integration of human factors into the design process of assistance systems.

Contribution	Design	Interface	Methods
Chen et al. [48]			
Yang et al. [49]			
Santochi and Failli [50]			
Romero et al. [51]	x		
Hold et al. [52]			
Ohtsuka et al. [53]			
Weiss et al. [54]			x
Soffker et al. [55]	x		
Pacaux-Lemoine et al. [56]			
Gorecky et al. [57]			
Paelke [58]		x	
Rauh et al. [59]		x	x

3. Integration of Human Factors into Human-Oriented Work Design in Cyber-Physical Production Systems

The results of the presented work can be separated in four parts: First, we derived crucial work design elements in CPPS (Section 3.1). Second, we developed a method for the integration of human factors into human-oriented work design in CPPS, which is based on the evaluation of the impact of work design elements on humans in a work environment (Section 3.2). Third, we developed an experimental platform for carrying out human factors studies as a part of the presented method (Section 3.3). Fourth, we applied and validated the method using the developed platform (Section 3.4).

3.1. Design Elements for Human–Machine Interfaces in CPPS

The findings presented in the previous sections on work in CPPS and on human factors in human–machine systems show the availability of a wide range of research work in this area. The consideration of work design in CPPS, however, shows only a few findings thus far. In addition, the applicability of many standards and design rules of work design for conventional production systems must be questioned. Since most of this research has been done before the introduction of cyber-physical work systems, it often involves an outdated understanding of work in production systems. Due to new tasks, a changed allocation of tasks between humans and machines as well as new assistance systems, cognitive work, in particular, becomes predominant, while the share of physical work decreases. As a result, an overview of new design elements was developed to meet the changing requirements of cyber-physical work systems. The basis is formed partly by research contributions on the current and future development of work in CPPS and partly by research on the human-oriented design of human–machine systems. Table 3 shows the areas of origin of these references and the contribution they provide to the finding of new design principles. We also refer to the corresponding sections of this publication, which contain detailed information on the specific areas. The subject of this discussion are always working systems in CPPS, which are designed as semi-automated systems.

Table 3. Basis of the new design elements concerning human factor-oriented work in CPPS and their corresponding contributions.

Area	Contribution	Reference
Cyber-physical production systems	technological capabilities, new products and processes	Section 1
Change of work	New requirements, changed division of labour, new work equipment	Section 1.1
Human–Machine-Interaction	Interface Design Principles for hybrid, semi-automated cyber-physical systems	Section 1.2.1
Human factors	Human needs and abilities as part of cyber-physical, socio-technical systems	Section 1.2.2

An overview of the new design principles is shown in Figure 1. It represents an excerpt of the aspects that work design in the sense of human-oriented work in CPPS must additionally take into account. Due to the thematic overlap of the subject areas, the design elements may originate from one or more of these areas. During their derivation, a maxim was set that they are of special relevance in the practical context of work systems in CPPS. In the context of this paper, they represent a basis for Phase 1 of the method (see Section 3.2), in which crucial design elements for the present work system are derived and selected for the subsequent phase steps, respectively. In the following, the design principles are referred to as design elements. They are explained briefly below:

Usability describes a facilitated usage of the interface for the worker. This is characterized by increased effectiveness, efficiency, and satisfaction when using the interface [60]. For example, usability can be increased by improving the readability of information or the simplicity of its presentation, as well as by placing dialog elements intuitively. The design element Attractive Design goes beyond the previously described aspect of usability and wants to create an additional positive experience for the working person when using the interface. Work tasks are perceived as easier if they can be performed with the help of attractive work equipment [61]. This can be achieved, for example, by exceptional design or the implementation of new technologies [26]. The Adaptability of the human–machine interface describes the possibility on behalf of the worker to individualize it. For example, special individual requirements can be taken into account by setting options such as font size, language, or contrast. Besides, adaptability can be used to personalize the use of existing assistance systems according to the skill level [22,62].

The design element Robustness describes the tolerance of the interface against wrong inputs and errors. Consequently, the design should aim to minimize the impact of errors on the outcome of the work. This can be done, for example, by automatic plausibility checks or the use of an "Undo" function [23,62]. Affordance describes the "invitation" of an object to use it for specific actions. For example, three-dimensionally designed buttons could be used, which seem to protrude spatially. Thus, they motivate the worker to click or press them, since they look like buttons on physical devices [23]. Constraints are (physical) restrictions of the interaction with an object [23]. For example, a drag and drop operation might only allow an object that has been picked up to be placed in the designated areas. This way, hints can be given on the intended processing of the work task, and errors are avoided. Mapping is a design element that provides a suitable assignment of real objects to the interface [23]. For example, with spatial mapping, the arrangement of production machines in production would be transferred in the same way to the arrangement of sensor data for these production machines on the interface.

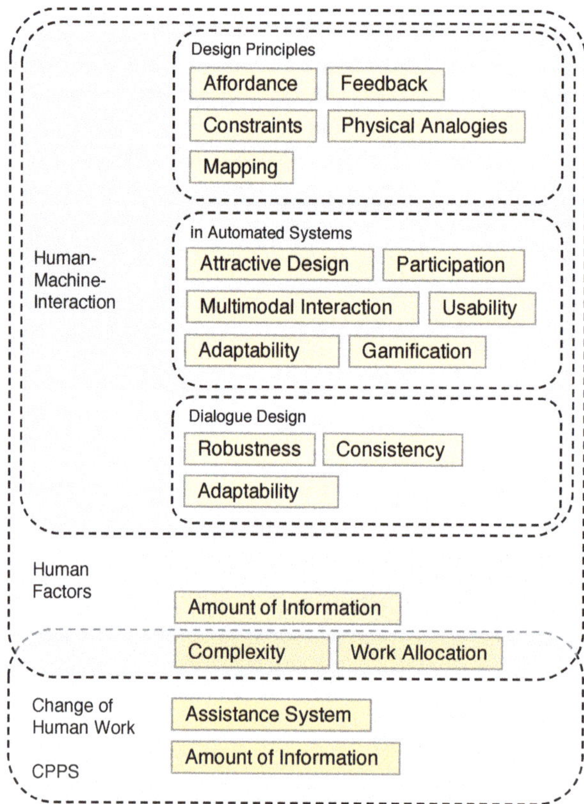

Figure 1. Design elements for human–machine interfaces in CPPS.

The element Consistency refers to the use of constant layouts, colors, or terminology for same and similar issues. For workers, this results in greater reliability in identifying functions and predicting their effects [23,62]. Physical analogies describe the behavior of buttons based on physics [63]. For example, activated buttons could be shown as pushed in buttons. Thus, the activation can be intuitively perceived by the worker. By using Assistance systems, workers should be supported in the processing their work tasks. A typical example is an augmented reality goggle that complements the perceived reality with helpful support. Such tools can represent the human–machine interface themselves but can also be designed as a function on an existing interface (e.g., on a tablet computer) and thus support the worker in different ways [64,65]. This kind of assistance system could, for example, be designed as color highlighting of valid areas when placing objects within a user interface.

The design element Amount of information describes the provision of the appropriate amount of information which is necessary and useful for solving the work tasks [66]. Only such information shall be displayed that contributes to an improved or simplified problem-solving process. ViaFeedback workers can be offered feedback both about the activities of the cyber-physical work system and about the quality of their work results. For example, this design element can range from confirmations or process status information to details on the user's goal achievement during or after task processing [22,23].

Gamification links work with elements from the field of computer games. For instance, it describes the use of virtual rewards (badges), a high score list, or different game levels in the remote context of a work system. This way, the intrinsic motivation of the workers can be increased, as the rewards lie in the work system itself and are not brought in externally [67,68]. The design element Multimodal interaction intends the use of several ways of communication between humans and machines. As a consequence, input and output should not only be carried out one way, for example by using a touch display but should also include other types of communication such as speech, gestures or facial expressions. This allows the human–machine interaction to be approximated to the natural communication between humans and their environment, and thereby to improve the information flow [69].

By Participation of the workers in the design process of interfaces, the interaction quality can be improved, and the efficiency of the design process can be increased. On the one hand, this participation can provide a complete picture of the work system during the requirements analysis and, on the other hand, can improve the acceptance of the interface at its introduction and operation [70,71]. Work allocation deals with the assignment of tasks within the human–machine system and allocates them to humans and automation [29]. This division of functions follows the present automation strategy and influences the quality of use and acceptance of CPPS [72]. The Complexity of a work task is a consequence of the required action and the available information and determines the work performance [73]. Complex conditions in terms of work design can be addressed by behavioral prevention (training) or structural prevention (reduction of complexity) [28].

Actions derived for the new design elements should be evaluated in terms of their effects on human work. Existing evaluation possibilities in the field of conventional ergonomics are somatographic aids. These are human-shaped models used to check the anthropometric design of workplaces. This procedure may also be implemented with the help of computers as an Augmented Reality application [74]. Furthermore, digital human models can be used in CAD models of workstations for vision or accessibility checks. These evaluation methods increase productivity and lead to cost reductions [75]. In addition to conventional ergonomics, usability engineering can be used in the evaluation of human–machine systems. It describes the process of continuously considering the usability and user-friendliness of work systems already during their design process [74]. Here, observations and user surveys are carried out frequently. The goal is to design a suitable user interface that avoids unnecessary complexity and maps the required functionalities of the work system [76]. In these types of tests, the users are confronted with representative work tasks. During and after processing, key data on effectiveness, efficiency, and user satisfaction can be collected [74].

To be able to check a large number of different design elements, a flexible way of conducting human factors studies is required. It should be usable for various derived actions. As a part of the method presented in the following section on the integration of human factors into the work design of CPPS, the development of an experimental platform is carried out.

3.2. Method for the Integration of Human Factors into Human-Oriented Work Design in CPPS

The method aims to support the understanding of the mechanisms between human factors and work in CPPS. It focuses on the area of work design. Here, the design of interfaces between humans and machines in manufacturing serves as a research subject. A method shall be designed to enable the assessment and evaluation of specific work design elements. The method should be capable of providing a first catalogue of rules regarding the interface design in CPPS.

We postulate that the integration of the new design elements presented in Section 3.1 has a significant effect on work performance and the perception of work. Our method is designed to assess and evaluate this expected effect. Its basic idea is to conduct experiments in which workers interact with cyber-physical systems via human–machine interfaces. The participants are asked to solve work tasks. Here, various key figures and inputs are recorded. At the same time, questionnaires are used during the experiments to gather personal, work psychological information about the test persons. Consequently, quantitative data on the task-specific achievement of objectives and personal status information is determined in the experiments.

If a design element is active when performing a first work task, but inactive when performing a second work task, the difference in the outcome variables can be linked to the design element. For example, a measure derived from a design element could be the illustration of a task description. In combination with suitable study design, this allows research to show the effects of design elements or draft work designs. These data enable a statistical analysis of the relationship between particular elements of work design, the perception of work, and the achieved performance. Thus, the hypotheses on essential design elements in CPPS can be tested and transformed into design principles that correspond to the particular effect.

The methodology provides a stepwise procedure in four phases:

- Phase 1: Hypothesis of key design elements;
- Phase 2: Design and implementation of representative work tasks as a human factors study;
- Phase 3: Execution and Evaluation of the human factors study; and
- Phase 4: Development of design principles for interfaces in CPPS.

Figure 2 summarizes a simple run of the method. A rectangle symbolizes an operation and an ellipse the end of the process. The method is presented in detail in the following.

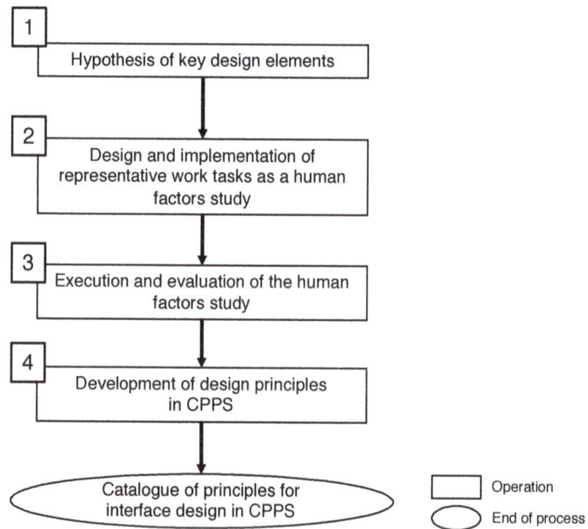

Figure 2. Method-schematic overview.

3.2.1. Phase 1: Hypothesis of Key Design Elements

Phase 1 of the method aims at identifying the key design elements in the present cyber-physical work system. At this point, the term "key" refers to the design elements that may have a significant effect on work performance or the perception of work. The selection of these elements will be initiated by a collection of potentially important design elements for human–machine interfaces in CPPS. This collection originated from previous research in the fields of human–machine interaction and human factors as presented in Section 3.1. The collection of possible key design elements is presented in the overview (see Section 3.1). These design elements will be evaluated in the next steps of the method.

3.2.2. Phase 2: Design and Implementation of Representative Work Tasks as a Human Factors Study

Phase 2 of the method implements the identified representative work tasks and the key design elements as a work design draft. This is carried out in a way that enables its implementation and subsequent use for a human factors study. Consequently, the design and preparation of the study is also part of Phase 2. First, the identified design elements and the representative work tasks are implemented on the experimental platform (see Section 3.3). As a representative work task, a model CPPS work system is used that does not originate from any concrete application case. The result is a work design draft that offers the possibility of modular integration of various design elements. This way, the design of experiments by a full factorial design is possible (see Section 2.1). Second, the design of the experiment is defined. This design of experiments includes the determination of the variables of the study and the preparation of a test plan. In the following third step, the work tasks realized on the experimental platform are transferred into a human factors study according to the design of experiments.

3.2.3. Phase 3: Execution and Evaluation of the Human Factors Study

In Phase 3 of the method, the human factors study is carried out, and the study results are evaluated statistically. As a first step, a sufficient number of adequate participants must be recruited for the study corresponding to the design of experiments. Within the study, the participants perform predefined work tasks. Here, the target variables (dependent variables) are automatically recorded by the experimental platform. Finally, the study results are provided for subsequent analysis. While carrying out the experiments, it is necessary to observe the test plan as well as the experimental conditions in order to achieve consistent and unbiased results. In a second step, the statistical evaluation of the study results takes place. Statistical analysis is performed to display significant differences in dependent variables caused by the design elements (independent variables or factors). The result provides the effect intensity of the design elements on the dependent variables (key figures).

3.2.4. Phase 4: Development of Design Principles for Interfaces in CPPS

Finally, Phase 4 aims to develop corresponding design principles for work design in CPPS based on the determined effect strengths and effect directions of the design elements. Therefore, differentiation between "Yes" and "No" rules will be made, which indicate whether an implementation of the respective design element can be recommended. The entirety of all key figures examined is explicitly considered. In the event of conflicting effects concerning individual factors, a division into "Yes" and "No" rules will also be made if a predominantly positive or negative effect can be identified. In all other cases, a "Maybe" rule is defined, which requires further investigations. Based on the study results, the effect sizes for each dependent variable are used as effect indicators for all examined design elements. Table 4 shows the generation of effect indicators for absolute or assessment variables. These are key figures where an increase or decrease are the possible changes, for instance, the processing time of a work task or the evaluation of the task difficulty on a rating scale. If the desired effect direction is a reduction (e.g., processing time), the effect indicator E^- is used; otherwise, E^+ is used for an increase. An observed reduction of the processing time by 12%, for example, would lead to an effect indicator of $+$ (see Table 4). The sums of the effect indicators are then compared. If a homogenous pattern of the effects can be identified for both aspects (work performance and perception of work), a "Yes" rule is assigned for a positive effect and a "No" rule for a negative effect. In the case of an indifferent pattern of effects, a "Maybe" rule is assigned.

Table 4. Generation of effect indicators for absolute and assessment variables.

Effect Size E^+/E^-	Effect Indicator E^+	Effect Indicator E^-
$E > +50\%$	$+++$	$---$
$+25\% < E \leq +50\%$	$++$	$--$
$+10\% < E \leq +25\%$	$+$	$-$
$-10\% < E \leq +10\%$	0	0
$-25\% < E \leq -10\%$	$-$	$+$
$-50\% < E \leq -25\%$	$--$	$++$
$E \leq -50\%$	$---$	$+++$

3.3. Conception and Development of an Experimental Platform

A platform needs to provide an interface to the investigator and the participant. The investigator configures the setup of the experimental platform to set up and configure the investigation. This includes the realization and customization of the work tasks, the design elements, as well as the set-up of the study. Once the experiment is finished, the experimental platform makes the recorded results available to the investigator. The test participant uses his or her interface to the system for task processing and answering of questionnaires. All entries and responses made are recorded. The described functionality of the interfaces is shown in Figure 3.

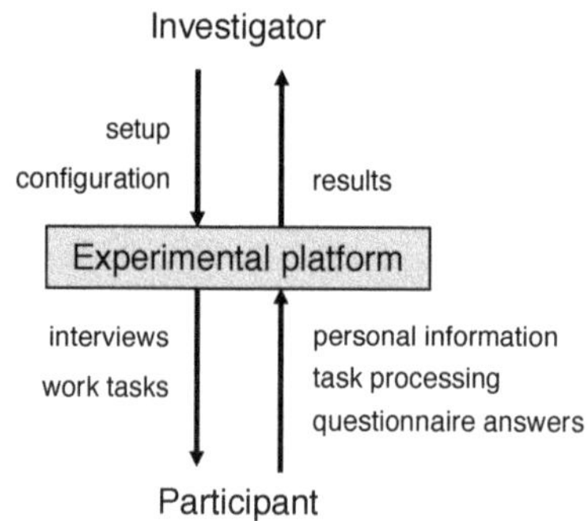

Figure 3. Functional diagram of the interfaces between the experimental platform and the users.

To meet the requirement that the experimental platform should be designed as flexible as possible to cope with different design elements, the setup and configuration of the experiments should be implemented as easy as possible. For example, including a new design element in a work task should be easy to integrate. In addition, there should be a set of modules which can serve as building blocks to sequentially assemble the experiment. The modules are implemented as pages (similar to a website or app page), which are displayed on a touch screen display and can be viewed by a test person in a sequence. A sequence consists of pages that can be used for task description, task processing, or interviews. There can also be additional pages for the information supply (e.g., title pages or introductory pages).

During task processing, the experimental platform can measure various key figures and record information. This information can be observable and non-observable. The observable variables are:

- the number of display touches;,
- the time used to solve a work task; and
- the outcome of the work task.

Non-observable variables are assessed using questionnaires. Depending on the chosen test design, questions can be asked about the perceived complexity of the task or the satisfaction with the own outcome. The collected data are recorded by the system and made accessible to the investigator.

The experimental platform is based on a Raspberry Pi 2 B single board computer combined with a 7-inch touch display. Further components are shown on the right side of Figure 4 as a cross-section of the system. Due to the use of common electronic components such as a Raspberry Pi 2 B or a USB power bank, a cost-effective implementation of the experimental platform is possible. The total price of a system is 233 Euro (as of 2019). The total price is in the range of commercially available tablet computers, which can be a possible alternative to this system. Due to its modular structure, however, the platform can be flexibly extended and adapted. The use of a Raspberry Pi, for example, provides extensive connection options (e.g., for operating the LED strip). The perspex box can be used to easily accommodate additional components. Thus, the use of the experimental platform is preferred over the use of a tablet. The left side of Figure 4 shows a photo of the system.

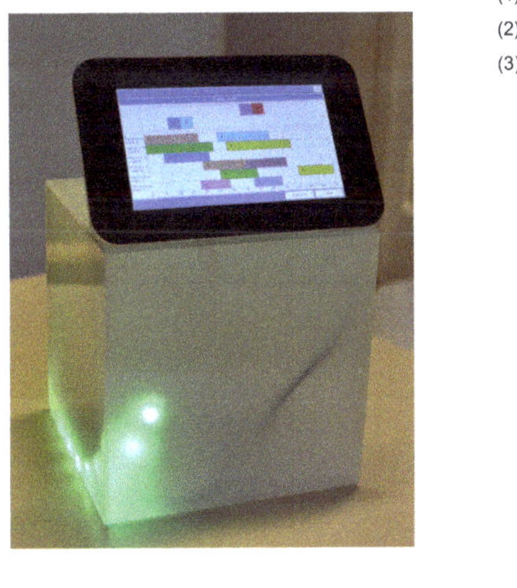
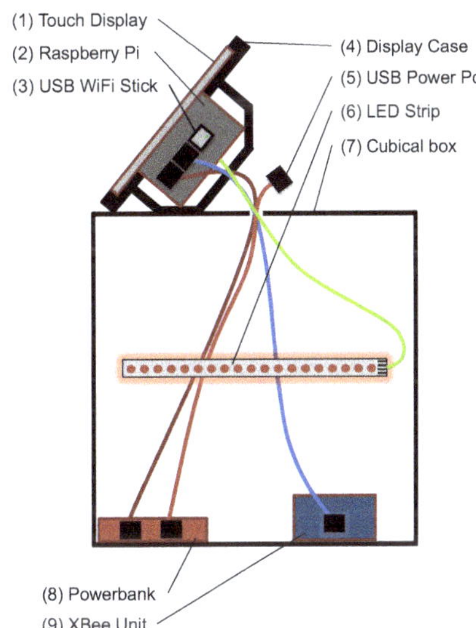

Figure 4. Photo and cross section of the experimental platform.

The software of the experimental platform is a self-developed Python 3 program. The operating system used is Linux for Raspberry Pi. Requirements for the software and its use cases from the perspective of the involved users were derived as follows [77]:

1. The system has to implement the representative work tasks and the design elements from Figure 1 and represent the interface between human and machine within a cyber-physical work system.
2. The system has to supply the necessary input and the required output regarding the representative work tasks to the workers.
3. The system has to facilitate the solution of the work task.
4. The system has to provide questionnaires regarding the query of work psychological indicators and personal information.
5. The system has to display feedback on the completed representative task.
6. The system has to independently record the desired target values of work performance and perception of work and make them available to the investigator.
7. The system has to be modular in order to be able to flexibly compose and adapt the experiment out of different modules (e.g., tasks, feedback, questionnaires) following the design of the experiment.

These requirements can be assigned to the users involved (participants and investigator). Requirements 2–5 represent the use cases of the participants, while Requirements 6 and 7 are attributed to the investigator. Requirement 1 represents the main function of the experimental platform and, therefore, cannot be assigned exclusively to the participants or the investigator. Figure 5 shows these use cases.

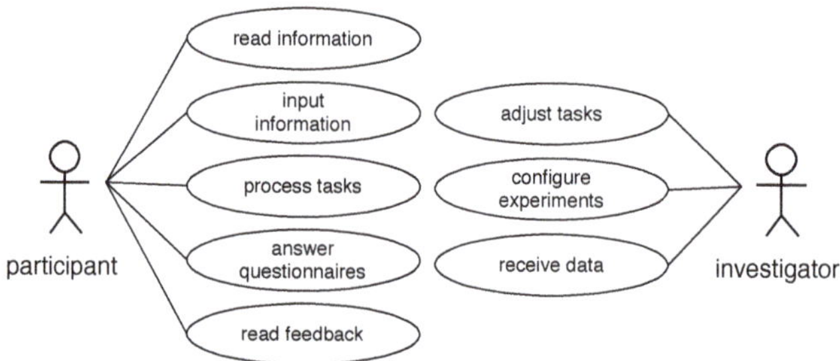

Figure 5. Use cases of the software of the experimental platform.

3.4. Application and Validation of the Method

The method developed in Section 3.2 was applied and validated (see below). For this purpose, the necessary parts, the overview of design elements in CPPS (see Section 3.1) and the experimental platform (see Section 3.3), were realized. Some design elements were selected from the overview provided and implemented on the experimental platform using representative work tasks from the field of machine scheduling. Subsequently, an extensive experimental study was conducted and analyzed, which was then used to evaluate the design elements. Finally, on this basis, initial design principles for interfaces in human–machine systems in cyber-physical work systems are proposed.

3.4.1. Phase 1: Hypothesis of Key Design Elements

The following three design elements from the overview in Section 3.1 were selected: "Feedback", "Assistance system", and "Usability". Since these design elements represent starting points for human-oriented design measures only, a further specification of these measures was necessary. The design element "Feedback" was implemented as a results feedback to the worker. For the integration of the design element "Assistance system", two different measures were used: firstly, a display of detailed information on incorrect actions by the worker and, secondly, visual support for operation and decision-making. The "Usability" design element was addressed by color differentiation of dialog elements. We postulate for all three design elements that the respective design measures would have a positive effect on work performance and perception (see Section 3.1). These hypotheses are examined below.

3.4.2. Phase 2: Design and Implementation of Representative Work Tasks as a Human Factors Study

The explanations in Section 1 of CPPS, the change of work and the existing research work on human-oriented work design results the following profile of a typical, semi-automated and cyber-physical work system:

- The worker performs a predominantly cognitive activity that takes place in interaction with automated systems.
- The human–machine interface is implemented as a cyber-physical assistance system.
- The task involves the understanding of a high process complexity resulting from the use of cyber-physical systems.

This profile was met by using a representative work task from the area of machine scheduling. The selected task confronted the participants of the study with a machine scheduling problem (a flexible flow shop problem) consisting of several production machines of different types that can be used to process several multi-step orders. The work objective was to complete all orders as soon as possible. We chose this scenario as it corresponds to the characteristic of a complex cognitive task that takes place in the manufacturing context. It was implemented on the developed experimental platform, which can be seen as an interface between human and machines within a cyber-physical assistance system. Other representative work tasks could also be considered for further applications of the method. The work tasks appeared in different versions: first, as an active task in which the machine scheduling problem is solved from scratch by the worker; and, second, as a reviewing and correcting task in which a predefined solution is evaluated and can be changed. This second version represents supervisory work tasks in CPPS, such as the monitoring of an automated production process.

The given jobs and their sets of operations (referred to as partial jobs in the following) served as input for the processing of the work task. Besides, the kind and characteristics of the production machines were mentioned (type, processing time). The predefined allocation for reviewing and corrective work tasks (referred to as corrective tasks in the following) varied according to different status: optimal, needs minor improvement, and needs major improvement. The expected output for both task versions consisted first of the capturing of the job and machine situation. This was followed by the machine scheduling (active) or the assessment of the need for changes and modifying the predefined scheduling (corrective). No other information or tools were provided to solve the tasks apart from the outlined input, and there is no time limit.

Design of Work Tasks on the Experimental Platform

The work tasks consisted both in the active and in the corrective version of two pages on the experimental platform: the task description page and the task processing page. This basis was supplemented by interview pages, which were used to record the perception of work and sometimes by a feedback page (depending on the design of experiments).

The task description page listed the existing jobs ($A1, A2, ..., An$) and the corresponding partial jobs (a, b, c, d). Each job contained at least one and a maximum of four partial jobs. A separate table showed the available production machines ($M1, M2, ..., Mn$) and indicated for which partial job they could be used and which processing time was required. Furthermore, the available machine network was represented by an illustration. Figure 6 shows a screenshot of the task description page.

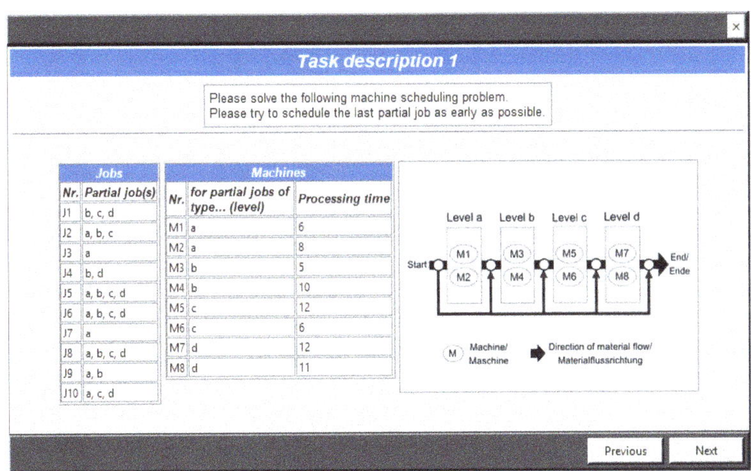

Figure 6. Task description page.

For active tasks, the task processing page contained all partial jobs of the jobs (for example, $A1.a$, $A1.b$, $A2.a$, etc.) as drag-and-drop elements. These could be picked up by the study participant and placed on the machine scheduling plan. For corrective tasks, the worker started here with predefined scheduling and had the option of making changes. On the task processing page, the processing time and the number of display touches made were measured and the final solution was saved. Figure 7 shows the task processing page.

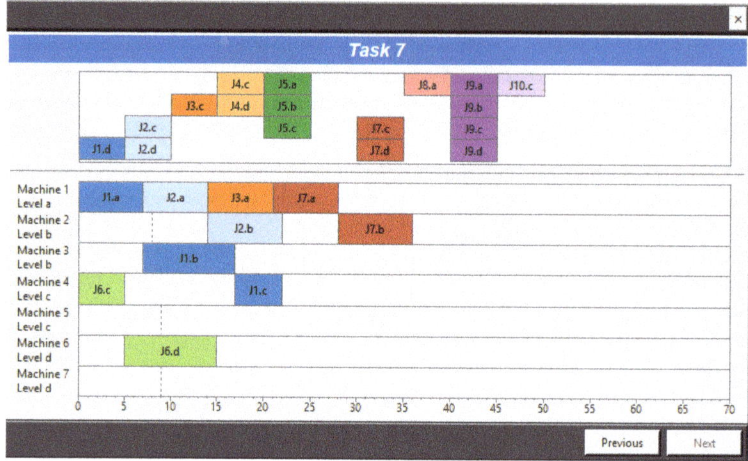

Figure 7. Task processing page.

The interview page was used to collect information about the participants' perception of work. For active tasks, the interview was conducted after task completion. The interview contained questions on the perceived difficulty of the problem, the satisfaction with the user's own final solution and the estimation of the quality of the solution. For corrective work tasks, a two-part questionnaire was presented. The first interview took place after the predefined scheduling plan had been assessed, and the second after the changes had been made. The first interview dealt with questions about the quality of the predefined solution and the certainty of this rating. The second interview was similar to the interview for active work tasks and contained questions on satisfaction with the solution and the evaluation of the quality of the solution.

The feedback page allowed a comparison of the participant's work results with an optimal solution. It compared the makespan of the achieved solution with that of an optimal solution. In addition, the machine scheduling plan for an optimal solution was displayed graphically. With the help of a button, the participant could switch it to the achieved solution to compare the placement of particular partial jobs.

The design elements selected in the previous phase or their respective measures were embedded in the four page types described above. The design element "Feedback" was implemented by using the feedback page for selected work tasks. If this design element was activated, the feedback page was displayed; otherwise, this page was skipped. Thus, the study participants only received feedback on the solution of a work task when this design element was activated. To implement the design element "Assistance system", a function was developed that displays details of the type of error in the event of incorrect actions during machine scheduling processing, e.g., "Production machine already occupied" or "Sequence of the work steps of the order not observed". Alternatively, visual support was provided to the operator for the placement of partial orders on the machine scheduling plan. The (re-)placement of drag-and-drop elements was supported by color highlighting. If an element was picked up, possible production machines on which this element can be placed (i.e., the appropriate machine type for the selected partial job) were marked by color shading. "Usability" was implemented by color coding of the jobs. All partial jobs that belong to a single job were represented by a color and were thus delimited from other jobs.

The machine scheduling problems were of comparable difficulty. For each of these tasks, CPLEX Optimizer was used to determine an optimal solution. In addition, predefined solutions were prepared in three levels: a solution without any need for improvement (no shorter completion time possible), a solution with medium improvement potential (approximately 10 percent) and a solution with high improvement potential (approximately 20 percent).

Experimental Design and Preparation of a Human Factors Study

Two experimental human factors studies were performed to illustrate the effects triggered by the design elements. Here, the three design elements selected in Phase 1 served as independent variables (factors), which were either set to "active" or "inactive". In Study 1, the display of error details was varied for the design element "Assistance system", while in Study 2 the visual operating support was varied. This resulted in eight experimental conditions of a factorial 2^3 design for both studies [32,33,35]. The dependent variables represented the aspects of work performance and perception of work to be examined.

The work performance was represented by:

- the number of display touches during task processing;
- the time required for task processing; and
- the quality of the solution (ratio of optimal makespan to achieved makespan).

The perception of work was represented by:

- the quality evaluation of the solution;
- the deviation between the solution quality and the quality evaluation;
- the satisfaction with the solution; and
- the evaluation of the task difficulty.

The dependent variables chosen are out of the areas of human–machine system performance and human subjective perception [32,78]. The solution quality in terms of the deviation from the optimal completion time can be seen as a direct indicator for work performance. The number of display touches and the time required for task processing can be seen as indirect indicators of work performance. They enable direct conclusions on the efficiency of task processing and the handling of the human–machine interface. The dependent variables concerning the perception of work were indirect indicators. Here, the quality evaluation, as well as the deviation between the solution quality and the quality evaluation, reflected the situation awareness of the participant. The evaluation of the task difficulty served as an indicator of cognitive support. Finally, the measurement of satisfaction with the user's own, final solution enabled an assessment of job satisfaction. The automatic and objective recording of the display touches and the time spent took place invisibly for the study participants in the background.

A combination of objective and subjective techniques was used for data recording. By a hidden, non-participating, and systematic observation, the objective measurement of the dependent variables on work performance took place [32,33,74,79]. Based on a questionnaire with standardized questions and answers, the subjective measurement of the dependent variables regarding the perception of work was carried out [80]. To determine the effects triggered by the design element "Feedback", the subsequent work task was used. The assignment of the study participants to the experimental conditions was performed as a within-subjects design with repeated measures [35]. Further, various randomizations were carried out: the problems generated occurred in random order for each study participant. Within these tasks, each experimental condition occurred at least once in a randomized sequence. Besides, there was a random sequence of active and corrective work tasks for each study participant. Finally, the order of the study participants was also randomized. The various randomizations were undertaken to minimize learning and fatigue effects [32,35].

3.4.3. Phase 3: Execution and Evaluation of the Human Factors Study

The experiments were carried out under constant environmental conditions to control external influences. Furthermore, the procedure and duration of participating in the experiment followed an identical pattern for all participants:

1. reception of the experimental platform;
2. start of the experiment and run of a test task;
3. processing of the tasks; and
4. end of the experiment after 45 min.

The studies were conducted as laboratory experiments to control the experimental conditions in the best possible way with the goal of determining basic rules for the design of human–machine interfaces in CPPS. A total of 68 persons participated in the studies ($n_1 = 33$, $n_2 = 35$). In a pilot study with nine participants, first insights were gained and problems regarding the experimental platform itself and the procedure of the experiment were identified [81]. These results were incorporated into the design of the main studies. Students from the University of Bremen and external participants were acquired as study participants. Due to their affiliation to different disciplines, professional heterogeneity of the group could be achieved, which mirrored the various application possibilities of the method. All study participants were assured an anonymous use of the results. Mostly automated execution of the experiments as well as the hidden data collection should avoid unintended external effects. In addition, the suitability of the empirical method was confirmed in advance based on various application examples (design of production cells [82], design of a multimodal user interface [83], and evaluation of different supply methods for material and work descriptions [84]).

The analysis was performed separately for each of the two studies (in the following, referred to as Study 1 and Study 2) as well as for each type of work task (in the following referred, to as Type A and Type B tasks). The studies differws regarding the chosen measure for the design element "Assistance system" (Study 1: error messages; Study 2: color highlighting) as well as the complexity of the work tasks. The work tasks occurred either as active tasks (Type A) or as corrective tasks (Type B). After completion of the studies, initially, data preparation was carried out. Here, we aimed to remove incomplete tasks from the results. All tasks in which the study participants spent less than 15 s on the assessment of the predefined solution and the task processing were removed. This threshold was set at approximately 10 percent of the average processing time. Then, 147 valid and completely solved Type A tasks and 135 Type B tasks remained for Study 1 and 106 and 68 for Study 2, respectively.

The statistical significance of the independent variables was assessed utilizing a subsequent three-factor variance analysis performed with the Minitab 18 software. Table 5 shows a selection of the resulting p-values. In the sense of a practical significance, all effects showing $p < 0.15$ were used for further consideration. If $p < 0.15$ is present, this entry is marked in gray color in Table 5. Only significant effects were used for further analysis and interpretation. A complete overview of all p-values of both studies can be found in t Appendix A.

Following the results of the analysis of variance in Table 5, the main effects of the factors are presented in Figure 8. They show the mean values of the dependent variables for each factor by level [35]. For example, the main effects graph at the upper left (processing time, Type A) shows a mean value of 190 s for the active factors "Feedback" (+1) and a mean value of 268 s for the inactive factor (−1). Consequently, there is an effect of the factor "Feedback" on the processing time of 78 s. Similarly, the factor "Assistance system" reduces the processing time by 66 s, while there is no significant effect for the factor "Usability" ($p = 0.61$, see Table 5).

Table 5. Selected Results of the analysis of variance (Study 1, Type A, significant results are marked with a gray background).

Dependent Variable	Usability	Assistance System	Feedback
Display touches	F(1,144) – 0.06, *p* – 0.802	F(1,144) = 0.43, *p* = 0.511	F(1,144) = 3.12, *p* = 0.079
Processing time	F(1,144) = 0.26, *p* = 0.61	F(1,144) = 4.37, *p* = 0.038	F(1,144) = 5.98, *p* = 0.016
Satisfaction	F(1,144) = 6.91, *p* = 0.009	F(1,144) = 0, *p* = 0.979	F(1,144) = 0.55, *p* = 0.459
Quality evaluation	F(1,144) = 3.25, *p* = 0.074	F(1,144) = 0.11, *p* = 0.735	F(1,144) = 0.75, *p* = 0.389

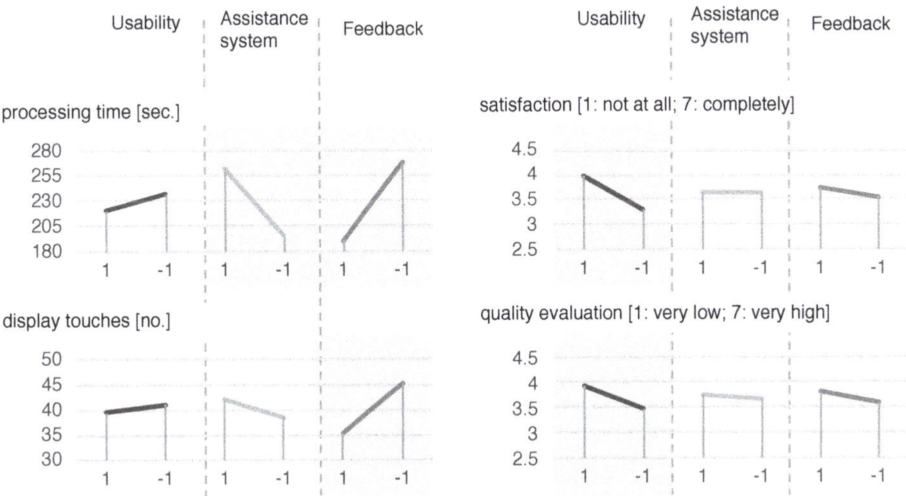

Figure 8. Selected main effects (Study 1, Type A); "+1" indicates an active design element, while "−1" indicates an inactive design element; significant results are marked with a gray background.

Furthermore, for the selected factors and dependent variables in Table 5 and Figure 8, further significant effects could be identified. There is an effect of the factor "Feedback" on the number of display touches of about 10 units. "Usability" led to an increase in the evaluation of the satisfaction with the user's own, final solution of approximately 0.7 steps on the seven-step rating scale. A similar effect was observed for the evaluation of the quality of the solution (approximately 0.5 steps). The size of these effects can be expressed as regression equations in addition to the main effect graphs. In the following, the equations are shown, again corresponding to the selected cases in Table 5 and Figure 8. Here, y_d refers to the number of display touches, y_p to the processing time, y_s to the satisfaction with the solution [1: very low; 7: very high], y_q to the quality evaluation of the study [1: very low; 7: very high], x_u to "Usability", x_a to "Assistance system", and x_f to "Feedback". Each consists of a constant baseline to which the respective effects of the independent factors are added or subtracted. For example, the combination of an active "Usability" (+1), an active "Assistance system" (+1) and an inactive "Feedback" (−1) would result in an expected number of 46.3 display touches ($= 40.25 - 0.70 \cdot (+1) + 1.82 \cdot (+1) - 4.93 \cdot (-1)$). A complete overview of all main effect graphs can be found in Appendix A of this paper.

$$y_d = 40.25 - 0.7 \cdot x_u + 1.82 \cdot x_a - 4.93 \cdot x_f$$

$$y_p = 229.3 - 8.1 \cdot x_u + 33.0 \cdot x_a - 39.0 \cdot x_f$$

$$y_s = 3.648 + 0.341 \cdot x_u + 0.003 \cdot x_a + 0.097 \cdot x_f$$

$$y_q = 3.713 + 0.225 \cdot x_u + 0.042 \cdot x_a + 0.108 \cdot x_f$$

In summary, all significant effects (with $p < 0.15$) of both studies for task Types A and B are presented in Figure 9. It can be seen that the design elements "Usability" and "Feedback" in particular induce several effects. For "Usability", nine significant observations could be made in the four areas examined (Study 1, Type A; Study 1, Type B; Study 2, Type A; and Study 2, Type B). It led to a reduction of the processing time, to a reduction of the number of display touches, to higher quality evaluation, a lower perceived difficulty and to an overestimation of the quality of the own solution compared to the actual solution quality (deviation of solution quality < 0).

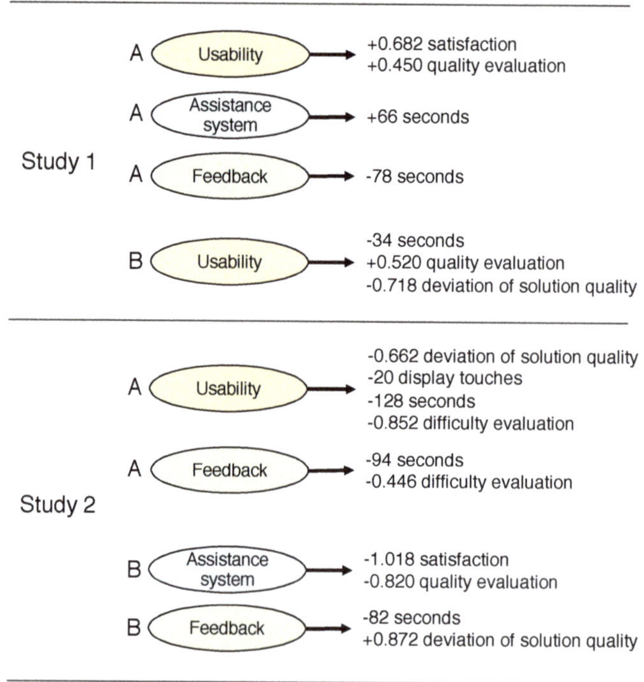

Figure 9. Summary of significant effects.

"Feedback" led to five significant observations consisting of a shorter task processing time and perceived task difficulty. The deviation in solution quality was positive at 0.872; consequently, the actual solution quality was underestimated. For the design element "Assistance system", three significant observations were made. Initially, an increase in the time required to solve the work tasks was observed. Additionally, the evaluation of satisfaction with the work result decreased, and the quality of the own solution was rated lower.

3.5. Phase 4: Development of Design Principles for Interfaces in CPPS

In the fourth phase of the method, the effect sizes and directions determined were transferred into design rules. Based on the effects summarized in Figure 9, the design rules were derived according to the scheme shown in Section 3.2.4.

In Table 6, the first step of this procedure is shown: the corresponding effect indicators were determined for all dependent variables for which significant effects were found. The existing scheme for deriving the effect indicators could be used for the key figures display touches, processing time, satisfaction, and difficulty rating, since these variables are absolute or rating figures. Here, the effect strengths E^+ and E^+ were calculated by dividing the main effects by the baseline values of the regression equations and then reading the corresponding effect indicators in Table 4. For the variable deviation of solution quality, however, a different scheme was used to determine the effect indicators. This variable is a relating figure where the possible change is a change in the distance between the related variables. A distance as small as possible represents the desired direction of change. Table 7 shows the schema we used here.

Table 6. Implementation: Effect sizes of the design elements.

	Work Performance			Perception of Work			
	Display Touches	Processing Time	Σ	Satisfaction	Evaluation of Difficulty	Deviation of Solution Quality	Σ
Study 1, Type A							
Assistance system	$--$	$--$					
Feedback	++	++					
Usability				+			+
Study 1, Type B							
Assistance system							
Feedback							
Usability	+	+				++	++
Study 2, Type A							
Assistance system							
Feedback		+	+		$-$		$-$
Usability	++	++	++++		$-$	++	+
Study 2, Type B							
Assistance system				$--$			$--$
Feedback		++	++			++	++
Usability							

Table 7. Caption on the formation of effect indicators for solution quality deviation.

Effect Size E	Effect Indicator
$E > 3.00$	$---$
$2.50 < E \leq 3.00$	$--$
$2.00 < E \leq 2.50$	$-$
$1.50 < E \leq 2.00$	0
$1.00 < E \leq 1.50$	+
$0.50 < E \leq 1.00$	++
$-0.50 < E \leq 0.50$	+++
$-1.00 < E \leq -0.50$	++
$-1.50 < E \leq -1.00$	+
$-2.00 < E \leq -1.50$	0
$-2.50 < E \leq -2.00$	$-$
$-3.00 < E \leq -2.50$	$--$
$E \leq 3.00$	$---$

Due to the mentioned differences between Study 1 and Study 2 as well as the task Types A and B, separate analyses were again carried out. By transferring the aggregated effect indicators, we can now compare the effects on work performance and perception of work in Table 8. This results in five "Yes" rules, in which either both aspects show positive effect indicators or only one area shows a positive effect indicator. The "Yes" rules here refer to the design elements "Usability" and "Feedback". For work tasks with Type A, there is an additional "Maybe" rule for "Feedback", since work performance (+) and perception of work ($-$) lead to a conflicting situation. Finally, two "No" rules result for "Assistance system"; here, analogous to the "Yes" rule, either one or two negative effect indicators occurred.

Table 8. Implementation: Determination scheme of design principles, Yes-Rules: green background, No-Rules: red background, and Maybe-Rules: yellow background.

	$\sum$ Work Performance	$\sum$ Perception of Work	Yes-Rule	No-Rule	Maybe-Rule
Study 1, Type A					
Assistance system	$--$			x	
Feedback	$++$		x		
Usability		$+$	x		
Study 1, Type B					
Assistance system					
Feedback					
Usability	$+$	$++$	x		
Study 2, Type A					
Assistance system					
Feedback	$+$	$-$			x
Usability	$++++$	$+$	x		
Study 2, Type B					
Assistance system		$--$		x	
Feedback	$++$	$++$	x		
Usability					

4. Discussion and Conclusions

Numerous approaches to human-oriented design are available from many disciplines associated with human factors [25]. In the present work, a linking of human-oriented design approaches to cyber-physical production systems enables us to focus on the cognitive elements which are essential for future work systems. The presented overview of work design elements, which we expect to be crucial for work design in CPPS, in Section 3.1 sets a starting point on the way to the development of work design principles for CPPS work systems, respectively. However, although the overview covers essential areas, it cannot be considered exhaustive. The selection, which has been made from the design elements (i.e., "Feedback", "Assistance system" and "Usability"), could be successfully implemented and converted into (first) work design principles using the method. Besides, we assume that, for further design elements of the presented overview, corresponding measures are conceivable as well.

The developed method provides a step-by-step procedure that keeps the applicability threshold for investigators low by subdividing the necessary steps into four phases. Further, necessary resources for the application of the method are available, since they were jointly developed within the scope of this research (i.e., starting points for work design elements in CPPS, the experimental platform). The method could meet the research goal and led to the desired quantification of work performance and perception of work (see also the following part regarding the application results).

The experimental platform was well received by the study participants and could thus contribute to the later acceptance of the human–machine interface to be developed. It led to a higher level of attention than would have been attained for a web application or similar. Besides, the modular design allows a variety of adaptations (e.g., use of other interfaces with multi-modal interaction, acoustic signals, classic input devices or conventional buttons). Thus, it can be seen as an enabler for future human factors studies, which might follow a different research goal and require a different setup, respectively. In addition, the platform is robust and could be used successfully and without issues across several studies.

With regard to the outcome of the two human factors studies conducted, significant effects of all three design elements investigated could be determined. We interpret the results in such a way that the participants received valuable information via "Feedback", which resulted in a quicker and more efficient way of finding a solution to the given tasks. The studies led to three significant observations of a faster task processing (Study 1, Type A: -78 s; Study 2, Type A: -94 s; and Study 2, Type B: -82 s) (see Figure 9). Since this effect occurred in both studies and both types of tasks, we assume, that "Feedback" should play a key role in CPPS work design. Besides, in terms of the effect indicators, two ++ and one + ratings support this conclusion (see Table 6). We also conclude that the differentiation by the color of the work orders ("Usability") supported the perception of information and thus helped to place the work orders in a meaningful way. Here, several significant effects could be observed: a higher evaluation of satisfaction (Study 1, Type A: +0.682), a shorter task processing (Study 1, Type B: −34 s; and Study 2, Type A: −128 s), and fewer display touches (Study 2, Type A). These effects have predominantly occurred more intensively in Study 2. We attribute this to the more complex work tasks given in this study. Again, regarding the effect indicators, two ++ and two + ratings support this finding (see Table 6). Finally, "Assistance system" led to two negative observations: first, it extended the processing time (Study 1, Type A: +66 s) and, second, the satisfaction evaluation was lower (Study 2, Type B: −1.018). This resulted in two −− ratings in terms of the effect indicators, respectively. Here, we assume that the intended idea of the worker assistance (Study 1: display of error details, Study 2: visual operating support) did not meet the user's requirements. A possible explanation is that the measures led to confusion.

Other observed significant effects for all design elements, as shown in Figure 9, do not point into a clearly positive or negative direction ("Usability", Study 1, Type A: +0.450 quality evaluation; Study 1, Type B: +0.520 quality evaluation; −0.718 deviation of solution quality; Study 2, Type A: −0.662 deviation of solution quality; −0.852 difficulty evaluation; "Feedback", Study 2, Type A: −0.446 difficulty evaluation; "Assistance system", Study 2, Type B: −0.820 quality evaluation; and "Feedback", Study 2, Type B: +0.872 deviation of solution quality). However, they can provide valuable information in combination with other effect observations. We consider the combined analysis of key performance indicators for work performance and work perception to be meaningful and necessary, as significant results could often only be obtained for a part of the dependent variables examined. Consequently, a consideration of a limited number of parameters would possibly lead to an inconsiderate rejection or confirmation of particular design elements. If, for example, a significant result was achieved as part of a work performance measurement, an equivalent effect on the perception of work cannot be assumed necessarily as well. Following this idea, we compared the effect indicators for work performance and work perception to derive the final design principles (see Table 8). Summarizing, we recommend the implementation of the design element "Usability". For "Feedback", we recommend a further examination in light of the "Maybe" rule. An implementation of "Assistance System" is not recommended.

These findings need to be considered in the context of the two studies with regard to the representative work tasks and the chosen measures and the limited number and variety of study participants. Thus, we do not claim general applicability to all CPPS work design use cases. Nevertheless, the results indicate effect sizes and directions of work design elements which can serve as starting points for future investigations.

As possible next research steps, we see the widening of the scope for deriving design elements through a focus extension. As mentioned previously, we do not expect the overview to be complete. Here, a broad review of other human factors related areas could reveal further design elements. Besides, we consider a variation of the method to enable a more practical research as very promising. This could be carried out such that the method does not start with a representative work task but with an industrial use case, which then will be transferred into a use case-specific representative task. Since human factors research is often used not only for basis research but also for applied research (e.g., to make design decisions out of several alternatives [45]), this variation could serve as a decision-making tool for SME practitioners. Additionally, a practical evaluation would provide information on a further development of the method as well. Here, experiences and insights could be used in terms of increasing the relevance of the results with regard to basic research and increasing the accessibility of the method with regard to applied research.

Further, collaborative processing of work tasks in the sense of human–machine-networks could open up promising new research possibilities. Here, the establishment of a communication link between the platforms (by using an already available XBee module, see Figure 4) as a basis for the implementation of collaborative tasks could lead to an investigation of design elements in a collaborative environment.

Finally, we see the conduction of further studies as another possible next research step. By using a different set of design elements according to Section 3.1, insights on other design elements could be achieved as well, which were not part of the presented studies. In addition, the use of other measures and representative tasks for the design elements already used in the presented studies could increase the generalizability of the results.

Overall, a set of results could be presented (work design elements, method, experimental platform, and first application results) which prove the functional capability of the procedure and expand the knowledge about human-oriented work design in CPPS. As presented in Section 3.2, the method aims to support the understanding of the mechanisms between human factors and work in CPPS and the design of interfaces between humans and machines. It can assess and evaluate specific work design elements to provide first design principles regarding the interface design in CPPS. We consider this as a valuable step for manufacturing companies to improve the integration process of CPPS ideas. This way, already at an early stage of the design and implementation process, unnecessary or even counterproductive work design measures can be minimized, and beneficial ones maximized instead at an early stage of the design and implementation process. This way, both social and economic sustainability of CPPS can be promoted.

Author Contributions: Conceptualization, H.S. and T.B.; methodology, H.S.; software, H.S.; validation, H.S. and T.B.; formal analysis, H.S. and T.B.; investigation, H.S.; resources, T.B.; data curation, H.S.; writing—original draft preparation, H.S.; writing—review and editing, T.B.; visualization, H.S.; supervision, T.B.; project administration, H.S. and T.B.; and funding acquisition, T.B.

Funding: This work was supported by the Institutional Strategy of the University of Bremen, funded by the German Excellence Initiative.

Conflicts of Interest: The authors declare no conflict of interest.

Abbreviations

The following abbreviations are used in this manuscript:

CPS	Cyber-physical system
CPPS	Cyber-physical production system
E	Effect size
y_d	Number of display touches
y_p	Processing time
y_s	Satisfaction with the solution
y_q	Quality evaluation of the study
x_u	Work design element "Usability"
x_a	Work design element "Assistance system"
x_f	Work design element "Feedback"

Appendix A

Appendix A.1. Main Effects Graphs

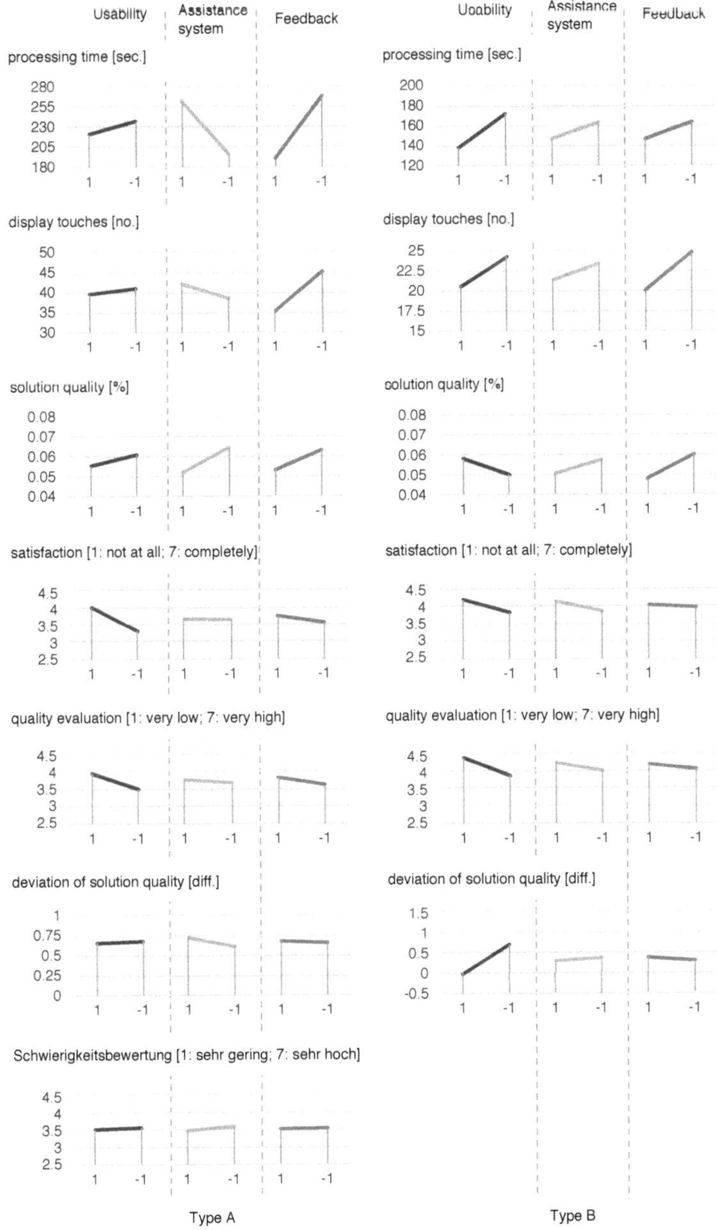

Figure A1. Main effects of Study 1; "+1" indicates an active design element, while "−1" indicates an inactive design element.

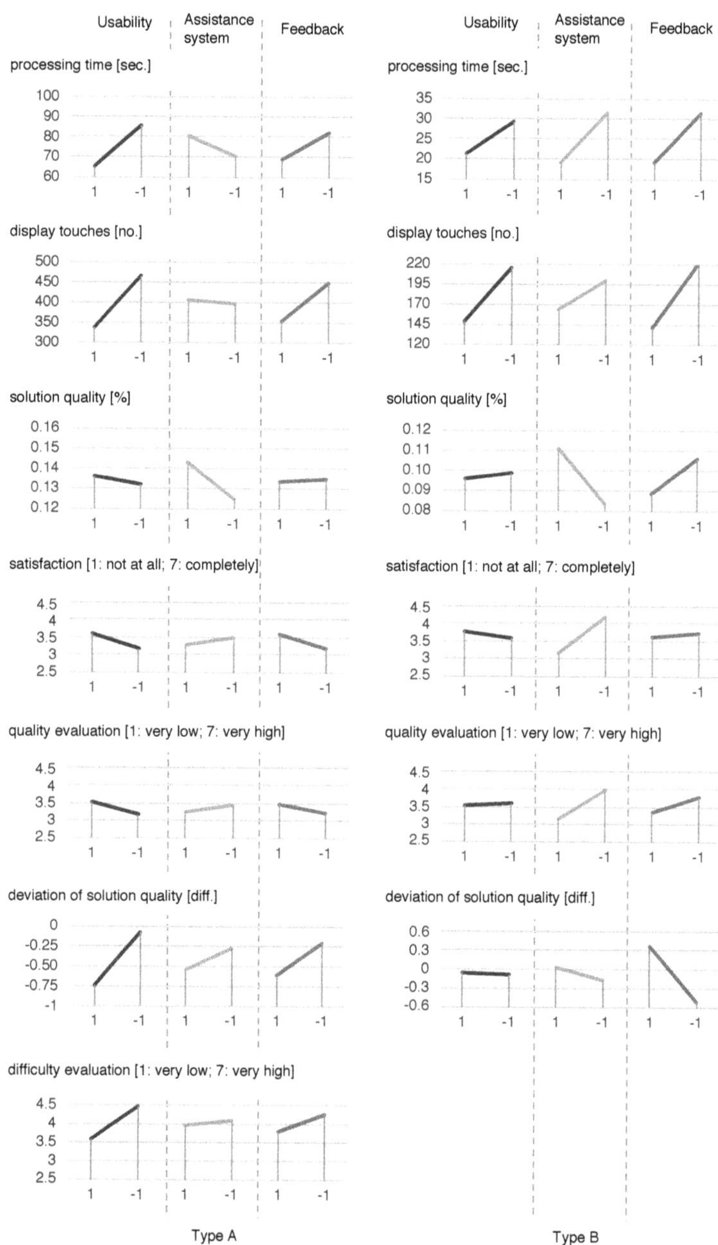

Figure A2. Main effects of Study 2; "+1" indicates an active design element, while "−1" indicates an inactive design element.

Appendix A.2. Results of the Analysis of Variance

Table A1. Results of the Analysis of Variance (Study 1, Type A).

Dependent Variable	Usability	Assistance System	Feedback
display touches	$F(1,144) = 0.06$ $p = 0.802$	$F(1,144) = 0.43$ $p = 0.511$	$F(1,144) = 2.12$ $p = 0.079$
processing time	$F(1,144) = 0.26$ $p = 0.61$	$F(1,144) = 4.37$ $p = 0.038$	$F(1,144) = 5.98$ $p = 0.016$
solution quality	$F(1,144) = 0.11$ $p = 0.737$	$F(1,144) = 0.6$ $p = 0.441$	$F(1,144) = 0.36$ $p = 0.551$
difficulty evaluation	$F(1,144) = 0.06$ $p = 0.808$	$F(1,144) = 0.29$ $p = 0.591$	$F(1,144) = 0.01$ $p = 0.938$
satisfaction	$F(1,144) = 6.91$ $p = 0.009$	$F(1,144) = 0$ $p = 0.979$	$F(1,144) = 0.55$ $p = 0.459$
quality evaluation	$F(1,144) = 2.25$ $p = 0.074$	$F(1,144) = 0.11$ $p = 0.735$	$F(1,144) = 0.75$ $p = 0.389$
deviation of solution quality	$F(1,144) = 0.00$ $p = 0.952$	$F(1,144) = 0.11$ $p = 0.745$	$F(1,144) = 0.00$ $p = 0.953$

Table A2. Results of the Analysis of Variance (Study 1, Type B).

Dependent Variable	Usability	Assistance System	Feedback
display touches	$F(1,132) = 0.76$ $p = 0.384$	$F(1,132) = 0.24$ $p = 0.627$	$F(1,132) = 1.2$ $p = 0.276$
processing time	$F(1,132) = 2.61$ $p = 0.109$	$F(1,132) = 0.64$ $p = 0.426$	$F(1,132) = 0.66$ $p = 0.418$
solution quality	$F(1,132) = 0.33$ $p = 0.569$	$F(1,132) = 0.23$ $p = 0.629$	$F(1,132) = 0.73$ $p = 0.395$
satisfaction	$F(1,132) = 1.64$ $p = 0.202$	$F(1,132) = 1.06$ $p = 0.306$	$F(1,132) = 0.04$ $p = 0.841$
quality evaluation	$F(1,132) = 2.98$ $p = 0.048$	$F(1,132) = 0.83$ $p = 0.365$	$F(1,132) = 0.26$ $p = 0.609$
deviation of solution quality	$F(1,132) = 2.60$ $p = 0.060$	$F(1,132) = 0.04$ $p = 0.851$	$F(1,132) = 0.05$ $p = 0.830$

Table A3. Results of the Analysis of Variance (Study 2, Type A).

Dependent Variable	Usability	Assistance System	Feedback
display touches	$F(1,103) = 4.63$ $p = 0.034$	$F(1,103) = 1.18$ $p = 0.28$	$F(1,103) = 2.01$ $p = 0.159$
processing time	$F(1,103) = 9.22$ $p = 0.003$	$F(1,103) = 0.05$ $p = 0.826$	$F(1,103) = 5.16$ $p = 0.025$
solution quality	$F(1,103) = 0.03$ $p = 0.867$	$F(1,103) = 0.58$ $p = 0.45$	$F(1,103) = 0$ $p = 0.956$
difficulty evaluation	$F(1,103) = 12.44$ $p = 0.001$	$F(1,103) = 0.26$ $p = 0.61$	$F(1,103) = 2.46$ $p = 0.066$
satisfaction	$F(1,103) = 1.46$ $p = 0.23$	$F(1,103) = 0.33$ $p = 0.569$	$F(1,103) = 1,3$ $p = 0.256$
quality evaluation	$F(1,103) = 1.09$ $p = 0.299$	$F(1,103) = 0.33$ $p = 0.564$	$F(1,103) = 0.57$ $p = 0.45$
deviation of solution quality	$F(1,103) = 2.56$ $p = 0.113$	$F(1,103) = 0.43$ $p = 0.512$	$F(1,103) = 1.00$ $p = 0.320$

Table A4. Results of the Analysis of Variance (Study 2, Type B).

Dependent Variable	Usability	Assistance System	Feedback
display touches	$F(1,63) = 0.67$ $p = 0.415$	$F(1,63) = 1.66$ $p = 0.203$	$F(1,63) = 1.52$ $p = 0.223$
processing time	$F(1,63) = 1.77$ $p = 0.188$	$F(1,63) = 0.51$ $p = 0.478$	$F(1,63) = 2.53$ $p = 0.116$
solution quality	$F(1,63) = 0.01$ $p = 0.915$	$F(1,63) = 1.08$ $p = 0.302$	$F(1,63) = 0.42$ $p = 0.518$
satisfaction	$F(1,63) = 0.2$ $p = 0.66$	$F(1,63) = 6.59$ $p = 0.013$	$F(1,63) = 0.07$ $p = 0.795$
quality evaluation	$F(1,63) = 0.03$ $p = 0.874$	$F(1,63) = 4.37$ $p = 0.041$	$F(1,63) = 1.14$ $p = 0.291$
deviation of solution quality	$F(1,63) = 0.00$ $p = 0.963$	$F(1,63) = 0.12$ $p = 0.729$	$F(1,63) = 2.18$ $p = 0.145$

References

1. Huber, W. *Industrie 4.0 in der Automobilproduktion*; Springer Fachmedien Wiesbaden: Wiesbaden, Germany, 2016.
2. Kagermann, H.; Wahlster, W.; Helbig, J. *Umsetzungsempfehlungen für das Zukunftsprojekt Industrie 4.0*; Forschungsunion Wirtschaft und Wissenschaft: Berlin, Germany, 2013.
3. Köhler-Schute, C. *Industrie 4.0: Ein praxisorientierter Ansatz*; KS-Energy Verlag: Berlin, Germany, 2015.
4. Baheti, R.; Gill, H. Cyber-physical Systems. *Impact Control Technol.* **2011**, *12*, 161–166. [CrossRef]
5. Andelfinger, V.; Hänisch, T. *Industrie 4.0 Wie Cyber-Physische Systeme die Arbeitswelt Verändern*; Springer: Berlin, Germany, 2017.
6. Mertens, P.; Barbian, D.; Baier, S. *Digitalisierung und Industrie 4.0 – eine Relativierung*; Springer Fachmedien Wiesbaden: Wiesbaden, Germany, 2017.
7. VDI/VDE-Gesellschaft. *Cyber-Physical Systems: Chancen und Nutzen aus Sicht der Automation*; Verein Deutscher Ingenieure e.V.: Düsseldorf, Germany, 2013.
8. Geisberger, E.; Broy, M. (Eds.) *AgendaCPS. Acatech Studie*; Springer: Berlin/Heidelberg, Germany, 2012; Volume 1, pp. 1–297. [CrossRef]
9. Spath, D.; Ganschar, O.; Gerlach, S.; Hämmerle, M.; Krause, T.; Schlund, S. *Produktionsarbeit der Zukunft-Industrie 4.0*; Fraunhofer Verlag: Stuttgart, Germany, 2013.
10. Hirsch-Kreinsen, H. Welche Auswirkungen hat "Industrie 4.0" auf die Arbeitswelt? *WISO Direkt* **2014**, *12*, 1–4.
11. Bainbridge, L. Ironies of automation. *Automatica* **1983**, *19*, 775–779. [CrossRef]
12. Vogler-Ludwig, K.; Düll, N. *Arbeitsmarkt 2030*; Bertelsmann Verlag: Gütersloh, Germany, 2013.
13. Prognos AG. *Arbeitslandschaft 2035*; Prognos AG: München, Germany, 2012.
14. Brossardt, B. *Arbeitslandschaft 2040 [Bericht]*; Prognos AG: München, Germany, 2015.
15. Wegge, J.; Wendsche, J.; Diestel, S. Arbeitsgestaltung. In *Lehrbuch Organisationspsychologie*, Schuler, H., Moser, K., Eds.; Huber: Bern, Switzerland, 2014; pp. 643–695.
16. Hackman, J.; Oldham, G.R. Motivation through the design of work: test of a theory. *Organ. Behav. Hum. Perform.* **1976**, *16*, 250–279. [CrossRef]
17. Luczak, H. *Arbeitswissenschaft: Konzepte, Arbeitspersonen, Arbeitsformen, Arbeitsumgebung*; Technische Universität Berlin: Berlin, Germany, 1991.
18. Rohmert, W. Aufgaben und Inhalt der Arbeitswissenschaft. *Die Berufsbild. Sch.* **1972**, *24*, 3–14.
19. Ulich, E. *Arbeitspsychologie*, 7th ed.; vdf, Hochschulverlag AG an der ETH Zürich: Zürich, Switzerland, 2011.
20. DIN6385. *Ergonomics Principles in the Design of Work Systems (ISO 6385:2016); German version EN ISO 6385:2016*; Beuth, Berlin, Germany, 2016.
21. DIN894. *Safety of Machinery—Ergonomics Requirements for the Design of Displays and Control Actuators—Part 1: General Principles for Human Interactions with Displays and Control Actuators; German Version EN 894-1:1997+A1:2008*; Beuth: Berlin, Germany, 2008.

22. DIN9241. *Ergonomics of Human-System Interaction—Part 392: Ergonomic Recommendations for the Reduction of Visual Fatigue From Stereoscopic Images (ISO 9241-392:2015); German Version EN ISO 9241-392:2017*; Beuth: Berlin, Germany, 2017.

23. Butz, A.; Krüger, A. *Mensch-Maschine-Interaktion*; De Gruyter Studium, De Gruyter: Berlin, Germany, 2017.

24. Peissner, M.; Hipp, C. *Potenziale Der Mensch-Technik-Interaktion für Die Effiziente Und Vernetzte Produktion Von Morgen*; Fraunhofer-Verl.: Stuttgart, Germany, 2013.

25. Badke-Schaub, P.; Hofinger, G.; Lauche, K. *Human Factors*; Springer: Berlin/Heidelberg, Germany, 2012.

26. Salvendy, G. *Handbook of Human Factors and Ergonomics*, 4th ed.; John Wiley & Sons, Inc.: Hoboken, NJ, USA, 2012.

27. Hollnagel, E. Handbook of Cognitive Task Design. In *Human factors and Ergonomics*; CRC Press: Boca Raton, FL, USA, 2010.

28. Hacker, W.; von der Werth, R. Denken-Entscheiden-Handeln. In *Human Factors*; Badke-Schaub, P., Hofinger, G., Lauche, K., Eds.; Springer: Berlin/Heidelberg, Germany, 2012.

29. Manzey, D. Systemgestaltung und Automatisierung. In *Human Factors*; Badke-Schaub, P., Hofinger, G., Lauche, K., Eds.; Springer: Berlin/Heidelberg, Germany, 2012.

30. Sheridan, T.B. Human Supervisory Control. In *Handbook of Human Factors and Ergonomics*, 4th ed.; Salvendy, G., Ed.; John Wiley & Sons, Inc.: Hoboken, NJ, USA, 2012.

31. Endsley, M.R. Toward a Theory of Situation Awareness in Dynamic Systems. *Human Factors*; Routledge: Abingdon, UK, 1995.

32. Jacko, J.A.; Yi, J.S.; Sainfort, F.; McClellan, M. Human Factors and Ergonomic Methods. In *Handbook of Human Factors and Ergonomics*; CRC Press: Boca Raton, FL, USA, 2012; pp. 289–329.

33. Wickens, C.D.; Gordon, S.E.; Liu, Y. *An Introduction to Human Factors Engineering*; Pearson: London, UK, 2014.

34. Miller, D.C.; Salkind, N.J. *Handbook of Research Design and Social Measurement*; SAGE Publications: Thousand Oaks, CA, USA, 2002.

35. Montgomery, D.C. *Design and Analysis of Experiments*, 7th ed.; Wiley: Hoboken, NJ, USA, 2009.

36. Fahrmeir, L.; Heumann, C.; Künstler, R.; Pigeot, I.; Tutz, G. *Statistik: Der Weg zur Datenanalyse*; Springer: Berlin/Heidelberg, Germany, 2016.

37. Edwards, J.R.; Scully, J.A.; Brtek, M.D. The nature and outcomes of work: A replication and extension of interdisciplinary work-design research. *J. Appl. Psychol.* **2000**, *85*, 860. [CrossRef]

38. Schaper, N. Arbeitsgestaltung in Produktion und Verwaltung. In *Arbeits-und Organisationspsychologie*; Nerdinger, F., Blickle, G., Schaper, N., Eds.; Springer: Berlin/Heidelberg, Germany, 2011; pp. 349–367.

39. Eklund, J.A. Relationships between ergonomics and quality in assembly work. *Appl. Ergon.* **1995**, *26*, 15–20. [CrossRef]

40. Zare, M.; Croq, M.; Hossein-Arabi, F.; Brunet, R.; Roquelaure, Y. Does Ergonomics Improve Product Quality and Reduce Costs? A Review Article. *Hum. Factors Ergon. Manuf. Serv. Ind.* **2016**, *26*, 205–223. [CrossRef]

41. Drury, C.G.; Hall, B. Human Factors and Quality: Integration and New Directions. *Hum. Factors Ergon. Manuf. Serv. Ind.* **2000**, *10*, 45–59. [CrossRef]

42. Goggins, R.W.; Spielholz, P.; Nothstein, G.L. Estimating the effectiveness of ergonomics interventions through case studies: Implications for predictive cost-benefit analysis. *J. Saf. Res.* **2008**, *39*, 339–344. [CrossRef] [PubMed]

43. Village, J.; Searcy, C.; Salustri, F.; Patrick Neumann, W. Design for human factors (DfHF): A grounded theory for integrating human factors into production design processes. *Ergonomics* **2015**, *58*, 1529–1546. [CrossRef] [PubMed]

44. Brauner, P.; Runge, S.; Groten, M.; Schuh, G.; Ziefle, M. Human Factors in Supply Chain Management. In Proceedings of the 15th HCI International 2013, Part III, LNCS 8018, Las Vegas, NV, USA, 21–26 July 2013; pp. 423–432._46. [CrossRef]

45. Brauner, P.; Ziefle, M. Human Factors in Production Systems. In *Advances in Production Technology*; Lecture Notes in Production Engineering; Brecher, C., Ed.; Springer International Publishing: Cham, Germany, 2015; Volume 30, pp. 187–199. [CrossRef]

46. Arning, K.; Ziefle, M. Different Perspectives on Technology Acceptance: The Role of Technology Type and Age. In *Lecture Notes in Computer Science (Including Subseries Lecture Notes in Artificial Intelligence and Lecture Notes in Bioinformatics)*; Springer: Berlin/Heidelberg, Germany, 2009; pp. 20–41.

47. Rauch, E.; Linder, C.; Dallasega, P. Anthropocentric perspective of production before and within Industry 4.0. *Comput. Ind. Eng.* **2019**. [CrossRef]
48. Chen, P.; Xia, P.J.; Lang, Y.D.; Yao, Y.X. A Human-Centered Virtual Assembly System. *Appl. Mech. Mater.* **2009**, *16–19*, 796–800. [CrossRef]
49. Yang, X.; Deines, E.; Lauer, C.; Aurich, J.C. A human-centered virtual factory. In Proceedings of the 2011 International Conference on Management Science and Industrial Engineering, MSIE 2011, Harbin, China, 8–11 January 2011; pp. 1138–1142. [CrossRef]
50. Santochi, M.; Failli, F. Sustainable Work for Human Centred Manufacturing. In *Green Design, Materials and Manufacturing Processes*; CRC Press: Boca Raton, FL, USA, 2013; pp. 161–166.
51. Romero, D.; Noran, O.; Stahre, J.; Bernus, P.; Fast-Berglund, Å. Towards a human-centred reference architecture for next generation balanced automation systems: Human-automation symbiosis. In *IFIP Advances in Information and Communication Technology*; Springer: Cham, Germany, 2015; Volume 460, pp. 556–566.
52. Hold, P.; Ranz, F.; Sihn, W.; Hummel, V. Planning Operator Support in Cyber-Physical Assembly Systems. *IFAC-Pap. OnLine* **2016**, *49*, 60–65. [CrossRef]
53. Ohtsuka, H.; Shibasato, K.; Kawaji, S. Experimental study of collaborater in human-machine system. *Mechatronics* **2009**, *19*, 450–456. [CrossRef]
54. Weiss, A.; Buchner, R.; Tscheligi, M.; Fischer, H. Exploring human-robot cooperation possibilities for semiconductor manufacturing. In Proceedings of the 2011 International Conference on Collaboration Technologies and Systems, CTS 2011, Philadelphia, PA, USA, 23–27 May 2011; pp. 173–177. [CrossRef]
55. Soffker, D.; Flesch, G.; Fu, X.; Hasselberg, A.; Langer, M. Assistance and supervisory control of operators in complex human-process-interaction. In Proceedings of the 2011 IEEE Jordan Conference on Applied Electrical Engineering and Computing Technologies, AEECT 2011, Amman, Jordan, 6–8 December 2011; pp. 1–6. [CrossRef]
56. Pacaux-Lemoine, M.P.; Trentesaux, D.; Rey, G.Z. Human-machine cooperation to design Intelligent Manufacturing Systems. In Proceedings of the IECON Proceedings (Industrial Electronics Conference), Florence, Italy, 23–26 October 2016; pp. 5904–5909. [CrossRef]
57. Gorecky, D.; Mura, K.; Arlt, F. A vision on training and knowledge sharing applications in future factories. In *IFAC Proceedings Volumes (IFAC-PapersOnline)*; Elsevier: Amsterdam, The Netherlands, 2013; Volume 12, pp. 90–97.
58. Paelke, V. Augmented reality in the smart factory: Supporting workers in an Industry 4.0. environment. In Proceedings of the 19th IEEE International Conference on Emerging Technologies and Factory Automation, ETFA 2014, Barcelona, Spain, 16–19 September 2014; pp. 1–4. [CrossRef]
59. Rauh, S.; Zsebedits, D.; Tamplon, E.; Bolch, S.; Meixner, G. Using Google Glass for mobile maintenance and calibration tasks in the AUDI A8 production line. In Proceedings of the IEEE International Conference on Emerging Technologies and Factory Automation, ETFA, Luxembourg, 8–11 September 2015; Volume 2015, pp. 1–4. [CrossRef]
60. Jordan, P.W. *An Introduction to Usability*; CRC Press: Boca Raton, FL, USA,1998.
61. Norman, D.A. *Emotional Design: Why We Love (or Hate) Everyday Things*; Basic Books: New York, NY, USA, 2004.
62. Zühlke, D. *Nutzergerechte Entwicklung von Mensch-Maschine-Systemen*; Springer: Berlin/Heidelberg, Germany, 2012; p. 250.
63. Jacob, R.J.K.; Girouard, A.; Hirshfield, L.M.; Horn, M.S.; Shaer, O.; Solovey, E.T.; Zigelbaum, J. Reality-based interaction: a framework for post-WIMP interfaces. In Proceedings of the SIGCHI conference on Human factors in computing systems, Florence, Italy, 5–10 April 2008; ACM: New York, NY, USA, 2008; pp. 201–210.
64. Delisle, S.; Moulin, B. User Interfaces and Help Systems: From Helplessness to Intelligent Assistance. *Artif. Intell. Rev.* **2002**, *18*, 117–157. [CrossRef]
65. Nee, A.; Ong, S.; Chryssolouris, G.; Mourtzis, D. Augmented reality applications in design and manufacturing. *CIRP Ann.* **2012**, *61*, 657–679. [CrossRef]
66. Wu, L.; Zhu, Z.; Cao, H.; Li, B. Influence of information overload on operator's user experience of human–machine interface in LED manufacturing systems. *Cognit. Technol. Work* **2016**, *18*, 161–173. [CrossRef]
67. Seaborn, K.; Fels, D.I. Gamification in theory and action: A survey. *Int. J. Hum. Comput. Stud.* **2015**, *74*, 14–31. [CrossRef]

68. Deterding, S.; Dixon, D.; Khaled, R.; Nacke, L. From Game Design Elements to Gamefulness: Defining "Gamification". In Proceedings of the 15th International Academic MindTrek Conference: Envisioning Future Media Environments, Tampere, Finland, 28–30 September 2011; ACM: New York, NY, USA, 2011; pp. 9–15. [CrossRef]

69. Turk, M. Multimodal interaction: A review. *Pattern Recognit. Lett.* **2014**, *36*, 189–195. [CrossRef]

70. Dombrowski, U.; Quack, S. Erfolgreiche Restrukturierungsprojekte durch Mitarbeiterpartizipation. *ZWF Z. Wirtsch. Fabr.* **2007**, *102*, 568–571. [CrossRef]

71. Kesting, P.; Parm Ulhøi, J. Employee-driven innovation: extending the license to foster innovation. *Manag. Decis.* **2010**, *48*, 65–84. [CrossRef]

72. Christoffersen, K.; Woods, D.D. How to Make Automated Systems Team Players. In *Advances in Human Performance and Cognitive Engineering Research*; Emerald Group Publishing Limited: Bingley, UK, 2002; pp. 1–12.

73. Wood, R.E. Task complexity: Definition of the construct. *Organ. Behav. Hum. Decis. Process.* **1986**, *37*, 60–82. [CrossRef]

74. Schlick, C.M.; Bruder, R.; Luczak, H. *Arbeitswissenschaft*; Springer: Berlin/Heidelberg, Germany, 2010.

75. Demirel, H.O.; Duffy, V.G. *Applications of Digital Human Modeling in Industry BT—Digital Human Modeling*; Springer: Berlin/Heidelberg, Germany, 2007; pp. 824–832.

76. Richter, M.; Flückiger, M.D. *Usability Engineering Kompakt*; IT Kompakt; Springer: Berlin/Heidelberg, Germany, 2013.

77. Sommerville, I. *Software Engineering*, 9th ed.; IT-Informatik; Pearson: München, Germany, 2012.

78. Sanders, M.S.; McCormick, E.J. *Human Factors in Engineering and Design*; McGRAW-HILL Book Company: New York, NY, USA, 1987.

79. Friedrichs, J. *Methoden Empirischer Sozialforschung*; VS Verlag für Sozialwissenschaften: Wiesbaden, Germany, 1990.

80. Raab-Steiner, E.; Benesch, M. *Der Fragebogen: Von Der Forschungsidee Zur SPSS-Auswertung*, 4th ed.; UTB 8607, Schlüsselkompetenzen; Facultas-Verl.: Wien, Austria, 2015.

81. Stern, H.; Becker, T. Development of a Model for the Integration of Human Factors in Cyber-physical Production Systems. *Procedia Manuf.* **2017**, *9*, 151–158. [CrossRef]

82. Hao, X.; Haraguchi, H.; Dong, Y. An experimental study of human factors' impact in cellular manufacturing and production line system. *Information* **2013**, *16*, 4509–4526.

83. Vitense, H.S.; Jacko, J.A.; Emery, V.K. Multimodal feedback: An assessment of performance and mental workload. *Ergonomics* **2003**, *46*, 68–87. [CrossRef] [PubMed]

84. Brolin, A.; Thorvald, P.; Case, K. Experimental study of cognitive aspects affecting human performance in manual assembly. *Prod. Manuf. Res.* **2017**, *5*, 141–163. [CrossRef]

Article

A Similarity-Based Hierarchical Clustering Method for Manufacturing Process Models

Hyun Ahn [1] and Tai-Woo Chang [2,*]

[1] Division of Computer Science and Engineering, Kyonggi University, Suwon, Gyeonggi 16227, Korea; hahn@kgu.ac.kr
[2] Department of Industrial and Management Engineering/Intelligence and Manufacturing Research Center, Kyonggi University, Suwon, Gyeonggi 16227, Korea
*   Correspondence: keenbee@kgu.ac.kr; Tel.: +82-31-249-9754

Received: 7 March 2019; Accepted: 1 May 2019; Published: 3 May 2019

Abstract: As the adoption of information technologies increases in the manufacturing industry, manufacturing companies should efficiently manage their data and manufacturing processes in order to enhance their manufacturing competency. Because smart factories acquire processing data from connected machines, the business process management (BPM) approach can enrich the capability of manufacturing operations management. Manufacturing companies could benefit from the well-defined methodologies and process-centric engineering practices of this BPM approach for optimizing their manufacturing processes. Based on the approach, this paper proposes a similarity-based hierarchical clustering method for manufacturing processes. To this end, first we describe process modeling based on the BPM-compliant standard so that the manufacturing processes can be controlled by BPM systems. Second, we present similarity measures for manufacturing process models that serve as a criterion for the hierarchical clustering. Then, we formulate the hierarchical clustering problem and describe an agglomerative clustering algorithm using the measured similarities. Our contribution is considered on the assumption that a manufacturing company adopts the BPM approach and it operates various manufacturing processes. We expect that our method enables manufacturing companies to design and manage a vast amount of manufacturing processes at a coarser level, and it also can be applied to various process (re)engineering problems.

Keywords: manufacturing process model; business process management; hierarchical clustering; similarity; BPMN

---

1. Introduction

Thanks to the evolution of manufacturing systems, manufacturing companies can efficiently plan, design, and produce their products. However, as the adoption of the internet of things (IoT) and manufacturing information systems increases, the complexity and interactivity of communications within information systems is expected to expand. Furthermore, there may be large volumes of data stored in a database, because IoT devices constantly transmit data of identified objects [1]. Because of these circumstances, it is still challenging for manufacturing companies to manage a vast array of their manufacturing data and processes. Although information and communication technologies have been introduced to promote technical support for manufacturing operations management (MOM), a more comprehensive methodology should be adopted to fully support manufacturing-process-centric management activities.

The business process management (BPM) approach can be a solution to tackle this hurdle by continually improving processes through automation and optimization. Manufacturing companies could also benefit from the well-defined methodologies and process-centric engineering practices of the BPM approach for optimizing their manufacturing processes. In particular, various analysis techniques which come from the BPM research field may be fruitful for efficiently managing a large set of manufacturing processes [2].

Based on the BPM approach, we propose a similarity-based hierarchical clustering method for manufacturing process models to facilitate managerial activities of manufacturing processes at a group level (e.g., the design and engineering tasks). To this end, modeling of manufacturing processes and enabling the measuring similarities must be performed beforehand. Therefore, this paper exploits the business process model and notation (BPMN) standard and its extension for modeling of manufacturing processes. The novel manufacturing process of modeling with the adaptation of the BPMN helps human understanding by a visual diagram and machine understanding is helped by a converted textual form that is executable in the BPM system. Measuring similarities, in this study, is performed by calculating the operation similarity and structural similarity between the models of manufacturing processes, respectively. The hierarchical clustering method this paper proposes operates based on the measured similarities between the models. We note that the contributions of this paper assume that a manufacturing company adopts the BPM approach and it operates a variety of manufacturing process models on a scale that is difficult to manage manually.

The remainder of this paper is organized as follows. In Section 2, we briefly introduce previous works related to this work. Next, in Section 3, we describe the adaptation of the BPM approach to the MOM area and its primary activities derived from this combined environment. Section 4 includes details of BPMN-based manufacturing process modeling. As the main work of this paper, Section 5 presents a similarity method for BPMN-based manufacturing process models, and Section 6 describes a hierarchical clustering phase of manufacturing process models based on the similarities we propose. Finally, this work will be concluded with future research in Section 7.

2. Related Works

The main elements of the method we propose in this paper are the similarity measures and clustering based on such similarity. We, therefore, organize the content of this section into two parts: process model similarity and clustering. Due to the lack of studies addressing similarity for manufacturing processes, a series of studies that propose similarity measures for business process models are mainly introduced and summarized in the similarity part. On the other hand, the clustering part includes related works applying clustering methods to the field of production research, and a comparison between them and this study.

Many similarity measures have been proposed in the field of BPM to handle a large collection of process models. Jung et al. [3] presented a similarity measure which is based on the execution probability and structural features of business process models. Dijkman et al. [4] investigated the similarity problem of process models centered on the structural aspect and proposed graph-matching algorithms for measuring similarities. They considered business process models as graphs and defined the structural feature between graphs, namely the graph-edit distance, as the criterion of similarity. Many different similarity concepts have been proposed from other studies, and a few comprehensive studies suggested that similarity measures between business process models can support the search and reuse of similar process models [5–7]. Despite a rich set of previous studies contributing to the process similarity research, there is still a lack of proper methods for manufacturing processes since the BPM approach has not been actively discussed in manufacturing industry. As slightly different applications, in [8], the similarity concept was applied to the problems of machine groupings for the design of manufacturing systems. Compared to our similarity, this study focused on relations between machines and components and therefore such measures are not process-centric.

On the other hand, clustering methods have been applied in manufacturing to address various manufacturing engineering problems including the machine grouping [9,10], CAD model grouping [11], product variants management [12], and product release planning [13]. The machine grouping problem has traditionally been the subject of manufacturing systems, especially cellular manufacturing. Dimopoulos and Mort [9] presented the machine-grouping technique based on the combination of genetic programming and hierarchical clustering. Likewise, Park and Suresh [10] addressed the part-machine grouping to identify families of parts with similar routing sequences. The hierarchical clustering method, considered as a solution of the grouping problem, is supported by using the Fuzzy ART neural network. Their experiments showed the compelling performance of hierarchical clustering for large data sets. These two studies provided prerequisites for the implementation of cellular manufacturing. As a different kind of grouping problem, Li and Xie [11] proposed a module partition approach with the hierarchical clustering of components to group CAD models into modules based on component dependencies. In addition, hierarchical clustering methods have also been applied to engineering applications, such as product variety management [12] and the optimization problem of deciding the contents of product releases [13].

As described above, hierarchical clustering methods have been applied for grouping elements that are part of a manufacturing process—such as machines, components, and CAD models—or to engineering applications that are irrelevant to the process grouping. This paper tries to combine the BPM approach to the field of MOM and proposes a hierarchical clustering method based on the similarity for grouping manufacturing process models, eventually enabling process group-level operations and engineering applications. Therefore, we believe that our work is distinguished from the previous works mentioned above.

3. Applying the BPM Approach to Manufacturing Operations Management

MOM is a methodology with the aim of optimizing the manufacturing operations in the stages including the creation, planning, production, and distribution of products. Because of the significance of each operation stage, there are well-established MOM systems (e.g., manufacturing execution systems) that have been developed to provide systematic support to the stage-level operations. In general, MOM systems are broadly classified into the following categories: production management system, performance analysis system, quality and compliance system, and human–machine interface system.

Despite the merits from such MOM methodology and systems, we believe that there is still a necessity to apply a process-centered approach to the manufacturing processes to concentrate the optimization efforts. Manufacturing companies have their own manufacturing processes as their core assets. Therefore, the management capability of such processes has a significant impact on the performance of manufacturing to a greater or lesser extent. It emphasizes that the collaboration capability supporting people-to-people, people-to-systems, and systems-to-systems interactions in performing manufacturing operations should be incorporated in order to meet technological requirements that emerge from trends in MOM systems.

The BPM approach [14] covers management methodologies and systems for business processes. However, it can also be an effective way to integrate and align the capabilities of manufacturing operations with the manufacturing processes. While existing MOM systems are intended to increase the efficiency of operations associated with the product lifecycle, BPM systems can be a valuable addition to MOM since they offer comprehensive capabilities for manufacturing process management. Therefore, manufacturing companies can achieve advantages of the BPM approach through the modeling and efficient implementation of their manufacturing processes.

The process model is a central concept of BPM, formed from the contents created during the process design, and can be executed automatically by the process engine. Relating to the manufacturing processes, Figure 1 represents details of the activities and expected outcomes based on manufacturing process models that correspond to each stage consisting in the BPM lifecycle.

- Design Stage—This stage includes a series of steps from the establishment of initial development plan for the product to the redesign of existing manufacturing processes. From the view of process, this stage specifies which manufacturing operations should be taken to produce the desired product, and which components are required for each operation. Basic production-related data—such as bill-of-materials (BOM), operation process charts, and other specifications—are the expected outcomes of this stage and will serve the modeling stage of manufacturing processes.

- Modeling Stage—Through the modeling of manufacturing processes, we can shift the manufacturing processes to the management area of BPM and ensure their automated execution. The outstanding standards for modeling processes, such as the BPMN and BPEL (Business Process Execution Language), can be applied in order to make manufacturing processes into the form of standard-compliant and executable models.

- Execution Stage—Modeled manufacturing processes can be automatically executed by BPM systems under the assumption that manufacturing execution and control capabilities (e.g., manufacturing execution systems) are integrated or communicated with the BPM system. Each task of a manufacturing process model implements an individual manufacturing operation on the shop floor. During executions of manufacturing processes, event log data containing detailed records of manufacturing operations are generated and stored. Collected event log data are used in various analytics stages (e.g., process monitoring and optimization).

- Monitoring Stage—As event logs are generated while manufacturing processes are being executed, it is possible to monitor the status of running manufacturing processes or process instances in real time [15]. The well-established data formats for event logging can support the monitoring stage. For example, the XES (eXtensible Event Stream) standard, which has been widely adopted in the BPM field, has extensible data schema to incorporate specialized attributes of processes from various industry domains. As the result of this stage, manufacturing process reports help process administrators to take appropriate actions.

- Optimization Stage—Based on event logs, process mining [16,17] and post-analysis techniques provide us with opportunities to investigate the quality and performance of completed manufacturing process executions and related operations and improve those inefficient parts. The derived requirements for manufacturing process improvement will be referenced in the process design or redesign stage of the next cycle.

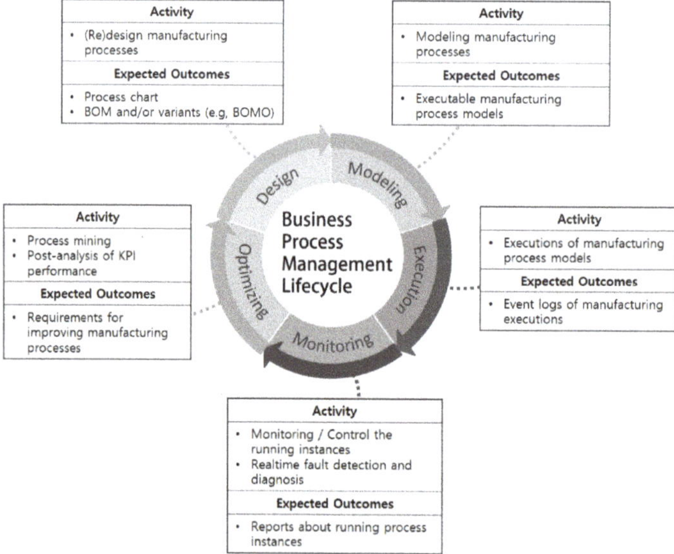

Figure 1. BPM lifecycle including the core stages in managing processes and details of activities and outcomes expected from the adoption of BPM into the MOM area.

Through continuous repetition of the BPM lifecycle applied into the MOM, manufacturing processes aligned with this management cycle can be expected to be constantly optimized to reflect the requirements that are derived by internal and external management factors. To take advantage of the convergence of these two management methodologies, modeling of the manufacturing process is a mandatory task. Therefore, the next section describes how to model the manufacturing process so that it complies with BPM-related standards from document-based manufacturing process data.

4. BPMN-Based Manufacturing Process Modeling

BPMN is one of the most widely used standards for modeling business processes. It provides a rich set of element types that can fully represent the context of a business process. Moreover, this standard can be easily extended, and it has been applied to modeling problems in various domains, such as wireless sensor network [18], healthcare processes [19], and manufacturing processes [20,21]. This section describes a construction procedure for manufacturing process models from the fundamental data of manufacturing processes.

Manufacturing companies conventionally possess operation process charts and BOM specifications to define and manage their manufacturing processes. In the Republic of Korea, the KS-A-3002 standard [22], which is for manufacturing process charts, has been used in manufacturing industries. However, it provides only a set of graphical notations, and there is no technical support for the modeling and automatic executions for manufacturing processes. Figure 2 shows an example of a manufacturing process chart for a thermocouple probe product.

A BOM specification is an essence of defining a manufacturing process that contains detailed data about the components (e.g., raw materials, parts, subassemblies, and end products) of each needed to manufacture a certain product. However, it is not sufficient to express the production flow of the manufacturing process. A manufacturing process consists of a set of manufacturing operations, which require a collection of components to manufacture and have prior relationships with other preceding and/or successive operations. In this regard, we additionally exploit the BOMO (Bill of Material Operations) concept, which is presented in [23] to define production-flow-oriented information of manufacturing process examples.

As shown in Table 1, each operation, such as wire welding, consumes a set of components and produces an intermediate component or end product. The preceding operation information provides an execution ordering of the operations within a manufacturing process. For example, the wire welding (OP1) and quartz tube winding (OP2) operations can be performed in parallel, but they must precede the wire injection operation (OP3) according to their ordering relationships. Through these basic ingredients of manufacturing process data (manufacturing process chart, BOM, and BOMO), we can organize structures for manufacturing processes.

Table 1. BOMO data of the product examples [2].

End Product	Operation		Component	Intermediate Component	Preceding Operation
	ID	**name**			
	OP6	Packaging	SUB-05, PACK-01, PACK-04, PACK-06, PACK-13, PACK-17	PROBE-01	OP5
	OP5	Insulation (cement)	SUB-04, MATL-11	SUB-05	OP4
PROBE-01	OP4	Housing	SUB-03, PART-08, PART-18, PART-19	SUB-04	OP3
	OP3	Wire injection	SUB-01, SUB-02	SUB-03	OP1, OP2
	OP2	Quartz tube winding	PART-15, MATL-02, MATL-04	SUB-02	-
	OP1	Wire welding	MATL-03, MATL-05, MATL-06	SUB-01	-

To create manufacturing process models, we apply the BPMN standard, with an extension of notations. The BPMN standard has a variety of its extensions, but it lacks modeling notations for the manufacturing domain. Although a few studies presented extensions for manufacturing processes [20,21], their extensions do not cover the full context of the manufacturing domain due to the absence of uniformity. Accordingly, we define a BPMN extension that has a minimal set of element types but suffices to model the examples we present in this paper.

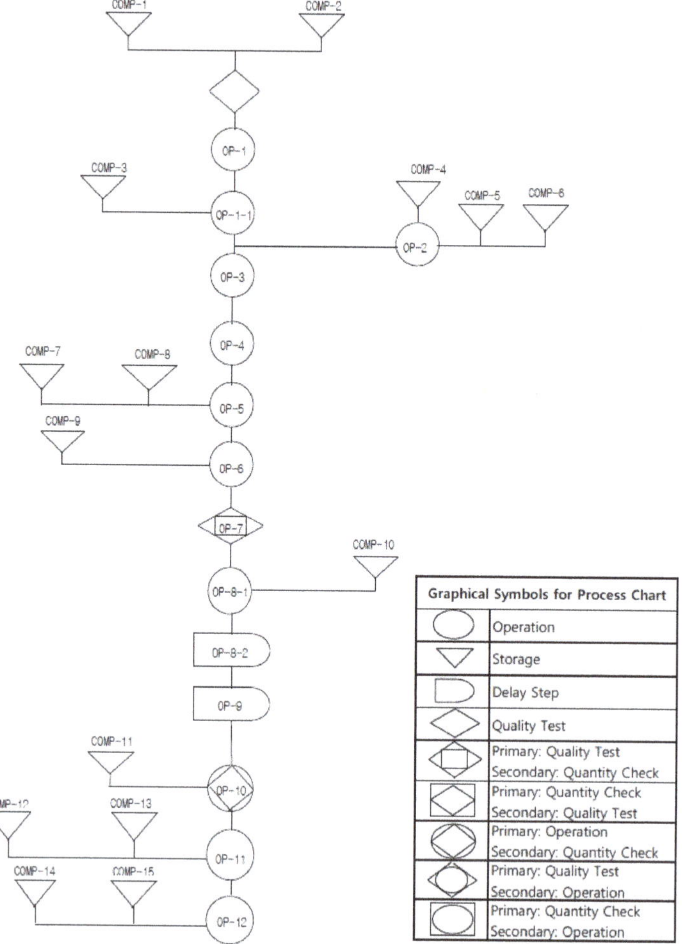

Figure 2. Operation process chart for a thermocouple probe product.

Every manufacturing process model essentially contains a start event and end event indicating their starting/terminating points. An operation refers to a primitive work within a manufacturing process. This type has precedence and/or successive relationships with other operations through connecting objects of sequence flows. Also, it is related to particular components through connecting objects of component associations, and production-related information such as operating cost. A component refers to an input of a particular operation and it can be a raw material, part, and subassembly.

Regarding the control-flow aspect, the notations of Table 2 are limited to focus on the examples including only sequential and parallel control-flow patterns. However, it is necessary to add extra notations for other patterns, such as selective, repetitive, and other complex patterns to facilitate modeling of advanced types of manufacturing processes and systems (e.g., flexible manufacturing systems and reconfigurable manufacturing systems). Figure 3 shows a result of the BPMN manufacturing process modeling for the thermocouple product specified as Table 1.

Table 2. BPMN notations for manufacturing process models [2]

Notation	Element type	Description
○	Start event	A 'start event' indicates where a particular manufacturing process will start.
◯	End event	An 'end event' indicates where a manufacturing process will terminate.
	Operation	An 'operation' is a generic term for manufacturing tasks. Each 'operation' can be performed by machines and/or human workers.
	Component	A 'component' is a generic term for raw materials, assemblies, and parts needed to manufacture a product.
◇⊕	Parallel gateway	A 'parallel gateway' is used to create and synchronize disjunctive flows which proceed in parallel fashion.
→	Sequence flow	A 'sequence flow' is used to show the order that operations will be performed in a manufacturing process.
┈┈►	Component association	A 'component association' is used to link components (e.g., material, part) and operations.

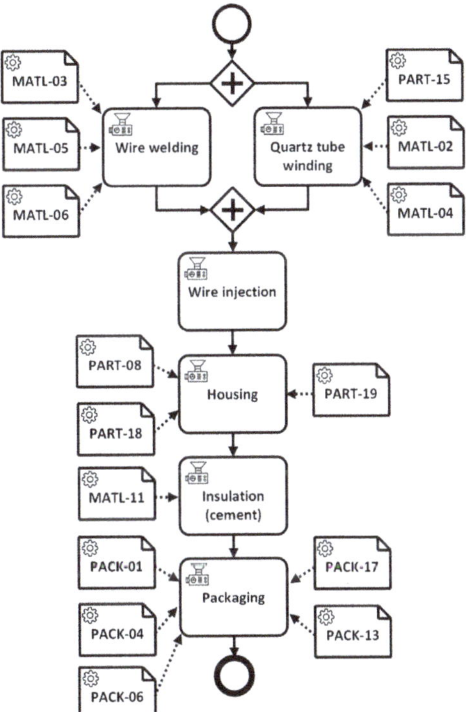

Figure 3. An exemplary model of a transformed BPMN manufacturing process.

The BPMN model is represented by a visual diagram (see Figure 3), while at the same time it is converted to a textual form (see Figure 4) that is executable by BPM systems. Figure 4 represents fragments of the textual content of the manufacturing process model example. This content represents detailed information, including definitions of operations, components, and connecting objects (e.g., gateway, sequence flow, and component association). When the manufacturing process model is instantiated, the execution engine of the BPM system interprets the information and creates a process instance to execute the manufacturing process with full control.

```xml
<?xml version="1.0" encoding="UTF-8"?>
<bpmn:definitions xmlns:xsi="http://www.w3.org/2001/XMLSchema-instance"
xmlns:bpmn="http://www.omg.org/spec/BPMN/20100524/MODEL">
  <bpmn:process id="_q1GEAP5u" name="BT Probe">
    <bpmn:componentSet id="btprobe_componentSet">
      <bpmn:component id="_q1xZcP5u" name="MATL-03"/>
      <bpmn:component id="_q3EZ8P5u" name="MATL-05"/>
      <bpmn:component id="_q5g7YP5u" name="MATL-06"/>
                          :
    </bpmn:componentSet>
    <bpmn:startEvent id="_q2AqAP5u" name="Start BT probe process"/>
    <bpmn:parallelGateway id="_q4J_EP5u" name="AND-Split Gateway-01"/>
    <bpmn:operation id="_q2VaIP5u" name="Wire welding">
      <bpmn:documentation>Cutting and welding a platinum wire.</bpmn:documentation>
    </bpmn:operation>
    <bpmn:operation id="_q4ogMP5u" name="Quartz tube winding">
      <bpmn:documentation>Bending a quartz tube using oxygen and LPG.</bpmn:documentation>
    </bpmn:operation>
    <bpmn:parallelGateway id="_q_9QMP5u" name="AND-Join Gateway-01"/>
    <bpmn:operation id="_q4eIIP5u" name="Wire injection">
      <bpmn:documentation>Injecting the welded wire into the quartz tube.</bpmn:documentation>
    </bpmn:operation>
                          :
    <bpmn:endEvent id="_q4UXIP5u" name="Terminate BT probe process"/>
    <bpmn:sequenceFlow id="_rAI2YP5u" name="OP1-to-G01" sourceRef="_q2VaIP5u" targetRef="_q_9QMP5u"/>
    <bpmn:sequenceFlow id="_rAUckP5u" name="OP2-to-G01" sourceRef="_q4ogMP5u" targetRef="_q_9QMP5u"/>
    <bpmn:sequenceFlow id="_rAUckP5u" name="G01-to-OP3" sourceRef="_q_9QMP5u" targetRef="_q4eIIP5u"/>
                          :
    <bpmn:componentAssociation id="_rBAZEP5u" name="MATL-03-to-OP1"
    sourceRef="_q1xZcP5u" targetRef="_q2VaIP5u"/>
    <bpmn:componentAssociation id="_rBN0cP5u" name="MATL-05-to-OP1"
    sourceRef="_q3EZ8P5u" targetRef="_q2VaIP5u"/>
    <bpmn:componentAssociation id="_rBEC2P5u" name="MATL-06-to-OP1"
    sourceRef="_q5g7YP5u" targetRef="_q2VaIP5u"/>
                          :
  </bpmn:process>
</bpmn:definitions>
```

Figure 4. An example of textual representation of the BPMN manufacturing process model.

5. Similarity Measure

Modeled manufacturing processes can be efficiently executed, tracked, and analyzed by the abundant toolsets in the BPM field. In this section, we describe similarity measures for manufacturing process models that serve as a basis of hierarchical clustering. Specifically, in order to capture the characteristics of manufacturing process, we define and explain two sub-types of similarity: operation similarity and structural similarity.

5.1. Preliminaries

We denote a set of manufacturing process models as $\mathbf{M} = M_1, \ldots, M_n$ with n indicating the number of manufacturing process models, a set of operations as $\mathbf{OP} = OP_1, \ldots, OP_m$ with m indicating the number of operations, and a set of components as $\mathbf{C} = C_1, \ldots, C_l$ with l indicating the number of components. We also introduce a mapping function $\delta(\mathbf{OP}) \rightarrow \wp(\mathbf{C})$ that maps from an operation OP_k^i to input components of OP_k^i which are a subset of total components, where $OP_k^i \in \mathbf{OP}$ is the operation OP_k in the process model M_i.

5.2. Operation Similarity

Although many similarity measures have been proposed for business processes, the manufacturing process has distinguishing features from business processes. In particular, production-related factors—which determine the characteristics of a manufacturing process—must be taken into account in measuring similarities.

The operation similarity we introduce in this paper is a similarity concept based on associations between operations and components that is one of these influential factors. Mostly, a manufacturing operation consumes a group of certain components to produce end products or intermediate components. Based on this feature, operation similarities between operations of the same type in different processes can be quantified. If two operations are the same type but have associations with different set of component types, these two operations are considered to have different characteristics. Mathematically, this similarity is based on the Jaccard coefficient, which is calculated by the division of the number of elements in the intersection set by the number of elements in the union set. Accordingly, the operation similarity between each operation OP_k in the process models M_i and M_j is calculated by Equation (1).

$$J\left(OP_k^i, OP_k^j\right) = \frac{\left|\delta\left(OP_k^i\right) \cap \delta\left(OP_k^j\right)\right|}{\left|\delta\left(OP_k^i\right) \cup \delta\left(OP_k^j\right)\right|} \tag{1}$$

For example, there are two 'Packaging' operations (OP_6) of the same type that is included in two different process models M_1 and M_2 as shown in Figure 5. The 'Packaging' operation in M_1 (Figure 5a) is associated with the set of input components $\delta\left(OP_6^1\right) = \{$PACK-01, PACK-04, PACK-06, PACK-13, PACK-17$\}$, that is different from the set of input components of the 'Packaging' operation in M_2 (Figure 5b), $\delta\left(OP_6^2\right) = \{$PACK-01, PACK-03, PACK-10, PACK-14, PACK-17$\}$. Accordingly, the operation similarity between these two operations is $J\left(OP_6^1, OP_6^2\right) = \frac{2}{8} = 0.25$.

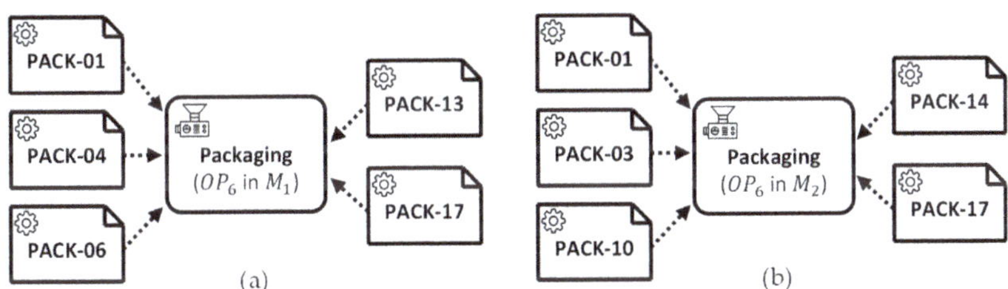

Figure 5. Operation–component associations.

We can measure operation similarities of the example models by calculating Jaccard coefficients based on the component association relationships of common operations (OP_1, OP_2, OP_3, OP_4, OP_5, and OP_6). In this regard, an operation similarity matrix including operation similarity measurements for all pairs of process models is defined as $X = (x_{rk}) \in \mathbb{R}^{\frac{n(n-1)}{2} \times m}$, where an element x_{rk} represents an operation similarity measure $J\left(OP_k^i, OP_k^j\right)$ between two operations of OP_k for the rth process model pair of the process models M_i and M_j. The index of the process model pair r is calculated by $r = \left(\sum_{a=1}^{i} n - 1\right) - (n - j)$, where $i < j$.

$$X = \begin{pmatrix} x_{1,1} & \cdots & x_{1,m} \\ \vdots & \ddots & \vdots \\ x_{\frac{n(n-1)}{2}, 1} & \cdots & x_{\frac{n(n-1)}{2}, m} \end{pmatrix} \tag{2}$$

Based on the above, let M_1 and M_2 be two process models as specified in Table 3. The operation similarity matrix of the example equals to the row vector represented by $X = [0.67, 1, 0, 0.40, 0, 0.25, \ldots, 0]$ since the example contains only two process models. Measured operation similarities affect the total similarity between two process models.

Table 3. Example of two process models including operation–component associations

Process Model	Operation	Components (without subassemblies)
Model 1 (M_1)	OP_1 (wire welding)	MATL-03, MATL-05, MATL-06
	OP_2 (quartz tube winding)	PART-15, MATL-02, MATL-04
	OP_3 (wire injection)	-
	OP_4 (housing)	PART-08, PART-18, PART-19
	OP_5 (insulation)	MATL-11
	OP_6 (packaging)	PACK-01, PACK-04, PACK-06, PACK-13, PACK-17
Model 2 (M_2)	OP_1 (wire welding)	MATL-05, MATL-06
	OP_2 (quartz tube winding)	PART-15, MATL-02, MATL-04
	OP_3 (wire injection)	-
	OP_4 (housing)	PART-02, PART-04, PART-18, PART-19
	OP_5 (insulation)	MATL-09, MATL-10, MATL-13
	OP_6 (packaging)	PACK-01, PACK-03, PACK-10, PACK-14, PACK-17
	OP_7 (fastening)	PART-10, MATL-01, MATL-07
	OP_8 (bar welding)	PART-09, PART-19
	OP_9 (sealing)	PACK-02, PACK-05

5.3. Structural Similarity

In this paper, we employ the similarity concepts of the activity vector and transition vector similarities, both are presented in [3]. We redefine these similarity concepts to fit manufacturing process models and call them structural similarity. As compared to the operation similarity based on the associations between operations and components, the structural similarity focuses on the existence of operations and the control dependency between operations. The structural similarity has two parts: operation vector similarity and transition vector similarity, and these concepts are slightly different from the similarity concepts for business process models in [3]. The total similarity is measured by putting these two similarities together.

5.3.1. Operation Vector Similarity

A manufacturing process comprises multiple operations that consume certain components and produce intermediate components or end products. Therefore, information indicating whether a specific operation is included in the process is the salient feature that characterizes manufacturing processes in terms of structural aspect. An operation vector v_i^O is an m-dimensional vector, where each element $v_{k,i}^O$ is a binary value (0 or 1) indicating whether the operation OP_k is included in the process model M_i.

$$v_i^O = \left[v_{1,i}^O,\ v_{2,i}^O,\ \ldots,\ v_{m,i}^O \right] \tag{3}$$

For example, Figure 6 represents two operation vectors v_1^O and v_2^O corresponding process models M_1 and M_2.

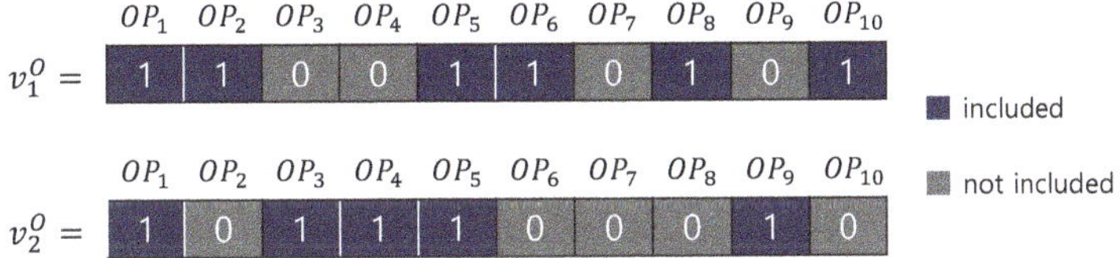

Figure 6. Representation of two operation vectors.

The operation vector similarity $sim_{ov}(M_i, M_j)$ is measured based on the Cosine coefficient and operation similarity. Given two operation vectors corresponding to two different manufacturing process models respectively (M_i and M_j), the following equation quantifies the similarity between these two operation vectors.

$$sim_{ov}(M_i, M_j) = \frac{\sum_{k=1}^{m} v_{k,i}^O \, v_{k,j}^O \, x_{rk}}{\sqrt{\sum_{k=1}^{m} \left(v_{k,i}^O\right)^2} \; \sqrt{\sum_{k=1}^{m} \left(v_{k,j}^O\right)^2}} \tag{4}$$

x_{rk} is the measured operation similarity of OP_k between M_i and M_j and it is a part of the numerator in the above equation. It implies that even if these vectors equal to each other, according to the operation similarities, the operation vector similarity varies from 1 to 0. For the example of Figure 6, the measured operation vector similarity is $sim_{ov}(M_1, M_2) \approx \frac{3.32}{5.48} \approx 0.61$.

In [3], activity vectors contain not only the inclusion of activity but also the execution probability that is varied based on the routing patterns in business process models. Fundamentally, there are four basic routing patterns: sequential, parallel, selective, and iterative routings. Among them, the selective routing patterns such as OR-split and XOR-split decide execution probabilities of activities.

On the other hand, in cases of manufacturing processes, we empirically found that the manufacturing processes we investigate are composed of only sequential routings and parallel routings (less frequently), and they intrinsically have no selective and iterative routing patterns. Conventionally, manufacturing processes proceed only according to the predefined operations and their stationary execution orderings without any decision points that cause splitting control-flows patterns. However, in state-of-the-art manufacturing systems (e.g., flexible manufacturing systems), where process routing and machine assignments are flexible to react in case of changes, it is necessary to support the concept of the routing patterns not considered in this paper and supplement the definition of operation vector.

5.3.2. Transition Vector Similarity

The transition is the fundamental property of all classes of process models which is formally represented as a directed acyclic graph. The causality of tasks (including operations) in a process model is established based on transitions between the tasks, and it is a main structural property of process models including manufacturing processes. For example, the starting task in a process model precedes all other tasks including a succeeding task directly following it. Therefore, for manufacturing processes, causal relationships between operations are quantified and weighted through calculations of distance weights, and these are represented as a transition vector.

Let v_i^T be a transition vector of manufacturing process model M_i. v_i^T is a row vector containing $m \times m$ elements for all pairs of operations, where each element $v_{kl,i}^T$ represents a causal relationship between OP_k and OP_l, measured by the reverse of distance weight $d_{kl,i}^T$ between OP_k and OP_l.

$$v_i^T = \left[v_{11,i}^T, \; v_{12,i}^T, \; \ldots, \; v_{mm,i}^T \right] \tag{5}$$

$$v_{kl,i}^T = \frac{1}{d_{kl,i}^T} \tag{6}$$

The Cosine coefficient-based transition vector similarity $sim_{tv}\left(M_i, M_j\right)$ is measured by the following equation.

$$sim_{tv}\left(M_i, M_j\right) = \frac{\sum_{k=1}^{m} \sum_{l=1}^{m} v_{kl,i}^{T}\, v_{kl,j}^{T}}{\sqrt{\sum_{k=1}^{m} \sum_{l=1}^{m} \left(v_{kl,i}^{T}\right)^2}\ \sqrt{\sum_{k=1}^{m} \sum_{l=1}^{m} \left(v_{kl,j}^{T}\right)^2}} \tag{7}$$

Based on the above equation, the result of measuring transition vector similarity between M_1 and M_2 represented in Figure 7 is $sim_{tv}(M_1, M_2) \approx \frac{4.9}{7.16} \approx 0.684$.

The total similarity between process models M_i and M_j is measured by putting these two vector similarities $sim_{ov}\left(M_i, M_j\right)$ and $sim_{tv}\left(M_i, M_j\right)$ together and adding a balancing parameter $\alpha \in [0, 1]$ to blend them.

$$sim\left(M_i, M_j\right) = \alpha \cdot sim_{ov}\left(M_i, M_j\right) + (1 - \alpha) \cdot sim_{tv}\left(M_i, M_j\right) \tag{8}$$

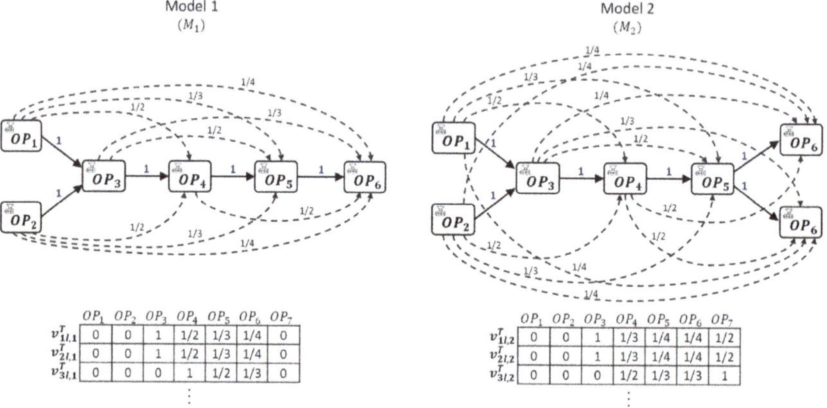

	OP_1	OP_2	OP_3	OP_4	OP_5	OP_6	OP_7
$v_{1l,1}^{T}$	0	0	1	1/2	1/3	1/4	0
$v_{2l,1}^{T}$	0	0	1	1/2	1/3	1/4	0
$v_{3l,1}^{T}$	0	0	0	1	1/2	1/3	0

	OP_1	OP_2	OP_3	OP_4	OP_5	OP_6	OP_7
$v_{1l,2}^{T}$	0	0	1	1/3	1/4	1/4	1/2
$v_{2l,2}^{T}$	0	0	1	1/3	1/4	1/4	1/2
$v_{3l,2}^{T}$	0	0	0	1/2	1/3	1/3	1

Figure 7. Representation of two transition vectors.

6. Similarity-Based Hierarchical Clustering

Based on the similarity measures, this section describes a hierarchical clustering algorithm of manufacturing process models. The hierarchical clustering is a useful technique that allows us to build a hierarchy of data clusters from similarities (or dissimilarities) between data. Hierarchical clustering methods generally fall into two types: the agglomerative (bottom-up) approach [24] and divisive (top-down) approach [25]. In this paper, we exploit an agglomerative approach, which is more common to the hierarchical clustering than the divisive approach, and we devise a clustering algorithm based on this approach.

6.1. Hierarchical Clustering Algorithm

The agglomerative clustering method builds a hierarchy from the n process models being clustered (see the algorithm below). Initially, each model is assigned to its own cluster (line 4–7) and then the clustering algorithm proceeds iteratively, at each step merging the two most similar clusters, continuing until all the models are merged into a single cluster (line 10–25).

Algorithm: Agglomerative Clustering using Process Similarity

Input: M, Set of manufacturing process models $(M_1, \ldots, M_n)$.
Output: H, Hierarchy of clusters $(\Lambda_1, \ldots, \Lambda_h)$.
1: **begin**
2: $l \leftarrow 1$;   //level of hierarchy
3: $\Lambda_l \leftarrow \varnothing$;
4: **for** $M_i \in \mathbf{M}$ //initialize a cluster set in level 1
5:     $\Lambda_l \leftarrow \varnothing$;
6:     $\Lambda_l \leftarrow \Lambda_l \cup M_i$;   //assign each model to its own cluster
7: **end for**
8: $\mathbf{H} \leftarrow \varnothing$;
9: $\mathbf{H} \leftarrow \mathbf{H} \cup \Lambda_l$;   //add the cluster set in level 1 to the hierarchy
10: **while** $(|\Lambda_l| > 1)$ //continue until all clusters are merged
11:     $\chi_m \leftarrow \varnothing$;   //a cluster for two clusters to be merged
12:     $max \leftarrow 0$;   //a variable for the global maximum similarity
13:     **for** $\chi_i \in \Lambda_l$ //find a maximum similarity between all pairs of clusters
14:       **for** $\chi_j \in \Lambda_l$, where $\chi_i \neq \chi_j$;
15:        **if** $sim^C_{max}(\chi_i, \chi_j) > max$ **then** //compare the global and local maximum similarities
16:          $\chi_m \leftarrow \varnothing$;
17:          $\chi_m \leftarrow \chi_i \cup \chi_j$;   //a set for the merged clusters
18:          $max \leftarrow sim^C_{max}(\chi_i, \chi_j)$;   //update the global maximal similarity
19:        **end if**
20:       **end for**
21:     **end for**
22:     $\Lambda_{l+1} \leftarrow \varnothing$;   //a cluster set in the next level
23:     $\Lambda_{l+1} \leftarrow \left((\Lambda_l - \chi_i) - \chi_j\right) \cup \chi_m$, where $\chi_i, \chi_j \subseteq \chi_m$;   //subtract χ_i and χ_j, and add their merged cluster χ_m
24:     $\mathbf{H} \leftarrow \mathbf{H} \cup \Lambda_{l+1}$;   //add the cluster set in level $l + 1$ to the hierarchy
25:     $l \leftarrow l + 1$;   //increase the level of hierarchy
26: **return H**;

There are linkage-criteria, each of which determines a way to compute similarities between clusters. The complete-linkage method [26] takes the lowest similarity value (or farthest distance) between data included in the pair of clusters as a cluster similarity. Conversely, the single-linkage method [26] merges two clusters based on the highest similarity value (or closest distance) between pairs of clusters. These two representative linkage methods have slight differences in terms of time and space complexities. We accordingly use the single-linkage method that can easily be adapted to agglomerative clustering problems. Therefore, the manufacturing clustering algorithm calculates local maximum similarities, each of which indicates the maximum similarity observed from a pair of model clusters (line 13–18).

$$sim^C_{max}(\chi_i, \chi_j) = \max\{sim(M_a, M_b) : (M_a, M_b) \in \chi_i \times \chi_j\} \tag{9}$$

In the next step, the two model clusters having the global maximum similarity are selected to be merged. For example, let us three of model clusters χ_1, χ_2, and χ_3 at the level l. The algorithm calculates local maximum similarities of all pairs of clusters, $sim^C_{max}(\chi_1, \chi_2)$, $sim^C_{max}(\chi_1, \chi_3)$, and $sim^C_{max}(\chi_2, \chi_3)$. In case of the cluster pair of χ_1, χ_2, a local maximum similarity of this pair can be obtained by finding a maximal value of similarity among all possible pairs of process models, $\forall(M_a, M_b) : M_a \in \chi_1, M_b \in \chi_2$, included in χ_1 and χ_2. If the cluster pair of χ_2 and χ_3 has the global maximum similarity, these two clusters are joined into a new cluster χ_4 for the next level of the hierarchy. In this way, the algorithm merges the two closest clusters at each level and eventually completes the entire hierarchy of clusters.

6.2. Running Example

To confirm the applicability of the hierarchical clustering method for manufacturing process models, we present a running example with manufacturing process models. These models are selected models from real-life manufacturing processes for thermocouple probe products (see Figure 8) and obtained by the BPMN modeling phase from process charts, BOM and BOMO data. The details of the model examples are partially omitted or simplified for the security reasons. Specifically, the running example contains 8 process models ($\mathbf{M} = M_1, \ldots, M_8$), 15 operations ($\mathbf{OP} = OP_1, \ldots, OP_{15}$), and 57 components ($\mathbf{C} = C_1, \ldots, C_{57}$). According to the characteristics of the routing patterns of manufacturing processes, all the models have only sequential and parallel routing patterns in terms of control-flow.

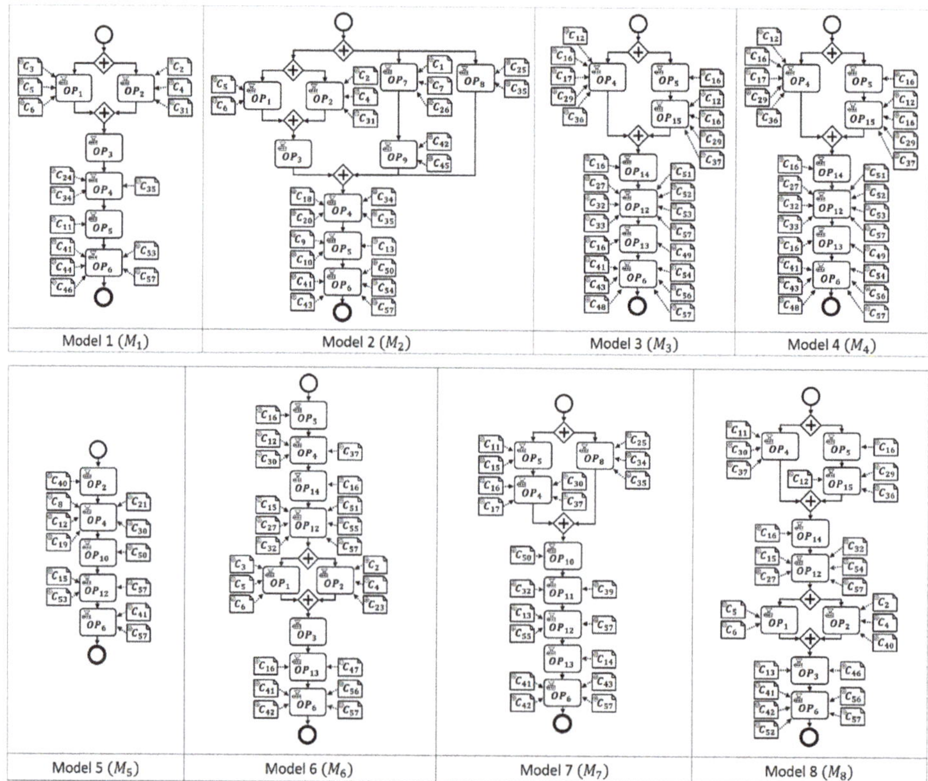

Figure 8. Manufacturing process models for the running example.

At the first step, an operation similarity matrix, $X = (x_{rk}) \in \mathbb{R}^{28 \times 15}$, between operations of the same type are calculated based on Equation (1). For example, the operation OP_4 is the operation of the same manufacturing task that is commonly included in all the process models. For the case of OP_4 in the model M_3, this operation is associated with the input components, $C_{12}, C_{16}, C_{17}, C_{29}$, and C_{36}. Due to the different operation–component associations from other process models, the measured operation similarities for OP_4^3 are

$$[0, 0, 1, 1, 0.11, 0.14, 0.29, 0]$$

Operation similarities affect the similarities between operation vectors, and Tables 4 and 5 represent the measured operation/transition vector similarities, respectively. The model pair (M_5, M_6) has the highest operation vector similarity, $sim_{ov}(M_i, M_j) = 0.6709$, while the two most similar models on the transition vector similarity are M_3 and M_8 with the topmost value of 0.828. Ultimately, the measured total similarities with the balancing parameter $\alpha = 0.5$ are shown in Table 6. This result shows that the two most similar models are M_5 and M_6 with the similarity value of 0.6998.

Table 4. Measured operation vector similarities.

Process model	M_1	M_2	M_3	M_4	M_5	M_6	M_7	M_8
M_1	-	0.4518	0.1000	0.0343	0.3791	0.3289	0.0730	0.1134
M_2		-	0.0454	0.0720	0.2725	0.3056	0.0596	0.1375
M_3			-	0.0955	0.1055	0.0794	0.2313	0.3945
M_4				-	0.4152	0.3916	0.1197	0.1103
M_5					-	0.6709	0.1768	0.1515
M_6						-	0.1839	0.1229
M_7							-	0.2914
M_8								-

Table 5. Measured transition vector similarities.

Process model	M_1	M_2	M_3	M_4	M_5	M_6	M_7	M_8
M_1	-	0.7792	0.0154	0.0769	0.2873	0.3206	0.1276	0.0346
M_2		-	0.0120	0.0354	0.2239	0.2502	0.0995	0.0336
M_3			-	0.3553	0.1906	0.0415	0.4119	0.8280
M_4				-	0.4492	0.5320	0.1317	0.3138
M_5					-	0.7287	0.0812	0.2638
M_6						-	0.0982	0.0447
M_7							-	0.3414
M_8								-

Table 6. Total similarities with $\alpha = 0.5$.

Process model	M_1	M_2	M_3	M_4	M_5	M_6	M_7	M_8
M_1	-	0.6155	0.0577	0.0556	0.3332	0.3247	0.1003	0.0740
M_2		-	0.0287	0.0537	0.2482	0.2779	0.0796	0.0855
M_3			-	0.2254	0.1480	0.0604	0.3216	0.6113
M_4				-	0.4322	0.4618	0.1257	0.2120
M_5					-	0.6998	0.1290	0.2077
M_6						-	0.1410	0.0838
M_7							-	0.3164
M_8								-

Through the hierarchical clustering, we can expect to build a hierarchy of clusters of manufacturing process models using the measured similarities. Figure 9 shows the hierarchical clustering result for the running example, visualized as a heat map with dendrograms. As mentioned above, the hierarchical clustering method operates in the bottom-up agglomerative way and finds and merges a pair with the highest similarity (as Equation (9)) among all model pairs at each stage. In addition, by cutting the hierarchy at a certain level, an intended number of clusters can be obtained. This means that it is possible to determine the desired scale in managing manufacturing processes. For example, at the level 5 of the hierarchy, there are four clusters $\chi_1 = \{M_1, M_2\}$, $\chi_2 = \{M_3, M_7, M_8\}$, $\chi_3 = \{M_4\}$, and $\chi_4 = \{M_5, M_6\}$. As clusters χ_3 and χ_4 have the global maximum similarity $sim_{max}^C(\chi_3, \chi_4) = 0.4618$ resulted from the model pair $(M_4 \in \chi_3, M_6 \in \chi_4)$, these two closest clusters are merged, resulting in a total of three clusters $\chi_1 = \{M_1, M_2\}$, $\chi_2 = \{M_3, M_7, M_8\}$, and $\chi_3 = \{M_4, M_5, M_6\}$ at the next level.

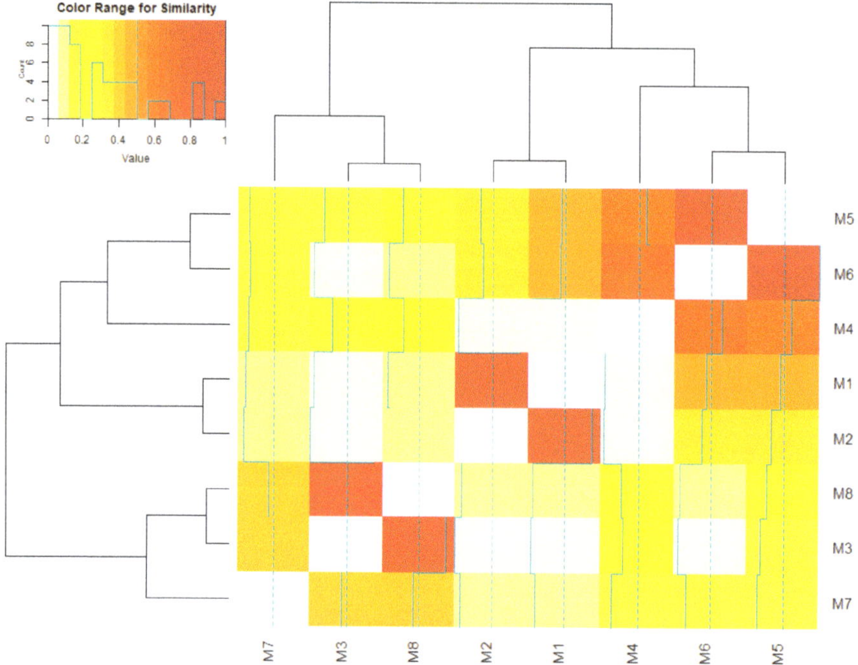

Figure 9. Hierarchical clustering result visualized as a heat map with dendrograms.

7. Conclusions

In the era of Industry 4.0, lots of manufacturing companies tend to have a vast array of manufacturing processes. With the attention of a need of manufacturing process management, this paper proposes a similarity-based hierarchical clustering method for manufacturing processes operated by manufacturing companies where the BPM methodology applied in. Specifically, the contributions of this paper are summarized as two-fold: (1) The manufacturing process modeling with the adaptation of the BPM approach. (2) The similarity measures for such manufacturing process models, and the hierarchical clustering based on the similarity. As the validation result of this work, the applicability of our clustering method was confirmed through the running example including real-life manufacturing processes for thermocouple products. In terms of manufacturing operation management, implications and possible applications are defined as follows:

BPM-compliant manufacturing process modeling—The process modeling standards generally provide visual notations and schema for storing textual information (e.g., BPMN). Thus, manufacturing process models that are visually standardized can help communications among related persons (e.g., process designers and stakeholders), at the same time, the textual information are interpreted by the BPM system to automatically execute manufacturing processes. Therefore, BPM-compliant manufacturing process models can be fully controlled in accordance with the BPM lifecycle. For example, process mining is an enabling technology to monitor status of entire manufacturing processes by analyzing event logs generated from the executions of process models.

Similarity-based hierarchical clustering of manufacturing process models—Clustering techniques aggregate similar objects into the same cluster. Thus, these facilitate group-level operations. For the manufacturing domain, operations in relevance with the production scheduling [27], resource planning and distribution can be performed effectively, centering on clustered manufacturing process groups. If there exist deficiencies such as bottleneck and low yield, which must be improved in the manufacturing process, the company may consider extending its reengineering task to the whole manufacturing processes of the same cluster from the corresponding single process. For unexpected situations, especially, the alternation operation of a manufacturing process or related resources (e.g., machines, components, and human resources) is also one viable application which is activated by manufacturing process clustering. Assuming a certain company operating a large-scale set of manufacturing processes that is the target environment of this paper, our clustering method can support in designing manufacturing processes for new products. By searching based on the similarity measures, process designers can find a matched process or partial information, valuable to be referred or reused.

In conclusion, we believe that our method will support manufacturing companies so as to design and manage a vast amount of manufacturing processes at a coarser level. However, in order for our method to be more realistic, the following points should be supplemented in a follow-up study. First, the BPMN modeling is insufficient to capture the other important characteristics of the manufacturing process. Machines and human resources are also important entities that steer the manufacturing process, therefore the modeling functionality supporting them should also be added. In addition, we need to review other similarity techniques (e.g., graph similarity) and apply it to make our method more reasonable and effective. Second, this work does not address the evaluation of how the proposed clustering method can contribute to quantitative and qualitative positive effects on manufacturing operation management and engineering problems. Therefore, our future works will cover the following subjects:

- Studying process modeling and similarity techniques for better capturing the characteristics of manufacturing processes.
- Conducting a case study applying the proposed clustering methods to a large set of real-life manufacturing processes and investigating its effects.
- Design and implementation of a system supporting decision-making in various engineering problems based on clustering results.

We are looking into implanting the BPM approach and systems into the manufacturing company of thermocouple products. In a broad sense, the ultimate objective of our future works is to establish realistic applications that provide an effective means supporting BPM-based manufacturing operation management.

Author Contributions: H.A. and T.C. did the initial design and analysis on the research; H.A. performed experiments; H.A. and T.C. analyzed the data; H.A. prepared the bulk of the manuscript and wrote the draft of the paper; while T.C. contributed write-up for the paper.

Acknowledgments: This work was supported by the GRRC program of Gyeonggi province. ((GRRC KGU 2017-B01), Research on Industrial Data Analytics for Intelligent Manufacturing).

Conflicts of Interest: The authors declare no conflict of interest.

References

1. Hwang, G.; Kang, S.; Dweekat, A.; Park, J.; Chang, T.-W. An IoT data anomaly response model for smart factory performance measurement. *Int. J. Ind. Eng. Theory Appl. Pract.* **2018**, *25*, 702–718.
2. Ahn, H.; Chang, T.-W. Measuring Similarity for Manufacturing Process Models. In Proceedings of the IFIP International Conference on Advances in Production Management Systems, Seoul, Korea, 27–30 August 2018; pp. 223–231.
3. Jung, J.-Y.; Bae, J.; Liu, L. Hierarchical Clustering of Business Process Models. *Int. J. Innov. Comput. Inf. Control* **2009**, *5*, 1349–4198.
4. Dijkman, R.; Dumas, M.; Garcia-Banuelos, L. Graph Matching Algorithms for Business Process Model Similarity Search. In Proceedings of the International Conference on Business Process Management, Ulm, Germany, 8–10 September 2009; pp. 48–63.
5. Ivan, A.; Akkiraju, R. Discovering Business Process Similarities: An Empirical Study with SAP Best Practice Business Processes. In Proceedings of the International Conference on Service-Oriented Computing, San Francisco, CA, USA, 7–10 December 2010; pp. 515–526.

6. Kunze, M.; Weske, M. Metric Trees for Efficient Similarity Search in Large Process Model Repositories. In Proceedings of the International Conference on Business Process Management, Hoboken, NJ, USA, 13–16 September 2010; pp. 535–546.

7. Schoknecht, A.; Thaler, T.; Fettke, P.; Oberweis, A.; Laue, R. Similarity of Business Process Models: A State-of-the-Art Analysis. *ACM Comput. Surv.* **2017**, *50*, 52. [CrossRef]

8. Gupta, T.; Seifoddini, H. Production Data Based Similarity Coefficient for Machine-Component Grouping Decisions in the Design of a Cellular Manufacturing System. *Int. J. Prod. Res.* **1990**, *28*, 1247–1269. [CrossRef]

9. Dimopoulos, C.; Mort, N. A Hierarchical Clustering Methodology Based on Genetic Programming for the Solution of Simple Cell-Formation Problems. *Int. J. Prod. Res.* **2001**, *39*, 1–19. [CrossRef]

10. Park, S.; Suresh, N.C. Performance of Fuzzy ART Neural Network and Hierarchical Clustering for Part-Machine Grouping Based on Operation Sequences. *Int. J. Prod. Res.* **2003**, *41*, 3185–3216. [CrossRef]

11. Li, B.M.; Xie, S.Q. Module Partition for 3D CAD Assembly Models: A Hierarchical Clustering Method Based on Component Dependencies. *Int. J. Prod. Res.* **2015**, *53*, 5225–5240. [CrossRef]

12. Daie, P.; Li, S. Managing Product Variety through Configuration of Pre-Assembled Vanilla Boxes using Hierarchical Clustering. *Int. J. Prod. Res.* **2016**, *54*, 5468–5479. [CrossRef]

13. Etgar, R.; Gelbard, R.; Cohen, Y. Presenting the Several-Release-Problem and Its Cluster-Based Solution Acceleration. *Int. J. Prod. Res.* **2017**. Available online: https://doi.org/10.1080/00207543.2017.1404657 (accessed on 2 May 2019). [CrossRef]

14. Weske, M. *Business Process Management: Concepts, Languages, Architectures*; Springer: Berlin/Heidelberg, Germany, 2012.

15. Stefanini, A.; Aloini, D.; Benevento, E.; Dulmin, R.; Mininno, V. Performance Analysis in Emergency Departments: A Data-driven Approach. *Meas. Bus. Excell.* **2018**, *22*, 130–145. [CrossRef]

16. Van der Aalst, W.M.P.; Reijers, H.A.; Weijters, A.J.M.M.; van Dongen, B.F.; Alves de Medeiros, A.K.; Song, M.; Verbeek, H.M.W. Business Process Mining: An Industrial Application. *Inf. Syst.* **2007**, *32*, 713–732. [CrossRef]

17. Zerbino, P.; Aloini, D.; Dulmin, R.; Mininno, V. Process-mining-enabled audit of Information Systems: Methodology and an application. *Expert Syst. Appl.* **2018**, *110*, 80–92. [CrossRef]

18. Sungur, C.T. Extending BPMN for Wireless Sensor Networks. Master's Dissertation, University of Stuttgart, Stuttgart, Germany, 2013.

19. Ruiz, F.; Garcia, F.; Calahorra, L.; Llorente, C.; Goncalves, L.; Daniel, C.; Blobel, B. Business Process Modeling in Healthcare. *Stud. Health Technol. Inform.* **2012**, *179*, 75–87. [PubMed]

20. Zor, S.; Leymann, F.; Schumm, D. A Proposal of BPMN Extensions for the Manufacturing Domain. In Proceedings of the 44th CIRP International Conference on Manufacturing Systems, Madison, WI, USA, 31 May–3 June 2011.

21. Bocciarelli, P.; D'Ambrogio, A.; Giglio, A.; Paglia, E. A BPMN Extension for Modeling Cyber-Physical-Production-Systems in the Context of Industry 4.0. In Proceedings of the 14th IEEE International Conference on Networking, Sensing and Control, Calabria, Italy, 16–18 May 2017; pp. 599–604.

22. KSA (Korean Standards Association). KS-A-3002 Standard: Graphical Symbols for Process Chart. Available online: https://www.kssn.net/en/search/stddetail.do?itemNo=K001010103692 (accessed on 2 May 2019).

23. Jiao, J.; Tseng, M.M.; Ma, Q.; Zou, Y. Generic Bill-of-Materials-and-Operations for High-Variety Production Management. *Concur. Eng.* **2000**, *8*, 297–321. [CrossRef]

24. Murtagh, F.; Legendre, P. Ward's Hierarchical Agglomerative Clustering Method: Which Algorithms Implement Ward's Criterion? *J. Classif.* **2014**, *31*, 274–295. [CrossRef]

25. Savaresi, S.M.; Boley, D.L. On the Performance of Bisecting K-means and PDDP. In Proceedings of the 2001 SIAM International Conference on Data Mining, Chicago, IL, USA, 5–7 April 2001; pp. 1–14.

26. Wilks, D.S. Cluster Analysis. *Int. Geophysics* **2011**, *100*, 603–616.

27. De Matta, R. Scheduling a Manufacturing Process with Restrictions on Resource Availability. *Int. J. Prod. Res.* **2017**, *56*, 6412–6429. [CrossRef]

Article

A Hybrid Methodology for Validation of Optimization Solutions Effects on Manufacturing Sustainability with Time Study and Simulation Approach for SMEs

Poorya Ghafoorpoor Yazdi [1,2,*], **Aydin Azizi** [2,*] **and Majid Hashemipour** [3]

[1] Department of Mechanical Engineering, Eastern Mediterranean University, 99628 Famagusta, Northern Cyprus

[2] Department of Engineering, German University of Technology, Muscat 130, Oman

[3] Department of Mechanical Engineering, Cyprus International University, 99258 Nicosia, Cyprus; mhashemipour@ciu.edu.tr

* Correspondence: poorya.ghafoorpoor@gutech.edu.om (P.G.Y.); aydin.azizi@gutech.edu.om (A.A.)

Received: 11 February 2019; Accepted: 5 March 2019; Published: 8 March 2019

Abstract: The properties of small- and medium-sized enterprises (SMEs) make them one of the most important categories of enterprises for the economics of challenging world. SMEs, in most countries, are still enterprises with marketing and financial challenges. In addition, most of these challenges are related to their production and product characteristics. On the other hand, SMEs should fulfil the costumer's demands. In order to reach these goals, SMEs must reach the highest level of production quality and quantity and successfully sustain them. Consequently, various manufacturing paradigms have been offered by an Industry 4.0 concept, which offers a variety of solutions to increase the productivity and enhance the performance of SMEs. It should be noted that implementation of these manufacturing paradigms for SMEs is quite difficult and sometimes risky for several reasons. Still, amidst all these difficulties and challenges, the benefits and idealism of the Industry 4.0 paradigms prevail. From productivity to market, it is difficult to deny that SMEs are frightened by the challenges they face and fleeing from the potential of overcoming them. This paper is an extended version of the research by Ghafoorpoor Yazdi et al. (2018) and conducts a hybrid methodology to satisfy the SMEs by validating and verifying any optimization idea before implementing the Industry 4.0 concept. To reach the study goals, an intelligent Material Handling System (MHS) with agent-based control architecture has been developed. The developed MHS has been utilized for auto parts distribution. The system performance has been evaluated, and some solutions have been provided to optimize the performance of system. To evaluate the target system's performance, an analytical time study method has been utilized. The time study has an Overall Equipment Effectiveness (OEE) standard approach to identifying the matters that need to be resolved and optimized to increase system performance. The other part of the methodology is generating a simulation model of the real system by use of ARENA® software to evaluate the system's performance before implementing the optimization idea and modifying the real system. Furthermore, as the sustainability strategies create many synergistic effects for SMEs, after evaluating the effects of the optimization ideas on OEE percentage, the influence of the OEE changes on manufacturing sustainability has been investigated. The results show that optimizing the OEE in SMEs with sustainability approaches can create competitive advantages, rather than simply focusing on reducing unsustainability.

Keywords: small and medium sized enterprises; material handling systems; simulation; ARENA®, time study; overall equipment effectiveness; manufacturing performance; Industry 4.0; manufacturing sustainability

---

1. Introduction

Nowadays, having businesses with increasing globalization and rapid technological changes requires more firms known as small- and medium-sized enterprises (SMEs), characterized by tight resources [1]. Industrial and commercial sectors, and large enterprises in particular, are changing due to globalization. These types of enterprises aim to shift and maintain their production, which are distributed to various regions abroad. For this reason, the impact of SMEs have become more prominent in an ever-changing globalized market [2,3]. SME firms need to produce admirable products with low cost, fast delivery to market, and appropriate quality. This strategy to meet these highly varying demands has made manufacturing systems more complex, dynamic, and demanding [4]. However, a manufacturing system including the mentioned capabilities requires proper control architecture, decision-making, and experts [5]. Moreover, it is difficult and quite risky for SMEs to utilize these systems before validation and verification of the possible advantages and disadvantages of these changes and its effects on their productivity [3]. Therefore, modelling, analysis, and simulation are essential aspects which can ensure the successful implementation and utilization of a new manufacturing system for SMEs [6].

One of the most important resources of a successful manufacturing system is the Material Handling System (MHS), which has a high level of influence on productivity [7]. Recently, automation of MHS has been focused on as an important feature of this type of system [8]. However, for a few SMEs which are currently utilizing automated/smart MHS, it would be risky and sometimes not feasible to apply any changes or optimization idea before validating and verifying their effects on different manufacturing aspects [9].

Many researchers have worked to develop a method to choose the right equipment and control architecture for MHS. Chan et al. (2011) proposed a model for improving MHS, called the Material Handling Equipment Selection Advisor (MHESA) [10]. Bhattacharya et al. (2002) also proposed another method, which is called the Multi-Criteria Decision-Making (MCDM) environment [11]. Shen et al. (2003) developed a new control architecture on their target MHS [12]. However, Fax and Murray (2004) proposed the use of a decentralized manufacturing control system which has the definition of an agent [13]. Durieux and Pierreval (2004) have proposed a study to design an automated manufacturing control system which is composed of machines working in parallel with MHS resources [14]. Schröder et al. (2016) developed an efficient way to manufacture products using automated MHS [15]. Johnstone et al. (2010) applied the concept behind the low-level control. This concept plays a huge role in preventing material collision in the considered MHS [16]. An operator-designed element-routing methodology for MHS was developed by Lau and Woo [17].

An increase in productivity and less rejections are the possible outcomes of increasing equipment effectiveness. The mentioned improvement is a major effect of the Industry 4.0 implementation on enterprises [18]. Considering the challenging market for SMEs, performance of the manufacturing system and reliable manufacturing equipment are currently considered the most important factors for increasing profitability [19]. Therefore, OEE is one of the best analytical methods for evaluating the performance of SMEs [20]. Generally, OEE focuses on the relationship between the loading time and the valuable time of the operations in manufacturing systems. Operating time can be defined as the time in which the equipment produces sufficient products, and loading time is the time in which the equipment needs to perform in a given period [21].

Many studies have been done about utilizing OEE to evaluate the performance of the enterprises. Nakajima (1988) introduced the total productive maintenance (TPM) concept. Overall Equipment Effectiveness (OEE) was the outcome of this concept as a quantative metric for measuring the productivity of individual equipment in an enterprise [22]. The quality rate, availability, and performance were the target aspects of OEE to be identified and measured. Huang et al. (2003) stated that the OEE concept has been widely used as a quantitative tool essential for the measurement of the manufacturing system's productivity [23].

Nakajima (1988) stated that the effectiveness of the equipment in manufacturing systems can be evaluated by OEE standards to detect the system's problems and limitations [22]. Jonsson and Lesshammar (1999) investigated the reason for losses during the production in manufacturing systems [24]. Muchiri et al. (2008) stated that there were six major losses available in manufacturing systems, which are downtime, Speed, Idling, and minor stoppage and losses due to speed reduction [25].

Calculation of performance, availability, and quality result in an obtainment of the OEE percentage [26]. The procedure for obtaining the percentage of each of these factors is illustrated in Figure 1 [27].

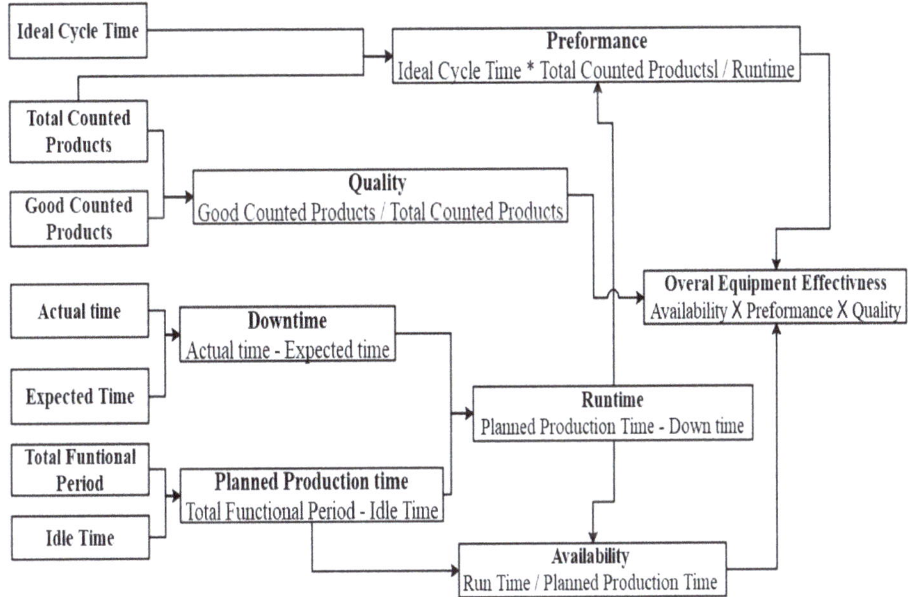

Figure 1. Overall Equipment Effectiveness (OEE) calculation procedure [27].

As stated previously, time is the most important factor when analyzing OEE. Consequently, a comprehensive time study can track each of the equipment and evaluate their behavior to identify the factors which might increase the system's performance. To reach this goal, the manufacturing process can be divided into sub-processes. A summation of the time study for each part of the process leads to an obtainment of the overall system's timing. Many studies have stated that utilization of a proper time study will identify the system's waste time and limitations which might occur during the production process [27–31].

There are several time study techniques and methods, and there are many studies available in regard to each of them. Patel (2015) performed a stopwatch time study to calculate the cycle time for labor tasks, Overall Equipment Effectiveness (OEE), and in regard to minimizing the cycle time of the overall shifts [32]. Elnekave (2006) conducted a time study in a textile factory to obtain the standard time for a printing machine to calculate the system performance [33]. Bon and Daim (2010) investigated time as a measuring tool for the performance evaluation of a company [34]. Longo and Mirabelli (2009) utilized the Method Time Measurement (MTM) to analyze the primary model of the assembly line. Since the study was based on a nonexistent assembly line, a simulation tool was used to model the manufacturing system [35]. Kuhlang et al. (2011) presented a study on an assembly line and production-logistic process which aims to create more added-value elements in the process within a fixed time [36].

Simulation has played a significant role in evaluating, validating, and verifying any optimizing idea or modifications for a manufacturing system's hardware, software, and layout design [37]. The operational performance of manufacturing systems could be obtained after any changes in simulation models before implementation on a real system [38,39]. There are different concepts in the investigation of the relationship between simulation and optimization, which are "simulation and optimization", "simulation-based optimization", and "simulation for optimization" [40–42]. ARENA simulation can be utilized to model different systems with different configurations to be subjected to optimization solutions and modifications. Therefore, ARENA is a proper simulation tool that can be used to validate, verify, and evaluate manufacturing systems which are modified by any optimization solution [43–45].

Manufacturing Sustainability Improvement

In order to evaluate the sustainability of the target manufacturing systems, the missions of the Organization for Economic Cooperation and Development (OECD) have been considered (Table 1). Based on OECD missions, a new business and production method, which is known as Sustainable Manufacturing, includes different types of green products and processes [46]. In addition, the concept of a sustainable manufacturing system is about decreasing the risk of business and increasing the chances which comes from the manufacturing system and process improvements [47].

Table 1. The Organization for Economic Cooperation and Development (OECD) requirements for sustainable manufacturing [27,46].

	Steps	Description
Preparation	Mapping impact and set priorities	In this step, the manufacturing, environmental impact of small- and medium-sized enterprises (SMEs) should be reviewed. Also, the priorities of each should be defined. Based on the detected and the priorities of these environmental impacts, sustainability objectives should be defined.
Preparation	Select useful performance indicators	Indicators that are essential for SMEs to increase the performance and learn about what data should be collected to help drive continuous improvement should be identified.
Measurement	Measure the inputs used in the production	The ways in which materials and components used in SMEs production processes influence environmental performance should be identified.
Measurement	Assess operations of the SMEs facility	The impact and efficiency of the operations in SME facilities, such as energy intensity, greenhouse gas generation, and emissions to air and water should be considered.
Improvement	Evaluate your products	Factors such as energy consumption in use, recyclability, and use of hazardous substances that help determine the sustainability of SME end-products should be identified.
Improvement	Understand measured results	Reading and interpreting the SME indicators and the methods for understanding trends in their performance should be learned.
Improvement	Take action to improve performance	Opportunities to improve SME performance and creating action plans to implement them should be chosen.

According to the OECD, there are eight key factors that have to be considered during any sustainability investigation (Figure 2).

According to Yazdi et al. (2018), Figure 2 can be simplified to the energy intensity relation with manufacturing sustainability. In the other word, out of the effective factors related to manufacturing sustainability, energy consumption of the system and its related resources can be focused on [27].

As mentioned previously, time is the main effective factor when evaluating the OEE percentage of manufacturing systems. Thus, the relationship between time and manufacturing sustainability for SMEs is one of the targets in this research. However, to have a comprehensive evaluation of the effectiveness of a manufacturing system and its energy efficiency, considering time as the only factor of focus is not enough. Therefore, to evaluate the energy efficiency of a manufacturing system, the relationship between the energy consumption of a manufacturing system in a specific period of time and its power requirements should too be considered, so that it is possible to conduct an action plan to reduce the time and energy consumption in parallel [48,49].

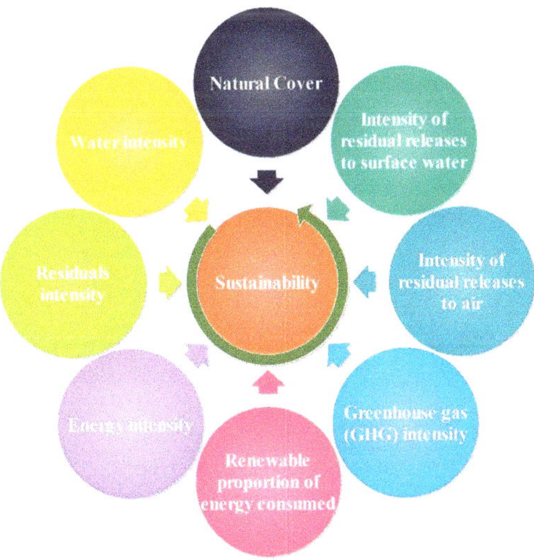

Figure 2. Considerable factors in manufacturing sustainability [27].

This study is predominantly about assessing the challenges and possible difficulties that SMEs might face during the implementation of Industry 4.0. However, SMEs have a fear of utilizing these solutions until the management layer, being convinced of the possible performance improvements and effects. On the other hand, for the category of SMEs in which the Industry 4.0 is adopted, they are likely to want to improve their current manufacturing system by new control architectures and optimization ideas and some other manufacturing paradigms. Hence, the proposed methodology in this research is likely to make SMEs able to evaluate the system before and after implementation of the new control architecture and optimization idea.

In general, the study had two stages and used a hybrid methodology for the possible industrial category target (SMEs). The Overall Equipment Effectiveness (OEE) standard analysis was conducted to determine an appropriate relationship between performance evaluation and optimization in such material handling systems for SMEs.

In the first stage of the research, the proposed hybrid methodology was implemented on an educational manufacturing system at the German University of Technology in Oman (GUtech) [50]. Before any changes were made, the system was a semi-automated material handling system for part/product distribution. In order to represent the essence and effect of the adoption of Industry 4.0 and the way in which the system's processes, operations, and overall performance could be improved, a new control architecture with a unique algorithm was developed. The reason behind this type of control architecture was to transform the target MHS from a semi-automated to an intelligent one, as this is significant for Industry 4.0 integration. To reach to an intelligent MHS, an agent-based control architecture was developed. The developed control architecture, including an algorithm for the target MHS, became a suitable solution due to the improvement in production planning and scheduling, as well as monitoring and control. Agent-based control architecture implementation resulted in improved decision-making and an effectively facilitated system.

In the second stage and in this paper, two categories of enterprises have been considered as the ones where the proposed method might be applicable for them. The first category included the enterprises which were going to initiate the production and which did not have a fixed control architecture, resources (hardware and software), or layout design. The other category was for the enterprises that were already producing and functioning, and where it was hard and risky to modify any of the hardware and software resources and layout design, especially the control architecture. Performance measurement of the system with new control architecture and the developed algorithm is indispensable to the target MHS, because if the efficiency of the control architecture cannot be measured, it cannot be properly modified and optimized. It is obvious that if the new proposed control architecture cannot improve the performance, some other hardware, software, and layout design problems will exist. Furthermore, these possible problems which may affect the system performance should be differentiated from the control architecture's possible problems and limitations.

The first stage of the study established that considering the second category of the SMEs, time, as the main performance indicator will determine most of the problems and limitations in relation to the hardware, software, control architecture, and layout design [27]. The study made use of timing as the main key factor for OEE. Quantitative required data were collected using "ProTime" Estimation software as the time study tool. The selected time study method is a combination of different time study techniques to obtain the required information for OEE calculation. Implementation of the proposed methodology for target MHS required a comprehensive overview of the system functionality and tasks, such that the MHS with new control architecture were separated into a different district, including the available tasks for each of them. OEE was calculated for the target MHS before any modification. The OEE result was analyzed, and the system hardware, software, and layout design obtained for each district and each task.

In the second stage and in this paper, out of the detected problems and limitations available in the system with new control architecture, several influencing factors have been identified. In order to optimize the system in regard to these three categories of problems, the distance between the devices and speed was considered as the optimization objective function. To investigate the effect of the optimization idea, the target intelligent MHS was modelled on an ARENA simulation to investigate any possible changes in system performance. The study's conclusion is that there were partial, as well as overall improvements in the systems. As the final stage of the research, the effect of the proposed methodology and optimization ideas on the sustainability of the target system were obtained.

2. Methodology

An intelligent material handling system, including the Master-Slave agent-based control architecture, was developed as the first stage of this research. The system layout and its agent-based control architecture and functionality has been described by Yazdi et al. in detail [27] (Figures 3 and 4).

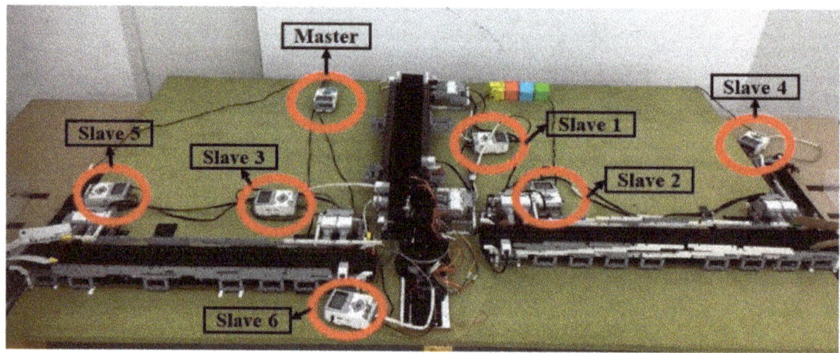

Figure 3. System overview [27].

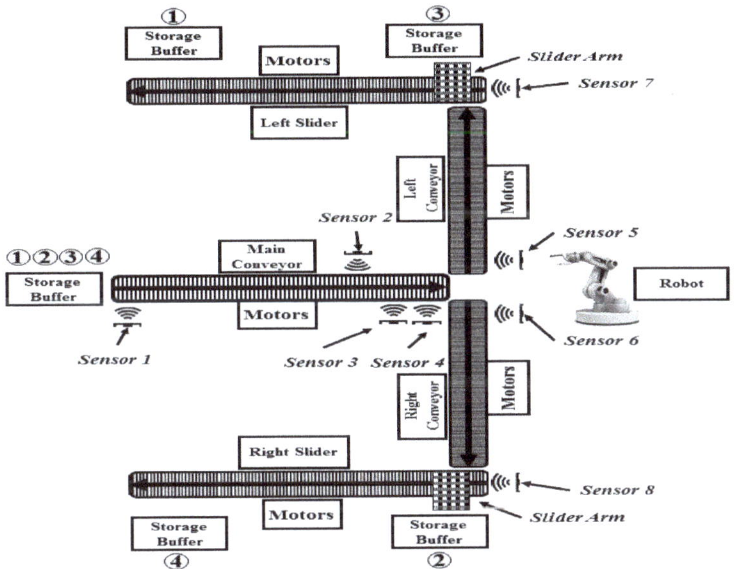

Figure 4. Layout design, top view [27].

Referring to Figures 3 and 4 and research done by Yazdi et al. [27], the system includes four phases. Phase 1 includes the storage, main conveyor, and its control unit (slave1). Phase 2 includes a robot arm and its control unit (slave6). Phases 3 and 4 include a slider with a control unit (Slave4 and 5), and a conveyor and its control units (Slave 2 and 3) and buffers [27].

A time study was done on the developed intelligent material handling system by Yazdi et al. [27] for each part of the process and the entire system. Detailed results of the time study in a previous part of the research is illustrated in Table 7, which is available in [27]. Obtaining the utilization time for all of the available resources in the system was the main target in doing the time study. In addition, with the selected time study technique, a comprehensive investigation of the difference between similar devices with the same functionalities was done. The time study results have been utilized to obtain the required factors for calculating the OEE percentages for every resource, as well as the entire system. Also, identification of the problems and limitations in hardware, software (control agents), and layout design was the result of the time study of the system. A combination of the outcomes of the time study and OEE evaluation led to the provision of proper optimization solutions for enhancing system performance. Due to the target system's characteristics as an educational manufacturing system, as well as its limitations, the time study evaluation was done for a short functioning period of the system, and expanded for a long period theoretically.

The time study results and its associated Gantt chart showed the overall system performance within the entire sections of the system, including the utilization time of each resource and the conflicts that occur when two or more resources are working at the same time. A more detailed Gantt chart and explanation is given by Yazdi et al. in [27].

Since part of the previous objectives of this research was to obtain methods for optimizing the system, it was essential to point out the issues and problems related to the system's performance and prepare a comprehensive overview of them. It has been stated by Yazdi et al. that a particular aspect which needs to be optimized in the system is idle time, since it affects the overall system utilization rate [27]. Thus, the authors presumed that there should be a direct relationship between a decrease in Busy time, Idle time, and overall production time [27]. As a result, all of the detected issues and possible solutions for enhancing system performance with the time study and visual observation point of view are listed in a Table in [27].

Once the problems and limitations and simulation model with functional specifications of the real system were clearly defined in the first stage of the research, it was possible to detect and gather the data, which was essential for verifying the simulation model with the same characteristics and properties of the real system. Once the necessary data were identified, determination of where the data came from was required. In utilizing the time study, the required data which needed to be collected manually was obtained.

In this research, the objective of the simulation was verification of the detected problems by time study and validation of the effects of implementation of the optimization method before execution on the real system.

When building the simulation model in ARENA, defining the primary aspects of the real manufacturing system plays an important role. These primary aspects include Resources, Variables, Attributes, and other elements. Once the fixed aspects have been defined, the simulation model's control logic can be generated by utilizing ARENA's flowchart-modelling methodology (Figure 5).

If the model's behavior makes sense and it is functioning like the real system—for instance, when entities are moving in the path that they should, and process steps are taking place as expected—we can be sure that the model has been verified. To achieve system verification, functional specification should be followed properly during the model-building phase.

Here, in the second stage of research, validation means users' approval about the output of the simulation model matching with the real system outputs. In this case, the model input may use real system data, and its output from the simulation model will be compared with the real system's results (time study results). In all cases, the changes in the existing system are amended to reflect on the simulation model.

To achieve behavior most similar to the real system by way of the simulation mode, it is mandatory to consider every hardware, software, and layout design specification in the model as they are in the real system. For this reason, the model was divided into different districts, which were the entrance of the system and the points at which parts were being loaded onto the system, main conveyor, Right and Left Conveyor, Right and Left Slider units, and Robotic arm (identical to the available resources in the first stage of research).

Visualization was applied in ARENA simulation that provided graphics and animation of the system's functionality. Through the use of graphics in the simulation, the user can gain better understanding of the system which has been modeled. As long as the implementation of any optimization and change in the existing real system which has been modeled is not possible before validation and verification, visualization of the system in ARENA provides one of the best methods for validating and verifying system optimization methods (Figure 6).

To generate an accurate model by ARENA and achieve similar behavior as it would be in a real system, Busy Time was considered as a target. Obtaining the Busy Time for each resource and related task/s required measurements of time, velocity, and distance between the resources for every section. For this reason, five participants with proper knowledge of the system and its functionality contributed to making this measurement have high-accuracy instrumentations. It is necessary to mention that the target factors required for creating the simulation model were obtained by human observation from the real system, meaning that human error may have affected the accuracy of the data. This means that as much as the observations, trials, and number of participants increases, the error will consequently be decreased.

Figure 5. ARENA simulation model.

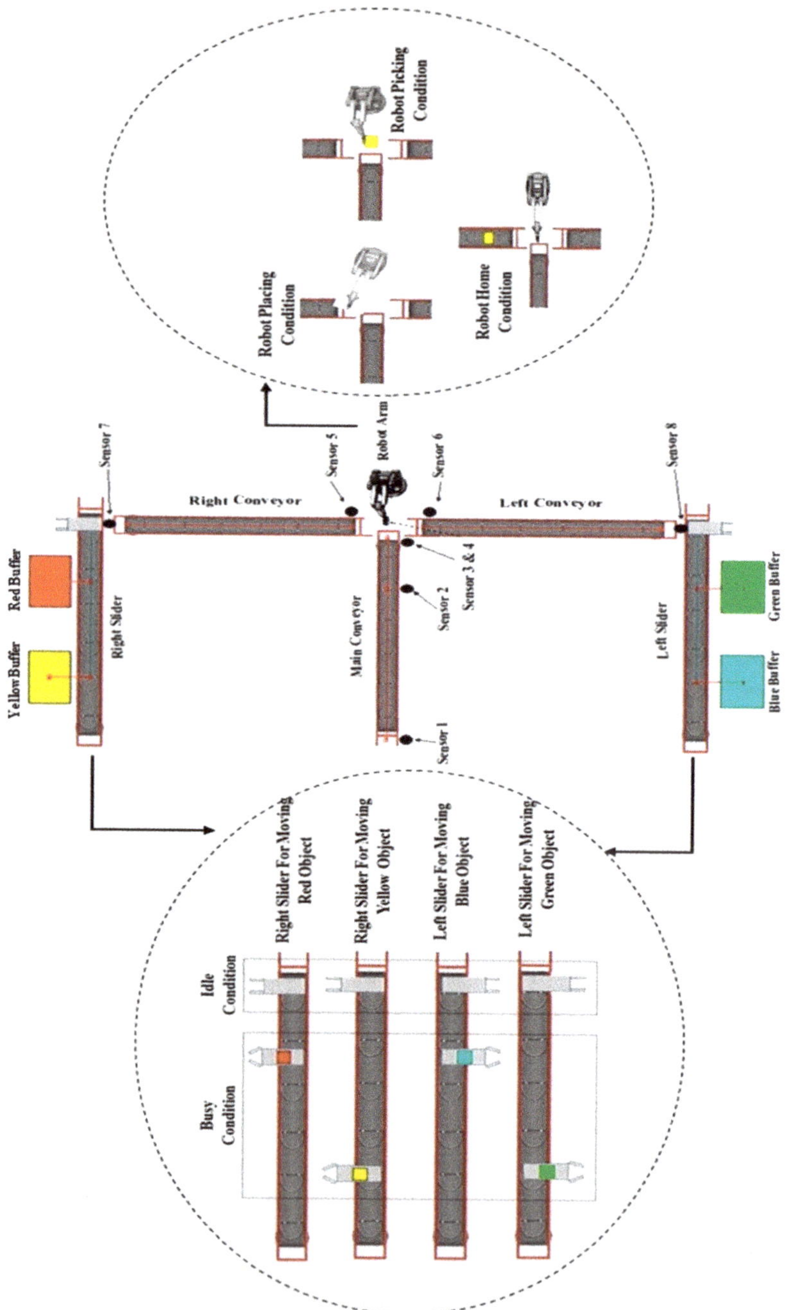

Figure 6. ARENA visualization of the real system.

The velocity of objects and resources was measured by utilizing a motion sensor. In this method, an electrostatic transducer in the face of the Motion Sensor transmitted a burst of 16 ultrasonic pulses with a frequency of about 49 kHz. The sensor measured the time between the trigger-rising edge and the echo-rising edge to measure the velocity. The distance between the resources and the displacement of the objects was measured with a laser distance meter to get the most accurate results. Each participant did the measurement individually and was asked to calculate the time while considering the velocity and measured distance (Table 2).

Table 2. Distance and speed measurements by participants.

From	To	Among	Resource	Distance (Free Unit)	Speed (Distance Unit Per Second)
Entrance/Sensor 1	Sensor 2	Main Conv.	Main Conv.	53	9.3
Sensor 2	Sensor 3&4	Main Conv.	Main Conv.	12	3.35
Sensor 3&4	Sensor 5	Main Conv. And Right Conv.	Robot Arm	34	12.9
Sensor 3&4	Sensor 6	Main Conv. And Left Conv.	Robot Arm	29	10.4
Sensor 5	Sensor 7/Right Slider Entrance	Right Conv. and Right Slider	Right Conv.	85	9
Sensor 7/Right Slider Entrance	Red Buffer	Right Slider	Right Slider	8	8
Sensor 7/Right Slider Entrance	Yellow Buffer	Right Slider	Right Slider	39	10
Sensor 8/Left Slider Entrance	Green Buffer	Left Slider	Left Slider	7	8
Sensor 8/Left Slider Entrance	Blue Buffer	Left Slider	Left Slider	39	10
Sensor 6	Sensor 8/Left Slider Entrance	Left Conv. and Left Slider	Left Conv.	75	9
Right Slider	Red Buffer	Right Slider and Red Buffer	Right Slider	1	2
Right Slider	Yellow Buffer	Right Slider and Yellow Buffer	Right Slider	1	3.13
Left Slider	Blue Buffer	Left Slider and Blue Buffer	Left Slider	1	3.13
Left Slider	Green Buffer	Left Slider and Green Buffer	Left Slider	1	2.39

The result was imported into the simulation model, just as it was measured in the real system. Table 3 shows the obtained average Busy time for each task measured by the participants. Table 3 also shows the difference between the real system's Busy time and average Busy time by the simulation, in which the model's input data is from the participants' measurements. The absolute time difference shows there is a 0.5 second difference between the Busy time in the simulation model and in the real system (Figure 7). Thus, the simulation model has the functional specifications of the real system, and is reliable for applying any modification.

Table 3. Time difference between "Busy" times of the real system and simulation model.

Task No.	Task Description	Busy Time (s)		
		Real System	Average of Participants Observation	Absolut Deference
1	Main conveyor handling Red object to Robot	9.81	9.50	0.31
2	Robot arm picking The Red Object from Main Conveyor	3.55	3.84	0.29
3	Robot arm placing Red object to Right Conveyor	3.56	3.66	0.10
4	Right conveyor handling Red object to Right Slider	9.22	9.40	0.18
5	Robot arm moves to its home position after placing Red object	2.62	2.50	0.12
6	Main conveyor handling Blue object to Robot	9.04	9.50	0.46
7	Right Slider transfers Red object to Red buffer	3.98	3.90	0.08
8	Right Slider unloading the Red object to Red buffer	4.5	4.50	0.00
9	Robot arm picking The Blue Object from Main Conveyor	3.88	3.55	0.33
10	Robot arm placing Blue object to Left Conveyor	3.82	3.66	0.16
11	Right Slider moves to its home position after unloading Red object	4.01	4.08	0.07
12	Left conveyor handling Blue object to Left Slider	7.94	8.33	0.39
13	Robot arm moves to its home position after placing Blue object	2.78	2.50	0.28
14	Main conveyor handling Yellow object to Robot	9.86	9.50	0.36
15	Left Slider transfers Blue object to Blue buffer	4.09	4.09	0.00
16	Left Slider unloading the Blue object to Blue buffer	4.42	4.40	0.02
17	Robot arm picking The Yellow Object from Main Conveyor	4.05	3.84	0.21
18	Left Slider moves to its home position after unloading Blue object	4.43	4.43	0.00
19	Robot arm placing Yellow object to Right Conveyor	3.63	3.66	0.03
20	Right conveyor handling Yellow object to Right Slider	9.3	9.40	0.10
21	Robot arm moves to its home position after placing Yellow object	2.57	2.50	0.07
22	Main conveyor handling Green object to Robot	9.62	9.50	0.12
23	Right Slider transfers Yellow object to Yellow buffer	0.98	0.99	0.01
24	Right Slider unloading the Yellow object to Yellow buffer	4.32	4.32	0.00
25	Robot arm picking The Green Object from Main Conveyor	3.84	3.55	0.29
26	Right Slider moves to its home position after unloading yellow object	1.25	1.25	0.00
27	Robot arm placing Green object to Left Conveyor	3.6	3.66	0.06
28	Left conveyor handling Green object to Left Slider	7.88	8.33	0.45
29	Robot arm moves to its home position after placing Green object	2.5	2.50	0.00
30	Left Slider transfers Green object to Green buffer	1.07	0.71	0.36
31	Left Slider unloading the Green object to Green buffer	4.33	4.32	0.01
32	Left Slider moves to its home position after unloading Green object	1.44	1.44	0.00

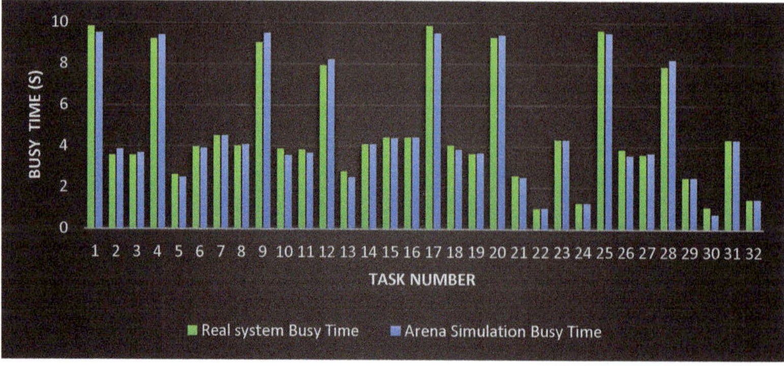

Figure 7. Time difference between "Utilization" time of the real system and simulation model.

The designed and implemented intelligent material handling system, with its related time study result and analysis, were evaluated by using the OEE standard to get the overall system performance. Detailed results of the OEE percentage for each resource and the entire system [27] have been illustrated by Yazdi et al. in [27], as well as in Table 5. This percentage provided an accurate view of how effectively the manufacturing process was running, and made it easier to track problems and improvements in the system over time. A detailed analysis of the OEE [27] results has been provided by Yazdi et al. in [27].

OEE Analysis after Optimization

After OEE analysis of the existing system and identifying the problems and limitations of the system which affected the OEE percentage, it was possible to modify the system to compensate and fix these issues and optimize the system. In order to reach this goal, an objective function was considered to evaluate the optimal methods and ideas.

For this reason, all of the system parameters which can be modified to optimize the system were formed as a parameter vector:

$$p = [p_1, p_2, p_3] \tag{1}$$

where p is the objective function, p_2 is the distance between the stations or segments, and p_3 is the devices' speed of motion.

p_1 is the production time, including Idle and Busy times:

$$T_i, T_b \in p_1 \tag{2}$$

where T_i is the Idle Time and T_b is the Busy time.

Time is influenced by the velocity and distance between the considered object, meaning that:

$$p_1 = \frac{p_2}{p_3} \tag{3}$$

$$p_{1,min} = \frac{p_{2,min}}{p_{3,max}} \tag{4}$$

It is obvious that by decreasing the tasks' durations (Busy time), the system's resource Idle times will be subsequently decreased. In addition, Idle time will be affected by many other parameters, such as the quality of the products and production planning. This means that idle time varies by changing different parameters. C_1 is the effectiveness coefficient of idle time, and shows that in different systems, Idle time has different effects. C_2 is the effectiveness coefficient of Busy time, which was evaluated internally for each resource and rarely affected by other parameters which are not related to the resources itself.

$$p_1 = \sum_{t=0}^{T} C_1 \, T_i + C_2 T_b \tag{5}$$

The proposed optimization methods in this study focused on time, speed, velocity, and layout design of the system. These parameters simultaneously affected Idle and Busy time. However, the influence of the changes in Busy time was more than the Idle time. Thus, to investigate the influence of each of the proposed optimization ideas, C_1 was assumed as 1 so as to only investigate the effects on Busy time.

$$C_1 = 1 \; \rightarrow \; p1 = \sum_{t=0}^{T} C_2 T_b \tag{6}$$

Since the aim of this study was to evaluate the optimization ideas and their effects on the OEE percentage, and as mentioned previously, time has been selected as the key affecting factor on OEE, optimization of timing was the outcome of the objective function. The time, p_1, which is the production time, should be minimized. As mentioned previously, to reach the minimum p_1, p_2 (Distance) should be minimized and p_3 (Velocity) should be maximized.

Following the instructions above and to verify the optimization ideas, the velocity of the resources was modified to the maximum possible value in the simulation model. The increasing of the velocity was based on the real resource's capacity, and the properties and limitation of the proposed control architecture. For instance, the velocity of the conveyors was increased to the point that the conveyor could handle the objects accurately with less vibration during the transportation. For this reason, without changing the system layout, several tests have been done practically on the similar conveyor to get the maximum velocity value.

The effects of the layout design and distance between the segments were investigated by changing the place of the sensors, instead of changing the length of conveyors in the simulation model. Table 4 and Figure 8 show the simulation model modifications for investigating the effect of the optimization ideas on the system resource's Busy times. In addition, a comprehensive comparison between the Busy times before and after implementing the optimization idea on the simulation model has been visualized in Figure 8.

Table 4. System modification and "Busy" times after the optimization methods.

Task Description	Real	Average of Participant Observation with Simulation	2/3 of the Real Distance Change	1/2 of the Real Distance Change	Speed Limit Change to Maximum
Main conveyor handling Red object to Robot	9.81	9.50	6.63	5.32	8.20
Robot arm picking The Red Object from Main Conveyor	3.55	3.84	3.84	3.84	3.25
Robot arm placing Red object to Right Conveyor	3.56	3.66	3.66	3.66	3.20
Right conveyor handling Red object to Right Slider	9.22	9.40	5.60	4.47	8.80
Robot arm moves to its home position after placing Red object	2.62	2.50	2.50	2.50	1.99
Main conveyor handling Blue object to Robot	9.04	9.50	6.63	5.32	8.20
Right Slider transfers Red object to Red buffer	3.98	3.90	2.64	2.03	3.70
Right Slider unloading the Red object to Red buffer	4.50	4.50	4.50	4.50	4.24
Robot arm picking The Blue Object from Main Conveyor	3.88	3.55	3.55	3.55	3.34
Robot arm placing Blue object to Left Conveyor	3.82	3.66	3.66	3.66	3.30
Right Slider moves to its home position after unloading Red object	4.02	4.08	2.68	2.06	3.90
Left conveyor handling Blue object to Left Slider	7.94	8.33	5.68	4.47	7.50
Robot arm moves to its home position after placing Blue object	2.78	2.50	2.50	2.50	1.99
Main conveyor handling Yellow object to Robot	9.86	9.50	6.63	5.32	8.20
Left Slider transfers Blue object to Blue buffer	4.09	4.09	2.72	2.00	4.02
Left Slider unloading the Blue object to Blue buffer	4.42	4.40	4.40	4.40	4.20
Robot arm picking The Yellow Object from Main Conveyor	4.05	3.84	3.84	3.84	3.25
Left Slider moves to its home position after unloading Blue object	4.43	4.43	2.95	2.27	4.06
Robot arm placing Yellow object to Right Conveyor	3.63	3.66	3.66	3.66	3.20
Right conveyor handling Yellow object to Right Slider	9.30	9.40	5.60	4.47	8.80
Robot arm moves to its home position after placing Yellow object	2.57	2.50	2.50	2.50	1.99
Main conveyor handling Green object to Robot	9.62	9.50	6.63	5.32	8.20
Right Slider transfers Yellow object to Yellow buffer	0.98	0.99	0.62	0.49	0.70
Right Slider unloading the Yellow object to Yellow buffer	4.32	4.32	4.32	4.32	4.10
Robot arm picking The Green Object from Main Conveyor	3.84	3.55	3.55	3.55	3.34
Right Slider moves to its home position after unloading yellow object	1.25	1.25	0.78	1.20	1.00
Robot arm placing Green object to Left Conveyor	3.60	3.66	3.66	3.66	3.30
Left conveyor handling Green object to Left Slider	7.88	8.33	5.68	4.47	7.50
Robot arm moves to its home position after placing Green object	2.50	2.50	2.50	2.50	1.99
Left Slider transfers Green object to Green buffer	1.07	0.71	0.62	0.49	0.66
Left Slider unloading the Green object to Green buffer	4.33	4.32	4.32	4.32	4.10
Left Slider moves to its home position after unloading Green object	1.44	1.44	1.03	0.82	1.07

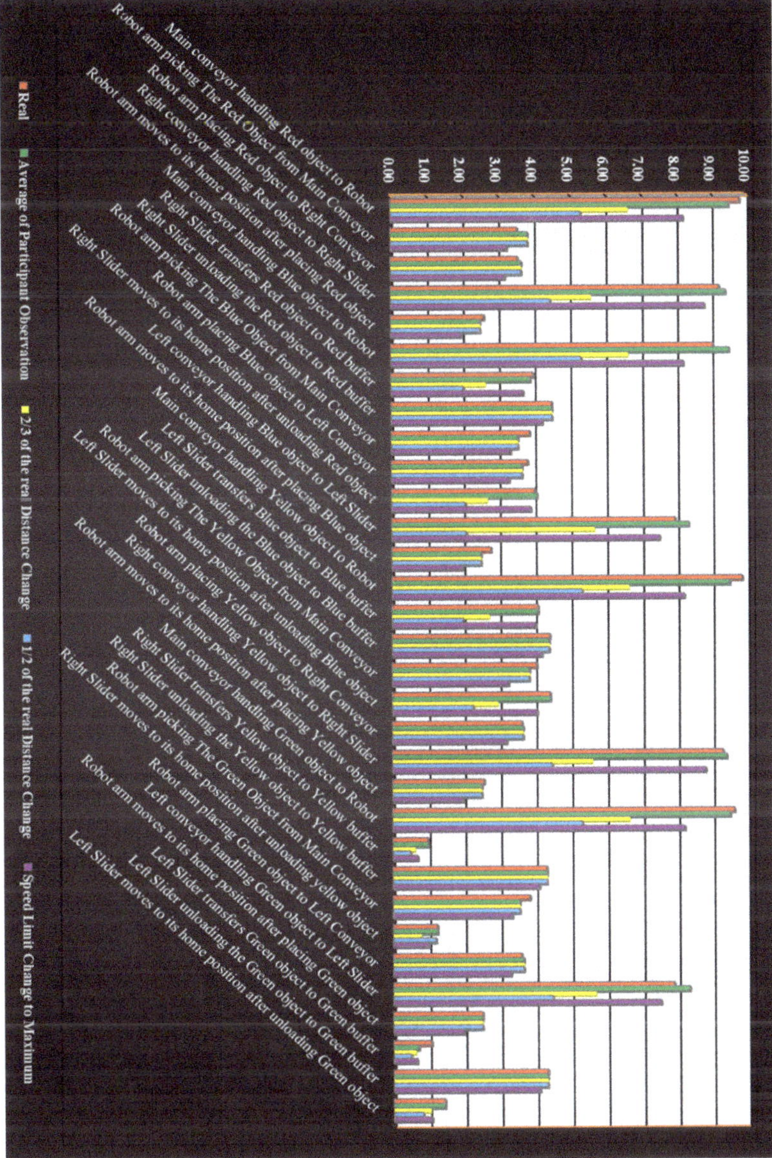

Figure 8. Comparison of "Busy" times before and after the optimization idea.

In order to investigate the effect of optimization solutions on OEE percentage, calculating the availability, performance, and quality are essential. As mentioned previously, the quality of the objects was considered as 100% because of the real system's characteristic (parts distribution). The factors required to obtain the Availability and Performance percentages were calculated individually for each system modification and each resource (Table 5).

Table 5. Overall equipment effectiveness (OEE) analysis before and after optimization.

Recourse	Actual Time	Expected Time	Down Time	Total Functional Period	Idle Time	Planned Production Time	Run Time	Availability %	Ideal Cycle Time	Total Count	Good Count	Performance %	Quality %	OEE %
						Speed to Maximum Limit								
Main Conveyor	32.80	32.50	0.30	58.36	25.56	32.80	32.50	99.09	8.00	4.00	4.00	98.46	100.00	97.56
Left Conveyor	15.00	14.50	0.50	78.88	62.88	16.00	15.50	96.88	7.50	2.00	2.00	96.77	100.00	93.75
Right Conveyor	17.60	17.50	0.10	78.33	60.73	17.60	17.50	99.43	8.00	2.00	2.00	91.43	100.00	90.91
Robot Arm	34.19	34.00	0.19	78.33	44.14	34.19	34.00	99.44	8.00	4.00	4.00	94.12	100.00	93.59
Right Sliders for Red	11.84	11.50	0.34	63.88	52.04	11.84	11.50	97.13	11.00	1.00	1.00	95.65	100.00	92.91
Right Sliders for Yellow	5.80	5.50	0.30	43.04	37.24	5.80	5.50	94.83	5.00	1.00	1.00	90.91	100.00	86.21
Left Slider for Blue	12.27	11.50	0.77	72.48	60.21	12.27	11.50	93.72	11.00	1.00	1.00	95.65	100.00	89.65
Left Slider for Green	5.83	5.50	0.33	13.25	7.42	5.83	5.50	94.34	5.00	1.00	1.00	90.91	100.00	85.76
						Distance 2/3								
Main Conveyor	26.52	26.00	0.52	56.23	29.71	26.52	26.00	98.04	6.00	4.00	4.00	92.31	100.00	90.50
Left Conveyor	11.36	11.00	0.36	75.09	63.73	11.36	11.00	96.83	5.00	2.00	2.00	90.91	100.00	88.03
Right Conveyor	11.20	11.00	0.20	75.09	63.89	11.20	11.00	98.21	5.00	2.00	2.00	90.91	100.00	89.29
Robot Arm	39.42	39.00	0.42	65.94	26.52	39.42	39.00	98.93	9.00	4.00	4.00	92.31	100.00	91.32
Right Sliders for Red	9.82	9.50	0.32	50.90	41.08	9.82	9.50	96.74	9.00	1.00	1.00	94.74	100.00	91.65
Right Sliders for Yellow	5.72	5.50	0.22	45.54	39.82	5.72	5.50	96.15	5.00	1.00	1.00	90.91	100.00	87.41
Left Slider for Blue	10.07	9.50	0.57	69.12	59.05	10.07	9.50	94.34	9.00	1.00	1.00	94.74	100.00	89.37
Left Slider for Green	5.97	5.50	0.47	29.79	23.82	5.97	5.50	92.13	5.00	1.00	1.00	90.91	100.00	83.75

Table 5. *Cont.*

Distance 1/2													
Recourse	Actual Time	Expected Time	Down Time	Total Functional period	Idle time	Planned Production Time	Run Time	Availability %	Ideal Cycle Time	Total Count	Good Count	Performance %	Quality %	OEE %
Main Conveyor	21.28	21.00	0.28	50.69	29.41	21.28	21.00	98.68	5.00	4.00	4.00	95.24	100.00	93.98
Left Conveyor	8.94	8.50	0.44	68.00	59.06	8.94	8.50	95.08	4.00	2.00	2.00	94.12	100.00	89.49
Right Conveyor	8.94	8.50	0.44	67.64	58.70	8.94	8.50	95.08	4.00	2.00	2.00	94.12	100.00	89.49
Robot Arm	39.13	39.00	0.13	60.41	21.28	39.13	39.00	99.67	9.00	4.00	4.00	92.31	100.00	92.00
Right Sliders for Red	9.13	9.00	0.13	43.96	34.83	9.13	9.00	98.58	8.50	1.00	1.00	94.44	100.00	93.10
Right Sliders for Yellow	5.51	5.50	0.01	42.52	37.08	5.44	5.43	99.82	5.00	1.00	1.00	92.08	100.00	91.91
Left Slider for Blue	8.67	8.50	0.17	62.37	53.70	8.67	8.50	98.04	8.00	1.00	1.00	94.12	100.00	92.27
Left Slider for Green	5.63	5.50	0.13	28.98	23.35	5.63	5.50	97.69	5.00	1.00	1.00	90.91	100.00	88.81

Before Optimization													
Recourse	Actual Time	Expected Time	Down Time	Total Functional period	Idle time	Planned Production Time	Run Time	Availability %	Ideal Cycle Time	Total Count	Good Count	Performance %	Quality %	OEE %
Main Conveyor	38.33	38.00	0.33	68.79	30.46	38.33	38.00	99.14	8.50	4.00	4.00	89.47	100.00	88.70
Left Conveyor	15.82	15.50	0.32	90.95	72.44	18.51	18.19	98.27	7.00	2.00	2.00	76.97	100.00	75.63
Right Conveyor	18.52	18.00	0.52	90.95	75.13	15.82	15.30	96.71	7.00	2.00	2.00	91.50	100.00	88.50
Robot Arm	40.4	40.00	0.40	90.95	50.55	40.40	40.00	99.01	9.00	4.00	4.00	90.00	100.00	89.11
Right Sliders for Red	12.49	12.00	0.49	68.81	56.32	12.49	12.00	96.08	11.00	1.00	1.00	91.67	100.00	88.07
Right Sliders for Yellow	6.55	6.00	0.55	52.33	45.77	6.56	6.01	91.62	5.00	1.00	1.00	83.19	100.00	76.22
Left Slider for Blue	12.94	12.00	0.94	84.11	71.17	12.94	12.00	92.74	11.00	1.00	1.00	91.67	100.00	85.01
Left Slider for Green	6.84	6.00	0.84	31.44	24.6	6.84	6.00	87.72	5.00	1.00	1.00	83.33	100.00	73.10

To obtain the effect of each optimization solution in each resource and choose the most effective one, Figure 9a–h have been provided to ease the investigation by visualizing the outcome of the system OEE and related factors after the optimization solution. Each part of Figure 9 shows the percentages of Availability, Performance, and OEE, respectively, for each resource.

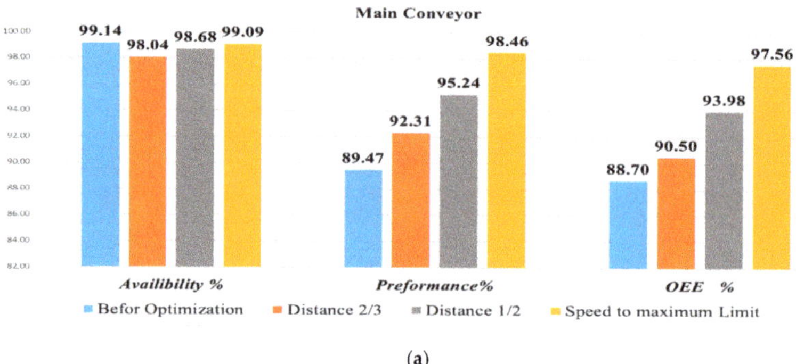

(a)

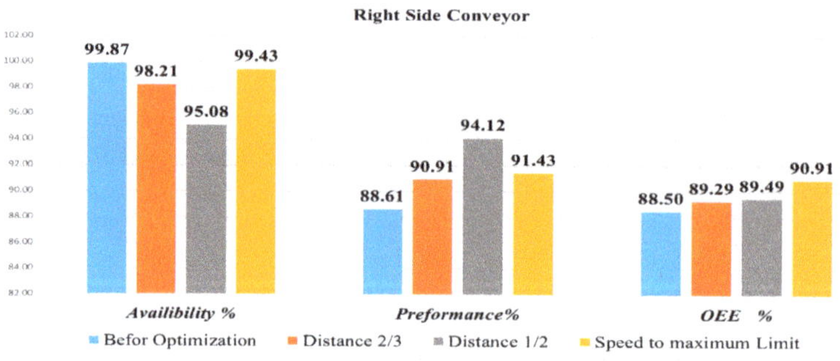

(b)

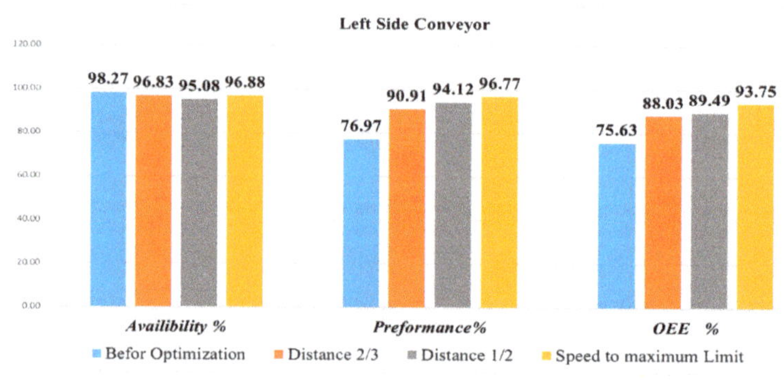

(c)

Figure 9. *Cont.*

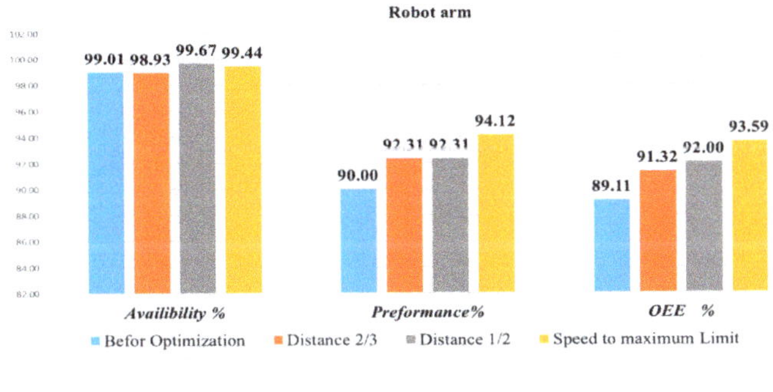

(**d**)

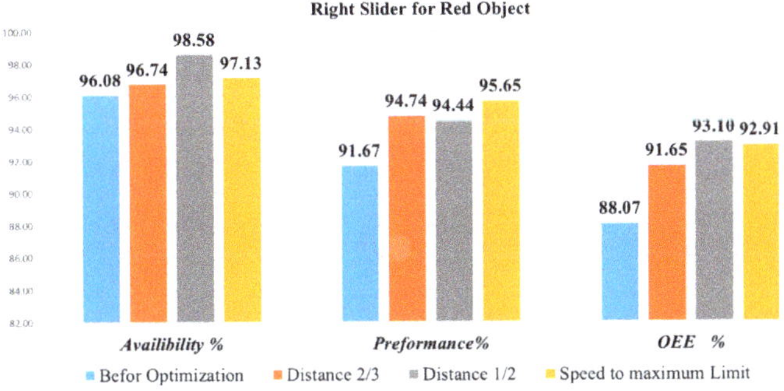

(**e**)

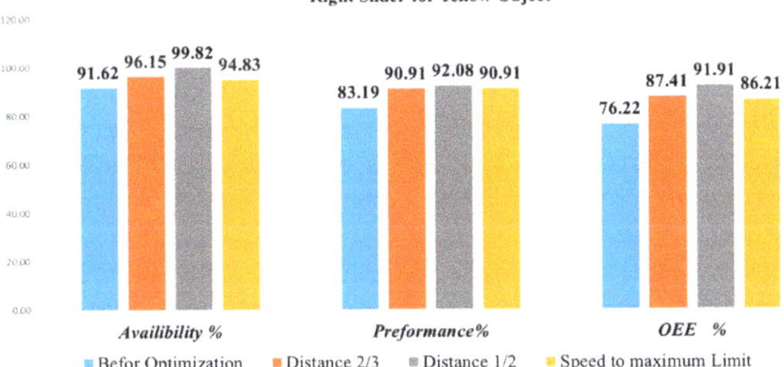

(**f**)

Figure 9. *Cont.*

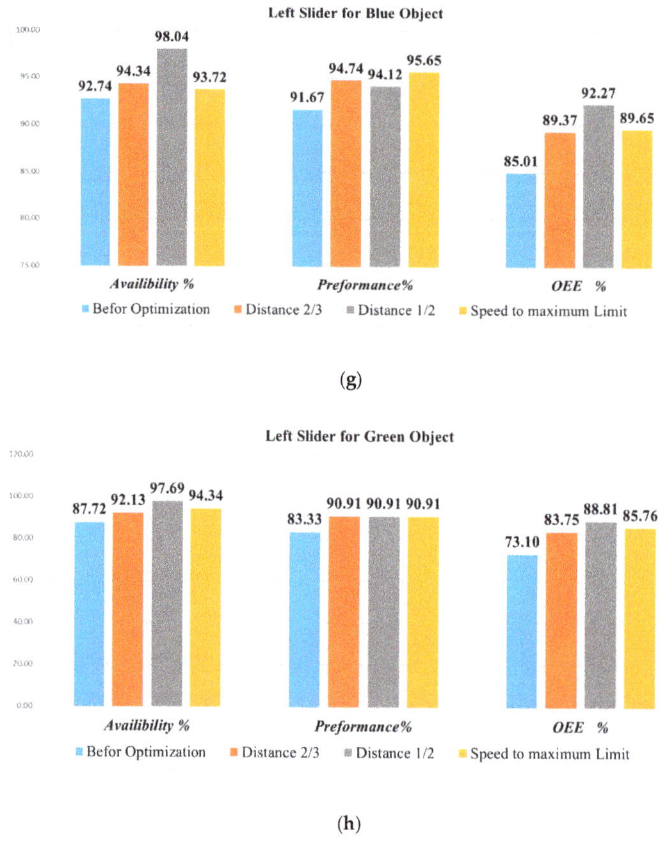

(g)

(h)

Figure 9. Comparison of OEE and its related factors' percentages.

Analyzing the outcome of the system modification on the simulation model verified that the overall equipment effectiveness (OEE) was significantly improved by each category of the optimization methods. However, this improvement has a direct relationship with system performance and availability. In almost all of the resources, all of the optimization methods increased the performance of the resources, but not equally.

Figure 9a–h shows that the layout design and changing the distance of the segments (decreasing the distances to $^2/_3$ and $\frac{1}{2}$ of the actual distance in the simulation model) by modifying the resources' layout design improves the OEE percentages of all of the resources which are connected, along with the conveyor units.

In the main conveyor, due to having two different segments (between sensors 1 and 2 and sensors 2 and 3) with different speeds, modifying the speed had the most effect on the OEE percentage. The OEE percentage of the conveyor was increased between 1.8% to 8.8%, in which the minimum effect was due to decreasing the distances to $^2/_3$ of the actual size, and the maximum effect was due to increasing the speed of the conveyor segments.

The same behavior of the OEE percentage changes was observed for the robot arm. However, it was not as significant, compared to the main conveyor. As long as the robot arm performance was affected by the main conveyor performance, it showed similar improvement to the main conveyor. Furthermore, the robot's motion speed is the only factor that can change the robot OEE percentage, but the speed depends on the robot's pick-and-place point distances to move the objects. This means that the distance and speed modification affected the Robot OEE percentages, and was between 2.21% and 4.48%.

The results which have been observed for the side conveyors are mostly like the main conveyor. The improvement in OEE percentage is between 0.79% to 2.41%, where the maximum improvement was the result of speed change. Negatively, the OEE percentage significantly improved in the left conveyor to between 12.4% to 18.12%. This considerable range of improvement is because of the equalization of the left conveyor length (this issue was discussed in the first stage of research).

Sliders have different behaviors in regard to the performance improvements by optimization methods. The percentage of OEE improves by changing the distance of the segments more than changing the speed. The right and left sliders are same in regard to the effectiveness of optimization methods for a short range of movement.

In the right slider, the OEE percentage of short-range transfer improved between 3.58% to 5.03%. The minimum and maximum values were the result of changing the distances to $^2/_3$ and $\frac{1}{2}$ of the actual distance. On the other hand, the OEE percentage improvement for the Left slider for a short range is between 4.64% to 7.26% for Speed and $\frac{1}{2}$ of the actual distance, respectively.

The left slider for short-range transfer had a 9.99% to 15.69% improvement for speed and $\frac{1}{2}$ of the actual distance change. Simultaneously, for the long range, there was a 12.66% to 15.71% improvement for modifying the speed to maximum limit and the distance to $\frac{1}{2}$ of the actual distance.

Table 6 shows the OEE percentage improvement in detail as the result of the optimization method, considering the most effective parameter as the mentioned objective function parameters.

Table 6. OEE improvement percentage for each resource.

Resource	OEE Improvement %			Priority of the Effectiveness		
	$\frac{1}{2}$ of the Actual Distance (p_{2-1})	$^2/_3$ of the Actual Distance (p_{2-2})	Speed to Maximum Limit (p_3)			
Main Conv.	5.28	1.30	8.86	p_{2-1} (2)	p_{2-2} (3)	p_3 (1)
Right side Conv.	0.99	0.79	2.41	p_{2-1} (2)	p_{2-2} (3)	p_3 (1)
Left Side Conv.	13.86	12.4	18.12	p_{2-1} (2)	p_{2-2} (3)	p_3 (1)
Robot arm	2.89	2.20	4.48	p_{2-1} (3)	p_{2-2} (2)	p_3 (1)
Right Slider for Red Obj.	5.03	3.58	4.48	p_{2-1} (1)	p_{2-2} (3)	p_3 (2)
Right slider for Yellow Obj.	15.69	11.19	9.90	p_{2-1} (1)	p_{2-2} (2)	p_3 (3)
Left Slider for Blue Obj.	7.26	4.36	4.46	p_{2-1} (1)	p_{2-2} (2)	p_3 (3)
Left Slider for Green Obj.	15.71	10.56	12.66	p_{2-1} (1)	p_{2-2} (3)	p_3 (2)

As the table above shows, all of the proposed optimization methods have improved the OEE percentage of the system simulation model. In addition, it is obvious that the Improvement percentages are not equal for all of the methods and neither for all of the resources. In order to generalize the solution to select the best method, a priority has been considered for all of the methods. This priority is based on the OEE percentage improvement amount.

As the general solution, to optimize the system for the main conveyor, right-side conveyor, left-side conveyor, and robot arm, the speed of the resources should be increased to the maximum limit, and there is no need to modify the distances between the segments due to the OEE percentage improvement being at the highest level. However, in some cases, increasing the speed is not a sufficient method because of the system control architecture limitations. This means that changing the distances between the segments will be the only solution, and as Table 6 shows, it is suggested to select the minimum distances between the segments by considering the layout design of the system.

This optimization of OEE could be generalized to all the systems, including resources which have material interaction by the conveyor and robot arm acting as the material handling system.

For slider units, or for the system generally containing a unit which slides a resource (i.e., Slider Robot arm), the optimization method will be different for a short and long range of motions. As Table 6 shows, for a short range of motions for both right and left sliders, changing the speed has the last priority. However, changing the distance between the segments further increases the OEE percentage. On the other hand, a long range of motion for the speed of both right and left sliders has second priority. This means that if the distance optimization is limited because of the layout design, hardware limitations make it impossible to decrease the distances to the minimum possible value, and changing the speed to the maximum limit is the most effective parameter to optimize the system.

It is worth mentioning that, as a general solution, it is possible to implement all the optimization methods if the enterprise has enough flexibility to handle all the modifications. However, in this research, SMEs are the target, where they are limited and less flexible to handle all of the proposed optimization methods, and it should be the chance for management to select the most effective method for increasing the OEE in the system.

The sustainability of the system is the next target to be evaluated after the investigation of the importance of time in the evaluation of OEE with utilizing simulation. To accomplish this task, the requirement to obtain sustainability has been illustrated. Identification of the environmental impacts on the target system is the first step toward the evaluation of sustainability. Yazdi et al. (2018) stated that out of the available environmental impacts (Table 7), energy consumption has the highest priority. The authors also stated that they selected this impact as the only effective one, due to the property of the target manufacturing system [27]. The authors also identified the most effective factors to consuming energy in each part (resource) of the target system (Table 8).

Table 7. Mapping impact and set priorities of the selected manufacturing system [27].

Mapping Impact and Set Priorities			
Impacts	**Definition**		**Priorities**
Water intensity	Consumption of water per unit of output	Not Selected	0
Residuals intensity	Generation of wastes per unit of output	Not Selected	0
Energy intensity	Energy consumed per unit of output	Selected	1
Renewable proportion of energy consumed	Used Energy from Sustainable Resources (%)	Not Selected	0
Greenhouse gas (GHG) intensity	GHGs produced during production per unit of output	Not Selected	0
Intensity of residual releases to air	Release of air emissions per unit of output	Not Selected	0
Intensity of residual releases to surface water	Release of effluents per unit of output	Not Selected	0
Natural Cover	The proportion of land occupied that is natural cover	Not Selected	0

Table 8. Measurement layer of the investigation of manufacturing sustainability [27].

Manufacturing System	Related Devices	Energy Consumption Reason	Energy Resource
Intelligent material handling system	Conveyors	Motors related to conveyor motions	Electricity
	Sliders	Motors related to Slider motions	
	Robot Arm	Motors Related to Robot Motions	
	Control units	processing the data	
	Sensors	excitation signal	

Results of the time study and properties of the target system shows that there is a direct relationship between the system efficiency and time. In other words, any improvement in the system with consideration of the proposed optimization ideas will directly affect the system times. It is again worthy of mentioning that there is a direct relationship between the energy consumption and system functional time. An OEE percentage evaluation of the system shows that consideration of the proposed control architecture and, in most cases, increasing the busy times and decreasing the idle time increases the OEE percentage. However, Table 6 shows that all the optimization ideas have different effects on increasing the OEE percentage. This means that, with considering the priority of the effectiveness on Table 6, the optimization methods which would further increase the sustainability of the system will be highlighted. In this case, in order to have more sustainable manufacturing, the amount of consumed electricity should be minimal (Figure 10). The way in which to reach this goal in the target system is by reducing the idle times and increasing the busy times. In addition, in most of the cases in which the idle time is not unpreventable, there should be some alternative solution. Running a device continuously on the system, when the devices are in idle time, drives up energy use and maintenance costs, which impacts on the sustainability of the manufacturing system.

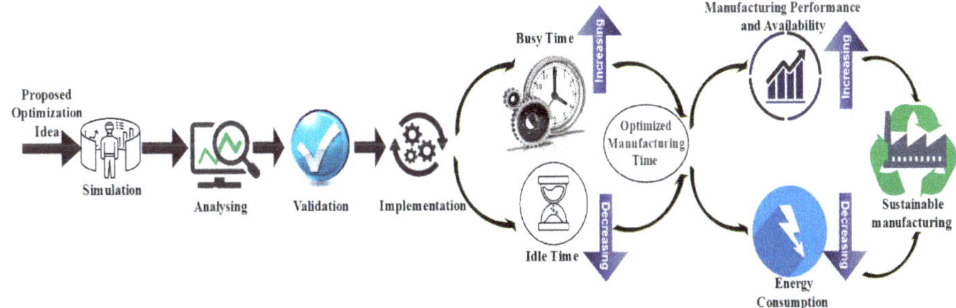

Figure 10. Manufacturing sustainability based on OEE and the Organization for Economic Cooperation and Development (OECD).

However, due to the property of the selected manufacturing system, and considering that as an example of SMEs and the energy resource which has been utilized, some limitations will show up. Calculations of energy consumption, OEE percentage evaluation, and its effects on sustainability of the target system require modification on the target system for it to perform for a longer time and consume more energy.

3. Conclusions

In the first stage of this research, an intelligent material-handling system for product differentiation was designed and implemented. A specific algorithm was created to deploy an agent-based control architecture across the system. The system was considered as the case study for this research and also as an example of SMEs with the same production cell. Different manufacturing aspects of the system were also evaluated. Time was focused on as the target manufacturing aspect. In order to evaluate the manufacturing system performance, OEE standards have been utilized. Due to the importance of time in the calculation of OEE percentages, the system was subjected to a proper time study. The OEE percentage was obtained for each resource available in the system, as well as the entire system. The results of the time study extracted some issues and limitations in the system during its performance. These problems and limitations were investigated in order to find a proper solution for them. They were categorized into different categories, and the effective parameters for each category have been achieved. Based on these achievements, some optimization methods have been proposed. In the second stage of the research, due to the SMEs being the target category of the enterprises, as well as their limitation and the difficulties which they might face in the implementation of the proposed methods, a simulation solution using ARENA software was considered. Each of the optimization methods were implemented on the simulation model to verify and validate the effects on the OEE percentage. A comprehensive solution was proposed for selecting the best optimization method and obtaining the best OEE percentage for SMEs. Finally, according to OECD definitions about the relationship between the time and energy consumption, manufacturing sustainability has also been investigated and described.

As mentioned previously, the system is a small-scale educational manufacturing system with a limited functionality period. Because of this selection, in the proposed methodology, system observation and data collection were conducted in a short period of time to prevent any system damage. As future work based on this research, the proposed methodology can be extended on such systems as a case study in order to execute all the evaluations over a long period of time.

Author Contributions: Conceptualization, G.Y.P. and A.A.; Methodology, G.Y.P.; Software, G.Y.P.; Validation, G.Y.P., A.A. and M.H.; Formal Analysis, G.Y.P.; Investigation, G.Y.P., A.A.; Resources, G.Y.P.; Data Curation, G.Y.P. and A.A.; Writing—Original Draft Preparation, G.Y.P.; Writing—Review & Editing, G.Y.P. and A.A.; Visualization, G.Y.P.; Supervision, A.A. and M.H.; Project Administration, A.A.

Funding: This research received no external funding.

Acknowledgments: This research was supported by German university of technology in Oman (GUtech). We thank our colleagues from GUtech who provided insight and expertise that greatly assisted the research.

Conflicts of Interest: The authors declare no conflict of interest.

References

1. Azizi, A.; Yazdi, P.G.; Hashemipour, M. Interactive design of storage unit utilizing virtual reality and ergonomic framework for production optimization in manufacturing industry. *Int. J. Interact. Des. Manuf. (IJIDEM)* **2018**, 1–9. [CrossRef]
2. Jamali, D.; Lund-Thomsen, P.; Jeppesen, S. Smes and csr in developing countries. *Bus. Soc.* **2017**, *56*, 11–22. [CrossRef]
3. Ahn, J.M.; Minshall, T.; Mortara, L. Open innovation: A new classification and its impact on firm performance in innovative smes. *J. Innov. Manag.* **2015**, *3*, 33–54.
4. Barbosa, J.; Leitão, P.; Adam, E.; Trentesaux, D. Dynamic self-organization in holonic multi-agent manufacturing systems: The adacor evolution. *Comput. Ind.* **2015**, *66*, 99–111. [CrossRef]
5. Hossain, M.; Kauranen, I. Open innovation in smes: A systematic literature review. *J. Strategy Manag.* **2016**, *9*, 58–73. [CrossRef]
6. Mourtzis, D.; Papakostas, N.; Mavrikios, D.; Makris, S.; Alexopoulos, K. The role of simulation in digital manufacturing: Applications and outlook. *Int. J. Comput. Integr. Manuf.* **2015**, *28*, 3–24. [CrossRef]
7. Esmaeilian, B.; Behdad, S.; Wang, B. The evolution and future of manufacturing: A review. *J. Manuf. Syst.* **2016**, *39*, 79–100. [CrossRef]
8. Michalos, G.; Makris, S.; Papakostas, N.; Mourtzis, D.; Chryssolouris, G. Automotive assembly technologies review: Challenges and outlook for a flexible and adaptive approach. *CIRP J. Manuf. Sci. Technol.* **2010**, *2*, 81–91. [CrossRef]
9. Islam, A.; Tedford, D.; Haemmerle, E. Risk determinants of small and mediumsized manufacturing enterprises (SMES)—An empirical investigation in New Zealand. *J. Ind. Eng. Int.* **2012**, *8*, 12. [CrossRef]
10. Chan, F.; Ip, R.; Lau, H. Integration of expert system with analytic hierarchy process for the design of material handling equipment selection system. *J. Mater. Process. Technol.* **2001**, *116*, 137–145. [CrossRef]
11. Bhattacharya, A.; Sarkar, B.; Mukherjee, S. A holistic multi-criteria approach for selecting machining processes. In Proceedings of the 30th International Conference on Computers & Industrial Engineering, Tinos Island, Greece, 28 June–1 July 2002; pp. 83–88.
12. Shen, W.; Norrie, D.H.; Barthès, J.-P. *Multi-Agent Systems for Concurrent Intelligent Design and Manufacturing*; CRC Press: Boca Raton, FL, USA, 2003.
13. Fax, J.A.; Murray, R.M. Information flow and cooperative control of vehicle formations. *IEEE Trans. Autom. Control* **2004**, *49*, 1465–1476. [CrossRef]
14. Durieux, S.; Pierreval, H. Regression metamodeling for the design of automated manufacturing system composed of parallel machines sharing a material handling resource. *Int. J. Prod. Econ.* **2004**, *89*, 21–30. [CrossRef]

15. Schröder, R.; Aydemir, M.; Glodde, A.; Seliger, G. Design and verification of an innovative handling system for electrodes in manufacturing lithium-ion battery cells. *Procedia CIRP* **2016**, *50*, 641–646. [CrossRef]

16. Johnstone, M.; Creighton, D.; Nahavandi, S. Status-based routing in baggage handling systems: Searching verses learning. *IEEE Trans. Syst. Man Cybern. Part C (Appl. Rev.)* **2010**, *40*, 189–200. [CrossRef]

17. Lau, H.Y.; Woo, S. An agent-based dynamic routing strategy for automated material handling systems. *Int. J. Comput. Integr. Manuf.* **2008**, *21*, 269–288. [CrossRef]

18. Ramesh, C.; Manickam, C.; Prasanna, S. Lean six sigma approach to improve overall equipment effectiveness performance: A case study in the indian small manufacturing firm. *Asian J. Res. Soc. Sci. Humanit.* **2016**, *6*, 1063–1072. [CrossRef]

19. Benjamin, S.J.; Marathamuthu, M.S.; Murugaiah, U. The use of 5-whys technique to eliminate oee's speed loss in a manufacturing firm. *J. Qual. Maint. Eng.* **2015**, *21*, 419–435. [CrossRef]

20. Yaşin, M.F.; Daş, G.S. Kobİ'lerde ekipman etkinliğinin iyileştirilmesinde tee tabanlı yeni bir yaklaşım: Bir ahşap işleme kuruluşunda uygulama. *J. Fac. Eng. Archit. Gazi Univ.* **2017**, *32*, 45–52.

21. Iannone, R.; Nenni, M.E. Managing oee to optimize factory performance. In *Operations Management*; InTech: London, UK, 2013.

22. Nakajima, S. *Introduction to TPM: Total Productive Maintenance*; Preventative Maintenance Series, Hardcover; Productivity Press: New York, NY, USA, 1988; ISBN 0-91529-923-2.

23. Huang, S.H.; Dismukes, J.P.; Shi, J.; Su, Q.; Razzak, M.A.; Bodhale, R.; Robinson, D.E. Manufacturing productivity improvement using effectiveness metrics and simulation analysis. *Int. J. Prod. Res.* **2003**, *41*, 513–527. [CrossRef]

24. Jonsson, P.; Lesshammar, M. Evaluation and improvement of manufacturing performance measurement systems-the role of oee. *Int. J. Oper. Prod. Manag.* **1999**, *19*, 55–78. [CrossRef]

25. Muchiri, P.; Pintelon, L. Performance measurement using overall equipment effectiveness (oee): Literature review and practical application discussion. *Int. J. Prod. Res.* **2008**, *46*, 3517–3535. [CrossRef]

26. Kumar, J.; Soni, V.; Agnihotri, G. Maintenance performance metrics for manufacturing industry. *Int. J. Res. Eng. Technol.* **2013**, *2*, 136–142.

27. Ghafoorpoor Yazdi, P.; Azizi, A.; Hashemipour, M. An empirical investigation of the relationship between overall equipment efficiency (oee) and manufacturing sustainability in industry 4.0 with time study approach. *Sustainability* **2018**, *10*, 3031. [CrossRef]

28. Stevenson, W.J.; Hojati, M.; Cao, J. *Operations Management*; McGraw-Hill/Irwin Boston: Boston, MA, USA, 2007; Volume 8.

29. Syska, A. Methods-time-measurement (mtm). In *Produktionsmanagement: Das A–Z Wichtiger Methoden und Konzepte für die Produktion von Heute*; Springer: Berlin/Heidelberg, Germany, 2006; p. 99.

30. Zandin, K.B. *Most Work Measurement Systems*; CRC Press: Boca Raton, FL, USA, 2002.

31. Heyde, G. Modular Arrangement of Predetermined Time Standards. Woodbridge, VA. 1996. Available online: www.modapts.com (accessed on 8 March 2019).

32. Patel, N. Reduction in product cycle time in bearing manufacturing company. *Int. J. Eng. Res. Gen. Sci.* **2015**, *3*, 466–471.

33. Elnekave, M.; Gilad, I. Rapid video-based analysis system for advanced work measurement. *Int. J. Prod. Res.* **2006**, *44*, 271–290. [CrossRef]

34. Bon, A.T.; Daim, D. Time motion study in determination of time standard in manpower process. In Proceedings of the 3rd Engineering Conference on Advancement in Mechanical and Manufacturing for Sustainable Environment, Kuching, Malaysia, 14–16 April 2010.

35. Longo, F.; Mirabelli, G. Effective design of an assembly line using modelling and simulation. *J. Simul.* **2009**, *3*, 50–60. [CrossRef]

36. Kuhlang, P.; Edtmayr, T.; Sihn, W. Methodical approach to increase productivity and reduce lead time in assembly and production-logistic processes. *CIRP J. Manuf. Sci. Technol.* **2011**, *4*, 24–32. [CrossRef]

37. Negahban, A.; Smith, J.S. Simulation for manufacturing system design and operation: Literature review and analysis. *J. Manuf. Syst.* **2014**, *33*, 241–261. [CrossRef]

38. Fui-Hoon Nah, F.; Lee-Shang Lau, J.; Kuang, J. Critical factors for successful implementation of enterprise systems. *Bus. Process Manag. J.* **2001**, *7*, 285–296. [CrossRef]

39. Kelton, W.D.; Smith, J.S.; Sturrock, D.T. *Simio & Simulation: Modeling, Analysis, Applications*; Learning Solutions: Sacramento, CA, USA, 2011.

40. Fu, M.C. Optimization for simulation: Theory vs. Practice. *Inf. J. Comput.* **2002**, *14*, 192–215. [CrossRef]
41. Juan, A.A.; Faulin, J.; Grasman, S.E.; Rabe, M.; Figueira, G. A review of simheuristics: Extending metaheuristics to deal with stochastic combinatorial optimization problems. *Oper. Res. Perspect.* **2015**, *2*, 62–72. [CrossRef]
42. Schriber, T.J.; Brunner, D.T.; Smith, J.S. Inside discrete-event simulation software: How it works and why it matters. In Proceedings of the 2014 Winter Simulation Conference, Savannah, GA, USA, 7–10 December 2014; pp. 132–146.
43. Alfieri, A.; Matta, A.; Pedrielli, G. Mathematical programming models for joint simulation–optimization applied to closed queueing networks. *Ann. Oper. Res.* **2015**, *231*, 105–127. [CrossRef]
44. Alfieri, A.; Matta, A. A time-based decomposition algorithm for fast simulation with mathematical programming models. In Proceedings of the Winter Simulation Conference, Berlin, Germany, 9–12 December 2012; p. 231.
45. Figueira, G.; Almada-Lobo, B. Hybrid simulation–optimization methods: A taxonomy and discussion. *Simul. Model. Pract. Theory* **2014**, *46*, 118–134. [CrossRef]
46. Toolkit, O.S.M. *Seven Steps to Environmental Excellence*; The Organization for Economic Co-operation and Development: Paris, France, 2014.
47. Roni, M.; Jabar, J.; Mohamad, M.; Yusof, M. Conceptual study on sustainable manufacturing practices and firm performance. In Proceedings of the International Symposium on Research in Innovation and Sustainability, ISoRIS, Malacca, Malaysia, 15–16 October 2014; pp. 1459–1465.
48. Lai-Ling Lam, M. Challenges of sustainable environmental programs of foreign multinational enterprises in china. *Manag. Res. Rev.* **2011**, *34*, 1153–1168. [CrossRef]
49. May, A.D. Urban transport and sustainability: The key challenges. *Int. J. Sustain. Transp.* **2013**, *7*, 170–185. [CrossRef]
50. Azizi, A.; Yazdi, P.; Humairi, A. Design and fabrication of intelligent material handling system in modern manufacturing with industry 4.0 approaches. *Int. Robot. Autom. J.* **2018**, *4*, 186–195. [CrossRef]

Article

Hit or Miss? Evaluating the Potential of a Research Niche: A Case Study in the Field of Virtual Quality Management

Albert Weckenmann [1], Ştefan Bodi [2,*], Sorin Popescu [2], Mihai Dragomir [2], Dan Hurgoiu [2] and Radu Comes [2]

[1] Friedrich-Alexander University Erlangen-Nürnberg; 4 Schloßplatz, 91054 Erlangen, Germany; aweckenmann@web.de

[2] Technical University of Cluj-Napoca, 103-105 Muncii Blvd., 400641 Cluj-Napoca, Romania; sorin.popescu@muri.utcluj.ro (S.P.); mihai.dragomir@muri.utcluj.ro (M.D.); dan.hurgoiu@muri.utcluj.ro (D.H.); radu.comes@muri.utcluj.ro (R.C.)

* Correspondence: stefan.bodi@muri.utcluj.ro; Tel.: +40-740-221-872

Received: 3 February 2019; Accepted: 4 March 2019; Published: 8 March 2019

Abstract: When knowledge is developed fast, as it is the case so often nowadays, one of the main difficulties in initiating new research in any field is to identify the domain's specific state-of-the-art and trends. In this context, to evaluate the potential of a research niche by assisting the literature review process and to add a new and modern large-scale and automated dimension to it, the paper proposes a methodology that uses "Latent Semantic Analysis" (LSA) for identifying trends, focused within the knowledge space created at the intersection of three sustainability-related methodologies/concepts: "virtual Quality Management" (vQM), "Industry 4.0", and "Product Life-Cycle" (PLC). The LSA was applied to a significant number of scientific papers published around these concepts to generate ontology charts that describe the knowledge structure of each by the frequency, position, and causal relation of associated notions. These notions are combined for defining the common high-density knowledge zone from where new technological solutions are expected to emerge throughout the PLC. The authors propose the concept of the knowledge space, which is characterized through specific descriptors with their own evaluation scales, obtained by processing the emerging information as identified by a combination of classic and innovative techniques. The results are validated through an investigation that surveys a relevant number of general managers, specialists, and consultants in the field of quality in the automotive sector from Romania. This practical demonstration follows each step of the theoretical approach and yields results that prove the capability of the method to contribute to the understanding and elucidation of the scientific area to which it is applied. Once validated, the method could be transferred to fields with similar characteristics.

Keywords: latent semantic analysis; virtual quality management; concept investigation; concept disambiguation; knowledge discovery; sustainable methodologies

---

1. Introduction

In the paradigm of the knowledge-based society, knowledge becomes an indispensable resource in the further development of the contemporary social and economic status. Hence, in recent years, for supporting the required advancements of modern society, the speed through which knowledge is generated increased exponentially. Yang Lu in Reference [1] associates this phenomenon with the Fourth Industrial Revolution and the emergence of the "Internet-of-Things", "Cyber-Physical Systems", and "Enterprise Integration"; however, the increasing computational capacity and connectivity of computers could be mentioned as facilitators. Often, this high dynamic in research (knowledge generation), especially in interdisciplinary fields, brings the need to express new specific contents, meanings or trends under a linguistic form. In this way, new syntagmas or concepts are brought to light and adopted by professionals as part of the specific jargon in the field. Even if their creators endowed them with a clear meaning at an incipient stage, when they become more popular in an emerging area, these concepts are quickly surrounded by a large amount of new knowledge that is developed with an amazing speed, enriching and enlarging their initial sphere.

The "virtual Quality Management" (vQM) concept could be a significant example for the circumstances described previously. It is born through a semantic operation, joining two established and mature concepts: "virtual" and "QM", thus it is representative for an area which is in a period of high dynamic development and of interest for companies preoccupied with sustainability from the perspective of operations management and organizational culture.

In this context in which the amount of information relating to new concepts quickly reaches unmanageable levels, regardless of the field, solutions that can analyze extended documentation with the purpose of disambiguating information and capturing the essentials, thus creating knowledge, become the focus of attention and gain in importance. Traditional solutions for that purpose lay in the literature review process, trying to collect, select, filter, and structure the existing and relevant information in a synthetic way. However, this solution has limits: on one hand the large amounts of data can lead to a decrease in the ability of a research team to handle that information and on the other, the selection and interpretation of the chosen references is exposed to strong subjectivity.

Natural Language Processing (NLP), a field of computational linguistics, can be considered as a possible alternative for the above stated issues making users capable of going through large amounts of information fast, retaining only important data. It deals with translating human language to computers, enabling them to summarize and essentialize inputted information from an unlimited number of references. The computer programs deployed for this task use various algorithms, each with their own advantages and disadvantages. Among these worth mentioning are: Part-of-Speech Tagging; Named Entity Recognition; Semantic Role Labeling; and Latent Semantic Analysis (LSA).

The current paper proposes a framework for identifying trends, based on LSA by analyzing the knowledge space resulting from the intersection of two knowledge structures given by the vQM and Industry 4.0 concepts.

The investigatory process for these two concepts was done by applying LSA to a significant number of scientific papers published around them, selected after they underwent an initial raw filtration process proposed by the authors. As a result of this analysis, the generated ontology charts defined the knowledge structures of the two concepts by the frequency, position in text, and causal relation of associated notions. Combining these notions with one another resulted in senseful descriptors. The common ones were retained and further used for identifying new technological solutions and industrial applications, which are considered to emerge from this particular intersection of concepts. They were finally sorted and ranked within the PLC stages using the affinity diagram (KJ diagram) and the matrix diagram.

The intended added value of this approach is to supersede the classic literature review in the state-of-the-art analysis with automated information processing tools, both classic (affinity diagramming, conceptual mapping and analysis—based on ontology charts) and modern (Latent Semantic Analysis) for investigating complex knowledge spaces born at the intersection of new concepts. The illustration of this endeavor was done with the help of the vQM and Industry 4.0 concepts, and its end result is the current and emerging practical Industry 4.0 applications related to vQM throughout the entire PLC stages, particularly in the automotive industry.

2. Objective of the Research

The main objective of the research presented in this paper is to develop an innovative and expedient approach to evaluate the result potential and dissemination outlook of a research topic at the beginning of an investigation endeavor, especially in emerging scientific domains that are characterized by a high degree of entropy and uncertainty. The presented methodology relies on the use of information processing tools, techniques, and methods (combing both classic and new) to address the most common concerns of research teams and help them arrive at a timely answer with an acceptable degree of validity to the issue concerning the relation between invested resources (time, staff, equipment, etc.) and foreseeable impact.

In this regard, the research approached in the current paper proposes, firstly, to analyze the knowledge density zone obtained by intersecting two knowledge structures, represented here by the vQM and Industry 4.0 axis, throughout the PLC stages (Figure 1).

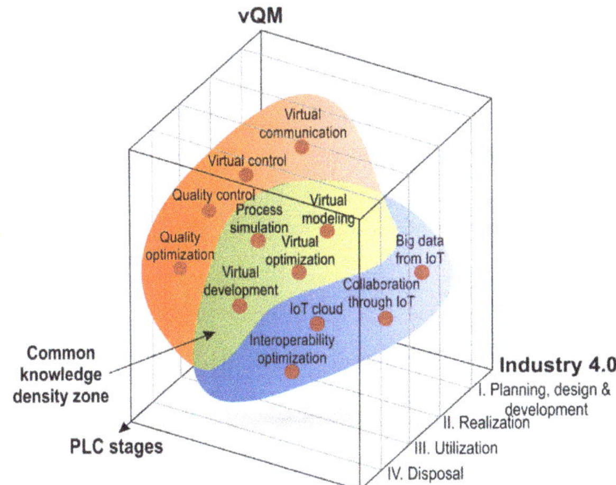

Figure 1. Knowledge density zone between vQM and Industry 4.0 along the PLC stages.

Secondly, in addition to concept investigation and disambiguation, it aims to identify and hierarchize industrial applications (or new technological solutions) that are expected to emerge.

Finally, the paper intends to validate the proposed methodology through an online survey.

The need to develop this algorithm appeared to the authors due to their involvement in high dynamic manufacturing research demarches, where they were compelled to generate practical solutions for companies before the theories were settled. As such, the validation included in this paper was also based on vQM and Industry 4.0 concepts but can be transferred to any domain.

Contributing to the achievement of these endeavors the following stages were established:

- In the first stage an identification of relevant papers and scientific references was carried out for each of the two main concepts individually: vQM and Industry 4.0. After that, a filtration process was performed for limiting as much as possible the amount of irrelevant information. The ontology charts, which were both similar in nature as two knowledge structures, both for vQM and Industry 4.0, were determined by applying LSA (using a dedicated software program) on the final sets of bibliographic references.

- The second stage was reserved for weighting notions that were in relation with the main concepts. Based on the ontology charts and other information provided by the software each notion relating to vQM and Industry 4.0 was ranked by calculating their CN (correlation number). Next, associations were made between them with the scope of obtaining syntagmas (descriptors) that characterized the two main concepts. A score was also calculated for each of them signifying their importance to both concepts.

- The third stage began with extracting the common meaningful descriptors between the two main concepts. Based on these expressions, applications were identified using search engines.

This way the focus was kept only on those applications that were in close reference to vQM and Industry 4.0.

- The fourth stage, firstly, was focused on identifying the product life cycle stages and secondly, on the construction and completion of the matrix diagram. It had, as inputs on the left, the product life cycle stages together with identified applications and above the common descriptors between vQM and Industry 4.0 (associated by their weighted importance). It illustrates not only the hierarchy of the main categories of applications throughout the product life cycle stages, but also the rank of every sub-category application within the main categories.

- Within the fifth stage, an online questionnaire was applied for checking results obtained from the matrix diagram against an average rating coming from experts from Romania's automotive industry.

The research methodology is summarized in the following figure (Figure 2) showing also inputs and outputs for all its phases in the form of an extended flowchart:

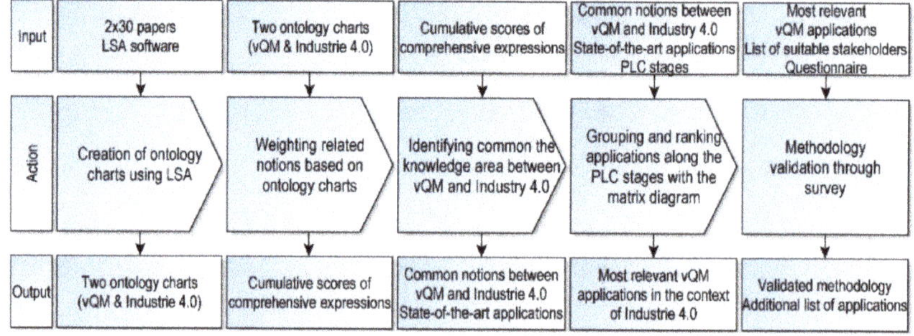

Figure 2. Proposed methodology overview.

3. The Review of the Scientific Literature

Considering the objectives set forth by the paper (detailed previously), in this section the authors assembled the elements relating to the background and literature review into two categories: aspects relating to the need to assist the literature review process with modern instruments and techniques; and background and aspects which related to the main concepts forming the analyzed knowledge space.

3.1. The Need for New Approaches

If the statement that the development of the current society is strongly linked to that of knowledge is already a truism, what seemed fascinating to scientists was and is its speed of growth. The study of knowledge growth has enjoyed attention starting with Price's theory on the exponential growth of science [2], passing through Buckminster Fuller's knowledge doubling curve [3], and arriving at recent works such as Reference [4] which summarizes mathematical models for knowledge growth or References [5,6] which focus on modern digital knowledge spreading channels, to mention just some of the countless publications on this topic in these last years.

At the communication level, new knowledge is asking usually for new concepts to formalize its significance. The emergence of new concepts or concept groups generate in their semantic proximity domain-specific or interdisciplinary knowledge spaces. Any connected new research pass by an exploratory phase when a preliminary investigation of the mentioned spaces is necessary to identify the state-of-the-art, directions or the trending elements in its area. This exploratory investigation is often carried out with the help of literature reviews, which are complex, laborious, and time-consuming tasks, whose results are strongly dependent on the author's critical thinking ability, experience, and competence in the domain. The unprecedented increase in the amount of information that needs to be processed to capture the relevant aspects of the knowledge space in question leads to either the reduction of the reference sample with the risk of losing important inferences from the analysis result, or the use as support of modern computer-assisted tools able to deliver a primary structure of the conceptual framework for the targeted knowledge spaces.

3.2. Literature Review Methods and Tools

A scientific literature review "has become the most common way of acquiring knowledge and oftentimes sets the direction for a study" [7]. Alongside the methods used and inputted constraints, the type of review can differ. The authors [8] identified a typology set of 14 main reviews for investigating and evaluating specialized literature, out of which worth mentioning are: "critical review"; "mapping review"; "rapid review"; "state-of-the-art review"; and "systematic search and review".

From the tools and techniques' perspective, two categories are identifiable: classic tools and those that use computer-assisted instruments. One might argue that, currently, literature reviews are carried out using tools from the latter category; however, some that have more of a classic background are still promoted: concept mapping (structuring the knowledge from a certain area based on main concepts in the form of highly-visual impact maps) [9,10]; affinity diagramming (structuring main ideas extracted from an initial literature search) [11]; snowball sampling (finding a network of professionals specific to a certain area by conducting interviews and asking them for other references) [12]; and literature tables used both for managing and ensuring the traceability of identified references, and for summarizing information and extracting main ideas.

The computer assisted tools are comprised of software programs, that help the user in making this review more efficient considering the following points:

- Researching and identifying relevant literature: Google Scholar; Distiller; arXiv; Citavi; Delve Health; LitAssist;
- Managing references: EndNote; Aigaion; Mendeley; Zotero; LiteRat; Bookends;
- Completing an automated text analysis and recognition (known as natural language processing (NLP)) for assisting the reviewer in classifying, tagging, parsing, semantic reasoning, and reducing the amount of information to a manageable size [13–15], but at the same time keeping its coherence, intelligibility, and meaning of the analyzed text: WordStat; QDA Miner; Carrot^2; RapidMiner; Tropes.

3.3. Natural Language Processing Approaches

Natural Language Processing (NLP), a field of computational linguistics, deals with translating human language to computers enabling them to understand, summarize, and essentialize inputted information from an unlimited number of references. An analysis of the specialty literature categorized mainly used models and algorithms (incorporated into software solutions) into ones that complete a basic text processing (part-of-speech tagging; named entity recognition), ones that use statistical language models (bag-of-words model; n-gram models; neural network language model), and ones that discern the text based on the semantical meaning of words within (semantic role labeling; latent semantic analysis). Based on these, the software assists the user by increasing the quality of the review and the amount of information subjected to the analysis, while it reduces the time in which the user can process individually that same amount of text (literature).

Part-Of-Speech (POS) Tagging assigns parts of speech to words from a text body, allowing words having the same written form, but different meanings in different contexts to be understood accordingly by the software. Developed by Reference [16] and later advanced by other contributors [17–19], the method is enjoying recent applications regarding the analysis of written information on social media websites [20–22], dealing with grammar related aspects [23,24], identifying plagiarism [25], or even acting as a human–machine interface [26].

Named Entity Recognition enables the localization and categorization of "important and proper nouns in a text" [27], helping the reviewer to focus on important concepts that characterize the text. According to Reference [28], "this is an important task because its performance directly affects the quality of many succeeding NLP applications such as information extraction". Its application recently gained popularity for processing semi-structured knowledge bases regarding entity disambiguation/mapping [29–31] and extracting/retrieving information [32] or for analyzing content generated on social media [33–35].

The Bag-of-Words Model identifies and represents by histograms the frequency of words appearing in the text [36] enabling the reviewer to conduct an initial text categorization. Although its application is grounded in NLP (e.g., opinion mining on information generated by social media users [37,38]), recent approaches prove its utility in other fields as well, such as image processing [39–41].

n-Gram Models help determine the probability of a sequence of words in a sentence or in a text. Their application varies from identifying patterns in text [42] to data extraction [43], automatic speech recognition, machine translation, and spell checking [44,45]. Neural Network Language models offer an improved version [46], both having the potential to be integrated into computer-assisted tools for supporting text reviewers.

Semantic Role Labeling traditionally uses a shallow parsing algorithm that identifies "arguments within the local context of a predicate" [47] enabling the software solutions, to a certain degree, to summarize information [48]. However, recent advancements, presented by References [49] and [50] outline the possibility for using this technique without syntactic parsing.

Latent Semantic Analysis (LSA) analyzes the relationship between terms contained in several documents and outputs a set of concepts that are semantically related to the terms and documents subjected to the analysis, and has multiple applications in the fields of information processing and knowledge discovery [51–53]. It uses Single Value Decomposition (SVD), a mathematical tool that allows the reduction of matrix rows, without seriously compromising the structure of columns. The matrix rows m (constructed out of the body of text) are the individual word types and the columns n represent the meaning of that word encoded into sentences and paragraphs. The cells that result contain information about the reoccurrence of a word in a paragraph [54]. After applying the SVD, and based on information obtained from it, an additional step is carried out, "sentence selection", in which the "most characteristic parts of text" are selected to illustrate ideas that are essential in the body of text [55].

3.4. Virtual Quality Management (vQM)

The expansion of applications in the virtual domain opened new horizons also in the case of "Quality Management", which evolved alongside the trends from modern industry. A new concept was thus born called "virtual Quality Management" (vQM).

The concept of vQM is based on, but not limited to, simulation studies, which are efficiently deployed for the sole purpose of "generating resilient knowledge and dimensioning quality techniques" [56] that can be applied either to products or processes, before they physically exist. By doing so, products and processes reach a certain level of maturity in the planning stage, so they can be introduced straight into production, having an increased level of performance compared to ones developed using conventional methods.

As identified from several references approaching the virtual area of quality management in manufacturing, simulation also plays an important role for vQM in developing models capable to foresee, adapt to, and optimize different scenarios by analyzing existing data [57], to increase efficiency and implement sustainable development [58] or even to reproduce measurement uncertainty due to temperature, humidity or other external influences [59]. Moreover, the approach proposed by Reference [60], regarding stochastic simulations for tolerance analysis "contributes to the improvement of virtual quality assessment", which in this case, takes place already in the design stage of a product's life cycle.

The combination between "simulation" and "virtual reality" allows operators to interact virtually with the fabrication process [61], which in term enables virtual control and even virtual inspection of the manufacturing flow [62]. Immersive virtual reality applications can also be used efficiently in problem-solving activities, as demonstrated by Reference [63].

The nature of vQM is highly innovative as it is capable to provide the necessary information for deploying quality management techniques before the actual start of the manufacturing processes [59]. Quality and process parameters are obtained with the help of advanced modeling and simulation tools, all contributing to the process design of a manufacturing system. The architecture of vQM, presented in Figure 3, illustrates precisely the above stated particularities.

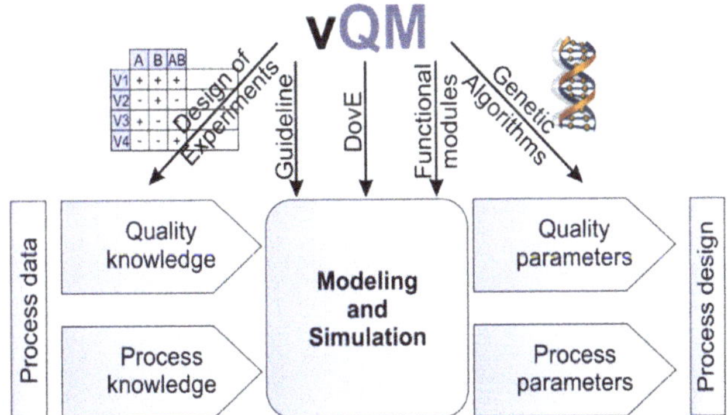

Figure 3. Architecture of vQM. Adapted from Reference [65].

Another important side of vQM is communication. The development of virtual channels (supported by the internet) brought several advantages in this field, such as improving the way we interact with each other or even with machines. They also increased the speed of information transfer, narrowed the gap between supplier and customer, and enabled product developers to provide their contributions remotely by forming "virtual teams" [64].

Although the definition of the concept is quite clear, the impact vQM has on current production engineering and quality management related concepts and how it interacts with them is yet to be found.

A possible research instrument that can exactly capture and contribute to the disambiguation of the vQM concept is LSA. This type of instrument is proposed because the clarification process should begin right at the semantic value of the expression, by analyzing it in the already used or possible contexts, put forth by the specialty literature of the latest years.

3.5. Industry 4.0

The forth industrial revolution (smart automation) is on the verge of unfolding and the driver of it will be the integration of physical objects in the information network. The main promoter of this leap in manufacturing industry is Germany, who introduced its own term "Industry 4.0" (or Industrie 4.0 in German) in 2011 at the Hannover Messe trade fair.

The year 2013 brought a clear definition of what the requirements are for the next industrial leap. In this sense, in Germany it was published the "Recommendations for Implementing the Strategic Initiative Industry 4.0" report, written by the German Communication Promoters Group of the Industry-Science Research Alliance, in collaboration with the National Academy of Science and Engineering and sponsored by the Federal Ministry of Education and Research. According to this report (also referred to as the Industry 4.0 National Working Group report), every physical object that is connected with the manufacturing process will be interlinked into a single network through the IoT. This network will incorporate everything starting from the factory floor to the delivery process, i.e., multiple systems that are outlaid as "Cyber-Physical Systems", forming the so-called "Smart

Factory". This sort of system has the capability to be self-aware, thus it can actively intervene in the manufacturing processes to prevent potential faults [66]. Its self-awareness is given by the sensory data collected within the system. Actions that are precisely ordered are based on previously stored information, this way making the system not only self-aware, but self-maintained and self-learning. Due to all these technological advancements, Industry 4.0 is viewed as one of the fastest developing trends in production engineering.

4. Research Questions, Methods, and Results

4.1. Research Questions and Rationale

Within this context, the current research demarche begins with the following set of questions that have been developed starting from the main premise that has been stated, that under severe time constraints complex and resource intensive scientific investigations should only be initiated after a preliminary analysis of available information arrives at the conclusion that there will be sufficient return on investment in the end. The authors laid out in a concentrated form the possible formulation of these questions:

- What is it and how can an ontology chart that depicts the layout of a concept be obtained?
- How can one evaluate the relationships between the main studied concept and associated notions?
- How can a knowledge density zone be defined by intersecting two or more highly volatile (new and ambiguous) concept domains over time?
- Which are the main areas in which this type of information processing can be capitalized upon?

4.2. Creation of Ontology Charts Using Latent Semantic Analysis

4.2.1. Theoretical Aspects (Identifying and Filtering the Documentation)

The input selection in the case of the two main concepts was made in the same way. Mostly, the authors focused on gathering references with the help of search engines that have access to large databases of scientific research that offer the possibility to access full-text articles (Elsevier's Scopus, Elsevier's Science Direct and Google Scholar). However, if additional documents were found with the help of traditional search engines (like Google or Yahoo) they were not disregarded but retained only if they were considered to be relevant after the first filtration process. This type of documentation was manually converted to a format supported by the LSA software (typically .pdf or .txt).

As search engines work with keywords, the challenge was to determine those words that accurately represent the studied concepts. In the first instance, the search was limited to identifying references in relation with the name of the concept, then it was extended to include certain keywords obtained from the top ten most cited scientific papers. These were restricted to six at the most for avoiding the display of insignificant documentation. The timeframe was also preset to the last 7 years for limiting outdated information. It was considered that documentation outside this framework cannot accurately capture state-of-the-art advancements in the studied fields. After a brief revision (the focus was kept on the abstract of the paper, but not limited to it) of displayed papers, a total number of 35 references were considered to be sufficient for illustration purposes for each main concept. They were downloaded and saved in separate folders. The brief review was chosen instead of an extensive literature browsing due to two reasons: the latter one would not serve to the proposed objective, which is quick identification of trending elements and concepts from a specific field of interest (meaning it was much more time consuming); and secondly, there were no guarantees that by reading the paper entirely it could be used in the semantic analysis, so there was the chance that after the review it would be disregarded. We point out the fact that the choices made here represent one of the limitations of the proposed methodology, that for making these decisions relating to the scope of the research it must rely on the experience and expertise of the reviewers, the quantity and quality of available literature, and on the relative positioning of the studied notions within the processed material when performing the first quantitative evaluation.

Making sure that every selected paper was indeed relevant for the analysis, the methodology included an additional filtration process performed by the review team. Each key word (mentioned above) was analyzed from its distribution perspective, throughout all the 35 papers. If the maximum number out of all keyword frequencies for each paper was not greater than five (viewed as a minimal threshold considering that all of the studied papers were more than five pages in length), the paper was considered to be irrelevant and it was removed from the analysis folder, where all were saved in the convened format. For this, the following condition had to be fulfilled:

$$\max_i(x_i) \leq 5 \tag{1}$$

where x represents the frequency of the i keyword (i = {1,8})

For illustration purposes, Figure 4 contains the frequency of mentioned keywords in a certain paper:

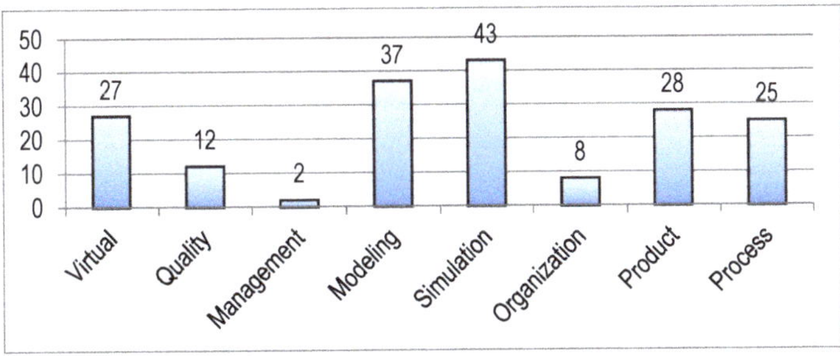

Figure 4. Occurrence of keywords in one identified paper.

The above figure contains the occurrence number of each keyword for one of the 35 papers. As it can be seen, the keywords "management" and "organization" are the two least frequent ones. Although "management" appears only two times it is not sufficient to consider this paper irrelevant; for a certain scientific reference to be entirely removed from the analysis it must be scarce in all key words (their maximum must be lower than or equal to five). It can happen that a certain paper is scarce in one key word, but rich in others; in this case, the paper is taken into account, as it is the case of this example, again appealing to the reviewers' competence, but to a reduced degree compared to the classic literature review. After this final filtration process a set of 30 papers remained, which were subject to the LSA.

The semantic analysis was carried out using a software program called Tropes v8.4 (English version) which is available free of charge and is capable of processing and perceiving, to a sufficiently advanced level, relationships between words, identifying or disregarding words specific to a domain of interest (based on input settings), counting the reoccurrence of words within different contexts, and retrieving neighboring concepts for certain notions or words.

4.2.2. Practical Implications (Defining the Ontology Charts)

As a direct result of the semantic analysis, an ontology chart of vQM was obtained and is presented in the following figure (Figure 5):

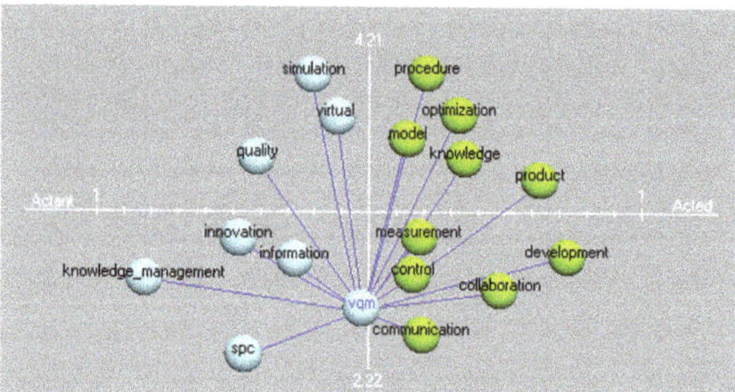

Figure 5. vQM ontology chart.

The vQM ontology chart (Figure 5) contains notions that are strongly related to quality management (measurement, control, quality, innovation, SPC). Some make references to instruments deployed through virtual means (such as simulation, modeling and optimization) and introduces "communication" and "collaboration" as notions linked with vQM (although having a low frequency; in sentences they were used closely together with vQM).

For "Industry 4.0", there appeared considerably more related notions on the ontology chart (Figure 6) than in the case of vQM, which can be explained by the fact that although the two concepts are considered to be relatively new, "Industry 4.0" is more developed and established in the scientific community as it is promoted to be the next generation of production engineering. Also, the known information on this subject reached a more mature level due to a higher number of publications.

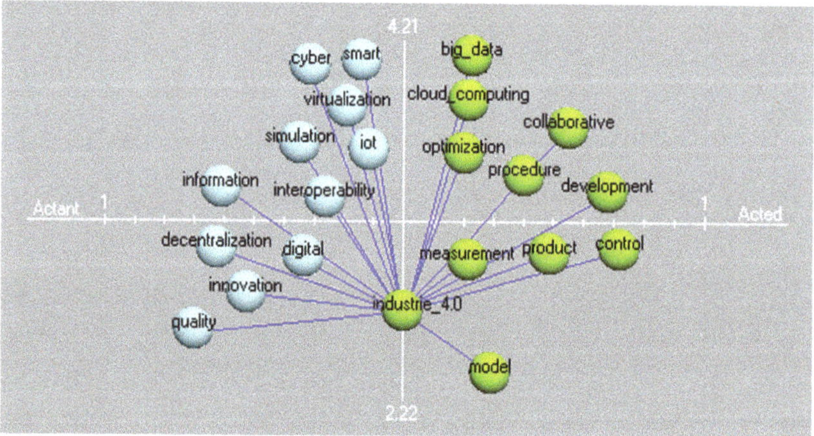

Figure 6. "Industry 4.0" ontology chart

The related notions in this chart include the main pillars of Industry 4.0, such as IoT (Internet of Things), cyber (making reference to "Cyber-Physical Systems"), "Big Data", and "Cloud Computing"; some design principles of Industry 4.0 as stated in Reference [67] (interoperability, decentralization, virtualization), and some notions which are also common to vQM.

The correlation degree of each current notion (denominated k) with regards to the main concept is reflected in its position on the chart. The position can be expressed as a function of the causal relation (the *x*-axis) and the concentration of relations—or frequency of association (the *y*-axis) [68]. Thus, the graph supports a visual comparison of the weight of relations between all concepts present on the chart with the main one.

On the *x*-axis, the number of words separating one concept from another in sentences can be considered as a direct reference for establishing their causal relation. Taking this into account, for quantifying the causal relation for both the "actant" (predecessor, enabler or source) and "acted" (successor, consequence or attribute) concepts, a scale from 0 to 1 was attributed from the center to both ends of the chart. As a certain *k* notion is further apart (closer to the value 1) on the chart from the main concept, it means that in sentences a significant number of words are placed between them. This value, called the "Causal Relationship Value" (CRV) takes values from 0 to 1:

$$CRV_k = [0 \dots 1] \tag{2}$$

On the *y*-axis, a fraction is represented between the total number of relations (or association) that each k concept has with the main one and the number of different relations. This value is called the "Concentration Value" or CV. The chart shows the minimum (bottom) and maximum (top) values, as they were calculated by the software program used to complete the semantic analysis. The formula for expressing the CV is:

$$CV_k = \frac{A}{A - B_k} \tag{3}$$

where A is the total number of relations and B_k is the number of relations that each concept has with the main one.

4.3. Weighting Related Notions Based on Ontology Charts

4.3.1. Theoretical Aspects

As a consequence of attributing the two scales for the "actant" and "acted" concepts their strength of the causal relation with the main concept is inversely proportional.

The CV and the inverse of CRV having completely different calculation rules but comparable in terms of magnitude, a correlation number (CN), obtained by the product of the two, was considered to be significant for expressing the importance of each concept in relation with the main one. The formula used for determining the CN is presented as follows:

$$CN_k = CV_k \times \frac{1}{CRV_k} \tag{4}$$

4.3.2. Practical Implications

The CN of each notion from the ontology charts are centralized in Tables 1 and 2, starting from the highest and descending to the lowest:

Table 1. Correlation number of "actant" and "acted" notions regarding vQM.

Actant	Score	Acted	Score
Virtual	37	Model/Modeling	24
Simulation	20	Procedure (referring to process)	20
Information	9	Control	17.4
Quality	8.3	Optimization	12.4
Innovation	5.8	Communication	12.1
Statistical Process Control (SPC)	4.4	Knowledge	9.4
Knowledge management	3	Measurement	5.8
		Collaboration	5.3
		Product	4.8
		Development	3.6

Table 2. Correlation number of "actant" and "acted" notions regarding "Industry 4.0".

Actant	Score	Acted	Score
IoT (Internet of Things)	31.4	Cloud_computing	17.9
Smart	27.6	Big_data	17.8
Virtualization	19.5	Optimization	17.0
Cyber	13.6	Measurement	12.7
Interoperability	11.5	Procedure	8.3
Simulation	10.0	Model	7.6
Digital	8.2	Collaborative	6.4
Information	5.3	Product	5.8
Innovation	5.0	Development	4.6
Decentralization	4.6	Control	4.0
Quality	3.6		

In the next step, combinations between the "actant" and "acted" notions were made for assuring increased homogeneity between all the terms of each concept. This action was taken for bringing the specific notions of each concept more closely together with the scope of finding common grounds between them.

A cumulative score can be calculated for each resulting association by multiplying the individual correlation values of the two forming notions, thus focusing on their conceptual intersections. This score can be used to determine the hierarchy between the obtained associations. However, only the meaningful associations were retained (thus becoming descriptors for each concept) and ordered in a descending manner in Tables 3 and 4:

Table 3. Cumulative scores of descriptors regarding vQM.

Descriptor	Cumulative Score	Descriptor	Cumulative Score
Virtual model/modeling	888.0	Innovation model	139.2
Virtual control	634.8	Virtual development	133.2
Simulation and modeling	480.0	Process innovation	116.0
Virtual optimization	458.8	Measurement through sim.	116.0
Virtual communication	447.7	Optimization information	111.6
Process simulation	400.0	Quality optimization	102.9
Virtual knowledge	347.8	Quality communication	100.4
Virtual measurement	214.0	Quality knowledge	78.0
Virtual collaboration	196.1	Innovation knowledge	54.5
Simulation knowledge	188.0	Measurement information	52.2
Information control	156.6	Product innovation	27.8
Quality control	144.4	Innovative development	20.9

Table 4. Cumulative scores of descriptors regarding "Industry 4.0".

Descriptor	Cumulative Score	Descriptor	Cumulative Score
IoT cloud	560.1	Cyber collaboration	87.0
Big_data from IoT	559.1	Process simulation	82.5
Smart cloud (computing)	494.0	Decentralized cloud	82.2
Smart measurement	352.1	Virtual(ized) control	78.4
Big_data virtualization	347.1	Decentralized optimization	78.3
Virtual(ized) optimization	331.1	Simulation & modeling	76.7
Virtual(ized) measurement	247.8	Measurement information	66.8
Cloud interoperability	205.0	Cyber development	62.3
Collaboration through IoT	199.6	Quality optimization	60.7
Interoperability optimization	195.2	Product simulation	58.2
Simulations on big_data	178.3	Cyber control	55.0
Smart collaboration	176.1	Interoperable develop.	52.3

Table 4. *Cont.*

Descriptor	Cumulative Score	Descriptor	Cumulative Score
Virtual(ized) modeling	149.3	Digital collaboration	51.8
Digital (big_)data	145.2	Development through sim.	45.6
Development through IoT	143.0	Process innovation	40.5
Digital optimization	138.4	Innovation model	37.6
Measurement through sim.	127.3	Digital development	37.1
Control through IoT	126.3	Digital control	32.8
Smart development	126.1	Collaborative innovation	31.2
Virtual(ized) collaboration	123.9	Decentralized collaboration	29.3
Smart control	111.4	Product innovation	28.5
Cyber model	104.8	Quality model	27.4
Digital measurement	103.6	Innovative development	22.4
Process interoperability	94.7	Information control	21.1
Cloud information	93.8	Decentralized develop.	21.0
Optimization information	89.3	Decentralized control	18.5
Virtual(ized) develop.	88.8	Quality control	14.4

In the case of the two concepts, the expressions obtained by pair-wise associations were filtered for redundancies and were verified if they have real-life applicability with the help of search engines. In some cases, connection words were added to further refine the expressions and to assure their coherence related to the main concept.

4.4. Features of the vQM x Industry 4.0 Knowledge Space

4.4.1. Theoretical Aspects

Based on the above steps, we can now characterize the previously ambiguous knowledge space, found at the intersection of the two notions, that are connected through a common goal, namely, sustainability. Overlapping the tables containing the descriptors of vQM and Industry 4.0 (Tables 3 and 4) it can be observed that there are common ones between them. These are used for determining common ground between two apposed concepts. Ranking them was done by calculating their weighted average (thus, still respecting their importance to each main concept) and consequently their weighted importance obtained in percentages.

4.4.2. Practical Implications

As seen in the table below, it can be the case that there is a considerable difference between the total scores of vQM and Industry 4.0. The weighted average provides an aggregate that accurately reflects the importance of each descriptor to both main concepts, regardless of their score difference. Although each common notion has its own LSA score depending on the main concept (vQM or Industry 4.0), the ranking presented in the above table was made according to the calculated weighted average of the score pairs.

These theoretical results can be further expanded for the use of practitioners by mapping onto the knowledge space the adequate concrete implementations that were found in industry. The proposed solution for identifying industrial applications uses the authors' expertise and is similar to identifying relevant papers for the LSA. Firstly, the top five most important common descriptors (obtained from the LSA—see Table 5) were entered into search engines followed by keywords extracted from the top 10 most cited "Industry 4.0" references together with the "application" keyword that was added to limit the amount of retrieved information. This way it is assured that the search was focused on practical applications and were related to "Industry 4.0" as its representative keywords were used. As "Industry 4.0" is considered to be the new trend in production engineering, applications related to it are also categorized as emergent. The retained applications were focused on those that had connections with—or can be used in relationship to vQM. A list of applications is obtained, but they are not sorted in any way.

Table 5. Common notions between vQM and Industry 4.0.

Common Descriptors	LSA Score		Weighted Average	Weighted Importance %
	vQM	**Industry 4.0**		
Virtual modeling	888.00	149.30	271.87	40.45%
Virtual optimization	458.80	331.10	78.95	11.75%
Simulation and modeling	480.00	76.70	152.74	22.73%
Process simulation	400.00	82.50	56.93	8.47%
Virtual measurement	214.00	247.80	62.27	9.27%
Quality optimization	102.92	60.70	7.35	1.09%
Measurement through simulation	116.00	127.30	11.83	1.76%
Virtual development	133.20	88.80	5.61	0.84%
Process innovation	116.00	40.50	16.88	2.51%
Innovation model	139.20	37.60	6.27	0.93%
Product innovation	27.84	28.50	0.88	0.13%
Innovative development	20.88	22.40	0.53	0.08%
TOTAL	3096.84	1293.2	672.10	100%

4.5. Grouping and Ranking Applications Along Time Stages (e.g., the PLC Stages)

4.5.1. Theoretical Aspects

Focusing on the theoretical connections uncovered before, this last step of the research methodology deploys the new structure of the two-fold common knowledge space into workable results that can be applied in companies. In other words, by using the proper applications in the proper manner, based on the guidelines below, firms can employ vQM tools to rip the competitive benefits of Industry 4.0. Taking into account the development (the third) dimension, to the two axes (vQM and Industry 4.0) another axis is added, the PLC, along which identified applications are grouped into its stages (see Figure 1 for an overview).

There are many views when it comes to defining precisely the product life cycle stages. For this reason, the authors synthesized information from an internationally accepted standard which inclusively covers all stages of a product life cycle: Reference [69]. The stages were grouped into main stages and sub-stages accordingly. These are presented in Table 6:

Table 6. Product life cycle stages, according to ISO/IEC/IEEE 15288:2015.

Main Stage	Sub-Stage
I. Planning, design, and development	Stakeholder requirement definition/analysis Product design and validation
II. Realization	Procurement Manufacturing Measurement and control
III. Utilization	Delivery Customer assistance Maintenance
IV. Disposal	Retirement Recycling

4.5.2. Practical Implications

The affinity diagram (also known as a KJ diagram) was applied for grouping the set of applications into the corresponding life cycle stages, defining both categories and sub-categories of applications. The matrix diagram (presented in Table A1 in the Appendix A Section) contains both these and the common notions with their weighted importance (expressed in percentages).

The cells resulted at the intersection between main or sub-category applications and the common notions were filled out according to the correlation between them, with a number on a scale of 0 to 3 (0 means that there is no relation between the two and 3 signifies a strong relation). The sum of products between these correlation numbers and the weighted importance of the common notions was calculated on each line, thus obtaining an aggregate number that can be viewed as the importance of each application. If a main category application contains more sub-category items, its score is provided by the average of them (indicated in bold in Table A1).

Having the importance scores for every sub- or main-category application, they can be hierarchized for all the stages of the product life cycle (in Table A1 the applications are already ranked). Thus, the matrix diagram contains a ranking of identified vQM applications in the context of Industry 4.0, that are or will become essential in each stage of the product life cycle. Where blank squares exist corresponding to a life cycle stage it means that there were no suitable applications identified that fit the previously mentioned search conditions (see Section 4.3). All identified applications grouped into the PLC stages are presented in Figure 7:

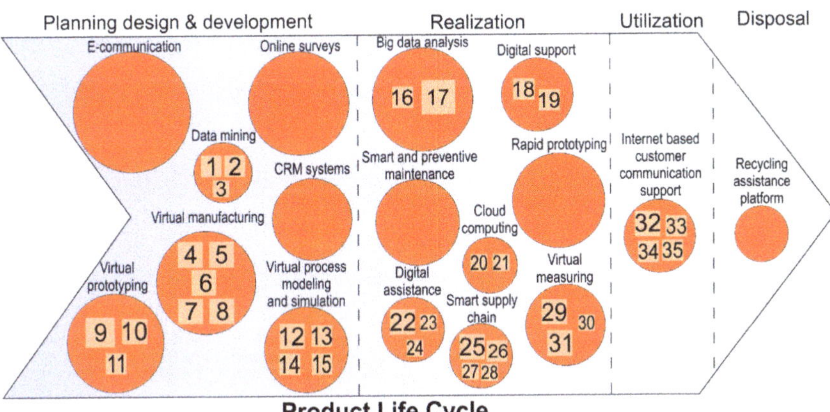

Figure 7. Industrial applications along the PLC stages.

1—Social networking data analysis
2—Google analytics for customer targeting
3—Google Trends analysis used for product benchmarking
4—Plant layout design used for production flow optimization
5—Manufacturability analysis
6—Work instructions design
7—Worker ergonomics planning and design
8—Work time study analysis
9—CAD used for product modeling
10—Kinematics simulation, analysis and optimization
11—Assembly and part finite element structural analysis
12—Simulation software used for process planning and flow analysis
13—Virtual commissioning used for equipment automation

14—Automated process optimization
15—Automated process robustness prediction
16—Preventive actions taken based on collected data
17—Automated real-time process optimization
18—Mobile device applications used for process monitoring and control
19—Smart glasses employing AR used for displaying product/process data
20—Remote process monitoring and control
21—Edge computing
22—Remote equipment control and assistance
23—Vision system used for part inspection
24—Smart glasses employing AR used for operator empowerment

25—Real-time inventory monitoring
26—Automated raw material order and purchase
27—Automated raw material approval
28—Smart supplier evaluation and selection
29—Virtual CMM used for evaluation of uncertainty estimates
30—Inline metrology used for fault, defect and deviation detection
31—Absolute multiline technology for online precision monitoring
32—Knowledge base used for problem solving for known issues
33—Remote access used for diagnostics and trouble ticket resolution
34—FAQ web page used for fault/defect identification and self-help
35—Automated online assistant used for tech support through live chat

The above representation combines PLC stages with the identified significant notions and their potential applicability degree to provide an overview for potential academic and industrial users. The circles represent the main categories of applications and their size is proportional with their importance as calculated in the matrix diagram (Table A1). The squares correspond to actual applications and their size also represents their priority in reference to the category they belong to. The denomination of each application is presented below the figure.

5. Research Implications

The main implication of the current research is uncovered by carrying out a validation stage of the proposed LSA-based methodology. This was done through an online questionnaire, applied to groups of professionals with relevant knowledge connected to the actual technological developments from the automotive industry, which is considered in Romania to have the highest degree of quality management and industrial practices proficiency and is also subjected to the most intense competition on a European scale.

The validation demarche raised the following challenges:

- covering the important companies from the automotive sector such that even if statistical relevance is not reached, this will be reflected by including within the survey all/most of the major actors from this field;
- selecting the target group such that it has access to relevant information;
- designing and constructing the questionnaire.

Statistically speaking, there are 510 automotive companies (NACE code 29) in Romania (according to the online database of Reference [70]), but in terms of our analysis not all of them present the same relevance. An analysis of the technological level of these companies clearly leads to the conclusion of an inhomogeneity of their relevance in relation to our analysis. One can expect not to obtain similar information on new technological developments regarding Industry 4.0/vQM from Renault or Ford (big players, global auto makers) and from a small firm that supplies low complexity parts. For this reason, the focus of this study was aimed at those few major manufacturers from this industry branch and questioning consultants, specialists, quality auditors or managers that draw their expertise from not just one, but several important companies and can provide answers with a high degree of relevance. This way, the answers of one individual can represent feedback from more than one company.

Considering the above-mentioned aspects, the companies were grouped into three main categories: the first two categories (A. automotive manufacturers and B. manufacturers of major components—e.g., Dacia–Renault, Ford, Bosch, Continental, Leoni, Rombat, Daimler, Pirelli, RAAL) represent manufacturers with international notoriety; all other companies were left for a third (C) category. The first category had 10 representatives, the second had around 150, and in the third category 350 companies were included.

The investigation was guided towards the first two categories, but there was also a concern about obtaining some samples as well from the third category companies, mainly from those which were preoccupied with new technologies. The reason behind this approach is that SME-type companies (from the C category) are mostly focused on concrete measures of physical QM due to resource constraints, lack of maturity, and limited openness towards adopting new technologies and due to the need to obtain quick and measurable results, with a noticeable impact on their process performance and bottom line.

The surveyed target group consisted of consultants, specialists, quality auditors or even managers that drew their expertise from not just one, but several important companies and could provide answers with a high degree of relevance. An important issue addressed was information confidentiality, as the research team made sure that the expertise and feedback of the focus group was captured but that company specific information was filtered out.

Within the survey, the abovementioned groups were provided with a list of 41 applications determined through the LSA (presented in the Matrix Diagram—Tables 2 and A1) and they had to rate them considering their implementation or future implementation into the companies whose manufacturing processes they were familiar with. Responders also had the possibility to name and rate additional applications that were not included, but were considered to be important by them (centralized in Table 3).

The scale for rating each application was kept the same as in the case of the matrix diagram, from 0 to 3, thus enabling an objective and direct comparison between the score calculated from the matrix diagram and the average score obtained from the survey (0–"it's not likely to be used by manufacturing companies"; 1—"only some manufacturing companies have or will implement it"; 2—"most of the manufacturing companies have or will implement it"; 3—"all manufacturing companies have or will implement it").

Table 7 includes the number of responses, the number of companies covered from each type and the percentage of coverage:

Table 7. Number of responses and percentage of coverage.

Target Group	Responses	Type A Companies Covered	Type B Companies Covered	Type C Companies Covered
Top management	6	0	10	9
Technical consultants	8	7	55	28
Specialists	5	2	35	11
Quality auditors	6	1	25	15
TOTAL	25	10 out of 10 (100% coverage)	125 out of 150 (83.3% coverage)	63 out of 350 (18% coverage)

As it can be seen, not all of the 510 companies were covered; however, all of the type A and most of the type B companies were reached within the study.

For analyzing the responses, the chosen reference values were the ones obtained from the questionnaire because they are considered to be the "voice" of the industry. A margin of error was calculated for each score obtained from the matrix diagram and findings showed that it was not more than ±8% (representing the deviation from the reference). The average margin of error was even better, being just under 4% (representing the mean of the deviation modulus values).

The list of applications alongside their rating and margins of error can be consulted in Table 2. It presents not only the accuracy of the research methodology approached within this paper, but also prioritizes trending applications that companies will need for their own development, thus defining the emerging research directions.

6. Conclusions

The current paper focused on analyzing the common knowledge space in specific areas by intersecting the knowledge structures obtained through a concept disambiguation process. The results of this analysis were further used to determine the technological solutions that are expected to emerge from that concept intersection space. Particularly, it was analyzed the possibility of identifying emergent industrial applications by intersecting the knowledge structures of two concepts: Industry 4.0 and vQM.

The investigation and disambiguation process were achieved with the help of a fairly new knowledge extraction method called latent semantic analysis (applied by a specialized software program), which is capable of analyzing large amounts of text and quickly identify the relationship, causality, and frequency of words, thus providing an interpretation of the state-of-the-art literature (in close connection with the two main concepts—vQM and Industry 4.0). By applying this technique, the common areas of interest (common knowledge space) were identified in the form of descriptors that were further used for identifying applications that are defined by or relate to them.

The set of applications was grouped into the stages of the product life cycle according to the ISO/IEC/IEEE 15288:2015 standard (with the help of a classic sorting tool—Affinity Diagram) and ranked considering their importance scores resulting from applying a matrix diagram. These calculated scores reflect the weighted importance of common descriptors between vQM and Industry 4.0, as they also served as inputs into the matrix diagram.

The research loop was closed with a survey that focused on obtaining information from experts (from the field of quality) with close ties to the automotive industry, that confirms the reliability of the presented methodology with an average margin of error of just under 4%.

As illustrated through the course of this paper, one can note the important advantages of this LSA-based methodology:

- Emerging technological trends can be identified from the common knowledge space of two or more concepts;
- The state-of-the-art knowledge discovery process (concept disambiguation) is based upon an exponentially greater number of scientific references and it is reduced to a few seconds;
- The outputs of the LSA can be further analyzed, sorted, and grouped by deploying various tools and techniques, thus obtaining more useful information.

Considering the abovementioned key points of the methodology, it can be stated that the resulting applications are the product of current research concerns in the two analyzed fields, vQM and Industry 4.0. The constructed matrix diagram also provides a useful preview of industrial applications throughout the product life cycle and ranks them to depict those that are more critical to quality based on their importance scores.

The current framework has a high degree of portability and it can be applied to other concept associations as well for determining common areas or emergent aspects in a specific context.

In a broader understanding, the model can make forecasts for applications that are more likely to become popular in the next period of time. Organizations that are capable to foresee them and focus on their development and implementation have a clear competitive advantage towards achieving sustainable growth and advancement. The applicability of the methodology and the results described in this paper are two-fold. From an academic point of view, they can provide the clarification of the concepts for the use in training and instruction as well as an approach to investigate the two domains in the future to gauge their development. On the other hand, industrial users could benefit from the identified ranking of potential applications to simplify and make more efficient their implementation projects, allowing them to keep the focus on the bottom line and on the concrete results that are needed for market success.

Author Contributions: Conceptualization, Ş.B., A.W., and S.P.; methodology, Ş.B., A.W., and S.P.; literature survey, Ş.B. and R.C.; investigation, Ş.B., D.H., R.C. and M.D.; validation, Ş.B., D.H., M.D.; visualization, R.C. and D.H.; writing—original draft, Ş.B., A.W., S.P. and M.D; writing—review and editing, Ş.B., M.D. and R.C.

Funding: This research received no external funding.

Acknowledgments: The authors would like to thank Dan Marius, the C.E.O. of S.C. Calitop S.R.L. and Călin Neamţu from the Technical University of Cluj-Napoca for their valuable expertise, discussions, and for facilitating connections with the industry.

Conflicts of Interest: The authors declare no conflict of interest.

Appendix A

Table A1. Matrix diagram for ranking the technological solutions.

Life cycle stage	Sub-stage	Category	Sub-category	Virtual Modeling	Simulation & Modeling	Virtual Optimization	Process Simulation	Virtual Measurement	Innovation Model	Virtual Development	Process Innovation	Measurement through Simulation	Quality Optimization	Product Innovation	Innovative Development	Sub-Category Score	Category Score
				0.40	0.12	0.23	0.08	0.09	0.01	0.02	0.01	0.03	0.01	0.001	0.001		
		E-communication (e-mail, e-conference, IM, file transfers)	-	3	3	3	3	2	1	3	1	2	1	2	2	2.823	2.823
	Stakeholder requirement definition/analysis	Online surveys and questionnaires	-	3	3	3	0	1	1	3	1	0	2	0	1	2.432	2.432
		Customer Relationship Management (CRM) systems	-	3	3	1	0	0	0	3	0	0	1	0	1	1.856	1.856
			Social networking data analysis	2	2	2	0	0	0	3	0	0	0	0	1	1.552	
		Data mining	Google analytics for customer targeting	2	1	2	0	0	1	3	0	0	0	3	3	1.451	1.419
			Google Trends analysis used for product benchmarking	2	1	1	0	0	1	3	1	0	3	0	1	1.255	
Planning, design, and development			CAD used for product modeling	3	3	2	1	2	2	3	1	1	3	3	2	2.432	
		Virtual prototyping	Kinematics simulation, analysis and optimization through software.	3	3	3	0	1	0	3	0	1	1	1	1	2.430	2.313
			Assembly and part finite element structural analysis	3	3	1	0	2	0	3	0	1	2	0	1	2.076	
			Simulation software used for process planning and flow analysis	3	3	2	3	1	1	3	1	3	2	1	2	2.536	
	Product design and validation	Virtual process modeling and simulation	Virtual commissioning used for equipment automation	2	2	3	1	1	0	3	2	1	2	0	2	2.018	2.059
			Automated process optimization	1	3	3	3	0	0	3	2	2	3	0	1	1.841	
			Automated process robustness prediction	1	3	3	3	0	0	3	2	2	3	0	1	1.841	
			Plant layout design used for production flow optimization	3	3	3	2	1	1	2	1	2	3	1	1	2.645	
			Manufacturability analysis	3	3	2	3	1	1	3	2	3	3	2	3	2.556	
		Virtual manufacturing	Work instructions design	2	3	3	3	2	0	3	2	2	3	0	1	2.431	2.417
			Worker ergonomics planning and design	2	3	3	3	1	0	3	1	2	0	0	0	2.301	
			Work time study analysis	1	3	3	3	3	1	3	1	3	3	1	3	2.150	

Table A1. *Cont.*

Life cycle stage	Sub-stage	Category	Sub-category	Virtual Modeling	Simulation & Modeling	Virtual Optimization	Process Simulation	Virtual Measurement	Innovation Model	Virtual Development	Process Innovation	Measurement through Simulation	Quality Optimization	Product Innovation	Innovative Development	Sub-Category Score	Category Score
				0.40	0.12	0.23	0.08	0.09	0.01	0.02	0.01	0.03	0.01	0.001	0.001		
	Procurement	Smart supply chain	Real-time inventory monitoring	1	2	3	3	2	1	0	3	1	3	0	3	1.852	
			Automated raw material order and purchase	1	1	3	0	1	0	1	3	2	3	0	3	1.420	
			Automated raw material approval	1	1	3	0	1	0	0	0	1	2	0	1	1.341	1.465
			Smart supplier evaluation and selection	1	1	2	1	1	1	2	2	0	3	0	1	1.245	
		Big data analysis	Preventive actions taken based on collected data	3	2	3	3	3	0	1	3	1	3	0	0	2.758	2.366
			Automated real-time process optimization	1	2	3	3	3	0	1	3	2	3	0	0	1.974	
		Rapid prototyping (Additive manufacturing)	3D printing used for product manufacturing	3	3	2	0	2	0	3	0	0	1	3	0	2.272	2.272
	Manufacturing	Smart and preventive maintenance	-	2	1	3	2	2	1	0	2	2	3	0	1	2.070	2.070
Realization		Digital assistance	Remote equipment control and assistance	2	1	2	1	2	1	1	2	0	3	0	1	1.725	
			Vision system used for part inspection	1	2	2	1	2	0	0	3	0	3	0	0	1.417	1.509
			Smart glasses employing AR for operator empowerment (displaying instructions, assigning tasks and facilitating information access)	2	1	1	0	2	0	1	1	0	2	0	0	1.384	
		Virtual measuring	Virtual CMMs used for evaluation of uncertainty estimates	2	2	3	1	3	1	2	1	3	2	0	2	2.238	
			Inline metrology used for fault, defect and deviation detection.	2	1	3	1	3	0	0	3	3	3	0	1	2.100	1.907
	Measurement and control		Absolute multiline techn. for online equipment precision monitoring	1	1	2	0	3	0	1	2	3	2	0	0	1.383	
		Digital support	Mobile device applications used for process monitoring & control	2	1	1	3	3	0	0	1	1	1	0	0	1.729	1.690
			Smart glasses employing AR used for displaying product/process data	2	2	1	1	3	0	0	1	0	1	0	0	1.652	
		Cloud computing	Remote process monitoring and control	1	2	2	0	2	0	1	0	1	1	0	0	1.331	1.315
			Edge computing	1	1	2	1	2	0	0	1	1	2	0	0	1.299	

Table A1. *Cont.*

Life cycle stage	Sub-stage	Category	Sub-category	Virtual Modeling	Simulation & Modeling	Virtual Optimization	Process Simulation	Virtual Measurement	Innovation Model	Virtual Development	Process Innovation	Measurement through Simulation	Quality Optimization	Product Innovation	Innovative Development	Sub-Category Score	Category Score
				0.40	0.12	0.23	0.08	0.09	0.01	0.02	0.01	0.03	0.01	0.001	0.001		
	Delivery	-	-	N/A	N/A	N/A	N/A	N/A	N/A	N/A	N/A	N/A	N/A	N/A	N/A	N/A	N/A
Utilization	Customer assistance	Internet based customer communication support	Knowledge base used for problem solving for known issues	2	2	3	1	0	3	3	3	0	3	2	1	1.952	1.250
			Remote access used for diagnostics and trouble ticket resolution	1	1	2	1	2	1	1	3	0	3	0	0	1.328	
			EAQ page used for fault/defect identification and self-help	0	0	3	0	1	2	0	3	1	3	1	1	0.877	
			Automated online assistant used for tech support through live chat	0	0	3	0	1	2	0	2	0	3	0	0	0.841	
	Maintenance	-	-	N/A	N/A	N/A	N/A	N/A	N/A	N/A	N/A	N/A	N/A	N/A	N/A	N/A	N/A
	Retirement	-	-	N/A	N/A	N/A	N/A	N/A	N/A	N/A	N/A	N/A	N/A	N/A	N/A	N/A	N/A
Disposal	Recycling	Recycling assistance platform providing info for customers for recycling	-	1	1	3	0	0	2	1	3	0	3	1	1	1.298	1.298

Table 2. List of rated applications.

Applications	Average of Responses	Calculated from LSA	Error (max. 8%)
Absolute multiline technology used for online precision monitoring	1.33	1.38	3.57%
Assembly and part finite element structural analysis	2.13	2.08	−2.70%
Automated online assistant used for tech support through live chat	0.87	0.84	−2.96%
Automated process optimization	2.00	1.84	−7.94%
Automated process robustness prediction	2.00	1.84	−7.94%
Automated raw material approval	1.33	1.34	0.57%
Automated raw material order and purchase	1.53	1.42	−7.42%
Automated real-time process optimization	1.87	1.97	5.44%
CAD used for product modeling	2.47	2.43	-1.40%
CRM systems used for requirement analysis and sorting	1.80	1.86	3.02%
E-communication: e-mail, e-conference, instant messaging, file transfers	2.87	2.82	−1.53%
Edge computing	1.33	1.30	−2.60%
FAQ page used for fault/defect self-identification and self-help	0.93	0.88	−6.08%
Google analytics for consumer targeting	1.47	1.45	−1.07%
Google trends analysis for product benchmarking	1.33	1.25	−5.91%
Inline metrology used for fault, defect and deviation detection	2.13	2.10	−1.56%
Kinematics simulation, analysis and optimization through software	2.27	2.43	6.71%
Knowledge base used for problem solving for known issues	2.07	1.95	−5.53%
Manufacturability analysis	2.53	2.56	0.88%
Mobile devices used for process monitoring and control	1.87	1.73	−7.40%
Smart glasses employing AR used for displaying product/process data	1.60	1.65	3.12%
Online surveys and questionnaires	2.47	2.43	−1.41%
Plant layout design used for production flow optimization	2.53	2.64	4.20%
Preventive actions taken based on collected data	2.67	2.76	3.31%
Rapid prototyping (3D printing)	2.27	2.27	0.22%
Real-time inventory monitoring	1.93	1.85	−4.20%
Recycling assistance platform providing info for customers for recycling	1.40	1.30	−7.26%
Remote access used for diagnostics and trouble ticket resolution	1.40	1.33	−5.14%
Remote equipment control and assistance	1.87	1.73	−7.59%
Remote process monitoring and control	1.33	1.33	−0.15%
Simulation software used for process planning and flow analysis	2.47	2.54	2.74%
Smart and preventive maintenance for equipment	2.20	2.07	−5.93%
Smart glasses used for operator empowerment	1.47	1.38	−5.66%
Smart supplier evaluation and selection	1.33	1.25	−6.59%
Social networking data analysis for requirement and stakeholder identification	1.47	1.55	5.50%
Virtual commissioning used for reliable equipment automation	2.00	2.02	0.89%
Virtual coordinate measuring machines	2.20	2.24	1.72%
Vision system used for part inspection	1.53	1.42	−7.59%
Work instruction design	2.40	2.43	1.28%
Work time study analysis	2.27	2.15	−5.16%
Worker ergonomics planning and design	2.27	2.30	1.51%
Average error			3.98%

Table 3. List of applications nominated by survey responders.

Applications	Average of Responses
Risk analysis applications	2.5
SAP Smart Data Access	2
Materials Management Operations Guideline/Logistics Evaluation—MMOG/LE	1.5
Design for Lean Six Sigma (DFLSS) used for product development	1
Eight Disciples (8D) problems solving	2

References

1. Lu, Y. Industry 4.0: A survey on technologies, applications and open research issues. *J. Ind. Inf. Integr.* **2017**, *6*, 1–10. [CrossRef]
2. De Solla Price, D.J. The exponential curve of science. *Discovery* **1956**, *17*, 240–243.
3. Fuller, R.B. *Critical Path*, 1st ed.; St Martins Press: New York, NY, USA, 1981.
4. Vitanov, N.K. *Science Dynamics and Research Production—Indicators, Indexes, Statistical Laws and Mathematical Models*; Springer: Basel, Switzerland, 2016.
5. Schmitt, U. Personal Knowledge Management Devices: The next Co-evolutionary Driver of Human Development. In Proceedings of the International Conference on Education and Social Sciences Abstracts & Proceedings (INTCESS14), Istanbul, Turkey, 3–5 February 2014; pp. 1081–1091. [CrossRef]
6. Petzold, T. *Global Knowledge Dynamics and Social Technology*; Palgrave Macmillan: Basingstoke, UK, 2017.
7. Onwuegbuzie, A.J.; Weinbaum, R. A framework for using qualitative comparative analysis for the review of the literature. *Qual. Rep.* **2017**, *22*, 359–372.
8. Grant, M.J.; Booth, A. A typology of reviews: An analysis of 14 review types and associated methodologies. *Health Inf. Libr. J.* **2009**, *26*, 91–108. [CrossRef] [PubMed]
9. Rosas, R.S.; Ridings, J.W. The use of concept mapping in measurement development and evaluation: Application and future directions. *Eval. Program Plan.* **2017**, *60*, 265–276. [CrossRef]
10. Wang, M.; Cheng, B.; Chen, J.; Mercer, N.; Kirschner, P.A. The use of web-based collaborative concept mapping to support group learning and interaction in an online environment. *Internet High. Educ.* **2017**, *34*, 28–40. [CrossRef]
11. Collins, H. *Creative Research: The Theory and Practice of Research for the Creative Industries*; AVA Publishing SA.: Lausanne, Switzerland, 2017.
12. Varela, A.R.; Pratt, M.; Harris, J.; Lecy, J.; Salvo, D.; Brownson, R.C.; Hallal, P.C. Mapping the historical development of physical activity and health research: A structured literature review and citation network analysis. *Prev. Med.* **2018**, *111*, 466–472. [CrossRef]
13. Bird, S.; Klein, E.; Loper, E. *Natural Language Processing with Python: Analyzing Text with the Natural Language Toolkit*; O'Reilly Media, Inc.: Sebastopol, CA, USA, 2009.
14. Sun, S.; Luo, C.; Chen, J. A review of natural language processing techniques for opinion mining systems. *Inf. Fusion* **2017**, *36*, 10–25. [CrossRef]
15. Hansen, S.; McMahon, M.; Prat, A. Transparency and deliberation within the FOMC: A computational linguistics approach. *Q. J. Econ.* **2018**, *133*, 801–870. [CrossRef]
16. Toutanova, K.; Manning, C.D. Enriching the knowledge sources used in a maximum entropy part-of-speech tagger. In Proceedings of the 2000 Joint SIGDAT Conference on Empirical Methods in Natural Language Processing and Very Large Corpora: Held in Conjunction with the 38th Annual Meeting of the Association for Computational Linguistics, Hong Kong, China, 7–8 October 2000; Association for Computational Linguistics: Stroudsburg, PA, USA, 2000; pp. 63–70. [CrossRef]
17. Shen, L.; Satta, G.; Joshi, A. Guided learning for bidirectional sequence classification. In Proceedings of the 45th Annual Meeting of the Association of Computational Linguistics, Prague, Czech Republic, 25–27 June 2007; Association for Computational Linguistics: Stroudsburg, PA, USA, 2007; pp. 760–767.
18. Søgaard, A. Simple semi-supervised training of part-of-speech taggers. In Proceedings of the ACL 2010 Conference Short Papers, Uppsala, Sweden, 11–16 July 2010; Association for Computational Linguistics: Stroudsburg, PA, USA, 2010; pp. 205–208.
19. Manning, C.D. Part-of-speech tagging from 97% to 100%: Is it time for some linguistics? In Proceedings of the International Conference on Intelligent Text Processing and Computational Linguistics, Tokyo, Japan, 20–26 February 2011; Springer: Berlin/Heidelberg, Germany, 2011; pp. 171–189. [CrossRef]
20. Owoputi, O.; O'Connor, B.; Dyer, C.; Gimpel, K.; Schneider, N.; Smith, N.A. Improved part-of-speech tagging for online conversational text with word clusters. In Proceedings of the 2013 Conference of the North American Chapter of the Association for Computational Linguistics: Human Language Technologies, Atlanta, GA, USA, 9–14 June 2013; Association for Computational Linguistics: Stroudsburg, PA, USA, 2013; pp. 380–390.

21. Gui, T.; Zhang, Q.; Huang, H.; Peng, M.; Huang, X. Part-of-speech tagging for twitter with adversarial neural networks. In Proceedings of the 2017 Conference on Empirical Methods in Natural Language Processing, Copenhagen, Denmark, 7–11 September 2017; Association for Computational Linguistics: Stroudsburg, PA, USA, 2017; pp. 2411–2420. [CrossRef]

22. Balusu, R.M.; Merghani, T.; Eisenstein, J. Stylistic variation in social media Part-of-Speech Tagging. In Proceedings of the 16th Annual Conference of the North American Chapter of the Association for Computational Linguistics, New Orleans, LA, USA, 1–6 June 2018; Association for Computational Linguistics: Stroudsburg, PA, USA, 2018; pp. 11–19. [CrossRef]

23. Fang, L.; Tuan, L.A.; Hui, S.C.; Wu, L. Syntactic based approach for grammar question retrieval. *Inf. Process. Manag.* **2018**, *54*, 184–202. [CrossRef]

24. Ganesh, B.R.; Gupta, D.; Sasikala, T. Grammar Error Detection Tool for Medical Transcription using Stop Words Parts-of-Speech Tags Ngram Based Model. *APTIKOM J. Comput. Sci. Inf. Technol.* **2017**, *2*, 47–56. [CrossRef]

25. Vani, K.; Deepa, G. Unmasking text plagiarism using syntactic-semantic based natural language processing techniques: Comparisons, analysis and challenges. *Inf. Process. Manag.* **2018**, *54*, 408–432. [CrossRef]

26. Aly, A.; Taniguchi, A.; Taniguchi, T. A generative framework for multimodal learning of spatial concepts and object categories: An unsupervised part-of-speech tagging and 3D visual perception based approach. In Proceedings of the Joint IEEE International Conference on Development and Learning and Epigenetic Robotics (ICDL-EpiRob), Lisbon, Portugal, 18–21 September 2017; pp. 376–383. [CrossRef]

27. Mohit, B. Named Entity Recognition. In *Natural Language Processing of Semitic Languages. Theory and Applications of Natural Language Processing*; Zitouni, I., Ed.; Springer: Berlin/Heidelberg, Germany, 2014; pp. 221–245.

28. Cho, H.-C.; Okazaki, N.; Miwa, M.; Tsujii, J. Named entity recognition with multiple segment representations. *Inf. Process. Manag.* **2013**, *49*, 954–965. [CrossRef]

29. Zhao, G.; Wu, J.; Wang, D.; Li, T. Entity disambiguation to Wikipedia using collective ranking. *Inf. Process. Manag.* **2016**, *52*, 1247–1257. [CrossRef]

30. Ni, J.; Florian, R. Improving Multilingual Named Entity Recognition with Wikipedia Entity Type Mapping. In Proceedings of the 2016 Conference on Empirical Methods in Natural Language Processing, Austin, TX, USA, 1–5 November 2016; Association for Computational Linguistics: Stroudsburg, PA, USA, 2016; pp. 1275–1284.

31. Das, A.; Ganguly, D.; Garain, U. Named entity recognition with word embeddings and Wikipedia categories for a low-resource language. *ACM Trans. Asian Low-Resour. Lang. Inf. Process. (Tallip)* **2017**, *16*, 18. [CrossRef]

32. Mehdi, M.; Okoli, C.; Mesgari, M.; Nielsen, F.Å.; Lanamäki, A. Excavating the mother lode of human-generated text: A systematic review of research that uses the wikipedia corpus. *Inf. Process. Manag.* **2017**, *53*, 505–529. [CrossRef]

33. Liu, X.; Wei, F.; Zhang, S.; Zhou, M. Named entity recognition for tweets. *ACM Trans. Intell. Syst. Technol.* **2013**, *4*, 3. [CrossRef]

34. Derczynski, L.; Maynard, D.; Rizzo, G.; van Erp, M.G.J.; Gorrell, G.; Troncy, R.; Petrak, K.; Bontcheva, K. Analysis of named entity recognition and linking for tweets. *Inf. Process. Manag.* **2015**, *51*, 32–49. [CrossRef]

35. Bellot, P.; Moriceau, V.; Mothe, J.; SanJuan, E.; Tannier, X. INEX Tweet Contextualization task: Evaluation, results and lesson learned. *Inf. Process. Manag.* **2016**, *52*, 801–819. [CrossRef]

36. Zhang, Y.; Jin, R.; Zhou, Z.H. Understanding bag-of-words model: A statistical framework. *Int. J. Mach. Learn. Cybern.* **2010**, *1*, 43–52. [CrossRef]

37. Baccianella, S.; Esuli, A.; Sebastiani, F. SentiWordNet 3.0: An enhanced lexical resource for sentiment analysis and opinion mining. In Proceedings of the 7th International Conference on Language Resources and Evaluation (LREC 2010), Valletta, Malta, 19–21 May 2010; European Language Resources Association: Paris, France, 2010; pp. 2200–2204.

38. Mukherjee, S.; Bhattacharyya, P. Sentiment analysis in Twitter with lightweight discourse analysis. In Proceedings of the 24th International Conference on Computational Linguistics (COLING 2012): Technical Papers, Mumbai, India, 8–15 December 2012; Indian Institute of Technology Bombay: Mumbai, India, 2012; pp. 1847–1864.

39. Song, J.; Song, X.; Liu, Y. Object recognition algorithm based on improved k-means bag-of-words model. *Univ. Politeh. Buchar. Sci. Bull. Ser. C Electr. Eng. Comput. Sci.* **2017**, *79*, 43–56.

40. Durou, A.; Aref, I.A.; Al-Maadeed, S.; Bouridane, A.; Benkhelifa, E. Writer identification approach based on bag of words with OBI features. *Inf. Process. Manag.* **2017**, *56*, 354–366. [CrossRef]

41. Sarwar, A.; Mehmood, Z.; Mehmood, T.; Qazi, K.A.; Adnan, A.; Jamal, H. A novel method for content-based image retrieval to improve the effectiveness of the bag-of-words model using a support vector machine. *J. Inf. Sci.* **2018**, *45*, 117–135. [CrossRef]

42. Gernot, A.F. *Markov Models for Pattern Recognition—From Theory to Applications*; Springer: London, UK, 2014.

43. Figueiredo, L.N.; de Assis, G.T.; Ferreira, A.A. DERIN: A data extraction method based on rendering information and n-gram. *Inf. Process. Manag.* **2017**, *53*, 1120–1138. [CrossRef]

44. Chelba, C.; Mikolov, T.; Schuster, M.; Ge, Q.; Brants, T.; Koehn, P.; Robinson, T. One billion word benchmark for measuring progress in statistical language modeling. In Proceedings of the 15th Annual Conference of the International Speech Communication Association, Singapore, 14–18 September 2014; Causal Productions Pty. Ltd.: Adelaide, Australia, 2014; pp. 2635–2639.

45. Vilares, J.; Alonso, M.A.; Doval, Y.; Vilares, M. Studying the effect and treatment of misspelled queries in Cross-Language Information Retrieval. *Inf. Process. Manag.* **2016**, *52*, 646–657. [CrossRef]

46. Xu, H.; Li, K.; Wang, Y.; Wang, J.; Kang, S.; Chen, X.; Povey, D.; Khudanpur, S. Neural network language modeling with letter-based features and importance sampling. In Proceedings of the 2018 IEEE International Conference on Acoustics, Speech and Signal Processing, Calgary, AB, Canada, 15–20 April 2018; pp. 6109–6113.

47. Osman, A.H.; Salim, N.; Binwahlan, M.S.; Alteeb, R.; Abuobieda, A. An improved plagiarism detection scheme based on semantic role labeling. *Appl. Soft Comput.* **2012**, *12*, 1493–1502. [CrossRef]

48. Christensen, J.; Soderland, S.; Etzioni, O. Semantic Role Labeling for Open Information Extraction. In Proceedings of the 1st International Workshop on Formalisms and Methodology for Learning by Reading (NAACL HLT 2010), Los Angeles, CA, USA, 5–6 June 2010; Association for Computational Linguistics: Stroudsburg, PA, USA, 2010; pp. 52–60.

49. Zhou, J.; Xu, W. End-to-end learning of semantic role labeling using recurrent neural networks. In Proceedings of the 53rd Annual Meeting of the Association for Computational Linguistics and the 7th International Joint Conference on Natural Language Processing, Beijing, China, 27–31 July 2015; Association for Computational Linguistics: Stroudsburg, PA, USA, 2015; pp. 1127–1137.

50. Marcheggiani, D.; Frolov, A.; Titov, I. A Simple and Accurate Syntax-Agnostic Neural Model for Dependency-based Semantic Role Labeling. In Proceedings of the 21st Conference on Computational Natural Language Learning (CoNLL 2017), Vancouver, BC, Canada, 3–4 August 2017; Association for Computational Linguistics: Stroudsburg, PA, USA, 2017; pp. 411–420.

51. Della Rocca, P.; Senatore, S.; Loia, V. A semantic-grained perspective of latent knowledge modeling. *Inf. Fusion* **2017**, *36*, 52–67. [CrossRef]

52. Zhang, P.; Essaid, A.; Zanni-Merk, C.; Cavallucci, D. Case-based Reasoning for Knowledge Capitalization in Inventive Design Using Latent Semantic Analysis. *Procedia Comput. Sci.* **2017**, *112*, 323–332. [CrossRef]

53. Zhu, G.; Iglesias, C.A. Exploiting semantic similarity for named entity disambiguation in knowledge graphs. *Expert Syst. Appl.* **2018**, *101*, 8–24. [CrossRef]

54. Babar, S.A.; Patil, P.D. Improving Performance of Text Summarization. *Procedia Comput. Sci.* **2015**, *46*, 354–363. [CrossRef]

55. Badry, R.M.; Eldin, A.S.; Elzanfally, D.S. Text Summarization within the Latent Semantic Analysis Framework: Comparative Study. *Int. J. Comput. Appl.* **2013**, *81*, 40–45. [CrossRef]

56. Bookjans, M.; Weckenmann, A. Virtual Quality Management: A holistic approach for the set-up of true-to-life process models and according virtual quality management tools. In Proceedings of the 1st International Conference on Quality and Innovation in Engineering and Management, Cluj-Napoca, Romania, 17–19 March 2011; pp. 17–22.

57. Mertins, K.; Rabe, M.; Gocev, P. Integration of Factory Planning and ERP/MES Systems: Adaptive Simulation Models. In *Lean Business Systems and Beyond*; Koch, T., Ed.; Springer: Boston, MA, USA, 2008; pp. 185–193.

58. Heilala, J.; Vatanen, S.; Tonteri, H.; Montonen, J.; Lind, S.; Johansson, B.; Stahre, J. Simulation-based sustainable manufacturing system design. In Proceedings of the 2008 Winter Simulation Conference, Miami, FL, USA, 7–10 December 2008; pp. 1922–1930. [CrossRef]

59. Bookjans, M.; Weckenmann, A. Virtual Quality Management—Validation of measurement systems by the use of simulation technologies. *Phys. Procedia Laser Assist. Net Shape Eng.* **2010**, *5*, 745–752. [CrossRef]

60. Stockinger, A.; Wittmann, S.; Martinek, M.; Meerkamm, H.; Wartzack, S. Virtual Assembly Analysis: Standard tolerance analysis compared to manufacturing simulation and relative positioning. In Proceedings of the International Design Conference (DESIGN 2010), Dubrovnik, Croatia, 17–20 May 2010; The Design Society: Glasgow, Scotland, 2010; pp. 1421–1430.

61. Dix, A. Human–computer interaction, foundations and new paradigms. *J. Vis. Lang. Comput.* **2017**, *42*, 122–134. [CrossRef]

62. Eschen, H.; Kötter, T.; Rodeck, R.; Harnisch, M.; Schüppstuhl, T. Augmented and Virtual Reality for Inspection and Maintenance Processes in the Aviation Industry. *Procedia Manuf.* **2018**, *19*, 156–163. [CrossRef]

63. Milella, F. Problem-Solving by Immersive Virtual Reality: Towards a More Efficient Product Emergence Process in Automotive. *J. Multidiscip. Eng. Sci. Technol.* **2015**, *2*, 860–867.

64. Dulebohn, J.H.; Hoch, J.E. Virtual teams in organizations. *Hum. Resour. Manag. Rev.* **2017**, *27*, 569–574. [CrossRef]

65. Weckenmann, A.; Bookjans, M. Zuverlässigere Prozesse durch Virtuelles Qualitätsmanagement. *Zeitschrift für wirtschaftlichen Fabrikbetrieb* **2008**, *103*, 448–451. [CrossRef]

66. Lee, J.; Kao, H.-A.; Yang, S. Service innovation and smart analytics for Industry 4.0 and big data environment. *Procedia CIRP* **2014**, *16*, 3–8. [CrossRef]

67. Kagermann, H.; Wahlster, W.; Helbig, J. *Recommendations for Implementing the Strategic Initiative Industrie 4.0—Final Report of the Industrie 4.0 Working Group*; Acatech: Munich, Germany, 2013.

68. Tropes US. Help Index: Classes and Relations graphs, Version 8.4.2 (build 0019). Released June 2014. Available online: https://www.semantic-knowledge.com/download.htm (accessed on 7 June 2016).

69. International Organization for Standardization: ISO/IEC/IEEE 15288:2015. Available online: https://www.iso.org/standard/63711.html (accessed on 5 January 2018).

70. List of Romanian Companies. Available online: https://www.romanian-companies.eu/ (accessed on 12 May 2018).

MDPI
St. Alban-Anlage 66
4052 Basel
Switzerland
Tel. +41 61 683 77 34
Fax +41 61 302 89 18
www.mdpi.com

Sustainability Editorial Office
E-mail: sustainability@mdpi.com
www.mdpi.com/journal/sustainability